T5-BAM-467

Phanerozoic Ironstones

Geological Society Special Publications

*Series Editor* K. COE

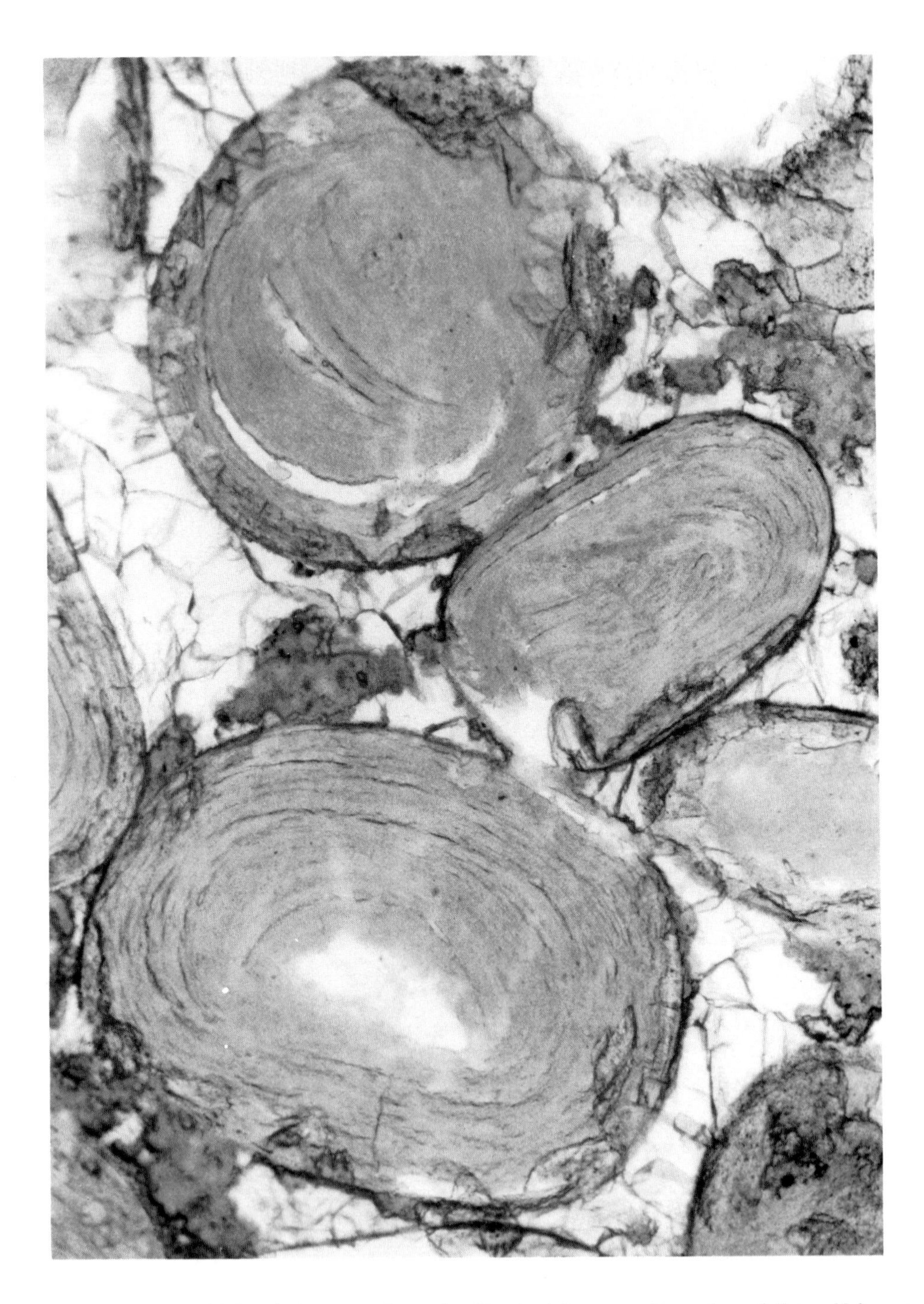

Thin section of an ooidal grain-ironstone from the Cleveland Ironstone Formation of Eston Nab, near Middlesborough, Teeside, England. This section was made in 1856 by Henry Clifton Sorby (1826 – 1908), the Sheffield scientist who established the modern science of petrography through the study of rocks in thin section. The section is housed in the collections of the Department of Geology, University of Sheffield. The specimen has berthierine ooids with, and partly replaced by, a siderite cement. Section viewed in plane-polarised light, ×180.

GEOLOGICAL SOCIETY SPECIAL PUBLICATION NO 46

# Phanerozoic Ironstones

EDITED BY

T. P. YOUNG
Department of Geology
University of Wales College of Cardiff

W. E. GORDON TAYLOR
Faculty of Applied Sciences
Luton College of Higher Education
Luton

1989

Published by

The Geological Society

London

# THE GEOLOGICAL SOCIETY

The Geological Society of London was founded in 1807 for the purposes of 'investigating the mineral structures of the earth'. It received its Royal Charter in 1825. The Society promotes all aspects of geological science by means of meetings, special lectures and courses, discussions, specialist groups, publications and library services.

It is expected that candidates for Fellowship will be graduates in geology or another earth science, or have equivalent qualifications or experience. All Fellows are entitled to receive for their subscription one of the Society's three journals: *The Quarterly Journal of Engineering Geology,* the *Journal of the Geological Society* or *Marine and Petroleum Geology.* On payment of an additional sum on the annual subscription, members may obtain copies of another journal.

Membership of the specialist groups is open to all Fellows without additional charge. Enquiries concerning Fellowship of the Society and membership of the specialist groups should be directed to the Executive Secretary, The Geological Society, Burlington House, Piccadilly, London W1V 0JU.

Published by the Geological Society from:
The Geological Society Publishing House
Unit 7
Brassmill Enterprise Centre
Brassmill Lane
Bath
Avon BA1 3JN
UK
(*Orders:* Tel. 0225 445046)

First published 1989



British Library Cataloguing in Publication Data

Phanerozoic ironstones
1. Phanerozoic strata. Ironstone
I. Young, Timothy Peter *1959 –* . II. Taylor, William Edward Gordon *1941 –*
III. Geological Society of London
IV. Series
551

ISBN 0-903317-43-5
Printed in Great Britain at the Alden Press, Oxford

# Contents

# Preface

This volume includes the papers presented at the International Symposium on Phanerozoic Ironstones held at the University of Sheffield in April 1987. Some sixty workers from eight nations attended the meeting and its accompanying field excursions. These field excursions visited Ordovician and Jurassic ironstones in North Wales, Yorkshire and Humberside. The symposium represented the starting point for the establishment of the International Geological Correlation Programme Project 277: Phanerozoic Oolitic Ironstones in 1988.

Research into iron-rich sediments is currently enjoying a renewal of involvement, as evidenced by the papers presented in this volume and by the enthusiasm displayed at the meeting. Much of the new interest has been generated by improvements in the technology for the investigation of these rocks and by the advancements in the geochemical background necessary for their interpretation. These influences are clearly to be seen in the following papers. We felt that the time was right to bring together those working in the field to discuss the various lines of research currently being pursued and the many different models for ironstone genesis proposed in recent years. The papers contained herein reflect the current diversity of opinion in the field. The scope of the symposium, and hence of the volume, was deliberately restricted to Phanerozoic examples of iron-rich sediments because Precambrian examples have had extensive recent discussion in other symposia (e.g., Trendall & Morris 1983).

Iron has been produced from Phanerozoic ironstones for several millenia, but, these days, the extensive use of Precambrian ores has now largely replaced the ironstones. For example, iron ore production in the UK reached a peak of 20.2m tonnes in 1942, when imports were restricted. Even in 1977 the production from the Frodingham Ironstone and the Northampton Sand Ironstone deposits contributed about 3.63m tonnes of the total UK production of 3.74m tonnes (Bowie *et al.* 1978). British ore production has continued to decline as a result of the availability of high grade–low phosphorus, imported ores and changes in iron and steel making technology which favour the use of high grade ores and concentrates. A recent European Communities' Working Document (Anon 1985) has, however, clearly indicated that some Phanerozoic ironstones, particularly those in West Africa, should be included as part of the Community raw materials supply policy.

The study of iron-rich sediments does, however, have real current economic relevance, as well as academic interest. Ironstones of Phanerozoic age are of great importance as a natural laboratory for the study of the diagenetic reactions involving iron, which have enormous implications for cement generation and kerogen maturation. They also figure prominently in the current discussion of sea-level changes during earth history, and the potential exploitation of these changes by seismic stratigraphy and its related fields. We hope that the papers presented in this volume will contribute towards the increased understanding of these sediments which have fascinated geologists for so long.

The volume has been organized into *four* major sections following a very necessary *Introduction* to the terminology of ironstones. The *Geochemical and Mineralogical Framework,* so important in the understanding of the formation of ironstones, sets the scene. The broader *Stratigraphic Patterns* are considered in a series of authoritative papers on a global and regional scale. Intriguing and complex aspects of the *Fabrics* of ironstones from the submicroscopic to the mesoscopic are then covered, and the volume is completed by a series of *Case Studies* from various locations in the world.

We would like to acknowledge the many people who have contributed to the symposium and to the production of this volume. In particular we would like to thank the academic, secretarial and technical staffs of both the Department of Geology, University of Sheffield, especially P. Mellor, S. Forster, P. Bootes, G. Mulhearn and M. Cooper, and the School of Applied Earth Sciences, Luton CHE, especially Mary Waller and Teresa O'Brien; also those who have shared with us the task of reviewing the papers presented in this volume: Charles Curtis, Charles Eccles, Colin Hughes, Anton Kearsley, Alan Spears, Robert Trythall and Franklyn Van Houten.

## References

ANON (Commission of the European Communities) 1985. *General Objectives: Steel 1990,* Commission Working Document Com, (85), 208.

BOWIE, S. H. U., KVALHEIM, A. & HASLAM, H. W. (ed.) 1978. *Mineral Deposits of Europe Vol 1: North-west Europe,* Institution of Mining & Metallurgy, London.

TRENDALL, A. F. & MORRIS, R. C. (eds) 1983. *Iron-Formation: Facts and Problems,* Elsevier, Amsterdam.

T. P. YOUNG, Department of Geology, UNCC, PO Box 914, Cardiff CF1 3YE, UK

W. E. G. TAYLOR, Faculty of Applied Sciences, Luton College of Higher Education, Park Square, Luton LU1 3JU, UK

# Phanerozoic ironstones: an introduction and review

## T.P. Young

SUMMARY: Ironstones, particularly ooidal ironstones, have long fascinated sedimentary geologists and have generated an enormous variety of interpretations, but have remained poorly understood. A recent upsurge in interest in this group of rocks has generated much new information derived from many different disciplines within the earth sciences.

The first part of this introduction examines recent advances, and attempts to produce a synthesis of them, with particular reference to the genesis of marine ooidal ironstones. Current models for the formation of marine oolitic ironstone-formations are discussed, and a working model based on various lines of recent research is proposed. This model invokes the intrasedimentary formation of berthierine ooids in marine environments during post-oxic diagenesis. The ooids may be mechanically modified and mineralogically transformed by reworking, as well as by later diagenesis. Goethite ooids may be oxidized berthierine ooids, reworked lateritic ooids or primary grains. The formation of oolitic ironstone-formations is favoured by a break in clastic sediment supply, continued supply of iron and physical reworking. Many sedimentary environments may supply these parameters, but changes in sea level leading to the flooding of land masses and the reworking of suitable terrestrial soils may be both an important source for the iron and provide a break in clastic sediment supply.

In the second part of this introduction the terminology and classification are discussed, and conventions proposed. The term *berthieroid* is introduced as a non-specific term for material of undetermined berthierine or chamosite composition. The term *ironstone* is proposed as a petrological term and as an informal lithostratigraphic term. A deposit composed of *ironstones* may be termed an *ironstone-formation.* It is recommended that a 'Dunham' style of classification is employed petrographically, with *-ironstone* replacing the *-stone* of the original carbonate classification. The nomenclature of allochems in ironstones is also discussed. A glossary of ironstone terminology is provided.

This introduction to the papers presented at the Geological Society International Symposium on 'Phanerozoic Ironstones', held at the University of Sheffield in April 1987 owes much to informal discussion with colleagues and friends during the meeting and field excursions.

Studies of ancient ironstones are dominantly of those with an ooidal texture, deposited with syndepositional iron enrichment in marine environments. Ironstones with texture and mineralogy similar to those of the marine examples are also found in non-marine contexts, and examples are described in the recent literature (e.g., Siehl & Thein 1989; Kearsley 1989). These non-marine examples have much to offer in the search for the conditions giving rise to iron-rich sediments.

Ironstones without ooidal textures (e.g., blackband ironstones, claystone ironstones and sphaerosiderites) are currently rather better understood than ooidal examples (Spears 1989).

It is the advance of the study of marine ooidal ironstones with which most workers in the field are primarily involved, and with which 12 of the 13 case studies in this volume are concerned. This review will therefore concentrate on the research into marine examples.

## Developments in the study of ironstones

Recent increased interest in the genesis of oolitic ironstone-formations is largely due to advances in the technology available for their examination and in the framework of sedimentology and diagenesis within which they can be interpreted.

In most ironstones the oxide, hydroxide and silicate phases are very fine-grained; classical optical petrographic technique has not been able to resolve microstructure, and has, therefore, fallen short of providing adequate basic description of the rocks. The major advances of recent years in techniques of high-resolution imaging and chemical analysis by electron beam instruments have provided a suite of important tools for petrographic and mineralogical investigation; high-resolution scanning electron microscopy, backscattered electron imagery, analytical transmission microscopy and electron probe micronanalysis have allowed a more detailed analysis of the textures and chemical composition of the mineral phases present in ironstones. The extremely fine-grained nature of the phases is highlighted by Hughes (1989), who was unable to obtain single-crystal energy

*From* YOUNG, T. P. & TAYLOR, W. E. G. (eds), 1989, *Phanerozoic Ironstones*
Geological Society Special Publication No. 46, pp. ix-xxv

dispersive spectroscopic chemical analysis for berthierine, despite using the technique (analytical transmission electron microscopy) which provides an analysis from the smallest volume currently possible. Clearly there is much more to be obtained through these areas of research. The small particle size of the phases in ironstones has similarly so far limited the investigation of their isotopic compositions, but the imminent capability for ultra-fine-scale measurement of the stable isotopes ratios of carbon, oxygen and sulphur will be of great significance for the study of ironstones.

The rather unusual sedimentological conditions during the formation of most oolitic ironstone-formations have, in the past, given rise to great problems in interpretation. The increased understanding of event stratification, cyclic sequences and episodic sedimentation has provided a new framework within which the conditions of deposition of these ironstones can be investigated.

A further major influence on the investigation of ironstones is the great advance in the study of diagenetic processes. In the last twenty years the broad framework for the understanding of the chemical processes involved in early diagenesis has been established. Although the study of the authigenesis of silicate minerals is not so advanced as that of carbonate or sulphide phases, much can now be said about the chemical conditions required for the formation of the mineral phases found in ironstones (Spears 1989).

## The genesis of ooidal ironstones

The problems associated with the origin of ooidal ironstones are threefold: why are these rocks iron-rich, why do they have the mineralogy that they do (particularly why do they have this specific silicate mineralogy?) and why do they have an ooidal texture? These problems are by no means solved, but advances have been made in our understanding of many areas. Any model for the origin of ironstones must simultaneously address the problems of the deposition of the ironstone-bearing units, of mineral genesis, and of the origin of the texture. Each of these areas will be discussed before the presentation of a working model.

### *The origin of oolitic ironstone-formations*

Much can be learnt about the controls on ironstone deposition from the study of the spatial and temporal distribution of oolitic ironstone-formations. The patterns of the temporal distribution of oolitic ironstone-formations have been analysed and show cyclicity on several scales (Van Houten & Arthur 1989). Two major peaks of ironstone production in the Jurassic and Ordovician have been interpreted by Van Houten & Purucker (1984) as being associated with high global stands of sea level and a high degree of continental dispersion, with large areas of peneplained terrain. These authors indicate a temporal relationship between the major periods of deposition of oolitic ironstone-formations and periods of repeated transgressions. Van Houten (1985) modified his view and indicated that 'neither dispersed continents nor major highstand of sea level was a necessary factor'. These conditions are clearly favourable, nevertheless, for sea-level stand and sea-level change control the ability of the continental areas to supply large amounts of iron to the marine environment during reworking of terrestrial weathering products.

Within these major peaks of ironstone deposition, periods of a few to a few tens of millions of years were conducive to enduring or recurring conditions of ironstone formation. This intermediate scale of temporal distribution has been attributed to hypothetical 32 Ma polytaxic oligotaxic cycles (Van Houten & Arthur 1989), but the great variation in the length of these events probably indicates that variations in local tectonic conditions, sea level and sediment supply may also be very important. The requirements for ironstone deposition may well locally be satisfied by such processes as a 'clastic trap', but only when many other parameters are also satisfied on a regional or global level.

Small-scale patterns, which may show a finer cyclicity on a scale of a few hundreds of thousands of years, have been tentatively attributed to 400 000 year Milankovitch cycles (Van Houten 1986; Van Houten & Arthur 1989) or to local tectonic controls. The low deposition rates required for the production of ironstones would make them very sensitive to minor environmental changes (Van Houten 1986).

The palaeogeographical distribution of ironstone deposits has been suggested as evidence for climatic control on their formation. The association of oolitic ironstone-formations with well vegetated land masses at moderately low latitudes, which may have had lateritic soils, has been claimed for the Jurassic (Taylor 1951; Bubinicek 1961; Hallam 1975; Hallam & Bradshaw 1979; Van Houten 1985), but some Devonian and Ordovician examples of ironstones occur at high palaeolatitudes inconsistent with tropical lateritic soils (Van Houten 1985).

Volcanic activity has been postulated as the iron source for the Devonian ironstones of Belgium (Dreesen 1989), while the nature of the high latitude Ordovician palaeosols, the reworking of which has been suggested for the abundant oolitic ironstones in the Ordovician of North Africa and Europe, is entirely unknown (Young 1989). Van Houten (1985) has suggested that a higher partial pressure of carbon dioxide in the Ordovician atmosphere may have produced more acidic groundwater at high latitudes than at present. Such acidic groundwater could have produced weathering rather similar to that seen in well vegetated tropical latitudes today.

The non-random distribution of ironstones through time, and the relationships between ironstone abundance and the estimates of global sea level allow two different interpretations of the genesis of ironstone units.

*The reworking of pedogenic ferruginous allochems into the marine environment.* The generation of iron-rich ooids and pisoids in terrestrial weathering profiles has been well documented (Adeleye 1973, 1975; Ambrosi & Nahon 1980; Nahon *et al.* 1980; Siehl & Thein 1989). Examples have been recorded of both igneous and sedimentary rocks forming the weathered substrate. Such ferruginous allochems are characterized by rather complex structures caused by the progressive invasion of a host, and by the repeated coating of earlier pisoids fractured by desiccation. The pisoids are typically of goethite with a high proportion of alumina.

The nature of pedogenic ferruginous ooids, pisoids and crusts implies that their reworking into secondary deposits, including marine ones, is possible, and is therefore almost certain to have occurred, at least locally. The problem is to ascertain whether the reworking of pedogenic allochems is the *usual* method of generation of marine ironstones (Siehl & Thein 1989).

Much of the debate over the non-marine origin of the ferruginous allochems in apparently marine deposits has centred on the well studied ironstones of the European Jurassic, particularly on the Minette Ironstone. The association of ironstone-formations with periods of rising sea level and with well vegetated land masses, probably with lateritic soils, in the Jurassic is ambiguous; it could imply that the supply of either chemical components or that of allochems was involved. If the reworking of pedogenic allochems was the *major* source of marine ironstones it seems curious that ferruginous ooids are not a more common component of clastic, particularly deltaic, sediments. The stratigraphically restricted occurrence of ferruginous ooids in muddy sediments of low clastic content, the common occurrence of marine ironstone formations, the common occurrence of ironstone-formations on offshore shoals, and their dependence on sea level, all provide circumstantial evidence that ferruginous ooids formed within the marine environment. These pieces of evidence can, however, also be explained by a non-marine model for ooid genesis, by controls on the processes of introduction of the allochems into the marine environment, by the processes of reworking the allochems within the marine environment, or by proposing previously unrecognized loci and periods of emergence and pedogenesis. It seems likely that both the supply of chemical components and final allochems was involved, with marine ironstone-formations being composed of reworked terrestrial ferruginous allochems in some cases (Siehl & Thein 1978, 1989), while others were generated from ferruginous granules formed in the marine environment from chemical components supplied by the reworking of terrestrial weathering products (Van Houten & Purucker 1984).

The investigation of whether there are chemical or textural criteria to distinguish between the ooids of non-marine origin and those of marine origin must be of high priority. The chemical evidence assembled to date (e.g., Siehl & Thein 1989; Myers 1989) does not apparently differentiate between terrestrial allochems and marine allochems formed by the reaction of ferruginous particulate matter derived from terrestrial environments.

*The generation of ironstones in the marine environment.* There are three broad models for the genesis of ironstones within the marine environment.

(1) The generation of the ferruginous allochems in nearshore, probably restricted lagoonal environments during periods of low sea level stand and their reworking into the basin of deposition by storms or during a subsequent transgressive event (Bayer 1989; McGhee & Bayer 1985).
(2) The development of ooidal ironstones on offshore swells which receive little clastic sediment, but on which may the sediments may be intensely reworked by wave activity (Hallam 1975).
(3) The *in situ* development of ooids on marine shelves during phases of sediment starvation, such as that produced by rising sea level (Young 1989).

These three models are not necessarily mutually exclusive, indeed they all represent methods of generating ironstones in areas of reduced clastic sedimentation. Well documented examples have been used as evidence for each of these models and it seems likely that they can all occur.

Firm evidence for the generation of ferruginous ooids in the marine environment is hard to obtain. There seems good reason to believe that this can occur, given the models whereby the textures and mineralogy could be generated in marine environments, and the problems in linking all ooidal ironstones with a non-marine allochem source.

*The formation of berthierine*

The nature and genesis of the minerals contained in ironstones have received much attention, and many models exist for their generation. The ferriferous silicate minerals involved in ironstones are usually berthierine and chamosite, although examples of nontronite ooids have been recorded both in the geological record (Petruk 1977; Petruk *et al.* 1977) and forming at the present day (Pedro *et al.* 1978). Chamosite in ironstones is believed to originate by transformation of berthierine at temperatures above about 120–160°C (Curtis 1985; Iijima & Matsumoto 1982). Goethite is a common mineral in many ironstones, sometimes it appears as an alteration product of berthierine, but often it appears to have been a primary mineral or a restructuring of an amorphous iron oxyhydroxide (Gehring 1989).

Several models for berthierine genesis in ironstones have been put forward.

*Direct precipitation from fluids.* The problem of what diagenetic environment is required for the authigenesis of berthierine has long troubled workers in this field. The problem is to reduce iron, but not to reduce sulphate, whilst producing an ooid which would apparently require turbulent, presumably oxic, conditions to form.

Following early attempts at a chemical approach to the problems of the genesis of iron minerals (e.g., Borchert 1965), Curtis & Spears (1968) provided the first detailed discussion of the chemical conditions under which various iron minerals would form. They showed that during early diagenesis under 'normal' conditions iron reduced in marine sediments would be quickly mopped up by the almost simultaneous reduction of sulphate to form iron monosulphides and then pyrite, whereas in freshwater environments, with very low sulphate activity, siderite could precipitate. The argument gained a very wide acceptance, with many authors proposing that ironstones formed during periods of reduced salinity. Most of the examples for which this has been proposed could simply not have had a sulphate activity lowered to such an extent as to inhibit sulphate reduction. Salinity variation in the marine environment seems incapable of providing the means for ironstones to form.

The common generation of berthierine as an authigenic cement phase raises the question of direct precipitation, as occurs in the authigenesis of other phyllosilicate cements. If berthierine is abundant as a neoformed phase in the cements of ironstones, then why should it not also be forming the material in ooids by precipitation from solution?

More recently the role of the organic component in the control of diagenetic reactions has been better appreciated. The prolonged reworking so evident in the sedimentology of many ironstones on transgressive unconformities or in condensed sequences gave rise to prolonged oxic and post-oxic reactions (Berner 1981; Maynard 1982). The products of any sulphate reduction which did occur would, by reworking into oxic environments, have become substrates for sulphide oxidizing bacteria. Many workers (Berner 1981; Spears 1989) now believe that in such an environment pre-sulphidic iron reduction allows the authigenesis of iron silicates and the quality of remaining organic matter is so refractory as to inhibit any subsequent sulphate reduction during burial diagenesis. A very similar environment has been proposed for the generation of phosphate minerals (Coleman 1985), and indeed these are frequently associated with ironstones.

*Precipitation from a gel.* The possibility of generating berthierine in the marine environment from a gel precursor has been described by Harder (1978, 1989). He proposes that a gel allows the preconcentration of the iron, alumina and silica which is otherwise very difficult to achieve in seawater. The formation of various minerals from such gels has been demonstrated experimentally.

*Transformation of a kaolinite/goethite mixture.* The model of berthierine formation from kaolinite/iron oxyhydroxide mixtures (Bhattacharyya 1983) has gained some acceptance (Bhattacharyya & Kakimoto 1982; Bhattacharyya 1989). It would appear however that berthierine is structurally rather dissimilar to kaolinite, and therefore that the reaction would have to proceed by dissolution and reprecipitation. The process may occur in some pedogenic environments (Siehl & Thein 1989), but seems unlikely for most marine ironstones, where the amount of kaolinite required would be very large. Evidence for marine highly kaolinitic sediments within ironstone bearing sequences, for instance the Jurassic of the UK, is rather sparse.

It is also possible that transformation of Si- and Al-rich iron oxyhydroxides could occur where the Si and Al are not present as kaolinite. Velde (1989) argues for the formation of berthierine in ironstones by the addition of silica and alumina to iron oxide ooids. These components are suggested to be derived by dissolution of kaolinite elsewhere in the sediment. Adeleye (1980) proposed a similar mode of origin for berthierine.

Many problems exist because there is no consensus on the sequence of origin of the minerals. Many authors have commented on the two main types of ooids in marine ironstones: the high-sphericity goethite ooids, and the low-sphericity berthierine-dominated ooids. This strongly suggests that berthierine (or possibly another phyllosilicate phase) was the dominant material in the ooids before any significant compaction had occurred, implying it is unlikely that burial diagenesis could account for berthierine formation in these cases. The early transformation of oxide ooids to berthierine could have occurred; the transformation of pedogenic ooids to berthierine does occur in the deeper, reducing parts of soil profiles (Siehl & Thein 1989).

### *The formation of ooids*

Many models for the origin of ferruginous ooids have been proposed, including the replacement of calcareous ooids (Sorby 1856; Cayeux 1909; Kimberley 1974, 1979a, b, 1980a, b, c), *in situ* growth as 'microconcretions' (Hallimond 1951; Hemingway 1974), crystallization from ferruginous gel precursors (Harder 1978), mechanical accretion of clays with subsequent transformation to iron-rich phases (Bhattacharyya & Kakimoto 1982; Van Houten & Purucker 1984), mineralization of calcareous microfossils (Champetier *et al.* 1987), biologically controlled environments (e.g., fungal mats, Dahannayake & Krumbein 1986) and derivation from lateritic soils (Siehl & Thein 1978). Only recently, however, has an adequate database of examples of well described ooid mineralogy and texture been built up, on which such models can be based. The contribution by Kearsley (1989) represents one of first attempts to examine a broad spectrum of ooid types and to discuss their relationships and genesis. The occurrence of non-marine ferruginous ooids, and the documentation of their structure (Kearsley 1989, Siehl & Thein 1989) has, in particular, provided much evidence for possible processes of ooid genesis.

Many of the previous arguments for modes of ooid genesis have been based on SEM imaging of the orientation of particles within the ooid cortex. Bhattacharyya & Kakimoto (1982) interpretated radial structures as primary precipitation and tangential as accretion. It is interesting, however, to note that the 'flakes' described by these and other authors from SEM images had already been described by Rohrlich (1974) as being polycrystalline aggregates; the orientation of the flakes may tell us nothing of the orientation of their component minerals (see also Hughes 1989). Even in circumstances where the structures resolved are individual minerals there are great problems with interpretation. Firstly the ooids studied by this technique have almost always been goethite ones: it is often uncertain whether the goethite in these ooids is primary, or a replacement of berthierine. Secondly the generality of the observations of Bhattacharyya & Kakimoto (1982) has been questioned; Siehl & Thein (1989) report tangential fabrics produced by the mechanical interference of growing ooids with other particles within the sediment, and Kearsley suggests rolling as a means of modifying a radial texture to a tangential one. Davies *et al.* (1978) have rather similarly attributed the tangentially structured Bahamian carbonate ooids to the modifying effects of turbulence, where all crystals on the surface of the ooid except tangentially oriented ones will be removed by attrition in turbulent conditions.

The timing of the generation of the sub-ooidal texture remains an important question. There is some evidence (Kearsley 1989) that much of the concentric structure may postdate the formation of the ooid. The sub-ooidal texture has often been interpreted as demonstrating particular modes of growth, but the possibility of post-formational textures, perhaps generated during the various proposed mineralogical transformations, must be carefully examined. In this light it seems that ultrastructural studies of chamosite or goethite ooids which were originally berthierine, or of berthierine ooids formed from goethite during diagenesis, may only provide evidence of recrystallization textures.

Many models for ooid generation have been proposed. The major models are outlined below.

*Mechanical accretion.* It is intriguing to note that the previously-favoured 'snow-ball' model of mechanical accretion has not been at all well documentated by the detailed textural studies (Kearsley 1989; Hughes 1989). The berthierine ooids examined by Hughes (1989) showed random orientation of berthierine within their shells; this would suggest against mechanical accretion of berthierine, but does not distinguish between direct preciptitation or precipitation from a gel. Chauvel and Guerrak (1989) demonstrate bizarre 'micro-armoured-mudball' textures which can be generated when physical accretion does play a part. Kearsley (1989) does, however, suggest that physical reworking may play a part in consolidating ooids by producing a tangential re-orientation of the crystals in the outermost cortical layer.

*Mechanical accretion of clays with subsequent transformation to iron-rich phases.* Mechanical accretion of kaolinite (Bhattacharyya 1983) as a mode of ooid genesis seems rather unlikely, for the kaolinite – berthierine transformation may not be a major mode of berthierine formation (see above). Most records of kaolinite ooids in ironstones probably refer to leached ferruginous ooids (subclass B4 of Kearsley 1989), with the generation of kaolinite occurring at various possible times from early diagenesis to sub-aerial weathering (Kearsley 1989). The generation of berthierine ooids from kaolinite seems problematical. If such a process occurred why do we not find primary kaolinite ooids? The evidence from pedogenic environments seems to suggest that either precipitation directly or from a gel seems more likely. Contrary to the assertations of Bhattacharyya (1983) and Van Houten & Bhattacharyya (1982), the structural reorganization needed to transform kaolinite to berthierine is not simple, and it seems the reaction would have to proceed by dissolution and reprecipitation. This would imply that primary accretion fabrics would be lost, if the model for kaolinite accretion and transformation were correct.

*Derivation from terrestrial soils.* The derivation of pedogenic ooids has been claimed as the major source of ironstone by Siehl & Thein (1989). Ferruginous ooids and pisoids can be generated in hydromorphic soils by alternate leaching of silicates and precipitation of iron and aluminium hydroxides. These processes allow the mobilization of iron and aluminium on a large scale. In the zone of permanent groundwater the ooids and pisoids may be transformed into siderite and berthierine. Various stages of the reworking of such soils are described by Siehl and Thein (1989), but it remains to be positively demonstrated that any of the major marine ironstone formation were generated from reworked pedogenic allochems.

*In situ growth as 'microconcretions'.* Whilst the argument of Siehl and Thein (1978, 1989) that ferruginous ooids are *usually* derived pedogenic ooids has not gained general acceptance, instrasedimentary growth, whether in a terrestrial or marine milieu, has great attraction (Hallimond 1951; Hemingway 1974). The concentrically varying proportions of chamosite/berthierine, haematite/goethite and apatites in marine ironstones has generally been attributed to reworking, with alternating periods of suspension and deposition. Clearly the texturally similar pedogenic ooids form under fluctuating chemical conditions *in situ.* This possibility must also be considered for marine ironstones. The potential variation in the chemical environment is very large:

(1) *Physical reworking between the sediment and the water column* is likely to be very short lived.
(2) *A variation in effects of bioturbation and/or removal/addition of sedimentary overburden* gives rise to fluctuations of reducing/oxidizing conditions *in situ.*
(3) *Physical reworking between one deposit and another, or between burial and exposure* gives rise to fluctuations of reducing/oxidizing conditions.

These changes in environment give opportunity for a variation in the phases precipitated. Reworking between oxic and post-oxic environments may lead to partial or total oxidation of neoformed berthierine to goethite. Some workers (e.g., Chavel & Guerrak 1989) claim that the concentric compositional variation seen in ironstone ooids reflects, at least in part, variations in the chemical environment of formation of the ooid. Kearsley (1989) suggests that much of the concentric variation in mineralogy may be due to the differential action of diagenetic processes in layers of the ooid with differing porosity.

*The replacement of calcareous ooids.* Since Sorby (1856) the replacement of calcareous ooids has been quoted as a possible mechanism for the generation of the ooids in ironstones. Kimberley revived the model with a series of papers (1974, 1979a, b, 1980a, b, c) proposing that ooidal ironstones are generated by the large-scale replacement of ooidal limestones by iron-rich porewaters. This model gained little general acceptance, with much critical discussion (Adeleye 1980; Binda & Moltzer 1979; Bradshaw *et al.* 1980), but it has been demonstrated (Kearsley 1989) that local replacement of carbonates by berthierine can occur, although it is unlikely to be capable of generating significant ironstone deposits.

*Crystallization from ferruginous gel precursors.* Harder (1978, 1989) proposed that crystallization from a gel would be a possible mechanism for generating iron-rich silicate minerals in early diagenetic environments. He suggested (1978) that the spherical bodies produced in some of the gel ageing experiments might be the precursors of ooids. The formation of ooids would thus be related to the mechanism of the mineral-forming reactions. Kearsley (1989) expresses doubts that ooids observed in ironstones could have been formed by such a process (class B3). Harder later described (1989), however, a model in which the ooids predate the generation of the silicate minerals; the ooid generating process therefore being independent of the role of gels in mineral genesis. Such a process might well be capable of generating the cortical shells of randomly oriented crystals as described by Hughes (1989).

*Mineralization of calcareous microfossils.* Champetier *et al.* (1987) suggested that ooids in the Jurassic Minette ironstone and in a Devonian ironstone from N. Africa were replacements of the tests of nubeculariid foraminifera. This suggestion has not been accepted by other workers (Chauvel & Guerrak 1989; Kearsley 1989). The material illustrated by Champetier *et al.* appear to show normal ironstone ooids. The central cavity described by them appears to be due to the loss of the ooid core during specimen preparation and the lunate divisions are the equatorial thickenings of cortical sheaths so common in berthieroid ooids (Siehl & Thein 1989). Chauvel & Guerrak (1989) also indicate that the Devonian example given by Champetier *et al.* predates the oldest known occurrence of nubeculariids.

*Biologically controlled environments (fungal mats).* Dahanayake & Krumbein (1986) argued that ferruginous ooids are not allochems, but are 'authigenically formed biogenic grains'. They presented figures claimed to demonstrate the 'stromatolitic layering' within the Minette ironstone of Lorraine. It is hard to reconcile their description with that of Teyssen (1984) who made a detailed analysis of the tidal controls of the sandwaves preserved in this ironstone-formation. Siehl & Thein (1989) comment on the SEM images discussed by Dahanayake & Krumbein, and present alternative interpretations; they also discuss the field evidence for the mats in which they ascribe the 'stromatolitic layering' of Dahanayake & Krumbein to true cross-lamination. It seems unlikely that this mode of origin produces ooids, although of course biological moderation of diagenetic reactions may be important within the other models for ooid genesis.

*Adsorption processes.* Gehring (1989) describes a model for the genesis of low Si and Al goethite ooids in the marine environment by adsorption of pre-existing goethite. This model provides an interesting alternative to the conventional models of mechanical accretion and *in situ* precipitation. The products of this process would appear to be very difficult to distinguish in practice from the products of other processes.

The problem of the allochthonous nature of the ooids in many ironstones has been addressed in many ways; it has been demonstrated that most marine ironstones involve at least a component of mechanical sorting and accumulation of the ferruginous components after their formation, making an assessment of the amount of transport involved very difficult. The association of the occurrence of ironstones with periods of low sedimentation rates renders them particularly susceptible to resedimentation. This reworking often entails winnowing of the ferruginous allochems into 'economic' deposits by storm activity (Dreesen 1989, Bayer 1989), tidal activity (Teyssen 1984, 1989), or other current activity (Bayer 1989). The attempts to identify sites of ooid generation (e.g., the lean oolites of Bhattacharyya (1989)) have not been able to do so entirely unambigously. The model (Spears 1989) of berthierine genesis in post-oxic conditions (*sensu* Berner 1981), after intense oxic zone reaction of organic material, may necessarily entail reworking of the sediment physically and biologically to maintain sufficient oxygenation of the upper layers of the sediment. This implies that the resultant ironstone will be dominated by reworked material, even if the ooids are being generated more-or-less *in situ*.

### The later diagenesis of ironstones

The rather particular chemical conditions of formation of ooidal ironstones mean that on burial the early formed mineral assemblages will often become quickly unstable. The earliest reactions may involve phosphatization of the components of the ironstone, and often the extensive generation of siderite. The sideritization process may involve the leaching of ooids (to leave kaolinite or opal relicts) and may give rise to a bulk iron enrichment of the ironstone (Spears 1989). Siderite generation probably occurs at various times from very soon after the deposition of the ironstone until late burial diagenesis. Early siderite cements are often strongly influenced by the presence of bioturbation.

In many marine ironstones calcite cements are common, and generally they postdate early siderite formation. Such calcite cements are particularly common in reworked ironstones with a grainstone texture and high initial porosity and permeability (e.g., the Frodingham Ironstone Formation, England).

The formation of pyrite as a relatively early diagenetic alteration of berthierine-rich ironstones is very common. Berthierine will be unstable even at low sulphide activity. Pyrite occurs particularly in the uppermost parts of ironstone-formations developed within mudstone sequences. The re-establishing of a sulphidic diagenetic zone after the formation of an ironstone allows the reaction of diffusing sulphide with the iron-rich phases in the ironstone. The temporary establishment of a sulphidic zone may allow the partial alteration of units within an ironstone-formation. These sulphide-bearing excursions in the ironstone facies often correlate laterally with intervals of resumed or accelerated sediment accumulation (e.g., the 'sulphur bed' of the Cleveland Ironstone Formation and the 'snap band' of the Frodingham Ironstone Formation). Ironstones showing such changes in sediment accumulation may have a reduced economic value because of their sulphide content.

Deep burial diagenesis of ironstones includes the transformation of berthierine to chamosite. This reaction is believed to occur at about 150–160°C and a depth of 3 km (Iijima & Matsumoto 1982; Curtis 1985). The reaction of siderite with kaolinite to give berthierine and quartz was also described by Iijima & Matsumoto (1982) at temperatures of 65–150°C and depths of 2–5 km. Reaction of early berthierine, hydrated iron oxides, kaolinite and siderite will tend to give chamosite, dehydrated iron oxides and quartz during burial and low-grade metamorphism. During low-grade metamorphism Velde (1989) has demonstrated that less iron-rich chlorites will form at the expense of chamosite. Higher grades of metamorphism (both regional and contact) are characterized by the development of magnetite and stilpnomelane as major iron-bearing phases.

## A model for the genesis of marine ooidal ironstones

The evidence that ferruginous ooids can form within marine environments is rather circumstantial, but since berthierine occurs widely in marine ironstones as an early diagenetic pore-fringing cement and as a replacement of calcareous bioclasts, then the genesis of this silicate must be feasible on a large scale within marine sediments; so why not in ooids? Ooids in marine ironstones may arise by a process similar to the formation of pedogenic ooids, under conditions of low sediment supply, adequate iron supply and intense reworking. The resultant authigenic mineral assemblage is a product of the post-oxic regime (Berner 1981). Post-oxic settings may involve fluctuating chemical conditions, particularly related to the physical and biological reworking evident in so many oolitic ironstone formations. Such fluctuating conditions may produce diagenetic micro-environments not dissimilar to those of hydromorphic soils. The formation of ironstones in the marine environment may be made possible by the introduction of ferruginous terrestrial weathering products. The exact paths of mineral formation have yet to be identified, but berthierine probably originated by direct precipitation from sea- or pore-waters. Goethite ooids occur both as the leached products of berthierine ooids, and as directly precipitated grains. Ooids presumably formed largely by intrasedimentary growth and not generally by mechanical accretion.

The sediment starvation required to produce these conditions may occur in restricted, protected lagoonal environments, but may also occur on open shelves. The starvation may be produced by autogenic means (channel and delta abandonment?) or perhaps more generally by allogenic means (tectonic control of swells, eustatic changes).

The reworking of lateritic and hydromorphic soils may also lead to the formation of ironstones as described by Siehl & Thein (1989). The ferruginous allochems may be reworked into terrestrial (fluvial or lacustrine) or marine environments.

The replacement of calcareous ooids seems not to be a major mode of ironstone formation. The replacement of calcite and aragonite has, however, been documented (Kearsley 1989) and may locally give rise to ooidal ironstones, but seems unlikely to account for any major deposits.

A priority for the immediate future of research in ironstones is the differentiation of marine and pedogenic ooids, and trying to establish the reaction pathways involved. The lack of actualistic examples of marine ironstones should not discourage research into the behaviour of authigenic iron silicates in recent marine sediments. We need to assemble more information about the mode of ooid formation in marine sediments by examining the geological record, and by being critical of observed microstructures. Further information on the nature of the load of modern rivers draining regions similar to those predicted as sources for the iron of ancient examples is also desirable. The fields of the organic geochemistry of ironstones also remains to be examined.

## Terminology

### Mineralogical nomenclature

Every effort has been made to ensure that mineralogical nomenclature in this volume conforms to international standards (Bailey 1980a, b; Newman & Brown 1987). These recommendations clearly state that *chamosite* should be used for a 2:1 trioctahedral chlorite (1.4 nm repeat) with $Fe^{2+}$ as the dominant divalent octahedral cation (with an end member formula of $(Fe^{2+}{}_5Al)(Si_3AL)O_{10}(OH)_8$), whereas *berthierine* has priority for an Fe-rich 1:1 type layer silicate of the serpentine group (0.7 nm repeat) having appreciable tetrahedral Al. The chemical composition of berthierine was reviewed by Brindley (1952). Identification of these minerals is often difficult, but a lax usage of the terms berthierine and chamosite should not be continued.

There is still a need for a usable field- and optical petrological-term which can be used where there has not been adequate determination of the iron-rich phyllosilicate involved. The use of *chamositic* in this context, in an analogous manner to the widespread use of *glauconitic,* has achieved some popularity (e.g., Van Houten & Purucker 1984, 1985). It is therefore appropriate to examine this analogy with a discussion of glauconite nomenclature.

The international recommendations for the term *glauconitic* (Bailey 1980a; Newman & Brown 1987) indicate that it is to be used to describe a mixture with an iron-rich mica as a major component. The use of the term *glauconitic* is based largely on Odin & Matter (1981), who distinguished a group of *glauconitic minerals* with end members *glauconitic smectite* (= '*green smectite*' of other authors) and *glauconitic mica* (= *glauconite* of other authors); they used *glaucony* as a facies term. The term '*glauconoid*' has been used recently as an adjective for grains optically similar to *glauconite sensu stricto,* but which have either undetermined mineralogy or are composed of minerals only distantly related to *glauconite,* such as iron-rich illites and smecities (K. Pye, personal communication).

The use of the term '*glauconitic* peloid' implies a peloid formed of species containing a significant proportion of iron-rich 1.0 nm components, whether these are present as discrete phases (e.g. *glauconite*) or as interstratifed layers in other phases (e.g. *glauconite-smectite*). On the other hand a '*chamositic* peloid' (*sensu* Van Houten & Purucker 1984) is a peloid composed entirely, or predominantly, of either *chamosite* or *berthierine.* There is no mineral or mineral structural unit which gives coherence to the *chamositic minerals sensu* Van Houten & Purucker (1984), although berthierine and chamosite are similar in chemical composition. The analogy between *chamositic* and *glauconitic* is, therefore, not a good one. Indeed the term *chamositic* would more reasonably mean *chamosite*-rich or *chamosite*-bearing (cf. common usage of calcitic, feldspathic, etc.). The use of quotation marks to distinguish *chamosite sensu stricto* from '*chamosite*' *sensu lato* (meaning either *berthierine* or *chamosite*) is prone to the problems of becoming misquoted and misinterpreted.

Van Houten & Arthur (1989) have suggested the use of the term *chloritic* instead of *chamositic sensu* Van Houten & Purucker (1984). This suffers from the problems that many chloritic rocks in a general geological sense would have no real relationship with the *chloritic* sediments they refer to, and that berthierine is not strictly a chlorite.

Berthierine would appear to be the dominant early diagenetic phase, and therefore a more desirable root for such a general term than chamosite, its late diagenetic/metamorphic equivalent. A useful derivative, by analogy with the gluaconite nomenclature, for a chamosite/ berthierine-like phase might be 'berthierinoid'. This could perhaps be shortened to *bertheroid,* to make it more easily pronouncable (again by analogy with *glauconoid*); it is therfore proposed that *berthieroid* should be used as an optical petrological and field term for these components. Reference may thus be made to '*berthieroid* ooids', or to 'ooids with alternating *berthieroid* and goethite sheaths'. *Berthieroidal* could be used as an adjective for material bearing, or rich in, *berthieroid* minerals, but known to contain others. Thus a mudstone containing *bertheroid* flakes as well as other material might be termed a *berthieroidal* mudstone.

A problem still exists for an adjective for a rock bearing, or rich in, berthierine, equivalent to the term *chamositic.* One informal suggestion has been made to use the term *berthieritic,* but this has a prior meaning of *berthierite*-bearing (*berthierite* is a sulphide with the composition $FeSb_2S_4$). Although rather difficult to pronounce *berthierinic* seems the best adjective here, although it may be preferable in some situations to use the full expressions 'berthierine-rich' or 'berthierine-bearing'.

### The 'ironstone' and 'iron-formation' problem

The use of the terms *ironstone* and *iron-formation* has vexed the nomenclature of iron-rich sediments for many years, aggravated by the general lack of communication between those studying the iron-rich sedimentary rocks of the Precambrian and Phanerozoic. A dual terminology has arisen despite the fact that lithological examples from the two periods may be identical.

The nomenclature workshop at the symposium debated the problem of the use of *ironstone* and *iron-formation,* and produced some recommendations for further discussion. These incorporate some ideas which are at variance with the current usage of the terms. The aim, however, is to produce terms which are better integrated into the scheme of petrological nomenclature, and which enable a rock to be allocated to a simple lithological classification regardless of age.

Those at the symposium strongly reaffirmed the proposition by Kimberley (1978) that *ironstone* should be a lithological term implying a rock of greater than 15 weight percent iron of any age. It was the opinion of the workshop, however, that *iron-formation,* a term applied to rock units of abnormally high iron content, should be modified to *ironstone-formation* (e.g. in *banded ironstone-formation*), since any such formation is characterized by sediments of high iron content (i.e. ironstones) rather than by the occurrence of elemental iron. The current tendency to use *ironstone* as a term solely for Phanerozoic examples, following James (1966), was rejected by the workshop.

The recent recommendation by Trendall (1983) that the use of *ironstone* be avoided altogether, and replaced by *iron-formation* even as a lithological term, was also strongly rejected by the workshop. Trendall argued that " 'limestone' and 'sandstone' among many others, effectively serve as both lithological and stratigraphic (formation) names". Although the International Stratigraphic Guide (Hedberg 1976) and the North American Stratigraphic Code (1983) do still permit such usage, some current codes of lithostratigraphic nomenclature (e.g., Holland *et al.* 1978) specifically state that formal lithostratigraphic rank should appear within the name of lithostratigraphic units wherever possible. Current lithostratigraphic practice, certainly in the UK, is to replace informal units bearing lithological names with formally named units incorporating rank terms, with an optional indication of the dominant lithology. In the same manner, therefore, that the 'Lincolnshire Limestone' has now been formally embraced by the Lincolnshire Limestone Formation (Ashton 1980), the 'Cleveland Ironstone' has now been formalized as the Cleveland Ironstone Formation (Howard 1985); the terms Limestone and Ironstone in these names being purely qualifiers indicating the dominant lithology. The form 'Cleveland *Iron-Formation*' (*Iron-formation sensu* Trendall 1983) would not be desirable as a formal lithostratigraphic term, because it contains no formal indication of rank, but does contain the word 'formation' suggesting to a reader that it is of formation rank; formal lithostratigraphic units with names such as 'Cleveland *Iron-Formation Member*' would be highly undesirable. Indeed, although *ironstone-formation* may be a useful term to describe a deposit formed of *ironstones,* in an informal reference to, for instance, the *ironstone-formation* of Lorraine, the argument of Trendall could be used to suggest the perpetuation of 'the Minette *ironstone*' as an informal lithostratigraphic term.

### The terminology of ooidal ironstone facies

Ooidal ironstones are often referred to as 'Minette'- or 'Clinton'-type ironstones. Most of those present at the terminology workshop felt that these terms were unsatisfactory. Whilst a facies term to distinguish the broad group of marine ooidal ironstones from other ironstones is desirable, the consensus was that it is probably premature to divide the group into facies types; this should await a broad review of the nature of the occurrence of Phanerozoic ironstones.

### Lithological nomenclature

The lithological nomenclature of ironstones has remained largely a matter for personal preference (see the various nomenclatures of Hallimond 1925; James 1966; Taylor 1949; Trendall 1983). The nomenclature workshop within the 'Phanerozoic Ironstones Symposium' sampled opinion on this subject, and guidelines resulting from the discussion are proposed here. The degree of consensus on the use of the various terms discussed varied considerably, although there was very strong support for a revision of nomenclature. The following discussion is based as far as possible on the majority opinion of the workshop.

The complex mineralogy of most ironstones

TABLE 1. *Recommended terminology of allochems in ironstones*

| | Recommended Terminology less than 2mm | above 2mm |
|---|---|---|
| | *Concentrically structured grains* | |
| grain | ooid | pisoid |
| adjective | ooidal | pisoidal |
| rock | oolite | pisolite |
| | *Plastically deformed concentrically structured grains* | |
| grain | spastolith | |
| adjective | spastolithic | |
| | *Grains without concentric structure* | |
| grain | peloid | |
| adjective | peloidal | |
| | *Grains of faecal origin* | |
| grain | pellet | pellet |
| adjective | pelletal | pelletal |
| | *Grains of probable cyanobacterial origin* | |
| grain | microoncoid | oncoid |
| adjective | microoncoidal | oncoidal |
| rock | microoncolite | oncolite |

Petrographic terms whose use is *not* recommended:

Oolith
Oolitoid
Ooloid (+ ooloid rock)
Oncolith
Oncoloid (+ oncoloid rock, microoncoloid, microoncoloid rock)
Pisolith
Pisoloid (+ pisoloid rock)

means that any petrographic nomenclature embracing all, or even just the common, phases soon become unwieldy. It is probably most useful in these rocks to stress the textural information in the nomenclature, leaving the user to add as much, or as little, mineralogical information as the context requires. The 'Dunham' style of classification, based on texture and originally erected for carbonates (Dunham 1962), has long been used as a basis for the nomenclature of ironstones. It is recommended that mineralogical and allochem qualifiers be used where appropriate. The use of such a classification in non-carbonate systems is well illustrated by its adoption for phosphorites (Cook and Shergold 1986). The one qualification proposed by the workshop was that in view of the common occurrence of ironstones and limestones in a single sequence, the use of '. . . stone' in the Dunham classification should be replaced by '. . .-ironstone' (e.g., mud-ironstone, ooidal pack-ironstone). The more complex terms which have sometimes been employed (such as 'chloritic sideritic cherty quartzose chlorite siderite quartz kaolinite magnetite oolite' (Weinberg 1973)) are probably of limited use; they are neither compact enough for practical lithological terms, nor complete enough to substitute for proper petrographic description. The use of a nomenclature based on the scheme for carbonates by Folk (1959, 1962) is usually inappropriate for ironstones because of the difficulty in distinguishing between cement and matrix.

The basic nomenclature for the allochems within ironstones follows that of limestones (see Table 1). Several recent schemes have used the inferred modes of origin of the allochems in their classification, but such schemes will necessarily be unstable when the mode of origin of the allochems is controversial. Instead, it is recommended that allochems are classified in a descriptive textural scheme without genetic inference. There remain several inconsistencies in the use of these allochem names. The use of -lite terminations for rocks and -id for the allochems is widespread. However the current use of adjectives is based almost entirely on -litic (pisolitic, oolitic, etc.), but an -idal termination would be more correct (pisoidal, ooidal, etc.). One exception to the allochem nomenclature are the deformed ooids known as spastoliths. The root (*spastos*: Greek meaning pulled or drawn) is an adjective, so the -lith ending for the granule should be used rather than the -id ending. In modern usage the diacritical mark ¨ is no longer required in the English language for the word ooid and its derivatives (Teichert 1970).

In a petrographic term which describes dominant allochem mineralogy as well as matrix/cement mineralogy the mineralogy should be given as a noun for the allochem and as an adjective for the groundmass if an adjective can be formed from the mineral name. If no adjectival form is available then the noun may be used. In all cases the sense is preserved by the word-order, with the allochem mineralogy given before the allochem type and the groundmass mineralogy before the textural term. A matrix-supported ironstone with 15% goethite ooids in a berthierine-rich matrix could be termed a 'goethite ooidal berthierine wacke-ironstone'.

The complex texture and mineralogy of ooidal ironstones requires the careful description of ooids. The problem has been addressed by Kearsley (1989) who has identified fifteen classes of ooid texture and mineralogy. Clearly the careful description of ooids is essential for the determination of their diagenetic history and for the evaluation of any evidence for their mode of origin. The petrographic description of an ironstone should include descriptions of ooid morphology and microstructure (Kearsley 1989) and, where appropriate, details of the ultra-structure of the cortical laminae themselves (Hughes 1989).

Dahanayake & Krumbein (1986) presented a classification of coated grains which involved classification by shape (regularity, continuity of laminae) and by origin (biogenic/abiogenic). The origin of all coated grains is highly controversial, and a genetic classification seems unjustifiable. The degree of variation of ooid morphology revealed by Kearsley (1989) clearly demonstrates that the divisions of Dahanayake & Krumbein would be arbitrary in practice. Their assertion that lateritic ooids can be distinguished from marine ones does not seem to be supported by the papers presented in this volume (Siehl & Thein 1989; Kearsley 1989). Their classification uses many recently erected terms which are etymologically unsound, as well as using old terms in a very different way from their common usage. In this review therefore a broader definition of the various coated grains is preferred, leaving authors free to provide full descriptions of the grains, instead of fitting them to narrow classifications.

# Glossary

*Berthierine*: an Fe-rich 1:1 type layer silicate of the serpentine group (0.7 nm repeat) having appreciable tetrahedral Al.

*Berthierinic*: *berthierine*-bearing, or *berthierine*-rich. Writing these expressions in full may be preferred.

*Berthieroid*: composed dominantly or entirely of minerals so that the whole is given an appearance in hand specimen, or in optical microscopy, similar to *berthierine*.

*Berthieroidal*: Composed of, or bearing, *berthieroid* material.

*Chained spastoliths: Spastoliths,* usually *egg-shell spastoliths,* where the outer *cortical layers* become confluent with those of adjacent *ooids,* because of collapse during early diagenesis. The chain is of a characteristic sigmoidal shape where the outer *cortical layers* of the lower side of one *ooid* become confluent with those on the upper side of an adjacent *ooid.*

*Chamosite*: a 2:1 trioctahedral chlorite (1.4 nm repeat) with $Fe^{2+}$ as the dominant divalent octahedral cation (with an end member formula of $(Fe^{2+}{}_5Al)(Si_3Al)O_{10}(OH)_8$).

*Chamositic*: *chamosite*-bearing, or *chamosite*-rich. Writing these expressions in full is to be preferred and may be necessary in making the same references to *berthierine.*

*Core*: the inner part of an *ooid* without concentric laminae, but which does not necessarily form an original *nucleus.* Refers particularly to a diagenetically altered central region of an *ooid.*

*Cortex*: The concentrically layered part of an *ooid* surrounding the *nucleus.*

*Cortical lamina:* a distinct concentric layer of an *ooid cortex.* Where such a lamina is present around most or all of an ooid, it can be termed a *sheath.* Sometime known simply as a cortical layer.

*Egg-shell spastolith*: A *spastolith* formed from a plastically deformed *ooid,* the outer *cortical layer(s)* of which have undergone brittle deformation. Common in berthierine *ooids* of which the outer layers have been phosphatized or sideritized in very early diagenesis.

*Ironstone*: 1. a rock of greater than 15 wt.% iron. 2. Combined with a geographical name as an informal lithostratigraphic name for a particular *ironstone-formation.*

*Ironstone-formation*: a rock unit or deposit of abnormally high iron content of any formal lithostratigraphic rank.

*Microoncoid*: used by Dahanayake & Krumbein (1986) for a grain of less than 2 mm, with irregular (wavy, discontinous) concentric structure, of biogenic origin.

*Microoncolite*: a rock largely or dominantly composed of *microoncoids* (*sensu* Dahanyake & Krumbein 1986).

*Microoncoloid*: used by Dahanayake & Krumbein (1986) for a grain of less than 2 mm, with irregular (wavy, discontinuous) concentric structure, of abiogenic origin. This use is *not recommended* here as the distinction from *microoncoid* is not purely textural.

*Microoncoloid rock*: used by Dahanayake & Krumbein (1986) for a rock composed of *microoncoloids, Not recommended*—see *microoncoloid.*

*Nucleus*: the central part of an *ooid* around which the *cortex* has been constructed.

*Oncoid*: grain of *pisoid* size, with concentric structure, believed to be of biogenic, usually cyanobacterial origin. Generally distinguished by irregular (often wavy), usually discontinuous, cortical laminae. Flügel & Kirchamyer (1962) have no lower size limits for oncoids.

*Oncoidal*: (adjective from *oncoid*) 1. pertaining to *oncoids.*
2. composed of *oncoids* (e.g. *oncoidal ironstone*).

*Oncolite*: 1. a rock largely or dominantly composed of *oncoids.*
2. Previously also used to mean *oncoid.*

*Oncolith*: previously used for *oncoid.*

*Oncolitic*: (adjective from *oncolite*). 1. pertaining to *oncolites.*
*2. composed of oncolites.*

*Oncoloid*: term used by Dahanayake & Krumbein (1986) to distinguish coated particles resembling *oncoids* but of an abiogenic or unknown origin. Its use is *not recommended* here since the distinction from *oncoid* is not purely textural.

*Oncoloid rock*: term used by Dahanayake & Krumbien (1986) for a rock composed of *oncoloids. Not recommended*—see oncoloid.

*Ooid:* 1. spherical or ellipsoidal grain, less than 2 mm in diameter, having regular concentric laminae. (Gk oion—egg, *oeides*—resembling; Kalkowsky 1908).
2. Dahanayake & Krumbein (1986) advocate its restriction to biogenic grains; such usage is subjective and controversial, so a purely textural definition is advocated here.

*Ooidal*: (adjective from *ooid*) 1. pertaining to *ooids.*
2. composed of *ooids* (e.g. *ooidal* ironstone).

*Oolite*: 1. informal term for a rock largely or dominantly formed of *ooids.*
2. Dahanayake & Krumbein (1986) advocate its restriction to rocks formed of biogenic grains (*ooids sensu* Dahanayake & Krumbein); such use is *not recommended.*

*Oolith*: previously used extensively meaning *ooid* (Rastall 1933; DeFord & Waldschmidt 1956), but also (originally) used meaning rock composed of *ooids. Not recommended* for use.

*Oolitic*: (adjective from *oolite*). 1. pertaining to *oolites.*
2. composed of *oolites* (e.g. *oolitic ironstone-formation).*

*Oolitoid*: has been used to refer to grains the size and shape of *ooids,* but lacking the concentric layering. Derivation ('*oolite*-like') is based on old usage of the word *oolite* and means this word is now inappropriate, and therefore *not recommended* in referring to grains. Fine-grained *oolitoid* (in its original meaning) granules are also known as *pseudo-ooids;* this latter term is to be preferred.

*Ooloid*: Dahanayake & Krumbein (1986) advocated this term for a biogenic concentrically structured coated grain; a biogenic *ooid.*
They also used it for such grains of unknown origin. Such usage is subjective and controversial, and is therefore *not recommended.* A purely textural classification is advocated here; ooid can be used for biogenic and abiogenic grains, as well as those of uncertain origin.

*Ooloid rock*: Dahanayake & Krumbein (1986) advocated this term for a rock composed of *oolids. Not recommended*—see *ooloid.*

*Pellet*: sub-spherical to cylindrical grains, usually 0.1–0.5 mm diameter, presumed to be of faecal origin.

*Pelletal*: (adjective from pellet) 1. pertaining to *pellets.*
2. composed of *pellets* (e.g. *pelletal ironstone*).

*Peloid*: grain of fine-grained material, up to several mm in diameter, without recognizable internal structure. *Peloids* of the same size and shape as accompanying *ooids* are often referred to as *pseudo-ooids.*

*Peloidal*: (adjective from *peloid*) 1. pertaining to *peloids.*
2. composed of *peloids* (e.g. *peloidal* ironstone).

*Pisoid*: 1. grain similar to an *ooid,* but greater than 2 mm in diameter, other than one of presumed biogenic origin (which is termed *oncoid*). (Gk. *pisos*—pea, *oeides*—resembling).
2. Dahanayake & Krumbein (1986) used this term for a regularly rounded coated grain, with continous concentric laminae, of biogenic origin. Such a usage is *not recommended* here; a purely textural use is advocated.

*Pisoidal*: (adjective from pisoid) 1. pertaining to *pisoids.*
2. composed of *pisoids* (e.g. *pisoidal ironstone*).

*Pisolite*: 1. a rock dominantly composed of *pisoids.*
2. Dahanayake & Krumbein (1986) used the term for a rock composed of *pisoids* in their usage, i.e. grains of biogenic origin. Such use is *not recommended* here.
3. previously used for *pisoid.*

*Pisolith*: previously used (e.g. Flügel & Kirchmayer 1962) for *pisoid.*

*Pisolitic*: (adjective from *pisolite*) 1. pertaining to *pisolites.*
2. composed of *pisolites* (e.g. *pisolitic ironstone-formation*).

*Pisoloid*: Dahanayake & Krumbein (1986) used this term for a regularly rounded coating grain, with continous concentric laminae, of abiogenic origin. Such use if *not recommended* here; use of the *purely textural term pisoid* is to be preferred.

*Pisoloid rock*: Dahanayake & Krumbein (1986) used this term for a rock composed of pisoloids. *Not recommend* here—see *pisoloid.*

*Protooid*: (sometimes spelled *proto-ooid*) an ooid bearing only a few cortical sheaths. Used by Van Houten & Bhattacharyya (1982) and Bhattacharyya (1989) in contexts where the authors believed ooids were preserved in the context of their original growth. The term may be synonomous with superficial ooid in cases where a distinct nucleus is present.

*Pseudo-ooids*: peloids of the same grain size and shape as *ooids* occurring with them.

*Sheath*: a persistent *cortical lamina* present over most or all of an *ooid.*

*Spastolith*: plastically deformed *ooid* (Rastall & Hemingway 1940). Often the outer *cortical lamina(e)* of *ooids* may have been replaced by siderite or a phosphate mineral prior to the deformation; this may undergo brittle deformation around the plastically deformed inner part of the *ooid* giving rise to a characteristic sigmoidal egg-shell *spastolith* (Kearsley 1989). (Gk. *spastos* —drawn, pulled, *lithos* —stone)

*Superficial ooid*: an *ooid* with a concentrically layered *cortex* thinner than the radius of its *nucleus* (Illing 1954).

ACKNOWLEDGEMENTS: This paper owes its origin to extensive discussion with many people, particularly during the symposium. However the particular contributions of Colin Hughes and Anton Kearsley must be acknowledged. The author also acknowledges the support of the Department of Geology, University of Sheffield, during the tenure of an NERC Research Fellowship, and during the organization and running of the Geological Society of London International Symposium on Phanerozoic Ironstones.

## References

ADELEYE, D.R. 1973. Origin of ironstones, an example from the middle Niger Valley, Nigeria. *Journal of Sedimentary Petrology,* **43,** 709–723.

——1975. Derivation of fragmentary oolites and pisolites from dessication cracks. *Journal of Sedimentary Petrology,* **45,** 794–798.

——1980. Origin of oolitic iron formations—discussion. *Journal of Sedimentary Petrology,* 50, 1001–1003.

AMBROSI, J.P. & NAHON, D. 1986. Petrological and Geochemical differentiation of lateritic iron crust profiles. *Chemical Geology,* **57,** 371–393.

ASHTON, M. 1980. The stratigraphy of the Lincolnshire Limestone Formation (Bajocian) in Lincolnshire and Rutland. *Proceedings of the Geologists' Association,* **91,** 203–233.

BAILEY, S.W. 1980a. Summary of recommendations of AIPEA nomenclature committee. *Clay Minerals,* **15,** 85–93.

——1980b. Structures of layer silicates. pp. 1– 123 *In:* BRINDLEY, G.W. and BROWN, G. (eds) *Crystal structures of clay minerals and their X-ray identification,* Mineralogical Society Monograph No. 5.

BAYER, U. 1989. Stratigraphic and environmental patterns of ironstone deposits, *In*: YOUNG, T.P. & TAYLOR, W.E.G. (eds) *Phanerozoic Ironstones,* Geological Society, London, Special Publication **46,** 105–117.

BERNER, R.A. 1981. New geochemical classification of sedimentary environments. *Journal of Sedimentary Petrology,* **51,** 359–365.

BHATTACHARYYA, D.P. 1983. Origin of berthierine in ironstones. *Clays and Clay Minerals,* **31** 173–182.

——1989. Concentrated and lean oolites: examples from the Nubia Formation at Aswan, Egypt, and significance of the oolite types in ironstone genesis. *In*: YOUNG T.P. & TAYLOR, W.E.G. (eds) *Phanerozoic Ironstones,* Geological Society, London, Special Publication, **46,** 93–103.

——& KAKIMOTO, P.K. 1982. Origin of ferriferous ooids: An SEM study of ironstone ooids and bauxite pisoids. *Journal of Sedimentary Petrology,* **52,** 849–857.

BINDA, P.L. & MOLTZER, J.G. 1979. Origin of oolitic iron formations—discussion. *Jounal of Sedimentary Petrology,* **49,** 1351–1353.

BORCHERT, H. 1965. Formation of marine sedimentary iron ores, pp. 159–204, *In*: RILEY, J.P. & SKIRROW, G. (eds), *Chemical Oceanography.* Academic Press, London.

BRADSHAW, M.J., JAMES, S.J. & TURNER, P. 1980. Origin of oolitic ironstones—discussion. *Journal of Sedimentary Petrology,* **50,** 295–304.

BRINDLEY, G.W. 1982. Chemical compositions of berthierines—a review. *Clays and Clay Minerals,* **30,** 153–155.

BUBINICEK, L. 1961. Recherches sur la constitution et la repartition du minerai de fer dans l'Aalenien de Lorraine. *Sciences de la Terre,* **8,** 5–204.

CAYEUX, L. 1909. *Les minerals de fer oolithiques de France. Tome 1: Minerais de fer primairies.* Imprimerie Nationale, Paris.

CHAMPETIER, Y., HAMBADOU, E. & HAMBODOU, H. 1987. Examples of biogenic support of mineralisation in two oolitic iron ores—Lorraine (France) and Gara Djebilet (Algeria). *Sedimentary Geology,* **51,** 249–255.

CHAUVEL, J. J. & GUERRAK, S. 1989. Oolitization processes in Palaeozoic ironstones of France, Algeria and Libya. *In*: YOUNG, T.P. & TAYLOR, W.E.G. (eds) *Phanerozoic Ironstones,* Geological Society, London, Special Publication, **46,** 165–174.

COLEMAN, M.L. 1985. Geochemistry of diagenetic non-silicate minerals: kinetic considerations. *Philosophical Transactions of the Royal Society of London,* **A315,** 39–56.

COOK, P.J. & SHERGOLD, J.H. 1986. Proterozoic and Cambrian phosphorites—an introduction. *In*: COOK, P.J. & SHERGOLD, J.H. (eds) *Phosphate deposits of the world, Volume 1: Proterozoic and*

Cambrian phosphorites, Cambridge University Press, 1–8.
CURTIS, C.D. 1985. Clay mineral precipitation and transformation during burial diagenesis. *Philosophical Transactions of the Royal Society of London,* **A315,** 91–105.
—& SPEARS, D.A. 1968. The formation of sedimentary iron minerals, *Economic Geology,* **63,** 257–270.
DAHANAYAKE, K., GERDES, G. & KRUMBEIN, W.E. 1985. Stromatolites, oncolites and oolites biogenically formed in situ. *Naturwissenschaften,* **72,** 513–518.
—& KRUMBEIN, W.E. 1986. Microbial structures in oolitic iron formations, *Mineralium Deposita,* **21,** 85–94.
DAVIES, P.J., BUBELA, B. & FERGUSON, J. 1978. The formation of ooids. *Sedimentology,* **25,** 703–730.
DEFORD, R.K. & WALDSCHIMDT, W.A. 1956. Oölite and oölith: *Bulletin of the American Association of Petroleum Geologists,* **30,** 1587–1588.
DREESEN, R. 1982. Storm-generated oolitic ironstones of the Fammenian (Falb–Fa2a) in the Vesdre and Dinant synclinoria (Upper Devonian, Belgium). *Annales de la Société Géologique de Belgique,* **105,** 105–129.
—1989. Oolitic ironstones as event-stratigraphical marker beds within the upper Devonian of the Ardenno-Rhenish Massif. *In*: YOUNG, T.P. & TAYLOR, W.E.G. (eds) *Phanerozoic Ironstones.* Geological Society, London, Special Publication, **46,** 65–78.
DUNHAM, R.J. 1962. Classification of carbonate rocks according to depositional texture. pp. 108–121 *In*: HAM, W.E. (ed.) *Classification of Carbonate Rocks,* Memoir of the American Association of Petroleum Geologists, 1.
FLÜGEL, E. & KIRCHMAYER, M. 1962. Zur Terminologie der Ooide, Onkoide and Pseudooide. *Neues Jahrbuch für Geologie und Paläontologie, Monatshefte,* 113– 123.
FOLK, R.L. 1959. Practical petrographic classification of limestones. *The American Association of Petroleum Geologists Bulletin,* **43,** 1–38.
—1962. Spectral subdivision of limestone types. pp. 62–84, *In*: HAM, W.E. (ed.) *Classification of Carbonate Rocks, Memoir of the American Association of Petroleum Geologists, 1.*
GEHRING, A.U. 1989. The formation of goethitic ooids in condensed Jurassic deposits in northern Switzerland. *In*: YOUNG, T.P. & TAYLOR, W.E.G. (eds) *Phanerozoic Ironstones,* Geological Society, London, Special Publication, **46,** 133–140.
HALLAM, A. 1975. *Jurassic Environments.* Cambridge University Press.
—& BRADSHAW, M.J. 1979. Bituminous shales and oolitic ironstones as indicators of transgressions and regressions. *Journal of the Geological Society of London,* **136,** 157–164.
HALLIMOND, A.F. 1925. *Iron ores: bedded ores of England and Wales. Petrography and chemistry.* Special Report on the Mineral Resources of Great Britain. Volume 29. Memoir of the Geological Survey.
—1951. Problems of the Sedimentary Iron Ores. *Proceedings of the Yorkshire Geological Society,* **28,** 61–66.
HARDER, H. 1978. Synthesis of iron layer silicate minerals under natural conditions. *Clays and Clay Minerals,* **26,** 65–72.
—1989. Mineral genesis in ironstones: a model based upon laboratory experiments and petrographic observations *In*: YOUNG, T.P. & TAYLOR, W.E.G. (eds) *Phanerozoic Ironstones,* Geological Society, London, Special Publication **46,** 9–18.
HEDBERG, H.D. 1976. *International Stratigraphic Guide: A guide to stratigraphic classification, terminology, and procedure.* John Wiley, New York.
HEMINGWAY, J.E. 1974. Jurassic. pp. 161–223, *In*: RAYNOR, D.H. & HEMINGWAY, J.E. (eds) *The Geology and Mineral Resources of Yorkshire.* Yorkshire Geological Society.
HOLLAND, C.H., AUDLEY-CHARLES, M.G., BASSETT, M.G., COWIE, J.W., CURRY, D., FITCH, F.J., HANCOCK, J.M., HOUSE, M.R., INGHAM, J.K., KENT, P.E., MORTON, N., RAMSBOTTOM, W.H.C., RAWSON, P.F., SMITH, D.B., STUBBLESFIELD, C.J., TORRENS, H.S., WALLACE, P. & WOODLAND, A.W. 1978. A guide to stratigraphical procedure. *Geological Society of London Special Report,* **10.**
HOWARD, A.S. 1985. Lithostratigraphy of the Staithes Sandstone and Cleveland Ironstone formations (Lower Jurassic) of north-east Yorkshire. *Proceedings of the Yorkshire Geological Society,* **45,** 261–175.
HUGHES, C.R. 1989. The application of Analytical Transmission Electron Microscopy to the study of oolitic ironstones: A preliminary study. *In*: YOUNG, T.P. & TAYLOR, W.E.G. (eds) *Phanerozoic Ironstones,* Geological Society of London, Special Publication **46,** 121–132.
IIJIMA, A. & MATSUMOTO, R. 1982. Berthierine and chamosite in coal measures of Japan. *Clays and Clay Minerals,* **30,** 264–274.
ILLING, L.V. 1954. Bahamian calcareous sands. *The American Association of Petroleum Geologists Bulletin,* **38,** 1–95.
JAMES, H.L., 1966. Chemistry of the iron-rich sedimentary rocks. *United States Geological Survey Professional Paper, 440*–W.
KALKOWSKY, E. 1908. Oolith and Stromatolith im norddeutschen Buntsandstein: *Deutsch. Geol. Gesell. Zeitschr,* **60,** 68–125.
KEARSLEY, A.T. 1989. Iron-rich ooids, their mineralogy and microfabric: clues to their origin and evolution, *In*: YOUNG, T.P. & TAYLOR, W.E.G. (eds) *Phanerozoic Ironstones,* Geological Society, London, Special Publication **46,** 141–164.
KIMBERLEY, M.M. 1974. Origin of oolitic iron ore by diagenetic replacement of calcareous oolite. *Nature,* **250,** 319–320.
—1978. Palaeoenvironmental classification of iron formations. *Economic Geology,* **73,** 215–229.
—1979a. Origin of oolitic iron minerals. *Journal of Sedimentary Petrology,* **49,** 110–132.
—1979b. Origin of oolitic ironstones—reply.

*Journal of Sedimentary Petrology,* **49,** 1352–1353.
——1980a. Origin of oolitic ironstones—reply. *Journal of Sedimentary Petrology,* **50,** 299–302.
——1980b. The Paz de Rio Oolitic Inland-Sea Iron Formation. *Economic Geology,* **75,** 97–106.
——1980c. Origin of oolitic ironstones—reply. *Journal of Sedimentary Petrology,* **50,** 1003–1004.
MAYNARD, J.B. 1982. Extension of Berner's "New geochemical classification of sedimentary environments" to ancient sediments. *Journal of Sedimentary Petrology,* **52,** 1325–1331.
MCGHEE, G.R. & BAYER, U. 1985. The local signature of sea-level changes. pp. 98–112, *In*: BAYER, U. & SEILACHER, A. (eds) *Sedimentary and Evolutionary Cycles,* Springer, New York.
MYERS, K.J. 1989. The origin of the Lower Jurassic Cleveland Ironstone Formation of North-East England: evidence from portable gamma-ray spectrophotometry. *In*: YOUNG, T.P. & TAYLOR, W.E.G. (eds) *Phanerozoic Ironstones,* Geological Society, London, Special Publication **46,** 221–228.
NAHON, D., CAROZZI, A.V. & PARRON, C. 1980. Lateritic weathering as a mechanism for the generarion of ferruginous ooids. *Journal of Sedimentary Petrology,* **50,** 1287–1298.
NEWMAN, A.C.D. and BROWN, G. 1987. The chemical constitution of clays, pp. 1–128, *In*: NEWMAN, A.C.D. (ed), 1987. *Chemistry of clays and clay minerals, Mineralogical Society Monograph No. 6.*
NORTH AMERICAN COMMISSION ON STRATIGRAPHIC NOMENCLATURE, 1983. North American Stratigraphic Code. *The American Association of Petroleum Geologists Bulletin,* **67,** 841–875.
ODIN, G.S. & MATTER, A. 1981. De glauconarium origine. *Sedimentology,* **28,** 611–641.
PEDRO, G., CARMOUZE, J.P. & VELDE, B. 1978. Peloidal nontronite formation in recent sediments of Lake Chad. *Chemical Geology,* **23,** 139–149.
PETRUK, W. 1977. Mineralogical characteristics of an oolitic iron deposit in the Peace River district Alberta. *Canadian Mineralogist,* **15,** 3–13.
——FARRELL, D.M., LAUFER, E.E., TREMBLAY, R.J. & MANNING, P.G. 1977. Nontronite and ferrunginous opal from the Peace River iron deposit in Alberta, Canada, *Canadian Mineralogist,* **15,** 14–21.
RASTALL, R.H. 1933. On the preparation of geological manuscripts. *Geological Magazine,* **70,** 481–488.
——& HEMINGWAY, J.E. 1940. The Yorkshire Dogger, 1. The Coastal Region. *Geological Magazine,* **77,** 257–275.
ROHRLICH, V. 1974. Microstructure and microchemistry of iron ooliths. *Mineralium Deposita,* **9,** 133–142.
SIEHL, A. & THEIN, J. 1978. Geochemische Trends in der Minetter (Jura, Luxembourg/Lothringen). *Geologische Rundschau,* **67,** 1052–1077.
——&——1989. Minette-Type Ironstones. *In*: YOUNG, T.P. & TAYLOR, W.E.G. (eds) *Phanerozoic Ironstones,* Geological Society, London, Special Publication **46,** 175–193.
SORBY, H.C. 1856. On the origin of the Cleveland Hill Ironstone. *Proceedings of the Geological and Polytechnic Society of the West Riding of Yorkshire,* **3,** 457–461.
SPEARS, D.A. 1989. Aspects of iron incorporation into sediments with particular reference to the Yorkshire Ironstones. *In*: YOUNG, T.P. & TAYLOR, W.E.G. (eds) *Phanerozoic Ironstones,* Geological Society, London, Special Publication **46,** 19–30.
TAYLOR, J.H. 1949. Petrology of the Northampton Sand Ironstone. *Memoir of the Geological Survey of Great Britain.*
——1951. Sedimentation problems of the Northampton Sand Ironstone. *Proceedings of the Yorkshire Geological Society,* **28,** 74–85.
TEICHERT, C. 1970. Oolite, oolith, ooid: discussion. *American Association of Petroleum Geologists Bulletin,* **54,** 1748–1749.
TEYSSEN, T.A.L. 1984. Sedimentology of the Minette oolitic ironstones of Luxembourg and Lorraine: a Jurassic subtidal sandwave complex. *Sedimentology,* **31,** 195–211.
——1989. A depositional model for the Liassic Minette ironstones (Luxembourg and France), in comparison with other Phanerozoic oolitic ironstones. *In*: YOUNG, T.P. and TAYLOR, W.E.G. (eds) *Phanerozoic Ironstones,* Geological Society, London, Special Publication **46,** 79–92.
TRENDALL, A.F. 1983. Introduction. *In*: Trendall, A.F. & Morris, R.C. (eds) *Iron-formation facts and problems.* Developments in Precambrian Geology, **6,** Elsevier 1–12.
VAN HOUTEN, F.B. 1985. Oolitic ironstones and contrasting Ordovician and Jurassic palaeogeography, *Geology,* **13,** 722–724.
——1986. Search for Milankovitch patterns among oolitic ironstones. *Palaeoceangraphy,* **1,** 459–466.
——& ARTHUR, M.A. 1989. Temporal patterns among Phanerozoic oolitic ironstones and oceanic anoxia. *In*: YOUNG, T.P. & TAYLOR, W.E.G. (eds) *Phanerozoic Ironstones,* Geological Society, London, Special Publication **46,** 33–50.
——& BHATTACHARYYA, D.P. 1982. Phanerozoic oolitic ironstones—geologic records and facies model. *Annual Reviews of Earth and Planetary Sciences,* **10,** 441–457.
——& PURUCKER, M.E. 1984. Glauconitic peloids and chamositic ooids—favourable factors, constraints and problems. *Earth-Science Reviews,* **20,** 211–243.
——&——1985. On the origin of glauconitic and chamositic granules. *Geo-marine Letters,* **5,** 47–49.
VELDE, B. 1989. Phyllosilicate formation in berthierine peloids and iron oolites. *In*: YOUNG, T.P. and TAYLOR, W.E.G. (eds) *Phanerozoic Ironstones,* Geological Society, London, Special Publication **46,** 3–8.
WEINBERG, R.M. 1973. *The petrology and geochemistry of the Cambro-Ordovician ironstones of North Wales.* Unpublished D.Phil. thesis. University of Oxford.

YOUNG, T.P. 1989. Eustatically controlled ooidal ironstone deposition: facies relationships of the Ordovician open-shelf ironstones of western Europe. *In*: YOUNG, T.P. & TAYLOR, W.E.G. (eds) *Phanerozoic Ironstones,* Geological Society, London, Special Publication **46**, 51–64.

T. P. YOUNG, Department of Geology, UNCC, PO Box 914, Cardiff, CF1 3YE, UK.

# Geochemical and Mineralogical Framework

# Phyllosilicate formation in berthierine peloids and iron oolites


## B. Velde



SUMMARY: Electron microprobe compositions of phyllosilicates formed in berthierine pellets and iron oolites suggest two different reaction paths which result in slightly different bulk chemical compositions.

In the case of berthierine peloids, for mineral aggregates formed and transformed at or very near the sediment–sea water interface, the starting materials seem to be most often dominated by kaolinite. The gradual incorporation of iron and magnesium produces the chlorite composition within the pellet. Systematic compositional zoning is apparent when pellet edges are compared to pellet centres. The compositions of the chlorite end-product have a rather large range of Fe/Fe + Mg (0.65–0.88). The general compositonal range of the transformed berthierine peloids is less aluminous than kaolinite suggesting a high silica activity in the solutions, which is independent of the dissolution of kaolinite.

In the case of berthierine/chamosite-bearing iron oxide oolites, the system is dominated by the presence of iron, as would be expected. The range of compositions observed using the electron microprobe on oolites which still contain iron oxides gives an Al/Si ratio of one, that of kaolinite. The formation of these chlorites is through a combination of divalent iron and kaolinite. These minerals have Fe/ Fe+Mg = 0.88–1.0. The bulk composition is $Si_{26}Al_{26}(Fe,Mg)_{48}$. Multiple point analyses in the same oolite or pellet indicate that the formation of chlorite occurs at various points in the mass with no specific zoning pattern.

When the iron oxide in a rock is completely reacted to form silicate and/or phosphate phases (the oolites are completely chloritized), the berthierine/chamosite mineral composition can become less aluminous due to a general chemical equilibrium established throughout the sample. In more metamorphic iron ore rocks, where the oolitic structure is absent and where new iron silicates are present (such as stilpnomelane) and iron oxides are present, the chlorite composition can be found to vary from that formed in the initial reaction, i.e. Si/Al = 1. They can be more silica-rich or more aluminous.

These analytical results indicate that the composition of the chlorites formed in high iron environments reflect the chemical system in which they form and the scale of that system, a pellet or a rock.


The question of iron berthierine/chamosite mineral formation in ironstones centres on the definition of diagenesis. In fact it is a process which occurs either at the sediment–surface interface (sedimentary) or after burial by sedimentation (diagenetic). These definitions might be somewhat unusual but they are at least clear in their geological meaning. However, this geological definition may or may not coincide with the chemical systems which are operative in the formation of minerals, berthierine/chamosites in the instance under investigation. The present study was designed to answer, if possible, the question of when, and thus where, do the iron-rich phyllosilicates form which are associated with the accumulations of iron which form the ironstones. The two geological expressions of iron accumulation which create the minerals are berthierine pellets and iron oxide oolite accumulations. The importance of the existence of a concentric structure to distinguish between the two groups of material has been noted by Van Houten & Purucker (1984) in their review paper. The problem of the geological environment of chlorite-type mineral formation is found in their question four (p.232): 'Why does berthierine (7Å mineral) form only in peloids today?'

A second problem which has at times confused the investigation of the origin of the iron-rich low-temperature phyllosilicates is one of structural form. The chlorite structure in the low-temperature facies can have two forms, one with a 7Å repeat distance in the $c \sin \theta$ direction and the other with 14Å spacings. These are the berthierine and chamosite forms respectively. Generally speaking, it is considered that the 7Å form occurs below 100°C in sedimentary rocks and the 14Å form occurs above this temperature limit (Velde 1985, p.174).

## Experimental methods

The electron microprobe was the almost exclusive investigative tool used in the study. X-ray diffraction


*From* YOUNG, T. P. & TAYLOR, W. E. G. (eds), 1989, *Phanerozoic Ironstones*
Geological Society Special Publication No. 46, pp. 3-8

of the whole rock samples indicated the presence of 7Å minerals but this is subject to caution because the 14Å band is often very weak (Brindley 1982) and would be masked by the high background counts encountered in the spectra obtained on the material investigated here. X-ray diffractograms of the sediment peloids do not show distinct characteristics which would allow the identification of a specific phyllosilicate mineral. It is not clear when kaolinite has disappeared and when berthierine is present. Therefore, little emphasis has been placed on the identification made using this method.

The microprobe used was a CAMECA Camebax machine operated at 15 KV accelerating voltage and 10 nA current intensity with counting times of 15 s on the peak intensities. These conditions gave statistical errors in the determinations of less than 1% of the amount of the element present for all elements (Mg, Al, Si, Fe) in the chlorites analysed. Given the range of compositions for many analyses, it is obvious that in many instances multiphase assemblages were present under the electron beam. The criterion used to indicate the presence of a chlorite phase was lack of Na, K, Ca, P (less than 0.2 wt%). The total of the cations should be near 20 and not above. The possibility of an intimate interlayering of kaolinite and chlorite, in a ditrioctahedral structure was not entertained here. Some authors (e.g. Bhattacharyya 1983) have proposed a more intimate substitution in the kaolinite to account for bulk compositions of transformed or transforming materials.

Another problem in the interpretation of electron microprobe data is the assignment of an oxidation state to the iron present in the mineral. Traditionally, berthierine and chamosite chlorites contain ferric and ferrous components (Brindley 1982). However the possibility of surface oxidation of the sample should possibly be considered. Unpublished observations (Beaufort 1986) indicate that a natural sample of chamosite can change oxidation state in a period of several years, the determinations being made by Mössbauer spectrometery. Beaufort (1986) has determined the $Fe^{2+}/Fe^{3+}$ values for iron-rich chlorites from metamorphic rocks using Mössbauer methods and finds that the proportion of the ferric component is usually low, between 10 and 25% of the iron present when this component is about 80% Fe/Fe + Mg. Since the exact proportion of ferric iron present is difficult to estimate and since it appears that this component of the iron in the chlorite is much less than half of the ions present, it will be assumed here that the iron is exclusively in the ferrous state in the plots of the analytical data gathered by the electron microprobe. It should be kept in mind that there will be a slight shift towards the trivalent ion pole (Al in the plots used) when more precise analyses are used. However, this error does not change the conclusions to any great extent.

The analytical results are plotted in two chemical frameworks, the $Si-Al-R^{2+}$ (Fe, Mg) and Al–Fe–Mg systems.

**Samples**

The data considered are either new or already summarized in Velde (1985). The berthierine peloids (no concentric ring structure apparent) are from recent sediments (Schellmann 1966; Giresse 1969) or sedimentary rocks (Velde *et al.* 1974; Leone *et al* 1975; Iijima & Matsumoto 1982). All analyses were done by electron microprobe. The oolite mineral analyses were from material described by Schellmann (1969) and new analyses presented here for material from the Falaise (Normandy) and from the Segre (Maine) areas of France. The samples from the Falaise area showed very apparent oolite ring structures. The mineral facies was one of hematite, siderite, chlorite. Samples from the Segre area showed more metamorphic minerals (stilpnomelane and magnetite) and rarely showed the oolite ring structure. The typical feature of the oolites investigated is the presence of a quartz or calcite grain in the oolite surrounded by concentric rings of iron oxide.

## Analytical results

Two major chemical substitutions are known in chlorites, MgSi = $A^{IV}Al^{VI}$ where the number of cations in the structure remains constant and the di-tri-octahedral substitution, MgSi = $Al^{VI}_{0.5}$ where a site vacancy develops in the octahedral sites. The first, iso-cation substitution, is represented on the diagrams $Si-Al-R^{2+}$ used here by a straight line from the serpentine composition to that of amesite. $R^{2+}$ is used for Mg and $Fe^{2+}$. Chlorites which have compositions above this line, the di-tri-octahedral series, are silica-rich with respect to the first series. Thus far, no evidence has been found for homogeneous, non-interlayered structures in the chlorites which form a continuous series between the di- and tri-octohedral mineral compositions. The compilation of chlorite compositions made by Velde (1973) indicates that there is a range from the tri-octahedral chlorites towards about 50% of the di-octahedral substitution. The minerals surveyed were from metamorphic and metasomatic rocks.

Microprobe analyses have been made on peloid minerals coming from sediments, i.e. found at the present-day sediment–sea water interface and on minerals found in sedimentary rocks which have been subjected to burial and probably burial diagenesis. Berthierine sediment peloids give compositions which vary within the individual grains as summarized in Velde (1985). The peloids often become more iron-rich and more aluminous in their mineralogic evolution

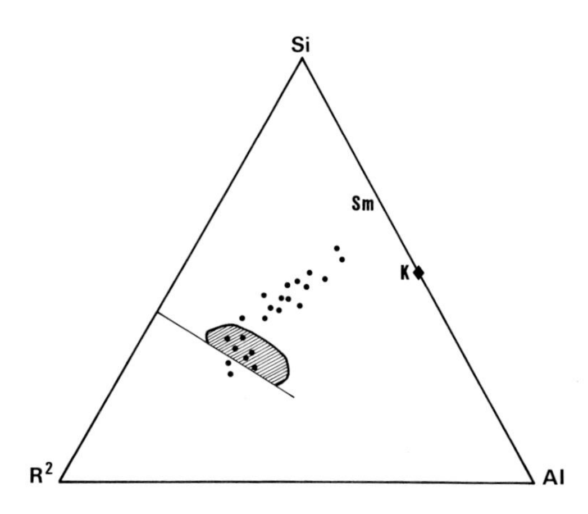


FIG. 1. Plot of peloid centre compositions for samples from the Gulf of Guinea and Gabon sediment surface. The line in the interior of the diagram indicates the strictly tri-octahedral substitution series in the chlorites. The shaded area is that for all berthierine compositions found by microprobe investigations. K = kaolinite, Sm = smectite compositions and $R^{2+}$ indicates the Mg, $Fe^{2+}$ pole. It is apparent that there is an enrichment for many pellets which incorporate $Fe^{2+}$ into the sedimentary pelletal material.

or in some cases they become more Si-rich. Where the composition is compatible with a chlorite the approximate composition $Mg_{2.00}\ Fe_{8.60}Al_{3.00}Si_{6.40}O_{20}(OH)_8$. The ratio of Al/Si is greater than one and the Mg/Mg + Fe ratio is near 0.8.

The peloids in sedimentary rocks which have experienced burial diagenesis seem to be homogeneous in composition throughout their volume. Their compositions are in general compatible with those for chlorites from other environments with a rather large scatter in Al/Si ratios and Si content. The berthierine (mineral) compositions given by Brindley (1982) which have a 7Å structure, are of variable composition also. However these minerals have a Si–Al ratio less than one. Analyses for clearly metamorphic or deep diagenesis of 7Å and 14Å phases described by Iijima and Matsumoto (1982) have compositions which cover a less aluminous field of compositions.

New data for berthierine/chamosites found in oolites coming from the low-temperature facies of Falaise for seven samples, and three samples from Echte, Germany (Schellman 1969), where hematite forming the oolite ring structure is present, are given in Fig. 1. A typical oxide-rich oolite is found to contain compositions which are between iron oxides and kaolinite. A chlorite composition is seen in many instances. It lies on the most silica-poor composition of the tri-octahedral chlorite solid solutions. The chlorites in the strictly tri-octahedral series give compositions which show an atomic total Si + Mg + Al + Fe of 20 when the analyses are calculated on a basis of $O_{10}(OH)_8$. When compositions in the oolites fall below the chlorite line in the Si–Al–$R^2$ diagram it is evident that the point analysed is multiphase and that there is a mixture of iron oxide and berthierine/chamosite. The oolite chlorite compositions are never silica rich nor do they vary much from the Si–Al ratio of one. Their composition is near $Mg_{0.74}Fe_{8.64}Al_{5.40}Si_{5.40}O_{20}\ (OH)_{16}$. A typical series of microprobe compositions from oolites containing oxides is given in Fig. 3.

In oolite samples where there is no longer obvious iron oxide present, cations total 20 or less. The chlorite compositions can be either more silica or alumina-rich than the previous types where the iron oxide is still present in the oolite. All analyses give a composition which is compatible with that of a chlorite. No compositions were found to lie between the limits of known magnesian chlorite compositions and kaolinite (i.e. exceeding 50% of the di-octahedral-type substitution. In samples where the presence of an oolite form was not apparent (higher grade of metamorphism, with stilpnomelane and magnetite facies samples) the same effect was seen: a wider variation in the mineral compositions. In rocks where

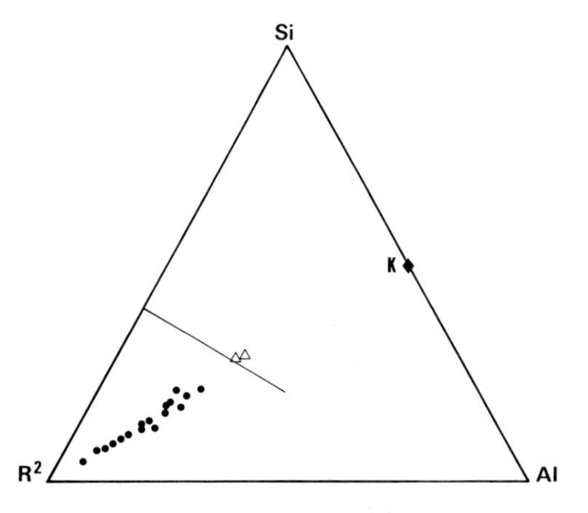


FIG. 2. Plot of microprobe compositions in iron oxide-bearing oolites in a single thin section. Dots show mixture of oxide and chlorite, triangles show pure chlorite composition which forms in the oolites. K = kaolinite composition. The points of the analyses form a continuous line between the oxide pole and that of the kaolinite mineral composition.

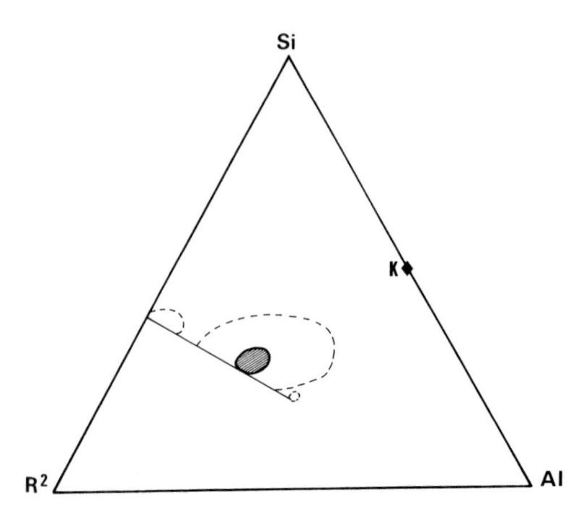


FIG. 3. Plot of the mineral compositions which occur in oolites which still contain iron oxide. Broken curves show the limits of compositions for chlorites of all compositions and origins summarized in Velde (1973).

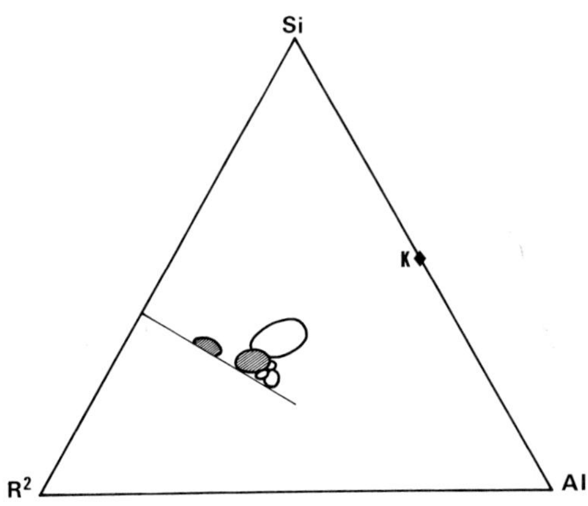


FIG. 4. Plot of compositions of phyllosilicate-filled oolites (shaded area) where no iron oxide remains and where the oolite now contains only the berthierine/chamosite phase. Circles show the compositional ranges of chlorites from higher-grade diagenesis or metamoprphism where stilpnomelane is present in the mineral facies in associated rocks or in the specimens examined. K = kaolinite composition.

stilpnomelane is abundant, the chamosites are more aluminous than in those where little or no stilpnomelane is found.

## Discussion

From the results given above, it appears that there are three types of iron-rich phyllosilicate-producing mechanisms in ironstones: one where the initial material evolves gradually in a confined volume (berthierine peloids) towards a siliceous, low-alumina composition, one where the combination of dissolved kaolinite and ferrous iron (oolite) produces a single composition phase and one where the berthierine/chamosites in peloids or oolites are brought into chemical equilibrium with the entire rock in which they are found. In the first instance, the general evolution of a peloid seems to be brought about through the recombination of a kaolinite-rich pellet in many instances with elements in the solution of deposition. The transfer of the elements necessary to form a iron-rich phyllosilicate phase is effected by migration into the peloid towards its centre. A concentration of the berthierine/chamosite is developed here. This effect is seen in peloids which have not been buried. The second mechanism is one where a pre-existing iron oxide oolite has been buried and transformed during burial diagenesis. This requires dissolution of the iron oxide, probably in the divalent state as this form is much more soluble than trivalent iron which tends to remain in the oxide form. The dissolved iron is immediately combined with Si and Al (plus minor Mg), present in equal quantities due, in all probability, to the dissolution of kaolinite. The resulting berthierine/chamosite, when it can be analysed with the electron microprobe in a pure state, is of the strictly tri-octahedral variety, i.e. silica-poor. This mineral transformation results in a chlorite with a very specific composition. The third type of chlorite-forming reaction is most likely one of recrystallization of re-equilibration of pre-existing berthierine/

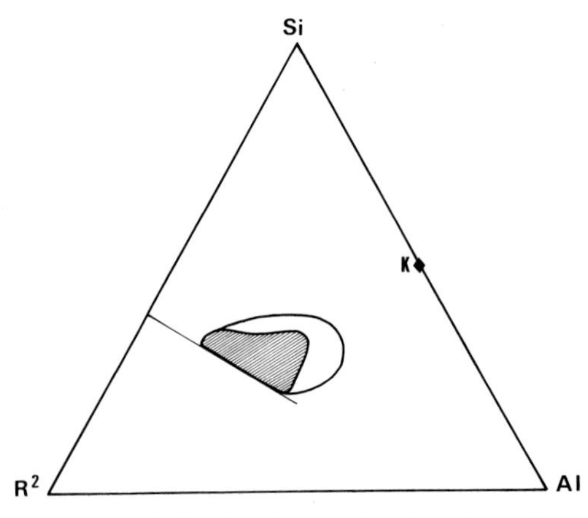


FIG. 5. Plot of all old and new mineral analyses of iron-rich minerals compared to the magnesian chlorites reported in Velde (1973).

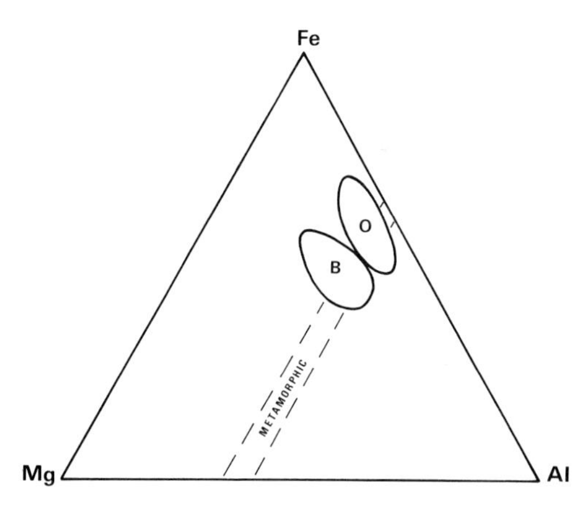


FIG. 6. Plot of oolite-origin mineral compositions (O) and berthierine peloid compositions (B) in Fe–Mg–Al coordinates. The limits of chlorites from metamorphic rocks where white mica is present are found between the broken lines on the diagram.

chamosite minerals formed by the two preceding mechanisms. Here the compositions of the minerals cover much of the range of the general chlorite compositional range as shown in Fig. 6 (present study: Brindley 1982; Klein & Fink 1976; Iijima and Matsumoto 1982; Bhattacharyya 1983). If the above analysis is correct, the post-burial crystallization and evolution of these iron-rich minerals can give phases with the 7Å structure and then the mineral berthierine is a sedimentary and burial diagenetic mineral.

When plotted in Fe–Mg–Al coordinates, the oolite origin and berthierine peloid origin minerals are separated by their iron–magnesium ratio. As one would expect, the oolite material gives a more iron-rich mineral. The berthierine and chamosite minerals of clearly metamorphic origin reported by Iijima & Matsumoto (1982) have compositions which are intermediate between the oolite origin and peloid material. The overall range in their composition is such that the berthierine and chamosites are of lower alumina content than that of the chlorites found in mica schists reported by Velde & Rumble (1977). The chlorites associated with white micas have a very restrained range of Al content (Fig. 6). The chlorites found in diagenetic shale sequences (i.e. coexisting with mixed layer illite/smectite clay minerals) tend to have compositions which are more alumina-rich than the metamorphic chlorites. Thus there seems to be a difference in the chemical parameters which form the diagenetic chlorites in a pelitic environment compared to one of ironstones. The chamosite and berthierine (14Å and 7Å forms) found in deep diagenetic coal measures (Iijima & Matsumoto 1982) have a composition between those of oolite-origin and peloid-origin chlorites. These minerals are of a high-temperature origin according to the description of the authors, clearly indicating that the transition of 7Å to 14Å structure phases occurs beyond the onset of burial diagenesis.

## Conclusions

The answer to the question posed by Van Houten & Purucker (1984) as to why no berthierine minerals formed in environments other than the peloid found at the sediment–seawater interface lies in the mechanism of their formation. Within the peloid micro-environment it must be possible to produce enough $Fe^{2+}$ to form a berthierine/chamosite from the silicates, the iron oxide present in the peloid and from the iron in aqueous solution. This is apparently not possible in the near-surface iron oolite environment where the iron must remain predominantly in the ferric state. In most cases the reduction of iron takes place in a metamorphic or burial diagenesis process. In the berthierines in iron oolite rocks and in the berthierine and chamosites found in coal measures in Japan, the mineral is clearly a product of post-burial processes. The change from the 7Å to 14Å structure is one which occurs beyond the initiation of the burial diagenesis berthierine-forming process. Thus berthierine and chamosite mineral names do not define a change in geologic environment but one of a change in physical conditions of burial, probably temperature.

ACKNOWLEDGEMENT: I wish to thank W. Schellmann for supplying samples.

## References

BEAUFORT, D. 1986. *Definition des equilibres chlorite—mica blanc dans le metamorphisme et le metasomatose.* These d'Etat Université de Poitiers, France.

BHATTACHARYYA, D. P. 1983. Origin of berthierine in ironstones. *Clays & Clay Minerals,* **31**, 173–182.

BRINDLEY, G. 1982. Chemical composition of berthierines—a review. *Clays & Clay Minerals,* **30**, 153–155.

GIRESSE, P. 1969. Etude des differents grains ferrugineux authigenes des sediments sous-marins au large du delta de l'Ogogue (Gabon). *Sciences de la Terre,* **14,** 27–62.

IIJIMA, A & MATSUMOTO, R. 1982. Berthierine and chamosite in coal measures of Japan. *Clays & Clay Minerals,* **30,** 264–274.

KLEIN, C & FINK, R. P. 1976. Petrology of the Sokoman Iron formation in the Howells River area at the western edge of the Labrador Trough. *Economic Geology,* **71,** 453–487.

LEONE, M, ALAINO, R. & CLADERONE, S. 1975. Genesis of chlorite pellets from Mesozoic bedded cherts of Sicily. *Journal of Sedimentary Petrology,* **45,** 618–628.

SCHELLMANN, W. 1966. Sekundare Bildung von Chamosite aus Goethit. *Erzmetal,* **20,** 302–305.

——1969. Die Bildungsbedingungen Sedimentarer Chamosit—und Hamatit-Eisenertze am Beispiel der Lagerstatte Echte. *Neues Jahrbuch Mineralogie Abhandlungen,* **111,** 1–31.

VAN HOUTEN, F. B. & PURUCKER, M. E. 1984. Glauconite peloids and chamositic ooids: favorable factors, constraints and problems. *Earth Science Reviews,* **20,** 211–243.

VELDE, B. 1973. Phase equilibria studies in the system $MgO-Al_2O_3-SiO_2-H_2O$: chlorites and associated minerals. *Mineralogical Magazine,* **39,** 297–312.

——1985. *Clay Minerals: A physico-chemical explanation of their occurrence,* Elsevier, Amsterdam.

——, RAOULT, J-M & LEIKINE, M. 1974. Metamorphosed berthierine pellets in mid-Cretaceous rocks from northeastern Algeria. *Journal of Sedimentary Petrology,* **45,** 618–628.

——& RUMBLE, D. 1977. Alumina content of chlorites in muscovite-bearing assemblages. *Carnegie Institution Washington Yearbook,* **76,** 621–623.

B. VELDE, Laboratoire de Géologie, CNRS 224, Ecole Normale Supérieure, 24 rue Lhomond, 75231, Paris, France.

# Mineral genesis in ironstones: a model based upon laboratory experiments and petrographic observations

**Hermann Harder**

SUMMARY: Although mineral assemblages of ironstones vary widely, with hematite and goethite being the main components of red deposits and berthierine and siderite predominating in green deposits, the bulk major and minor elemental chemical compositions for both types of deposits are very similar. This indicates that the source material and physico – chemical conditions at the site of deposition must have been almost identical and that the different mineral associations probably developed during early diagenesis.

Data from experimental studies including iron clay mineral synthesis throw light upon the post-depositional processes. It is demonstrated that berthierine can be synthesized from iron hydroxides in solution with initial $SiO_2$ concentrations of 5–9ppm, low concentrations of Mg and K, and with ferrous iron dominant.

Decay of organic matter causes changes of Eh towards reducing conditions, which destabilizes the iron-rich particles and leads to both increased concentration of ferrous ions in the pore solution and the release of silicic acid ions. pH and concentration of silicic and carbonic acid controls whether berthierine or siderite is formed. In red deposits the concentration of organic material is low, suggesting that the presence and amount of organic material is a controlling influence.

Although the mineral composition in oolitic ironstones can vary widely, the bulk major and minor elemental chemical composition is broadly similar (Tables 1 and 2).

Most ironstones contain three different types of iron minerals: oxides and oxyhydroxides, silicates and carbonates. The iron oxides–oxyhydroxides are represented by hematite, goethite and minor amounts of magnetite; berthierine (7Å) and chamosite (14Å) are the main iron silicates, predominating over glauconite; and siderite is the main iron-bearing carbonate mineral. These minerals occur in the ironstones in widely varying proportions and in most cases these mineral assemblages are not equilibrium assemblages. In addition, these minerals seldom represent the pure end members of a solid solution series. Aluminium is present in many goethites and hematites in solid solution; berthierine may contain appreciable amounts of aluminium.

Other minerals associated with the iron minerals like quartz, feldspars, various clay minerals, calcite and apatite may form during diagenesis. In some ironstones terrigenous siliciclastic material is virtually absent. In addition to the crystalline phases, X-ray amorphous or poorly crystallized material can be present in the form of iron-oxyhydrates or iron-silicates. Moreover, the content of organic material in ironstones can be very variable.

A major difference between the various iron ores lies in the oxidation state of the iron. In the minette iron ores there are red, brown, green and black coloured deposits, frequently arranged in alternating layering, and such layering can even be observed within single ooids (Berg 1944; Taylor 1949; Harder 1951). The red to purple

TABLE 1. *Bulk chemical composition of iron ores of the minette type*

| | Average values in weight % | | | |
|---|---|---|---|---|
| | Hematite (goethite) ores | Magnetite – berthierine ores | Berthierine – siderite ores | Berthierine rich ores |
| ΣFe | 35 | 45 | 35 | 30 |
| $Al_2O_3$ | 9 | 5 | 7 | 12 |
| $SiO_2$ | 16 | 7 | 12 | 20 |

Data obtained from several hundred chemical analyses of minette iron ores from different locations from all over the world.

*From* YOUNG, T. P. & TAYLOR, W. E. G. (eds), 1989, *Phanerozoic Ironstones*
Geological Society Special Publication No. 46, pp. 9-18

TABLE 2. *Trace element content of 10 samples of Lias iron ores from Dögerode near Echte/Germany (Harder 1964a)*

| | Average values in weight % | |
|---|---|---|
| | hematite ore | berthierine-rich ore |
| Ti | 0.3 | 0.4 |
| Mn | 0.15 (0.08–0.20) | 0.18 (0.027–0.45) |
| P | 0.6 | 0.7 |
| V | 0.06 | 0.09 |
| Cr | 0.08 | 0.1 |
| Zn | 0.014 | 0.014 |
| Cu | 0.02 | 0.014 |
| Pb | 0.003 | 0.015 |
| Co | 0.001 | 0.004 |
| Ni | 0.011 | 0.02 |

colours indicate that iron is present mainly in its oxidized state, as the trivalent ion, in the green ores divalent iron predominates, while in the blackish, dark greenish and dark red deposits both valencies are present in varying quantities. Thus the different states of oxidation lead to different iron ore facies. What causes these differences? Two explanations may be considered as foremost:

(1) The change of the physico-chemical conditions during deposition,
(2) the roles of the diagenetic processes.

If the present mineralogy of the iron ores represents the originally deposited mineral association, this would imply either variable supplies of ferrous iron to the site of deposition, to be precipitated there as ferrous iron species, or fluctuation of the Eh at the site of deposition, causing alternate formation of ferrous and ferric iron minerals (Smith 1892; Taylor 1949; Borchert 1952; James 1966; Maynard *et al.* 1986). These explanations are hardly consistent with observations of bioturbation or of organic remains in life-like position, both of which indicate effective aeration in the area of deposition.

If ferrous and ferrous-ferric iron minerals form diagenetically after deposition (Sorby 1856; Schneiderhöhn 1924; Correns 1942; Harder 1951, 1964a; Strakov 1959a; Bubenicek 1960) then oolitic iron ores would be primarily deposited in agitated, well oxygenated coastal waters in the presence of organisms such as echinoderms, brachiopods, burrowing animals, boring algae, fungal mats (Dahanayake & Krumbein 1986) and bacteria (Harder 1919). A further argument for the formation of the oolitic iron ores in an oxygenated environment is the uniform content of trace elements in the ferric- and ferrous-rich iron ores (Table 2). If the ferrous- and ferric iron ores were formed under different depositional Eh conditions, substantial differences in the trace element contents would be expected, especially for elements that occur in more than one valency (V, Cr, Mn) and the chalcophile elements (Pb, Zn, Cu, Ni).

Although sedimentary ironstones have been studied for a long time, there is still no generally accepted mechanism for their formation. Various sources for the iron supply have been discussed, and only a little is known about its chemical composition. Geological, petrographical and geochemical data give very little information about the physico-chemical conditions of the environments in which the ores were formed, about enrichment mechanisms and mineral formation processes. Since pore water analyses are lacking, a reconstruction of diagenetic reactions is difficult. Under these circumstances syntheses of minerals, especially iron silicates, under surface conditions might contribute to the understanding of the enrichment mechanisms. They can give some detailed information about the chemical composition of the solutions in which the different iron minerals can form. The data from experiments must always be supplemented by observations and data of iron mineral occurrences in recent sediments.

Iron can be enriched in sediments by detrital accumulation or chemical precipitation. For the formation of ironstones relatively high iron concentrations are needed. Intensive weathering can produce iron-enriched accumulations. A supply of mobilized iron to the site of deposition where it is chemically precipitated is another possibility.

Mudstones with gel-like textures seem to be the primary iron-rich sediments. Reworking of such 'muddy' ironstones may lead to the formation of oolitic iron ores, possibly with further enrichment of iron. Often the ooids are cemented by material similar in composition to that of the mudstones. Iron ores with the typical composition and oolitic texture are frequently called minette iron ores.

## Silica-bearing hydroxides: precursors in iron mineral formation

Chemical analyses indicate that oolitic ironstones contain substantial amounts of silica and aluminium, even in the case of hematitic and goethitic ores. Why are ores with such a different mineralogy so similar in their chemical composition? Were aluminium, silica and also the trace elements deposited simultaneously with the iron? If this latter situation were the case, how can these elements then be transported to the site of deposition together? Did they occur in true solution, as colloids or as detrital minerals? Only very little iron can be transported in true solution in water with pH of 5–8 and positive Eh; e.g. in river water iron is predominantly transported in colloidal form, aluminium behaves similarly, but what about silica?

Data derived from both experimental work (Harder 1965; Flehmig 1967) and investigation on several post-volcanic spring waters (Harder 1964a, b) showed that hydroxides can remove silica even from very dilute solutions by co-precipitation or chemisorption. To study these processes more thoroughly two sets of experiments were carried out:

(1) precipitation of iron hydroxides in silica-bearing solutions,
(2) aged iron hydroxides were put into silica-bearing solutions.

## Experimental methods

The concentration in the initial solutions was kept between 50 ppm and to 0.1 ppm; in this concentration range silicic acid is present in its monomeric form and results can be compared with natural situations. The solutions were undersaturated with respect to amorphous silica (solubility at 20°C and pH 5–8 ≈ 100 ppm $SiO_2$) and even in some experiments with respect to quartz (solubility at 20°C and pH 5–8 ≈ 10 ppm $SiO_2$).

The iron solution was freshly prepared from ferrous sulphate, aluminium was added as sulphate. Reducing conditions were produced mainly by addition of Na-dithionite stabilized by a nitrogen atmosphere. Iron hydroxides were precipitated either by slowly changing the Eh or the pH. Following the precipitation Eh and pH were kept constant.

The fixation of the silica into the hydroxides occurs as a reaction between colloids and the ions in the solution. The particles of the iron hydoxide sols are generally positively charged and attract the negatively charged silicic acid anions. Excess of silicic acid may even lead to charge reversal connected with aggregation phenomena.

The precipitates were aged in the solution from a day to several months, then filtered, carefully washed and dried (if necessary under a nitrogen atmosphere).

Results of experiments of the first type are summarized in Table 3. They show the enrichment of silica in the precipitates under various conditions. Since the formation of iron

TABLE 3. *Chemisorption of silica by iron hydroxides under various Eh conditions (Harder 1978)*

| Reducing cond.: % Na-dithionite | pH | Temperature (°C) | Concentration in initial solution | | | Conc. in final solution (ppm Fe) | Composition of the precipitates (ignited to 1000°C) % | | |
|---|---|---|---|---|---|---|---|---|---|
| | | | (ppm) Fe | (ppm) $SiO_2$ | % $Al_2O_3$ | | $Fe_2O_3$ | $SiO_2$ | $Al_2O_3$ |
| – | 7 | 3 | 20 | 18 | | – | 70 | 30 | |
| 0.3 | 8 | 3 | 20 | 20 | | 16 | 68 | 32 | |
| 1 | 8 | 3 | 20 | 20 | | 20 | no precipitate | | |
| – | 7 | 22 | 20 | 18 | | – | 74 | 26 | |
| 0.1 | 8 | 20 | 20 | 20 | | 6 | 76 | 24 | |
| – | 9 | 20 | 4 | 20 | | – | 64 | 36 | |
| 0.03 | 8 | 20 | 4 | 20 | | 0.4 | 68 | 32 | |
| 0.1 | 8 | 20 | 4 | 20 | | 1.8 | 40 | 60 | |
| – | 7 | 22 | 0.2 | 18 | | 0.04 | 44 | 56 | |
| 0.1 | 8.5 | 20 | 0.3 | 13 | | 0.3 | no precipitate | | |
| 0.1 | 8.5 | 20 | 0.3 | 13 | 0.3 | 0.2 | 5 | 80 | 15 |

hydroxides is strongly dependent on the Eh and pH conditions in the solution, the precipitates may contain ferrous and/or ferric iron hydroxides. These silica-bearing iron hydroxides are relatively stable in contrast to 'pure' iron hydroxides. The latter are initially also amorphous in X-ray studies, but after a short time they commence crystallization into goethite and/or magnetite losing in the process much of the initial chemisorption capacity, but still retaining part of their ability to remove silica from solutions (Flehmig 1967).

Similar results were obtained with hydroxides of other elements such as aluminium and magnesium, and with mixed hydroxides (Al + Mg, Al + $Fe^{2+}$, $Fe^{2+}$ + Mg). With potassium present in the initial solution potassium–silica-bearing hydroxides were formed (Harder 1965; Flehmig 1967; Kurze 1971). Depending on the initial concentration of the cations and the silicic acid in the solutions, the changes of pH and Eh during the experiments and the temperature, precipitates were formed which contained all elements necessary for clay mineral formation. It was reasonable, therefore, to assume that under certain circumstances they may act as precursors for clay mineral formation. Further experiments on the ageing of these precipitates became necessary to demonstrate this.

## Synthesis of iron clay minerals under natural conditions

Conditions favourable for iron clay mineral synthesis were primarily obtained from recent occurrences. In the experiments the conditions were varied until the silica-hydroxide gels possessed the right composition so that, during the process of ageing, crystallization of clay minerals could take place.

All experiments were carried out in the same way: the hydroxides were precipitated from solutions that contained iron, silicic acid and in some cases also aluminium, potassium and magnesium, either by changing the pH or the Eh. Following precipitation Eh and pH were kept constant, except in some experiments in which the hydroxides were precipitated under oxidizing conditions and then aged in a reducing environment. The time of ageing varied from a day to several months. After careful filtration, washing and drying, the precipitates were investigated by X-ray diffraction and chemical analysis. Because most of the synthesized products were only poorly crystallized, Debye Scherrer cameras were used. The interpretation of the powder diffraction photographs were based on the following scheme:

(a) the *d*(001) values were used to differentiate between the clay mineral groups as follows:

| | | |
|---|---|---|
| 14–16Å | air dry, expanding after glycol treatment | iron-rich smectite (nontronite) |
| 14Å | no reaction with glycol | iron-rich chlorite (chamosite) |
| 10Å | no expansion after glycol treatment | iron-rich illite (glauconite) |
| 7Å | | iron-rich mineral of the kaolin–serpentine group (berthierine, greenalite). |

(b) The *d*(060) values were used to distinguish between di- and tri-octahedral minerals as follows:

| | |
|---|---|
| 1.49–1.52 Å | di-octahedral, $Fe^{3+}$ and $Al^{3+}$ are the main cations in octahedral position; |
| 1.52–1.54 Å | tri-octahedral, $Fe^{2+}$ and $Mg^{2+}$ cations occupy the octahedral positions. |

In Table 4 data are summarized from experiments, as examples of the synthesis of berthierine, nontronite and glauconite. The results of these investigations show that:

(1) (to allow synthesis) the chemical composition of the precipitate must be similar to that of the appropriate clay mineral;
(2) di- or tri-octahedral, two- or three-layer iron clay minerals form only under reducing conditions;
(3) in solutions with a higher pH (8–9) and low Eh values the iron clay minerals crystallize more fully and at faster rate;
(4) concentrations of 300 ppm KCl in the initial solution are necessary for the synthesis of glauconite.

Besides the formation of clay minerals, minerals like quartz or alkali feldspars can develop within the silica-containing hydroxides. If the silica concentration is too low in the precipitates a transformation into goethite, hematite and quartz (Harder & Flehmig 1970) will occur. The newly formed quartz crystals had grain sizes between 0.01 and 1 mm and rarely showed an idiomorphic habit, more often they were irregularly shaped, sometimes with inclusions of hydroxides. This observation must be taken into account in any attempt to differentiate between

TABLE 4. *Syntheses of berthierine, nontronite and glauconite, at 20° C, pH 7–9*

| Composition of initial solution | | | | | | Composition of precipitates (ignited to 1000° C) | | | | | X-ray data | |
|---|---|---|---|---|---|---|---|---|---|---|---|---|
| Na-dithion (%) | $SiO_2$ (ppm) | Fe (ppm) | Al (ppm) | Mg (ppm) | KCl (%) | $SiO_2$ (%) | $Fe_2O_3$ (%) | $Al_2O_3$ (%) | MgO (%) | $K_2O$ (%) | *d*(A) (001) | (060) |
| – | 20 | 10 | | | | | | | | | amorphous | |
| –[1] | 20 | 10 | | | | | | | | | 7.9 | 1.54 |
| 0.1 | 20 | 20 | | | | | | | | | 7.9 | 1.50 |
| 0.1 | 10 | 0.6 | 0.3 | | | | | | | | 7.9 | 1.50 |
| 0.1 | 20 | 7,5 | 1.5 | 1290 | | | | | | | 7.9 | 1.54 |
| 0.1 | 20 | 20 | | 1290 | | 31 | 60 | | 8 | | 7.9 | 1.55 |
| 0.1 | 20 | 10 | 5 | 1290 | | 34 | 23 | 13 | 30 | | 8.5 | 1.54 |
| – | 20 | 4 | | | | 28 | 72 | | | | amorphous | |
| 0.03 | 20 | 4 | | | | 49 | 51 | | | | 11–16 | 1.515 |
| 0.3 | 20 | 4 | | | | 47 | 53 | | | | 11–16 | 1.53 |
| 0.1 | 16 | 1 | 0.3 | | | 51 | 32 | 17 | | | 11–16 | 1.508 |
| 0.1 | 13 | 0.6 | 0.3 | 300 | 0.1 | 57 | 16 | 13 | 6 | 6 | 10.0 | 1.51 |

[1] Aged under reducing conditions 0.1% Na-dithionite at pH 8→Eh – 0.6 V.

detrital and authigenic quartz in sediments such as ironstones.

As a rule one can say that in the $Fe^{2+}-SiO_2$ system the silica concentration in the initial solution controls the mineral formation in the precipitates as follows:

| **$SiO_2$ concentration in initial solution** | **mineral formed** |
|---|---|
| <5 ppm | quartz |
| 5–9 ppm | berthierine |
| >10 ppm | nontronite, glauconite |

The formation of minerals within the hydroxides may be likened to a process similar to the gel technique as the voluminous precipitates also contain a large amount of pore solution (initially greater than 90%). The chemical compositions of these pore solutions are different from that of the bulk solutions, and there can be no doubt that the composition changes during the process of ageing, but unfortunately little is known about this. One can only speculate that, for example, in the case of quartz formation within hydroxides an initial solution undersaturated with respect to quartz, the pore solution of the hydroxides would need to have a silica concentration which is equal to or slightly higher than that known from quartz solubility measurements (10 ppm $SiO_2$ at 20°C and pH 5–8).

But how can these experimentally obtained data help in the understanding of the genesis of ironstones? It is necessary first to study natural processes and their mechanisms as far as possible and then try to find similarities.

## Source of major and minor elements for the minette iron ores

An understanding of the sedimentary processes that led in the past to the formation of ironstones is not possible without studying present-day processes. The origin of iron and its transport to a site of deposition is of great importance, but this should only be considered in the context of the source and transport mechanisms of the aluminium, silica, titanium and other trace elements. Volcanic activity and weathering are two possible sources.

In the vicinity of active volcanoes (e.g., Etna, Italy; Santorini, Greece) and also in areas of post-volcanic activity (e.g., Eifel, Germany) spring waters relatively rich in ferrous iron occur. Low-temperature (10–30°C) thermal waters generally have pH values about 6, controlled by carbonic acid. They contain alkali ions, alkali earth ions, silica and manganese but very little aluminium and titanium (Table 5). When these waters become oxidized, ferric iron hydroxides precipitate. The composition of such precipitates is also given in Table 5. These data indicate that iron ores of this origin can be rich in silica and manganese but low in aluminium, titanium and phosphate. The iron ores of the Lahn Dill type are assumed to have been formed by exhalative sedimentary processes and their chemical composition is similar to that given for recent precipitates.

The iron-rich hot brine sediments in the Red Sea, consisting dominantly of amorphous to poorly crystallized iron oxide, amorphous silica

TABLE 5. *Chemical composition of spring waters and precipitates from such low-temperature thermal waters (Harder 1964b; James 1969)*

| | Spring water Wehrer Kessel Eifel Germany (ppm) | | Precipitates from spring waters Wehrer Kessel (%) | Santorini (Greece) (%) | Hot brine sediments Red Sea (%) |
|---|---|---|---|---|---|
| Fe | 18.8 | $Fe_2O_3$ | 64 | 58 | 43.2 |
| $SiO_2$ | 60.8 | $SiO_2$ | 8 | 18 | 39.1 |
| Mn | 4.1 | MnO | 0.3 | 0.15 | 2.7 |
| Al | 0.3 | $Al_2O_3$ | 0.8 | 0.05 | 5.7 |
| Ti | 0.005 | $TiO_2$ | 0.02 | 0.01 | <0.05 |
| $HPO_3$ | 0.27 | $P_2O_5$ | 0.17 | 0.3 | 0.4 |

and nontronite, are also genetically related to hot springs, but these waters are highly saline. The deposits, precipitated from the brine water column, generally have higher aluminium but very low titanium content (Table 5).

From areas of subaerial weathering under tropical to subtropical climates very little iron and aluminium is transported by rivers in true solution (e.g., Amazon: 0.002–0.1 ppm Fe, 0.02–0.06 ppm Al, 9.1–12.4 ppm $SiO_2$; from Gibbs 1967). Most is carried in colloidal form, as detrital particles or in organic complexes. Such iron-rich colloidal phases were collected during a wet season from the river Nile (Harder 1964b). Iron oxyhydrates and silica were the main components, but the colloidal phases also contained up to 15% $Al_2O_3$, 1.8% $TiO_2$, 1.7% MnO, and 0.25% $V_2O_5$. The overall chemical composition of the suspension load of rivers is certainly somewhat dependent on the petrography of the source area.

In areas of higher latitude so called 'bog iron ores' in lakes and swamps (Von Gehlen & Harder 1956) are still forming. In the subaqueous facies of humic and podzolic soils and peat bogs iron can be transported as soluble humates or bicarbonate and it will be precipitated in response to rise in Eh, when the waters become aerated, or by an increase of pH. Precipitates and ores, formed from such solutions are relatively poor in aluminium, titanium and certain trace elements (Table 6).

Submarine remobilization of iron under reducing conditions from sediments and transport as ferrous iron to the site of deposition should result in iron ore compositions similar to that of the bog ores, in so far as the aluminium, titanium and vanadium contents are also expected to be very low. The above observations and investigations indicate that an adequate supply of iron can be introduced via rivers from lateritic weathering areas to the sites of deposition of ironstones.

## A discussion of depositional and post-depositional processes in the ironstones

Minette iron ores are generally interpreted as having formed close to the shoreline in well oxygenated brackish to marine waters. The iron-rich particles, pre-concentrated in aluminium, silica and titanium, were either derived from lateritic weathering areas close to the shoreline by erosion or transported into this environment by rivers. In zones of increasing salinity the colloidal particles were flocculated and accumulated together with siliceous and calcareous organisms. The ferric iron oxyhydrates, geothite, lepidocrocite and hematite, perhaps with small amounts of ferriferous minerals of detrital origin, were deposited either as mudstones, from which by reworking oolitic ironstones developed, or ooids formed directly from the material supplied by the rivers. With the simultaneous deposition of organic matter, sediments with varying quality and quantities of organic carbon content were produced, dependent upon factors such as organic productivity and rate of sedimentation.

TABLE 6. *Composition of a recent iron-rich precipitate from bog water and of a recent bog iron ore, rich in siderite ('Weisserz') (Gehlen & Harder 1964)*

| | Precipitate (%) | 'Weisserz' (%) |
|---|---|---|
| FeO | n.d. | 41.8–49.6 |
| MnO | 0.1 | 0.5–0.7 |
| $Al_2O_3$ | 0.8 | 0.0–1.5 |
| $SiO_2$ | 5.3 | n.d. |
| $P_2O_5$ | 0.3 | 0.7–2.2 |
| $TiO_2$ | 0.01 | n.d |

Generally, soon after deposition, the decay of organic matter and reactions of the solid with the aqueous phase led to changes in the chemistry of the pore solutions. Lower Eh values result from anaerobic bacterial decomposition of organic material. In marine sediments, containing abundant dissolved sulphate, this involves reduction of sulphate to $H_2S$ (Schneiderhöhn 1944). In fresh water sediments, low in dissolved sulphate, the anaerobic bacterial decay may enable the attainment of a low Eh and high $P_{CO_2}$. Because of the almost complete absence of pyrite in the oolitic iron ores it has to be assumed that either the sediments were deposited in a more brackish environment low in dissolved sulphate, or the oxidation potential was low enough to convert $Fe^{3+}$ to $Fe^{2+}$ but not so low as to allow the reduction of sulphate to $H_2S$. Thus, the early diagenetic environment may be described according to the geochemical classification of Berner (1981), as post-oxic (weakly reducing).

In the early stages of diagenesis of the ironstones the bacterial decomposition and microbiological fermentation of organic matter may result in a change of the redox potential towards reducing conditions and influence the pH of the pore solutions, depending on the amounts of $CO_2$, organic acids and nitrogeneous bases formed. In addition the organic matter serves as a reducing agent, destabilizing ferric iron hydroxides to $Fe^{2+}$ and thus increasing the concentration of dissolved ferrous iron in the interstitial solutions. Simultaneously, ions such as silicic acid and aluminium, preconcentrated in the sedimented iron-rich particles become mobilized. An increase of dissolved silicic acid in the pore solution would also be achieved through dissolution of siliceous skeletal debris. With increasing overlying sediment load, pore waters probably moved upward into less reducing environments. By reaction of such pore solutions with the solid phases, accompanied by progressive alteration and replacement of the substrate, the various mineral assemblages of the ironstones crystallized in the pores; but what were the specific geochemical conditions? An attempt to gain an insight into the conditions of these mineral formations within the ironstones can be made by combining field data from modern sediments with the laboratory data.

In modern marine sediments authigenic berthierine and glauconite occur predominantly in the form of pods such as fecal pellets or test fillings. The formation of berthierine seems to be restricted to water depths of less than 80 m, probably related to organic activity, while glauconites are found in temperate latitudes in waters as shallow as 10 m, but in tropical climates in depths up to 100 m. It is assumed that berthierine and glauconite develop by early diagenetic anoxic transformation of a precursor such as lateritic iron-rich colloidal particles or iron-rich coatings around detrital grains (von Gärtner & Schellmann 1965), and that the glauconitization process takes place close to the water/sediment interface since free access to seawater seems to be necessary in this process. Berthierine in the form of fecal pellets and envelopes around sand grains is also described from recent sediments of the low salinity (23–27ppm) Loch Etive in Scotland. At temperatures of about 8°C berthierine is formed diagenetically within the sediment (Rohrlich *et al.* 1969).

These observations suggest that the environment of deposition is less critical and that the diagenetic reactions within the sediments, controlled by the inorganic phases, organic matter and interstitial waters, play a far more important role not only in the formation of berthierine and glauconite but also of magnetite and siderite. Since little is known about these reactions the experiments may contribute some information. The investigation of the formation of greenalite can be used as an example. Greenalite was synthesized in anoxic solution that contained initially 20 ppm $Fe^{2+}$ and 20 ppm $SiO_2$. Under slightly alkaline conditions precipitation of ferrous hydroxide decreased the ferrous iron concentration in the solution to about 5 ppm Fe and the concentration of the silicic acid as a result of its coprecipitation to about 10–15 ppm $SiO_2$. This solution was in equilibrium with a precipitate having an $FeO/SiO_2$ ratio favourable to act as precursor for greenalite.

In pore solutions of brackish to fresh water sediments it is more likely that berthierine forms during early anoxic diagenesis with magnesium from the entrapped water and ferrous iron, silica and aluminium released from the iron-rich particles. However, berthierine will develop only when the metal oxide/$SiO_2$ ratio in the hydroxide gels is about two, while ratios of about one favour the formation of the 2:1 clay minerals such as nontronite, or, if enough magesium and potassium is available and the redox potential less reducing, glauconite crystallizes in the precipitates (Harder 1980). Conditions for the formation of glauconite can only be realized close to the sediment/water interface in a marine environment, where sea water components can easily migrate into the pore solution and where the redox potential allows a high $Fe^{3+}/Fe^{2+}$ ratio in the pore

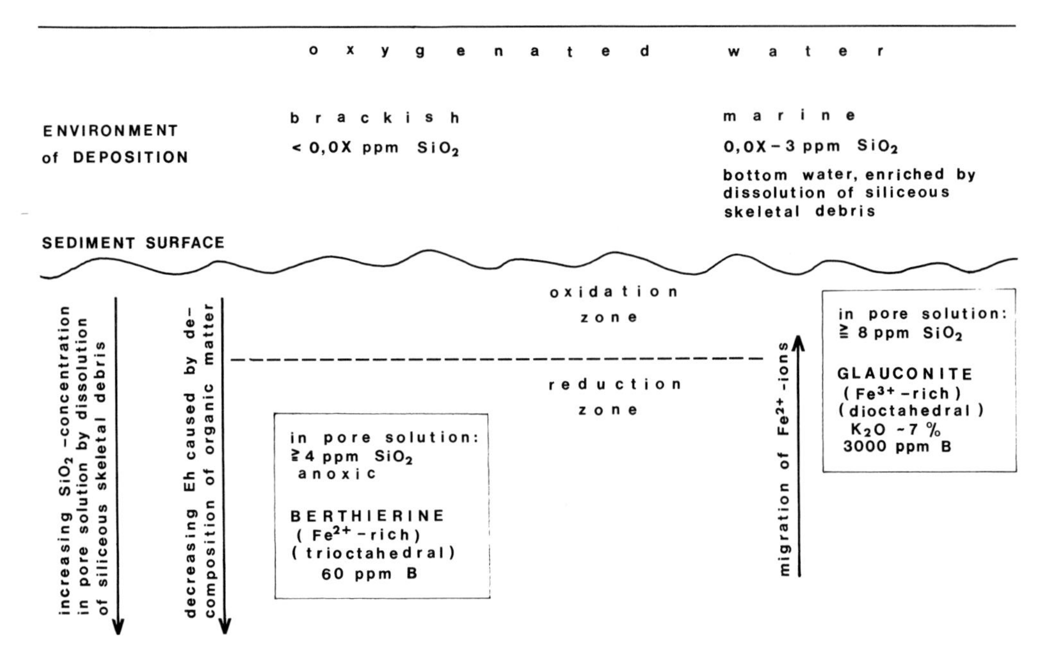


FIG. 1. Diagenetic iron clay mineral formation.

solution. The high boron content of glauconite (up to 3000 ppm; Harder 1974) points towards an origin in marine sediments. Glauconites are almost entirely absent in ironstones and thus indicate that these processes were of minor importance in the genesis of the ironstones, and the low boron contents of berthierines (50—150 ppm B; Harder 1951, 1974) supports the hypothesis of ironstone deposition in a more brackish milieu. Some criteria for the formation of berthierine glauconite are summarized in Fig. 1.

Siderite is diagenetically formed in an anoxic milieu with high $CO_2$ concentration, caused by the decay of organic matter. Siderite can also replace calcite when the concentration of dissolved ferrous iron in the pore solution is higher than that of calcium. The amount of siderite formed depends on the $CO_2$ content and the ferrous iron concentration available for this reaction, but it should be realized that ferrous iron ions may participate in more than one reaction process at a time. This explains why berthierine and siderite may be found together in ironstones. To reconstruct the various physico-chemical conditions favourable for the different mineral assemblages is very difficult because of the very complicated pore water chemistry with the complex interactions, (e.g., between metal ions and dissolved organic substances).

## Conclusions

The mineral genesis of ironstones can be schematically illustrated (Fig. 2) according to the above concepts. The source of the iron is lateritic weathered areas. Iron-rich particles, pre-concentrated in aluminium, silica, and titanium, are transported from there via rivers to the site of deposition. Here they accumulate together with organic material.

In sediment layers initially low in organic matter, goethite and hematite will be stable, some recrystallization may take place and silica, released by dissolution processes, will precipitate as quartz. The increase of organic matter in sediment layers causes a change of Eh towards reducing conditions and a decrease of the $Fe^{3+}/Fe^{2+}$ ratio in the pore solutions. Ferrous iron can be enriched in the solution to such an extent that magnetite crystallizes; hematite and possibly goethite will still be present because of their low solubility and reactivity. In this environment quartz may also form. With still more organic matter available Eh decreases further to levels where $Fe^{2+}$ is the dominant iron ion in the pore solution. Depending on pH, concentration of silicic acid and $CO_2$, berthierine or siderite will form, but it is also possible that the physico-chemical conditions will favour the formation of berthierine + siderite or

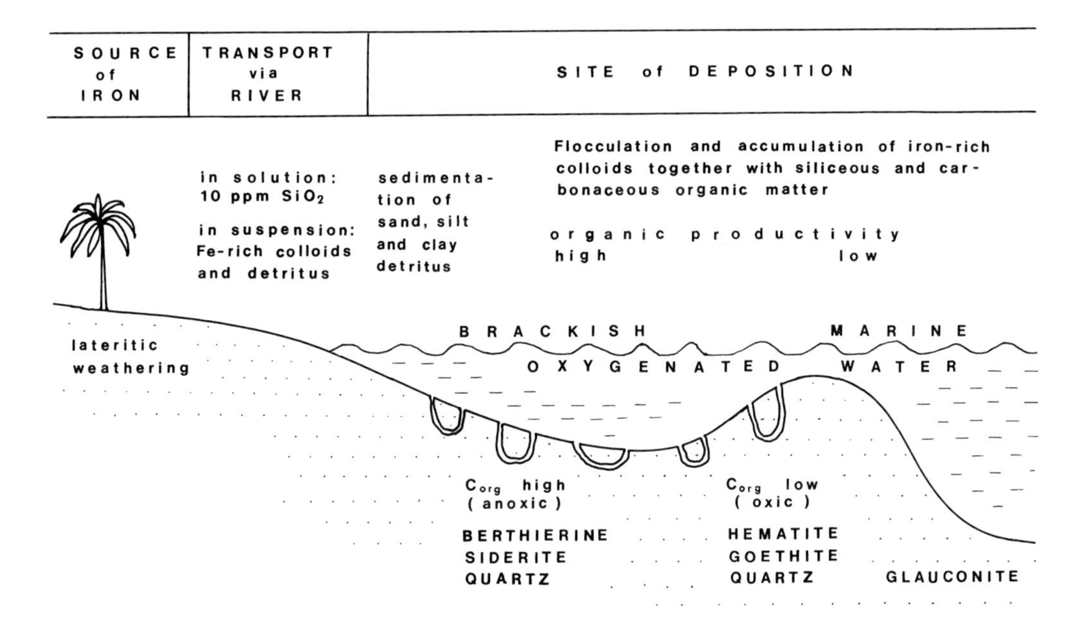


FIG. 2. Weathering, transport and diagenesis in iron mineral formation.

siderite + quartz. Because of their low reactivity and solubility, hematite and magnetite may also be present.

For the formation of glauconite it is assumed that iron-rich particles were deposited in marine environment with bottom water possibly enriched in silica by dissolution of siliceous skeletal debris. Because of the low initial content of organic matter Eh within the sediment will drop only slightly thus creating $Fe^{3+}/Fe^{2+}$ ratios favourable for glauconite formation. The high silicic acid concentrations required for glauconite formation were derived from dissolution of siliceous skeletal debris within the sediment, from entrapped sea water (together with the magnesium and potassium required) and by exchange processes at the sediment/water interface.

With this concept the alternating layering of oxidized and reduced iron ore beds can be explained, as can the absence of glauconite in the ironstones. Subsequent to these very early diagenetic processes, further reactions can take place in the iron ores and chamosite can mainly be interpreted as having been formed during late diagenetic to low-grade metamorphic processes.

Chemical weathering near or at the earth's surface will somewhat reverse the above discussed reactions. Siderite and berthierine become unstable under oxidizing conditions, leaving geothite and hematite as dominant iron minerals in the deposits together with some kaolinite or smectite.

ACKNOWLEDGEMENT: I am gratefully indebted to Dr A. Heydemann for his critical review of the text and for improving the English, and to the reviewers.

## References

BERG, G. 1944. Vergleichende Petrographie oolithischer Eisenerze. *Archives Lagerstattenforschungen,* **76**, Berlin.

BERNER, R.A. 1981. Authigenic mineral formation resulting from organic matter decomposition in modern sediments. *Forschungen Mineralogie,* **59**, 117–135.

BORCHERT, H. 1952. Die Bildungsbedingungen mariner Eisenerzlagerstätten. *Chemie der Erde,* **16**, 49–74.

BUBENICEK, L. 1960. *Rechercles sur la Constitution et la Repartion des Minerais de fer dans L'Aalanien de Lorraine.* These Docteur, Nancy.

CORRENS, C.W. 1942. Der Eisengehalt der marinen Sedimente und seine Entstehund. *Archives Lagerstattenforschungen,* **75**, 47–57.

DAHANAYAKE, K. & KRUMBEIN, W.E. 1986. Microbial stuctures in oolithic iron formations. *Mineralium Deposita,* **21,** 85–94.

FLEHMIG, W. 1967. . *Zur Erklarung des Kieselsauregehaltes in Nadeleisenerzooiden.* Dissertation, Munster.

VON GEHLEN, K. & HARDER, H. 1956. Zur Genese der kretazischen Eisenerze von Auerbach (Oberpfalz). *Heidelberg Beitrage zur Mineralogie und Petrographie,* **5,** 118–138.

VON GARTNER, H.R. & SCHELLMANN, W. 1965. Rezente Sedimente im Kustenbereich der Halbinsel Kaloum, Guinea. *Mitteilungen der Mineralogie Petrogrophie,* **10,** 349–367.

GIBBS, R.J. 1967. The Geochemistry of the Amazon River System: Part 1. The Factors that Control the Salinity and the Composition and Concentration of the Suspended Solids. *Bulletin of the Geological Society of America,* **78,** 1203–1232.

HARDER, E.C. 1919. Iron-Depositing bacteria and their geologic relations. *Professional Paper 113 of U.S. Geological Survey.*

HARDER, H. 1951. Uber Den Mineralbestand und die Entstehung einiger sedimentrer Eisenerze des Lias. *Heidelberg Beitrage der Mineralogie und Petrographie,* **2,** 455–476.

—1964a. On the diagenetic origin of Berthierine (Chamositic) Iron Ores, *12th International Geological Congress, New Delhi, Part V,* 193–198.

—1964b. The use of Trace elements in distinguishing different genetic types of marine sedimentary iron ores. *12th International Geological Congress, New Delhi, Part V,* 551–556.

—1965. Experiment zur "Ausfällung" der Kieselsaure. *Geochimica et Cosmochimica Acta,* 29, 429–442.

—1974. Boron *In:* Wedepohl, K. H, and others Handbook of Geochemistry. Vol. 2/4, Springer-Verlag, Berlin.

—1978. Synthesis of iron layer silicate minerals under natural conditions. *Clays and Clay Minerals,* **26,** 65–72.

—1980. Syntheses of Glauconite at surface temperatures. *Clays and Clay Minerals,* **28,** 217–222.

—& FLEHMIG, W. 1970. Quarzsynthese bei tiefen Temperaturen. *Naturwissenschaften,* **60,** 517.

JAMES, H.L. 1966. Chemistry of the iron-rich sedimentary rocks. *Professional Paper, 440-W of the U.S. Geological Survey.*

—1969. Comparison between Red Sea Deposits and Older Ironstone and Iron-formation. *In:* DEGENS, E.T. & ROSS, D.A. (eds) *Hot Brines and Recent heavy Metal Deposits in the Red Sea,* Springer Verlag, Berlin.

KURZE, R. 1971. *Synthese von Illit bei tiefen Temperaturen* Thesis, Universitat Gottingen.

MAYNARD, B.J. 1986. Geochemistry of Oolitic Iron Ores, an Electron Microprobe Study. *Economic Geology,* **81,** 1473–1483.

ROHRLICH, V., PRICE, N.B. & CALVERT, S.E. 1969. Chamosite in the recent sediments of Loch Etive, Scotland. *Journal of Sedimentary Petrology,* **39,** 624–631.

SCNEIDERHOHN, H. 1924. Untersuchungen uber die Aufbereitungs-moglichkeiten der Eisenerze des Salzgitterhohenzuges aufgrund ihrer mineralogisch-mikroskopischen Beschaffenheit. *Mitteilungen der Kaiser Wilhelm Institut Eisenforschungen Dusseldorf,* **5,** 79–108.

—1944. *Erzlagerstatten.* Gustav Fischer Verlag, Jena.

SMITH, H. 1892. On the Clinton iron ore. *American Journal of Science,* **43,** 487.

SORBY, H.C. 1856. On the origin of the Cleveland Hill Ironstone. *Proceedings of the Geological and Polytechnic Society of the West Riding of Yorkshire,* **3,** 457–461.

STRAKOV, N.M. 1959 Schema de la diagenese des depots marins. *Ecologae Geologiae Helvetiae,* **51,** 761–767.

TAYLOR, J.H. 1949. *Petrology of the Northampton Sand Ironstone Formation. Memoir of the Geological Survey of Great Britain,* His Majesty's Stationary Office, London.

H HARDER, Sedimentpetrographisches, Institut der Universitat Gottingen, D-3400 Gottingen, Goldschmidtstrassel, Federal Republic of Germany.

# Aspects of iron incorporation into sediments with special reference to the Yorkshire Ironstones

**D.A. Spears**

SUMMARY: Ironstones of former economic value occur in both Coal Measures and Jurassic sequences in Yorkshire. The distribution of pyrite and siderite in the Coal Measures sediments is explained by the stability relationships between the two minerals and the controls on the rate of pyrite formation, particularly the sulphate availability, but also involving the amount and quality of metabolizable organic matter and the iron availability. In marine black shales early diagenetic pyrite formation may have immobilized much of the available iron, whereas in non-marine sediments iron was available over much longer diagenetic periods to form siderite in the mudrocks and pyrite in the coals. The qualitatively more important siderite forms clay ironstones by a process of diagenetic redistribution and concentration of original iron. In the marine Jurassic sediments pyrite and siderite again show a cyclic distribution. The same explanation holds for pyrite in the marine black shales, but in the grey shales loss of metabolizable organic matter under oxic conditions mainly allows post-oxic reactions, which includes reduction of iron but little sulphate reduction. In the oolitic ironstones efficient loss of organic matter in an important oxic zone is seen as a key factor. The oolitic ironstones do represent an original enrichment of iron in contrast to the other ironstones considered. Post-oxic reactions are thought to be important and produced berthierine and siderite. The possibility of some diagenetic enrichment of iron in the oolitic ironstones should be recognized.

The ironstones in Yorkshire are no longer of economic value, but they nevertheless remain of great historical importance and of considerable academic interest.

The ironstones were worked from both Carboniferous (Westphalian) and Jurassic (mainly Pliensbachian) sources. The Coal Measures ores were worked during the Roman occupation and from the twelfth century exploitation was in the hands of the great monastic houses. In the early nineteenth century mining activity increased and the Coal Measures ironstones were an essential ingredient of the developing iron and steel industry. The peak output of 798 000 tonnes was recorded in 1868; thereafter production declined and ceased after World War I. A detailed account and bibliography are given in Dunham (1960), who estimated total output at over 200 million tonnes. The decline in the extraction of these ores corresponded with an increase in the production of Jurassic ores; particularly from the Cleveland Main Seam in North Yorkshire. In 1876 peak production of 6 667 000 tonnes was reached and for the next fifty years production seldom fell below 5 million tonnes per annum. Output declined after this period and by the mid-1950s was well below 1 million tonnes per annum. Production ceased in 1964 without significantly adding to the total given in Dunham (1960) of 366 million tonnes.

The ironstones from the Coal Measures are the clay-ironstones so named because of the detrital sediment intimately associated with the siderite. The siderite typically occurs as nodules, 5–15 cm in diameter, flattened in the plane of the bedding, with economic accumulations occuring in the *communis* and *modiolaris* zones. The compositions of some of the main ores are given in Dunham (1960). The siderite deviates from an end member composition with Ca, Mg and Mn in solid solution and Dunham (1960) noted that the carbonates warranted a fuller study than they had hitherto received.

The stratigraphy and petrography of the Jurassic ironstones in Yorkshire were reviewed by Dunham (1960) and Hemingway (1974). The ores are described as sideritic berthierine oolites, with or without berthierine in the matrix, grading both laterally and vertically by loss of the oolite fraction to either a siderite mudstone or a siderite–berthierine mudstone. The Cleveland Main Seam was one of the first rocks to be examined in thin section (Sorby 1856), and the general conclusion reached was that, at first, the Cleveland Hill Ironstone was a kind of oolitic limestone, interstratified with ordinary clays containing a large amount of the oxides of iron and also organic matter, which by their natural reaction, gave rise to a solution of bicarbonate of iron—that this solution percolated through the ironstone, and, removing a large part of the

*From* YOUNG, T. P. & TAYLOR, W. E. G. (eds), 1989, *Phanerozoic Ironstones*
Geological Society Special Publication No. 46, pp. 19-30

carbonate of lime by solution, left in its place carbonate of iron; and not that the rock was formed as a simple deposit at the bottom of the sea. However, Hallimond (1925) established to the satisfaction of most subsequent British workers the primary nature of the berthierine ooliths. Nevertheless the evidence presented by Sorby (1856) and its interpretation bears further investigation. One of the main points made by Sorby was that the ironstones did not accumulate in their present form, and that diagenetic alteration was important. The involvement of organic matter in the diagenetic reactions was recognized. In thin section partial and total replacement of shell material by siderite was noted. Similar alteration of the oolites was also recorded. He also argued, quite correctly, that if replacement of the larger grains by siderite could be demonstrated, alteration of the fine matrix could be assumed. Sorby also noted that ferric oxide was stable in modern oceans, hence the necessity for diagenetic reduction and also the lack of any modern analogues which showed comparable iron enrichment to the ironstones.

In later work Sorby (1879) returned to the Cleveland Hill Ironstone. He demonstrated experimentally that replacement of calcite by siderite was possible. He also commented on the composition of the ooliths as follows: 'In some cases peroxide of iron appears to have formed an original constituent of the oolitic grains, having been collected on them mechanically or by contemporaneous or subsequent chemical replacement, and afterward reduced to black oxide.' There is clear recognition here that an iron mineral could be a primary constituent of the oolites, which casts further doubt on the subsequent interpretation of Sorby's views on the origin of ironstones as being simply replacement of oolitic limestones.

One of the aims of the present paper is to consider iron mobility during diagenesis and the potential for iron enrichment, possibly of an already iron-rich unit. Replacement of an oolitic limestone is one of the possibilities and an example, which may have this origin, will be described from the Yorkshire Jurassic. Another aim is to consider the controls on pyrite and siderite formation, hence to account for their distributions, and in so doing to provide a framework in which the other iron minerals in the oolitic ironstones should be examined.

## Iron minerals in coal measures sediments

Pyrite and siderite are the main precipitate iron minerals in these sediments. The interrelationships are reasonably well established and an understanding of their distributions is important if more complex associations, such as those in the Jurassic, are to be understood. In the Coal Measures abundant pyrite occurs in some of the marine shales and microscopic and macroscopic siderite characterizes the non-marine shales.

The equilibrium Eh-pH-$P_{S_2}$-$P_{CO_2}$ conditions for the stability of pyrite and siderite are well known (Garrels & Christ 1965). A low Eh is required for the formation of both minerals but siderite will form in preference to pyrite in the presence of dissolved carbonate only if the dissolved sulphide concentration is extremely low. It is logical, therefore, to first consider the limiting conditions of pyrite formation. Critical factors determining the rate of pyrite formation are the rates of supply of decomposable organic matter, disolved sulphate and reactive detrital iron minerals (Berner 1984).

### Organic matter and sulphate reduction

The input of organic matter into marine shales is shown schematically on Fig. 1. The normal marine, non-euxinic situation is shown in which the bottom waters are oxygenated and the upper sediment is bioturbated. Dissolved oxygen is therefore present in the upper sediment as a result of physical mixing, pore fluid movement and molecular diffusion. In the presence of dissolved oxygen the major microbial process is indicated by reaction 1 (Fig. 1). The more reactive organic matter is utilized first and in general land-derived organic matter is less reactive or, in other words, more refractory. Only when the dissolved $O_2$ concentration

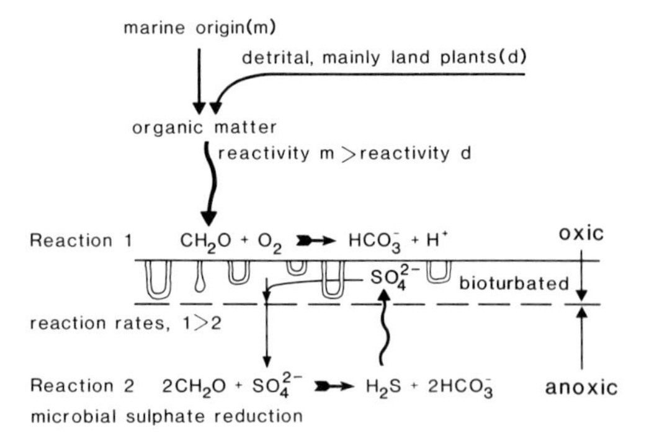


FIG. 1. Organic input and microbial destruction in marine muds.

approaches zero do the anaerobic micro-organisms become important leading to sulphate reduction (reaction 2, Fig. 1) and other similar reactions. The rate of sulphate reduction, and thus the rate of sulphide precipitation given an excess of $Fe^{2+}$, is a function of the amount and reactivity of the organic matter and the dissolved sulphate concentration. The effect of sulphate concentration in modern marine sediments is considered important only at values less than approximately 5 mM, which were attained at depths <0.9 m in one detailed study (Berner 1984). In normal marine sediments a positive relationship between %C and %S is observed (Leventhal 1983; Berner & Raiswell 1983; Raiswell & Berner 1986), of the form %C = 2.9 %S with zero intercept, which clearly demonstrates the control of organic matter on the sulphide precipitation. However, the relationship is complex because firstly up to 90% of the $H_2S$ may be reoxidized to sulphate (Jørgenson 1978; Berner & Westrich 1985) and secondly the organic C is that remaining after reaction. The evidence of the positive S against C relationship implies the organic matter did have approximately the same initial composition and that the loss of the reactive fraction prior to the onset of sulpate reduction was also relatively constant. A further complication in marine shales is that organic reactions do not stop at the zone of sulphate reduction but continue into deeper diagenetic zones. This influence on C–S values is considered in the work of Raiswell & Berner (1986).

In euxinic marine shales sulphate reduction takes place within the water column as well as in the sediment. The S–C plot may show (a) an intercept on the S axis indicating increased sulphide retention (Leventhal 1983) or (b) lack of correlation stemming from reactions in the water column not necessarily closely linked to reactions within the analysed samples, that is organic C and Fe depostion are decoupled (Berner 1984). In Coal Measures marine bands the *Gastrioceras–Pectinoid* faunal phase (Calver 1968) is probably euxinic based on absence of benthonic fauna and bioturbation—the shales are finely laminated. Other marine faunal phases may represent normal marine conditions, i.e. non-euxinic, but there are lateral variations in the composition of the sediment, related to grain size and rate of sedimentation which all complicate the C–S relationship.

## Pyrite

The precipitation of pyrite via monosulphides is shown on Fig. 2 for marine sediment. At

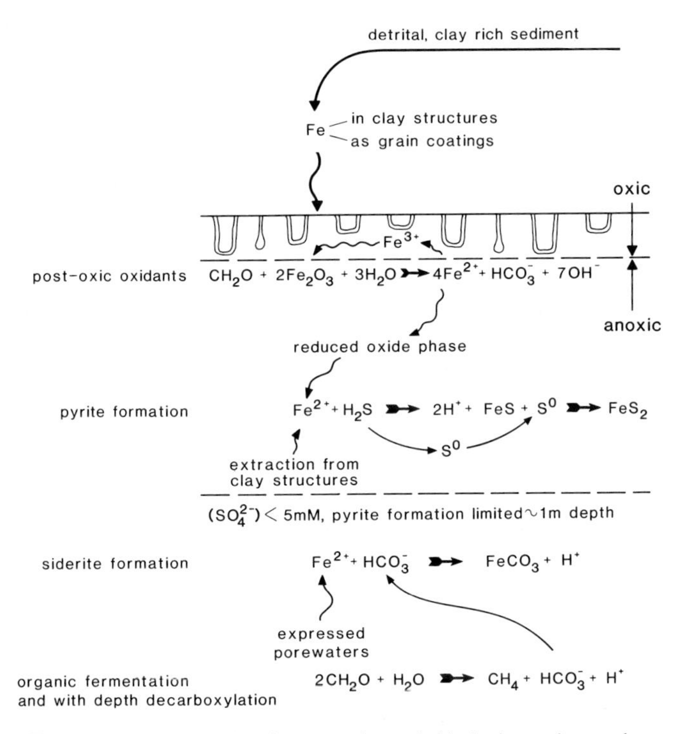


FIG. 2 Iron input and fixation as pyrite and siderite in marine muds.

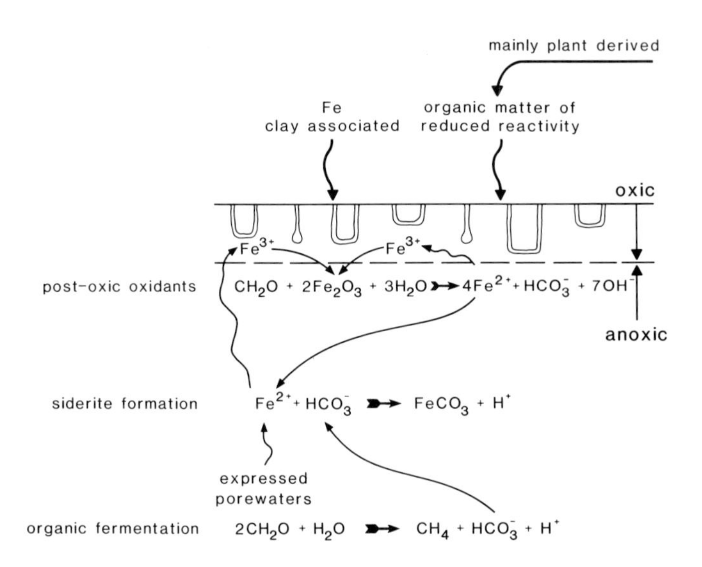


FIG. 3 Organic and iron input in non-marine shales and iron fixation

depth the sulphate availability becomes a limiting factor on pyrite formation and at greater depths the activity of reduced sulphur species falls progressively and siderite precipitation is possible. Analyses of pyrite in marine shales of the Upper Carboniferous (Pearson 1979; Spears & Amin 1981a, b) contain very little siderite which suggests that in these shales there was little iron available after pyrite formation.

In Coal Measures the major contrast is the concentration of pyrite in the marine shales and siderite in the non-marine. The non-marine environment is shown on Fig. 3. The limiting factor on pyrite formation is sulphate availability and this is much lower because of the decrease in salinity; hence the dominance of siderite in these sediments. The major difference between figs 2 and 3 therefore, is the decreased importance of sulphate reduction.

Pyrite is also present in the coals. High pyrite contents associated with marine roofs are well documented and reflect sulphate availability (e.g. Williams & Keith 1963). However, in addition to the early framboidal pyrite, euhedral, concretionary and vein infill pyrite are present in coal seams independent of any marine influence. The association with the one rock type suggests an organic S source for the reduced sulphur species (Berner 1980; Spears & Caswell 1986). In the mudrocks where organic C contents are generally less than 10%, this direct influence of the organic matter is not so readily discernible. The time of formation of this coal pyrite spans pre- and post-compactional stages and demonstrates not only the continued presence and production of reduced sulphur species but the availability of $Fe^{2+}$ for reaction. The veins are vertical joint (cleat) fractures and indicate movement of solutions through the coal and hence through the succession as a whole (Spears & Caswell 1986).

## Siderite

It should be clear from the above account that siderite formed in Coal Measures sediments only because pyrite precipitation was restricted. Both $Fe^{2+}$ and $HCO_3^-$ are generated within the system and siderite formation may therefore cover a long time span. This has been demonstrated by serial analysis of porosity, clay orientation, and chemical and isotope composition (Oertel & Curtis 1972; Pearson 1979; Matsumoto & Iijima 1981; Curtis & Coleman 1986). Initially pore fluid movement would be essentially upwards, but with time the lateral component of movement would become more important. Flow would also be concentrated through fracture planes and along more permeable horizons providing the opportunity for mixing and increased saturation. The siderite concretions, and those that coalesce to form bands are believed to be primary diagenetic siderite precipitates which developed from a diagenetic redistribution of original detrital iron. These are the clay ironstones.

In addition to the macroscopic concretions, siderite occurs as microscopic rhombs in the non-marine mudrocks (Pearson 1979, Table 5). The distribution is similar to that of the pyrite in the marine shales and an early diagenetic formation may be suggested, possibly equivalent to the earliest stage of nodule formation. Iron-rich carbonates, mainly ankerite but including siderite, occurs as bands. Again these are due to a diagenetic concentration of iron but replacing a primary $CaCO_3$ based on the preservation of original textures and structures, such as shells and cone-in-cone structures (Taylor & Spears 1967). A spherulitic form of siderite also occurs in Coal Measures sediments (Deans 1934). The siderite is characteristically associated with seatearths and also differs from the clay ironstones in forming modules less regular both in overall shape and in surface detail due to projecting spherulites. Although having a radial and concentric texture and a size of about 1 mm these are not the typical ooids of the oolitic ironstones. *In situ* formation is indicated by the interfering and displacive crystal growth. The siderite could be primary, but is certainly diagenetic.

### *Blackband ironstones*

Not to be confused with the clay ironstones are the blackband ironstones. These ironstones are finely laminated and are associated with a spore-rich carbonaceous or cannel coal matrix indicative of a lacustrine environment (Boardman 1978). The blackband ironstones, which are relatively rare, are interpreted by Boardman as fossil bog iron ores which were converted to siderite early in diagenesis. The primary ferric iron enrichment was due to acid, low Eh solutions containing ferrous iron emerging from the peat and precipitating as ferric iron in the lake. The blackband ironstones thus represent a primary iron enrichment at the time of sedimentation but triggered by a diagenetic leaching and redistribution of iron already available with the system.

## **Iron availability and clay ironstones**

In the study of ironstones it is important to establish the extent of original and secondary iron enrichment in the ironstones, for on this will hinge the sedimentological interpretation. The clay ironstones represent a diagenetic redistribution of original iron and it also is important to understand the diagenetic controls in order to explain the presence of siderite in some sequences and absence in others. In the work of Berner (1984) and Raiswell & Berner (1986) attention is focussed on the role of organic matter in sulphate reduction and sulphide precipitation. It may be that iron availability was not a limiting factor on sulphide precipitation in fine-grained detrital sediments, but nevertheless sufficient depletion of iron may have occurred that subsequent diagenetic events were influenced. This possibility is explored further.

Part of the iron necessary for the growth of pyrite in Coal Measures sediments originated in a mobile, probably oxidate phase and the same source supplied all the siderite iron (Pearson 1979). The remaining iron in the pyrite was extracted from clay structures, probably chlorite and/or iron rich smectites (Pearson 1979). The Mg–Fe exchange noted by Drever (1971) in modern sediments would appear to be limited and extraction of iron may have involved greater structural modifications (Spears 1973). This loss of iron does suggest a very low activity of $Fe^{2+}$ in the pore solutions. Possibly this depletion of reactive iron is the explanation for the lack of siderite, noted earlier, post-dating the pyrite in the marine shales. In the non-marine siderite rich sediments, similar extraction of iron from the clay structures has not been observed (Pearson 1979).

Further evidence of restricted iron availability is provided by a comparison of the iron contents of Carboniferous marine and iron-marine shales (Table 1). The marine shales have a high $Fe_2O_3$ content reflecting the abundance of pyrite (pyrite iron is expressed as $Fe_2O_3$ because the $HF-H_2SO_4$ digestion for $Fe^{2+}$ leaves pyrite unattacked). In non-marine shales the higher FeO content is due primarily to siderite, although FeO in the marine shales is also lowered by pyrite extraction of iron from the clay structures. Although the iron distribution differs in marine and non-marine samples the total iron contents are broadly similar, which again suggests that the iron availability during pyrite formation was limited. It is also probable that the average iron content of the non-marine shales has been underestimated because of the inhomogenous distribution of siderite (i.e. as micron-sized grains and concretions).

Finally, the Namurian sequence contains turbidites which originated either from the instability of pro-delta clays or direct from river discharge and therefore should reflect the detrital input into the basin. The rapid rate of sedimentation would favour a closed system and lead to retention of the original iron. In the analyses of the Mam Tor sandstones (Spears & Amin 1981b) both FeO and $Fe_2O_3$ average 2.5%. Expressing this as $Fe_2O_3$ and recasting to a comparable clay basis gives a total iron content of 10.5% $Fe_2O_3$ which is similar to the analyses in Table 1. In the turbidites loss of mobile iron may have been prevented by rapid sedimentation and in the marine and non-marine shales by pyrite and siderite precipitation respectively.

There is therefore some evidence to suggest that in the marine black shales the iron availability was limited during sulphide precipitation and there was depletion of reactive iron. Reactive iron was therefore fixed in pyrite thereby limiting the iron source potential of the sediment for later diagenetic minerals either to be precipitated *in situ* or elsewhere in the sequence. Pyrite therefore has the potential to limit the formation of other iron minerals not only by more favourable equilibrium stability conditions but also by fixing and thus removing reactive iron.

## Iron minerals in Jurassic sediments

The ironstones worked in Yorkshire were siderite mudstones and oolitic ironstones. The former are the clay ironstones of the Coal Measures but comparable ores occur in the Jurassic (Hemingway 1974). In the Lower Jurassic there are cycles of sedimentation closely involving the iron minerals (Hemingway 1951; Hallam 1967, 1978; Hallam & Bradshaw 1979; Maynard 1983). The cycle is from black shale to grey shale, to sandstone and terminating in oolitic ironstone.

TABLE 1. *Average iron contents in Namurian Shales from the Tansley Borehole*[1]

| | Non-marine shales | Marine Shales |
|---|---|---|
| $Fe_2O_3$ | 3.07 (±1.32) | 7.97 (±1.78) |
| FeO | 3.82 (±4.40) | 0.96 (±1.34) |
| Total Fe ($Fe_2O_3$) | 7.31 | 9.04 |

[1]Spears & Amin (1981a) (from Table 1)

Bituminous, pyritic shales, such as the Jet Rock occur at or near the base of transgressive units. The shales are laminated, probably due to the presence of varves, and marine fauna lived either at or just above the water–sediment interface. Conditions were ideal for the formation of pyrite in an analogous manner to the Carboniferous marine shales. Concretions in the pyrite rich shales, such as the Jet Rock, are calcareous, demonstrating limited activity of $Fe^{2-}$ because of sulphide precipitation. These concretions show notable pyrite rims and zones with the fauna well preserved. Zoned dolomite rhombs, 5–20 μm in size, occur in the Jet Rock (Pye 1985). The outer zone is iron rich and indicates some iron availability after pyrite formation.

The upwards passage from pyrite shales is to shales which are characterized by less pyrite and organic matter, bioturbation and siderite bands and nodules. Pye & Krinsley (1986) noted that the grey mudstones contained siderite in addition to pyrite, as randomly dispersed rhombs, irregularly shaped grains and in the more silty sediment patches of intergranular cement. The upwards sequence from pyrite to siderite is reminiscent of Coal Measures, but the sequence is marine. In the earlier work of Curtis & Spears (1968) anion availability, and that of sulphate in particular, was considered to be a major control. In order to explain the limited sulphate reduction in a marine sequence a rapid rate of sedimentation was proposed (Curtis & Spears 1968) which rapidly took the sediment below the zone of sulphate diffusion (Fig.1). Maynard (1983) has argued that such a kinetic approach is probably not correct, and we would certainly agree. The upwards passage from black shales to grey shales signifies a changing environment in which bioturbation was more important. This increased the importance of the oxic zone and lead to greater loss of organic matter and thus loss of sulphate reducing capacity in the sediment. The rate of reaction of the residual organic matter would be much slower and the reaction products available over a longer time period leading to the formation of siderite. This argument is equally applicable to the oolitic ironstones.

The complexity of the organic reactivity was not appreciated by Curtis & Spears (1968), nor was the sequential reduction and utilization of energy sources in early diagenesis (Froelich *et al.* 1979). Maynard (1983) favours this thermodynamic explanation for siderite in marine shales, noting that siderite does have a stability field on an Eh–pH diagram in the presence of dissolved S if FeS and $Fe(OH)_3$ are chosen as primary phases. In a geochemical classification of environments, particularly those pertaining during diagenesis, Berner (1981) has coined the term post-oxic, rather than sub-oxic, to describe the environment in which all the dissolved oxygen is consumed by organic decomposition and nitrate, iron and manganese reduction takes place. The environment contains very little metabolizable organic matter and the energetically less favourable reaction of sulphate reduction is inhibited although $SO_4^{2-}$ is available.

The siderite present in the grey Jurassic shales differs from that in non-marine Coal Measures shales in that sulphate availability was not the key factor. It is thought that rate of sedimentation was important, but not by a rapid rate limiting sulphate diffusion as was originally suggested, but rather the rate being sufficiently slow that the major part of the reactive organic matter was destroyed in an oxic environment in which the fauna played an important role. The black shales may have accumulated at an even slower rate but conditions were dominantly anoxic. In the grey shales the residual organic matter was utilized in post-oxic reactions leading to relatively early siderite precipitation as suggested by Maynard (1983). In addition, or alternatively, the residual organic matter, although less suitable for sulphate reduction because of reaction rate, nevertheless provided an energy source for reactions taking place over a much longer time period. In this view only part of the siderite need be an ealy diagenetic precipitate. In the marine Gammon shale, U.S.A., Gautier (1982) concluded that the siderite in the concretions did not start to precipitate until after the onset of methane generation. In rapidly deposited sediments concretions formed at shallow depths (less than 10m) and at greater depths (greater than 200 m) in more slowly deposited sediments. The variation within nodules was less important than that between nodules.

Siderite beds in the Yorkshire Lias were studied by Sellwood (1971). Some prominent siderite rich horizons are associated with boundaries of faunal zones, which together with extensive evidence of faunal activity within the bed suggested a phase of slower sedimentation. Fossils are preserved whole, indicating early iron precipitation and pyrite is often contained within the fossil. In life the fauna promotes an oxic environment through the upper part of the sediment, but on death the reactive organic matter provides a micro-environment in the sediment for sulphate reduction and sulphide formation. The macro-environment was clearly one in which organic matter was destroyed restricting sulphate reduction, as outlined above, and siderite precipitation resulted. Sellwood (1971) suggested the slow rate of sedimentation lead to iron precipitation in the upper sediment by upwards movement of ferrous iron and this oxyhydroxide material converted to siderite below the zone of sulphate diffusion. Less attention would now be paid to the latter factor as discussed earlier. The idea of iron enrichment by redistribution and linked to sedimentological events is attractive (Sellwood 1971). Hemingway (1974 p 180) too has noted the lateral continuity of concretion rich horizons, with individual beds traceable with little change over a few scores of kilometres which suggested some form of bedding control.

### *Millepore Bed*

In the Middle Jurassic in Yorkshire both marine and non-marine sediments occur. The Millepore Bed is a marine intercalation, in a dominantly non-marine sequence, which passes in a northerly direction from an offshore oolitic limestone to an inshore sideritic sandstone (Hemingway 1974 pp 198–199). The latter could be considered an ironstone and its formation an example of iron metasomatism of an oolitic limestone as considered by Sorby (1856). In thin section many of the ooids in the limestone contain quartz as a nucleus and all stages of oolitic coating of quartz grains occur. Bioclastic grains are also important, particularly brachiopod and echinoderm debris. The siderite sandstone, on the other hand, consists of quartz grains floating in a siderite matrix of typical micron sized rhombs. Large, bioclastic grains of calcite are also present, but these are partially replaced by siderite. The presence of residual bioclastic debris suggest that ooids were also originally present. Textural evidence for the ooids is lacking except for the quartz grains which are comparable in size, shape and possibly distribution in thin section to the quartz grains in the oolitic limestone. This may be an example of an oolitic limestone which was replaced during diagenesis by $Fe^{2+}$, as Sorby suggested (1856), but textures are not preserved and the replacement by siderite is destructive. A case can therefore be made for siderite replacement of an oolitic limestone as Sorby (1856) originally suggested, but the result is not, however, an oolitic ironstone. In the limestone quartz is abundant and it is possible that the inshore facies originally lacked ooids and was a calcareous sandstone. I would argue against this on the evidence of lack of packing exhibited by the quartz grains in the siderite matrix and the presence of residual biogenic debris.

TABLE 2. *Major element analyses of representative samples of Millepore Bed*

| | Offshore (grid. ref. TA 084844) | Inshore (grid. ref. TA 020952) |
|---|---|---|
| $SiO_2$ | 25.16 | 28.84 |
| $TiO_2$ | 0.11 | 0.36 |
| $Al_2O_3$ | 1.38 | 2.66 |
| $Fe_2O_3$ | 2.35 | 2.59 |
| FeO | 0.73 | 29.9 |
| MnO | 0.15 | 0.28 |
| MgO | 0.41 | 2.08 |
| CaO | 37.96 | 5.55 |
| $Na_2O$ | $<0.10$ | 0.12 |
| $K_2O$ | 0.41 | 0.36 |
| $P_2O_5$ | 0.05 | 0.52 |
| $SO_3$ | $<0.02$ | 0.25 |
| $H_2O^+$ | 0.74 | 0.84 |
| $CO_2$ | 30.0 | 24.5 |
| Organic C | 0.66 | 0.54 |
| Total | 100.1 | 99.4 |
| Calculated carbonate % (with correction for CaO in apatite) | | |
| $CaCO_3$ | 67.7 | 8.1 |
| $MgCO_3$ | 0.9 | 4.3 |
| $FeCO_3$ | 1.2 | 48.2 |
| Theorectical $CO_2$ | 30.7 | 24.5 |
| Measured $CO_2$ | 30.0 | 24.5 |

The chemical compositions of representative samples of Millepore Bed are given in Table 2 to demonstrate the contrast in carbonate composition from calcite to siderite with residual calcite. The analyses were made by X-ray fluorescence spectrometry using the fused disc method of Norrish & Hutton (1969). The silica content is mainly due to quartz as feldspar and/or clay minerals are of minor importance (low $Al_2O_3$ and $K_2O$ contents). The silica content demonstrates the importance of quartz. Approximately comparable concentrations in both rock types do support the concept of iron metasomatism of an oolitic limestone. It is not realistic to compare more closely by making mass adjustments of losses and gains because the original lateral variations in carbonate-detrital proportions are not known.

The Millepore Bed is a good example of iron metasomatism during diagenesis, as described by Sorby (1856) but whether or not ooids were replaced is debatable. If they were, original textures are not preserved however, and that part of Sorby's hypothesis which proposed formation of oolitic ironstones by replacement is not supported. The extent of iron addition to the Millepore Bed should be noted. This iron was derived from adjacent non-marine sediments in which, by virtue of only minor pyrite formation, iron availability would be predicted during diagenesis. Similarly if the adjacent sediments had been marine grey shales the evidence provide by siderite concretions demonstrates iron availability and the potential to alter the oolitic limestone to form a siderite rich bed. In pyritic, marine black shales, on the other hand, early immobilization of iron would have led to the preservation of adjacent calcareous beds.

### Oolitic ironstones

The petrography and problems of formation of these ironstones are dealt with at length by a number of authors in this volume. The emphasis in this paper is not on the oolitic ironstones, but rather the relationships of the iron minerals to depositional and diagenetic processes in the sequence as a whole, which nevertheless have an important bearing on ironstones, such as the Cleveland Main Seam.

The depositional environment in which the oolitic ironstones accumulated was marine, shallow-water, rich in a macrofauna with bioturbation trace fossils and evidence of reworking and erosion on many scales (Hemingway 1974). The environment was near shore and was adjacent to an extensive low-lying well vegetated land mass experiencing a warm humid climate, where preconcentration of iron by intensive leaching took place (Hallam & Bradshaw 1979). It is generally accepted that the sediment was iron rich and this is attested by the composition of the ooids and the non-siderite component of the matrix. Erosion of mature soil, probably due to a sea level change, would provide fine-grained detritus with a clay mineralogy dominated by kaolinite and associated with Fe–Al oxyhydroxides. The main processes thought to have operated are shown diagramatically on Fig. 4. There is an original iron enrichment and thus these ironstones differ from those considered earlier in the paper. Nevertheless diagenetic redistribution and concentration are effective processes and the sediments with which the oolitic ironstones are associated would provide a source of iron. Secondary, diagenetic enrichment of iron in the ironstones would therefore be predicted. The depositional environment for the oolitic ironstones was one of slow net sedimentation with an important oxic zone within the sediment in which the rate of organic production was balanced by the rate of destruction. Sufficient reactive organic matter was generally not available for sulphate reduction and pyrite formation was therefore

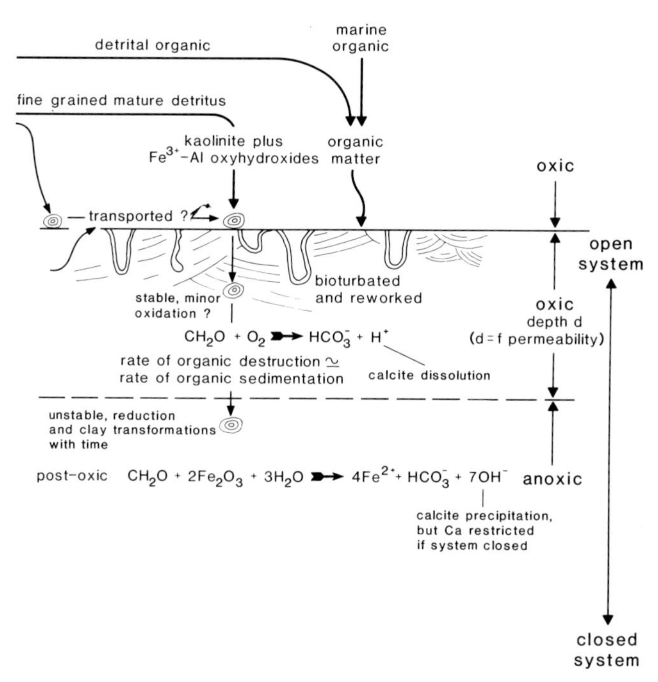


FIG. 4 Organic and iron inputs in marine ironstones and some diagenetic reactions. Note the increased importance of the oxic zone compared with text (Figs. 1–3).

limited except locally where sufficient organic matter passed through the oxic zone to promote major sulphate reduction, as presumably happened in the case of the Sulphur Band within the Cleveland Main Seam. The survival of reactive organic matter below the oxic zone was, however, exceptional. The loss of this organic matter before sulphate reduction is seen as a key factor in oolitic ironstone formation (irrespective of whether or not the ooids formed *in situ*). Extensive pyrite formation would have limited iron availability and precluded other iron minerals as anything other than minor phases.

Elimination of dissolved $O_2$ in porewaters and the initiation of post-oxic reduction lead to the precipitation of siderite. How early in diagenesis siderite first developed is debatable. Siderite in the ooids is generally replacive and siderite bands within ooids are rare (Dunham 1960). If siderite replacement took place at a shallow depth within the zone of reworking discrete siderite bands within ooids would be common. Siderite precipitation and replacement must in general take place below the limits of reworking and a very early formation is not supported by the petrography. By analogy with concretions, siderite formation may have covered a long time period. Furthermore, the siderite is visible in thin section as rhombs forming a texture which has been described as microspar and which would not be considered typical of early diagenesis. In the oolitic ironstones and associated facies the siderite presumably formed from an iron-rich precursor. The $HCO_3^-$ may have been generated from organic matter within the ironstone, but probably the $HCO_3^-$ potential was greater in the clastic beds associated with the ironstone. Also the possibility of some iron movement into the ironstone to form siderite should not be overlooked. Replacement of calcium carbonate grains and cement is one possibility, as considered earlier, but movement of both $Fe^{2+}$ and $HCO_3^-$ to the site of precipitation from an external source is also possible. Quantatitive chemical analyses on a micro and macro scale, coupled with petrography, are required to resolve such possibilities.

Not detailed in this paper is the process of ooid formation and the stability of the primary iron phases. These key questions are addressed by other speakers at this meeting. However, in this paper the importance of the oxic zone has been stressed. Thus sulphate reduction and pyrite precipitation are of minor importance and siderite formation is thought to take place at depth in the sediment, that is below the oxic zone. The intimate association of $Fe^{3+}-Al$ oxyhydroxides and clay minerals probably provide the ingredients for berthierine formation, but again below the oxic zone.

## Conclusions

In the Coal Measures.

(1) Pyrite is associated with the marine shales and siderite with the non-marine mudstones. Sulphate availability, controlled by salinity, is the limiting factor on pyrite formation. Siderite precipitation is only possible if the concentration of reduced sulphur species is very low.

(2) For marine shales, pyrite formation may have utilized all the available iron based on total iron concentrations and the insignificance of siderite post-dating pyrite. (The complex C–S relationship could include a relationship between organic and iron reactivities).

(3) For non-marine sediments, iron was available over a much longer period of diagenesis based on siderite concretions in non-marine mudstones and pyrite pre- and post-dating compaction in the coals (organic S source for reduced sulphur species).

(4) Siderite concretions forming clay ironstones represent a diagenetic redistribution of original detrital iron. Siderite (and ankerite) also occurs, replacing original $CaCO_3$. Blackband ironstones differ in that there was an original iron enrichment, as bog iron ore, which converted to siderite during diagenesis. The original enrichment, however, results from the escape of diagenetic pore fluids from a raised peat swamp.

In the Jurassic sediments.

(5) Pyrite is similarly associated with marine black shales whereas siderite is more common in the grey shales which, however, are also marine. Restricted sulphate availability as a function of sedimentation rate is not the factor responsible for siderite precipitation. Loss of metabolizable organic matter in the oxic zone is thought to be responsible.

(6) The oxic zone is also thought to be very important in the oolitic ironstones in that post-oxic precipitation of siderite was possible. Siderite formation probably covered a long period of diagenesis but the earliest precipitation was in the post-oxic stage which may have been below the limit of reworking.

(7) An oolitic ironstone, although representing a depositional enrichment of iron, accompanied by a kaolinite dominated clay assemblage, was probably not a closed system during diagenesis and both cations and anions may have been added to the system.

(8) The Millepore Bed in the Middle Jurassic could represent an iron metasomatized oolitic limestone, but the oolitic texture is not preserved.

(9) Notwithstanding (8), non-preservation of calcareous ooids, other aspects of oolitic ironstones as described by Sorby (1856) are largely substantiated. In particular the necessity for diagenetic reactions to precipitate the ferrous minerals (siderite) the involvment of organic matter in those reactions, the mobility of ions within the sequence, and the possibility of the mudrocks acting as an ion source.

ACKNOWLEDGEMENTS: The useful comments of anonymous referees and Dr. Raiswell are gratefully acknowledged. The author, however, accepts responsibility for the views expressed in the paper.

## References

BERNER, R.A. 1980. *Early diagenesis: A theorectical approach.* Princeton University Press.

——1981. A new geochemical classification of sedimentary environments. *Journal of Sedimentary Petrology,* **51**, 359–365.

——1984. Sedimentary pyrite formation–an update. *Geochimica et Cosmochimica Acta,* **48**, 605–615.

——& RAISWELL, R. 1983. Burial of organic carbon and pyrite sulpur in sediments over Phanaerozoic time: A new theory. *Geochimica et Cosmochimica Acta,* **47**, 855–862.

——& WESTRICH, J.T. 1985. Bioturbation and the early diagenesis of carbon and sulfur. *American Journal of Science,* **285**, 193–206.

BOARDMAN, E.L. 1978. The blackband ironstones of the North Staffordshire Coalfield. *North Staffordshire Journal of Field Studies,* **18**, 193–206.

——1988. Coal Measures (Namurian and Westphalian) Blackband iron formations—fossil bog iron ore. *Sedimentology.*

CALVER, M.A. 1968. Distribution of Westphalian marine faunas in Northern England and adjoining areas. *Proceedings of the Yorkshire Geological Society,* **37**, 1–72.

CURTIS, C.D. & COLEMAN, M.L. 1986. Controls on

the precipitation of early diagenetic calcite, dolomite and siderite concretions in complex depositonal sequences. *In:* GAUTIER, G.L., (ed) *The relationship of organic matter and mineral diagenesis* & Special Publication of the Society of Economic Palaeontologists and Mineralogists.

——& SPEARS, D.A. 1968. The formation of sedimentary iron minerals. *Economic Geology,* **63,** 257–270.

DEANS, T. 1934. The spherulitic ironstones of West Yorkshire. *Geological Magazine,* **71,** 49–65.

DREVER, J.I. 1971. Early diagenesis of clay minerals, Rio Ameca Basin, Mexico. *Journal of Sedimentary Petrology,* **41,** 982–994.

DUNHAM, K.C. 1960. Syngenetic and diagenetic mineralisation in Yorkshire. *Proceedings of the Yorkshire Geological Society,* **32,** 229–284.

FROELICH, P.N., KLINKHAMMER, G.P., BENDER, M.L., LUEDTKE, N.A., HEATH, G.R., CULLEN, D. & DAUPHIN, P. 1979. Early oxidation of organic matter in pelagic sediments of the eastern equatorial Atlantic: Suboxic diagenesis. *Geochimica et Cosmochimica Acta,* **43,** 1075–1090.

GARRELS, R.M. & CHRIST, C.L. 1965. *Solutions, minerals and equilibria.* Harper and Row, New York.

GAUTIER, D.L. 1982. Siderite concretions: Indicators of early diagenesis in the Gammon Shale (Cretaceous). *Journal of Sedimentary Petrology,* **52,** 859–871.

HALLAM, A. 1967. An environmental study of the Upper Domerian and Lower Toarcian in Great Britain. *Philosophical Transactions of the Royal Society of London, Series,* **B252,** 393–445.

——1978. Eustatic cycles in the Jurassic. *Palaeogeography Palaeoclimatology Palaeoecology,* **23,** 1–32.

——& BRADSHAW, M.J. 1979. Bituminous shales and oolitic ironstones as indicators of transgressions and regressions. *Journal of the Geological Society, London.* **136,** 157–164.

HALLIMOND, A.F. 1925. Iron ores: bedded ores of England and Wales. Petrography and chemistry. *Special Report Mineral Resources, G.B., Geological Survey,* **29,** London.

HEMINGWAY, J.E. 1951. Cyclic sedimentation and the deposition of ironstone in the Yorkshire Lias. *Proceedings of the Yorkshire Geological Society,* **28,** 67–74.

——1974. Jurassic. *In:* RAYNER, D.H. & HEMINGWAY, J.E. (Eds) The Geology and Mineral Resources of Yorkshire. Yorkshire Geological Society, Leeds, 161–223.

JØRGENSEN, B.B. 1978. A comparison of methods for the quantification of bacterial sulphate reduction in coastal marine sediments. III. Estimation from chemical and bacteriological field data. *Geomicrobiology Journal,* **1,** 49–64.

LEVENTHAL, J.S. 1983. An interpretation of carbon and sulphur relationships in Black Sea sediments as indicators of environments of deposition. *Geochimica et Cosmochimica Acta,* **47,** 133–137.

MATSUMOTO, R. & IIJIMA, A. 1981. Origin and diagenetic evolution of Ca-Mg-Fe carbonates in some coalfields of Japan. *Sedimentology,* **28,** 239–259.

MAYNARD, J.B. 1983. *Geochemistry of sedimentary ore deposits.* Springer-Verlag.

NORRISH, K. & HUTTON, J.T. 1969. An accurate X-ray spectrographic method for the analysis of a wide range of geological samples. *Geochimica Cosmochimica Acta,* **33,** 431–453.

OERTEL, G., & CURTIS, C.D. 1972. Clay-ironstone concretion preserving fabrics due to progressive compaction. *Bulletin of the Geological Society of America,* **83,** 2597–2606.

PEARSON, M.J. 1979. Geochemistry of the Hepworth Carboniferous sediment sequence and origin of the diagenetic iron minerals and concretions. *Geochimica et Cosmochimica Acta,* **43,** 927–941.

PRYOR, W.A. 1975. Biogenic sedimentation and alteration of argillaceous sediments in shallow marine environments. *Bulletin of the Geological Society of America,* **86,** 1244–1254.

PYE, K. 1985. Electron microscope analysis of zoned dolomite rhombs in the Jet Rock formation (Lower Toarcian) of the Whitby area, U.K. *Geological Magazine,* **122,** 279–286.

——& KRINSLEY, D.H. 1986. Microfabric, mineralogy and early diagenetic history of the Whitbian Mudstone Formation (Toarcian), Cleveland Basin, U.K. *Geological Magazine,* **123,** 191–203.

RAISWELL, R. & BERNER, R.A. 1986. Pyrite and organic matter in Phanerozoic normal marine shales. *Geochimica et Cosmochimica Acta,* **50,** 1967–1976.

SELLWOOD, B.W. 1971. The genesis of some siderite beds in the Yorkshire Lias (England). *Journal of Sedimentary Petrology,* **41,** 854–858.

SORBY, H.C. 1856. On the origin of the Cleveland Hill Ironstone. *Proceedings of the Geological and Polytechnic Society of the West Riding of Yorkshire.* **3,** 457–461.

——1879. Anniversary address of the President: On the structure and origin of limestones. *Quarterly Journal of the Geological Society of London,* **35,** 56–95.

SPEARS, D.A. 1973. Relationship between exchangeable cations and palaeosalinity. *Geochimica et Cosmochimica Acta,* **37,** 77–85.

——& AMIN, M.A. 1981a. Geochemistry and mineralogy of marine and non-marine Namurian black shales from the Tansley borehole, Derbyshire. *Sedimentology,* **28,** 407–417.

——&—— 1981b. A mineralogical and geochemical study of turbidite sandstones and interbedded shales, Mam Tor, Derbyshire. U.K. *Clay Minerals,* **16,** 333–345.

——& CASWELL, S.A. 1986. Mineral matter in coals: Cleat minerals and their origin in some coals from the English Midlands. *International Journal of Coal Geology,* **6,** 107–125.

TAYLOR, R.K. & SPEARS, D.A. 1967. An unusual carbonate band in the East Pennine Coalfield (England). *Sedimentology,* **9,** 55–73.

VAN HOUTEN, F.B. & PURUCKER, M.E. 1984. Glauconitic pecoids and chamositic ooids—favourable factors, constraints and problems. *Earth Science Reviews,* **20,** 211–243.

WILLIAMS, E.G. & KEITH, M.L. 1963. Relationship between sulphur in coals and occurrence of marine roof beds. *Economic Geology,* **58,** 720–729.

D.A. SPEARS, Department of Geology, University of Sheffield, Sheffield S1 3JD, UK.

# Stratigraphic Patterns

# Temporal patterns among Phanerozoic oolitic ironstones and oceanic anoxia

**F. B. Van Houten & M. A. Arthur**

SUMMARY: The stratigraphic distribution of Phanerozoic oolitic ironstones corresponds to temporal patterns recorded in common detrital deposits. These patterns, in which ironstones share significant associations with black shales, comprise successively smaller-scale sequences, each defined mainly by major transgressions and regressions, changing rates of erosion and sedimentation, and inter-regional unconformities.

Two 300 Ma episodes of continental dispersal and Pangaeic assemblage of cratons encompassed the maximum rise and fall of sea level. Deposition of ironstones and black shales was essentially limited to 150–170 Ma phases (Ordovician–Devonian; Jurassic–Palaeogene) marked by extensive epicontinental seas and ineffective oxygenation of deep-water masses.

Smaller-scale tectonically distinct phases of cratonic submergence, several to many tens of Ma long, were separated by briefer erosional intervals, or by longer episodes of oscillating conditions during which several interregional unconformities developed. Widespread submergence generally favoured major production of ironstones and black shales. Within this framework both were especially common during the Ordovician and Jurassic periods. A hypothetical quasiperiodic (c. 32 Ma) pattern of global climate change was accompanied by recurring relatively high sea level and expansion of oxygen-depleted water masses. Many of these phases led to widespread development of both ironstones and black shales.

Episodes of fluctuating coastal onlap one to several Ma long reflect variations in sea level, sediment supply, and/or subsidence. Major successions of ironstones repeated on a similar time scale were limited to ten tectonic provinces comprising foreland basins, cratonic margins and unstable cratons, and intracratonic basins. During the favourable phases black shales and ironstones were deposited alternately in some successions.

In many of the major successions an ironstone developed between each of several repeated small-scale sequences in a cadence of several hundred thousand years. Ironstones in these asymmetric units record a long lapse in normal sedimentation during the initial stage of renewed transgressions that commonly spread organic-rich mud across the shelf. The repeated association of ironstones and black shales with sequences of common sedimentary rocks helps constrain speculation about conditions controlling their origin, and suggests that ironstones reflect the more local development of productive conditions.

The record of Phanerozoic ferric oxide-chloritic[1] oolitic ironstones comprises temporal patterns of several different scales or orders. Moreover, in them (Fig. 1) ironstones share significant stratigraphic associations with black shales[2]. Each of these distinctive deposits was relatively abundant during two long episodes of extensive epicontinental seas, relatively ineffective deep-oceanic circulation, and scattered cratonic blocks — the 150–170 Ma first-order depositional sequences of Vail *et al.* (1977), and both facies recurred on a rather similar time-scale (Cook & McElhinny 1979) as well as alternately in some stratigraphic sections (McGhee & Bayer 1985; Hallam & Bradshaw 1979). These similarities and associations suggest a possible genetic relation of the sort Jenkyns (1980) reconstructed for shallow marine glauconitic greensands, phosphorites, and hardgrounds with coeval deeper-water organic-rich mud during Cretaceous anoxic events.

In order to investigate the possible relationships we review relevant stratigraphic data and then introduce the basic question—in what ways may oolitic ironstones and black shales have had a common cause in oceanic anoxia? Because there are no well-developed modern analogues we have had to rely heavily on the stratigraphic record. In this endeavour we assume that the ages in the Geologic Time Scale (Palmer 1983) are precise enough for our purpose; yet we are mindful that many of the stratigraphic intervals involved are poorly constrained. Moreover, the

[1] Chloritic—a generic term for the several kinds of Fe-rich chlorite (2:1) and serpentine–kaolin (1:1) clay minerals.

[2] Including mudstones containing considerable organic carbon and those with scattered pyrite and little organic matter.

*From* YOUNG, T. P. & TAYLOR, W. E. G. (eds), 1989, *Phanerozoic Ironstones*
Geological Society Special Publication No. 46, pp. 33-49

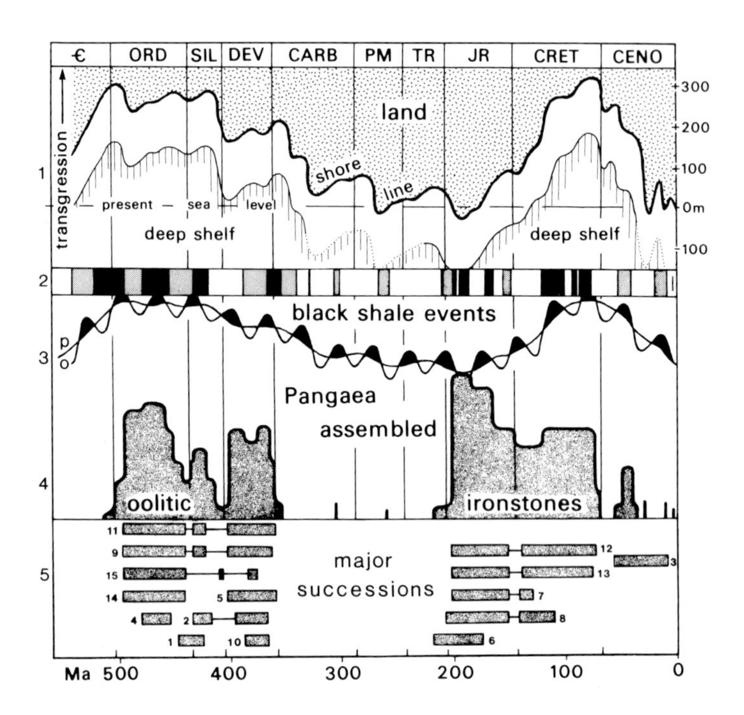


FIG. 1. Phanerozoic temporal patterns. 1. Estimated sea level and shoreline changes after Wilde & Berry (1984). 2. Widespread black shale events (black) with poorly-developed intervals dotted. 3. Hypothetical 32 Ma cycles. P-polytaxic (black), O-oligotaxic phases after Fischer & Arthur (1977) plotted along a sea-level curve after Vail *et al.* (1977). 4. Global abundance. 5. Selected major successions of oolitic ironstones numbered as in Table 1.

record of ironstones is limited, but it is not so sporadic that it reflects only random development. For convenience we use the term, oolitic ironstone, in a very general way without reference to the amount of ooids present.

## Oolitic ironstones

### General statement

Most of the approximately 175 known oolitic ironstones accumulated on shallow shelves after waning of normal sedimentation that had built shoaling-upward detrital (siliciclastic) sequences a few metres to a few tens of metres thick (Fig. 2). In this reconstruction most of the ironstones were associated with a widespread sediment-starved hiatus developed after progradation and at the beginning of renewed transgression (Bayer *et al.* 1985; Van Houten 1986). Recognition of this motif facilitates identification of relevant stratigraphic patterns. Although the ferruginous ooids were commonly concentrated in near-shore bars and sand waves, many were swept offshore and spread in sheets across distal muddy facies (Fig. 2); some also accumulated *in situ,* in the coastal environment where they had formed (Bhattacharyya, 1989).

### Major successions

During the two congenial Phanerozoic episodes (Ordovician through Devonian and Jurassic through Palaeogene time) the worldwide abundance of ironstones waxed and waned (Fig. 1) in a cadence corresponding to a hierarchy of common depositional sequences (Busch & West 1987) generally construed as reflecting global rise and fall of sea level. Within this pattern ironstones developed repeatedly in about 15 major successions preserved in only about ten sedimentary basins throughout the globe (Table 1: Figs 1 and 3). Among these tectonic frameworks long enduring or recurring favourable conditions prevailed for a few to several tens of millions of years. Unusually long episodes (Table 1, Nos 2,7,8,9,11,12,13,15) comprise composite successions of ironstones interrupted by non-productive intervals a few tens of million years long.

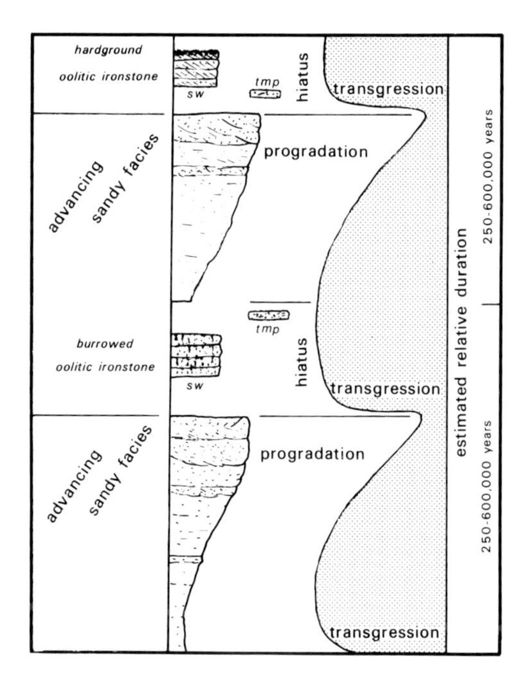


FIG. 2. Facies model of coarsening (shoaling)-upward regressive detrital sequence commonly 10m to several tens of metres thick, with capping oolitic ironstone developed during hiatus. Reconstruction portrays rapid transgressive and development of ironstone before renewed detrital influx. Dotted area suggests the relative extent and duration of transgression and regression. SW—sand wave; tmp—tempestite.

Successive ironstones of this sort, commonly with a correlative offshore black shale facies, developed most readily during Jurassic time (Fig. 4), especially along the margin of post-Variscan swells and basins produced in the extensional regime of middle and northwestern Europe (Bayer *et al.* 1985; Hallam & Sellwood 1976; McGhee & Bayer 1985; Mouterde & Tintant 1980; Sellwood & Jenkyns 1975). Ironstones also flourished in the Ordovician Period (Fig. 5) along the margins of north-western Gondwana and on Armorican blocks and other peri-Gondwanide terranes (Van der Voo 1982; Van Houten 1985). In addition, notable successions of ironstones developed during Silurian time along the proximal margin of the Appalachian foreland basin in eastern North America (Fig. 7), as well as during Middle and Late Devonian time on the unstable post-Caledonian Ardennes shelf of Belgium and adjacent France (Dreesen 1982, 1989) and on the South China craton (Table 1., No. 10; Liao 1958, 1964; stratigraphic details not available). In the following discussion we review selected sequences associated with black shale (Table 1, Nos 2,7,8,9,14; Figs. 4–7) which have a well constrained stratigraphic record.

## Organic-rich black shales

### General statement

During Phanerozoic time widespread development of a black shale facies in ocean basins (Fig. 1) was commonly accompanied by recurring shoreward spread of black mud (Arthur 1979; Wilde & Berry 1984). Preservation of organic carbon in these deposits is generally attributed to intensification and expansion of oceanic anoxia (Berry & Wilde 1978; Fischer & Arthur 1977; Jenkyns 1980; Schlanger & Jenkyns 1976). Secular changes in the carbon and sulphur isotopic composition of ancient oceans, recorded in carbonates and evaporites (Holser 1977, 1984; Berner & Raiswell 1983), provide constraints for reconstructing the cycling of organic carbon and reduced sulphur (OCRS) from oceans to muddy sediments (Arthur *et al.* 1984; Berner 1987; Berner & Raiswell 1983; Garrels & Lerman 1981, 1984; Mackenzie & Pigott 1981). Meaningful interpretation of this record requires distinguishing between terrestrial and marine organic matter in ancient sediments, as well as recognizing that not all marine black shales are a record of global oceanic stagnation (Waples 1983). Moreover, anoxic water that did develop may have been only a thin layer above the sea floor (Kauffman 1978), rather than an expanded oxygen-minimum zone or the result of deoxygenation of an entire ocean basin. A brief review of the record of Phanerozoic black shales and evidence of oceanic anoxia provides a background for discussing their relation to oolitic ironstones.

### Palaeozoic record

Widespread oceanic anoxia during much of early Palaeozoic time is implied by the prevalence of light carbon and heavy sulphur isotopes (Berner & Raiswell 1983) and by widespread Cambrian to Silurian black shales (Fig. 1, Table 2). Extensive development of these deposits in Late Cambrian and Early Ordovician, Middle Ordovician, and Early Silurian time was associated with major transgressions (Leggett 1980; Wilde & Berry 1984). Increased rates of burial of organic carbon during these intervals may be reflected in positive excursions in the record of lighter carbon isotopes, but because of modification by

TABLE 1. *Major successions of repeated oolitic ironstones*

| Sequence | Tetchonic framework | Stratigraphic age | Number of episodes | Duration (Ma) |
|---|---|---|---|---|
| Foreland Basins | | | | |
| 1 | Southern Appalachians<br>Alabama, Georgia | Late Ordovician –<br>Early Silurian | 5 | 20 |
| 2[1] | Central Appalachians<br>Virginia to New York | Early Silurian –<br>Late Devonian | 8 – 9 | 65 |
| 3 | Northern Andes<br>NE Colombia, W Venezuela | Early Eocene –<br>Late Miocene | 5 | 45 |
| Cratonic margins – unstable cratons | | | | |
| 4 | Southwest Baltica<br>Southern Sweden | Middle Ordovician –<br>Late Ordovician | 5 | 22 |
| 5 | Ardennes Shelf<br>S Belgium, NE France | Early Devonian –<br>Latest Devonian | 7-8 | 45 |
| 6 | Sverdrup Basin<br>NE Arctic Canada | Late Triassic –<br>Middle Jurassic | 6 | 40 |
| 7[1] | Variscan Europe<br>Great Britain | Early Jurassic –<br>Early Cretaceous | 13-14 | 73 |
| 8[1] | Variscan Europe<br>France, Germany | Early Jurassic –<br>Middle Cretaceous | 23 | 90 |
| 9[1] | Northwestern Gondwana<br>S Morocco, W Algeria | Early Ordovician –<br>Late Devonian | 18 | 125-30 |
| Stable cratons | | | | |
| 10 | South China Craton<br>C and W South China | Middle Devonian –<br>Late Devonian | several | 20 |
| 11[1] | Central North Africa<br>NW Libya | Early Ordovician –<br>Late Devonian | 10 | 125-30 |
| 12[1] | Northeast Africa<br>Levant, Egypt, S Arabia | Early Jurassic –<br>Late Cretaceous | 7 | 110 |
| 13[1] | Russian Platform<br>W USSR | Early Jurassic –<br>Late Cretaceous | 6 | 120 |
| Amorican blocks | | | | |
| 14 | Bohemian Massif<br>W Czechoslavakia, SE Germany | Early Ordovician –<br>Late Ordovician | 12 | 45 |
| 15[1] | Iberian Massif<br>NW Spain, Portugal | Early Ordovician<br>Middle Devonian | 7 | 100 |

[1] Composite successions with non-productive intervals a few tens of Ma long.

controls such as diagenesis the resolution of the relevant secular curve of $\delta^{13}C$ is not adequate for reliable interpretation.

Mid-Silurian through Early Devonian time apparently was characterized by relatively low rates of accumulation of OCRS. Then another prolonged episode of black shale and oceanic anoxia occurred in Middle Devonian to early Carboniferous time (Krebs 1979). Increased rate of burial of OCRS in Late Devonian marine deposits is reflected in sharp changes in the $\delta^{13}C$ and $\delta^{34}S$ curves (Arthur *et al.* 1984).

During the Carboniferous and Permian periods most of the preserved organic carbon was in vascular land plant debris buried in both non-marine and marine deposits (Anderson *et al.* 1982; Berner & Raiswell 1983). Brief intervals of widespread accumulation of marine black shales on cratons occurred in Late Carboniferous (Middle Pennsylvanian) and Middle to Late Permian time (Fig. 1; Table 2). These episodes are not reflected in the available stable isotope curves, however.

**Mesozoic record**

No episodes of widespread black shale deposition have been recognized in the relevant Triassic stratigraphic record. In addition, insignificant accumulation of OCRS during much of the period is suggested by the available carbon and sulphur isotope data. Increased burial in Late Triassic time is reflected in positive excursions in the curves, however. During most

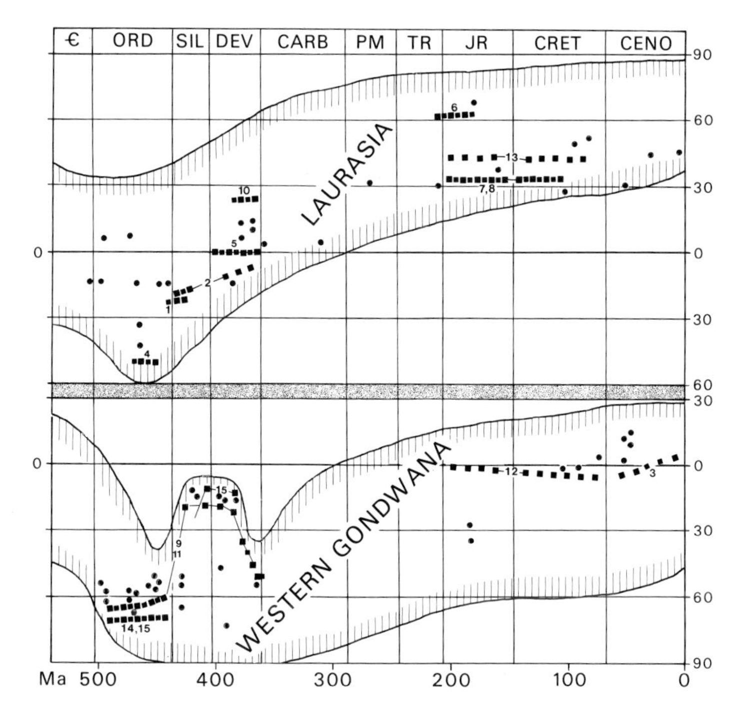


FIG. 3. Time–latitude distribution of cratonic blocks of Laurasia and western Gondwana (including peri-Gondwanide Armorican blocks), with plot of Phanerozoic oolitic ironstones. ■ — major successions of repeated ironstones numbered as in Table 1. ● — other oolitic ironstones.

of Jurassic times (Figs. 1 and 4) rates of burial of OCRS apparently were relatively high, recording major oceanic anoxic events (Jenkyns 1980, 1985) in the Toarcian and Callovian-Oxfordian epochs, as well as minor ones during Hettangian and Kimmeridgian times (Morris 1980).

In the Cretaceous Period enhanced burial of OCRS was associated with widespread transgression and oxygen depletion in oceanic mid- to deep-water masses (Arthur 1979; Arthur *et al.* 1985; Jenkyns 1980; Ryan & Cita 1977; Scholle & Arthur 1980). Relatively high rates of burial prevailed during Aptian–Albian time, at the Cenomanian–Turonian boundary, and during Coniacian–Santonian times (Fig. 1).

## Cenozoic record

There is no evidence of widespread oceanic anoxia during the Cenozoic Era. Nevertheless, Eocene and mid-Miocene deposits record some features commonly associated with reduced ventilation of the oceans (Arthur 1982). Examples of more local stagnation, such as the Middle Pliocene Pontian sediments of Paratheths, were related to the existence of isolated euxinic basins.

## 32 Ma Polytaxic–Oligotaxic cycles

A model of the atmosphere and oceans fluctuating between contrasting conditions with a rhythm of about 32 million years (Fig. 1; Table 2) has been proposed by Fischer & Arthur (1977). During their polytaxic modes latitudinal climate gradients were low, sea level was relatively high, the oceans convected sluggishly, or at least were not effectively oxygenated, and mid-water oxygen-minimum zones apparently expanded and intensified, fostering accumulation of organic matter on the sea floor. The converse of these conditions prevailed in the alternating oligotaxic phase of a cycle. In a recent analysis of Devonian and Jurassic stratigraphic records Bayer & McGhee (1986) noted evidence suggesting a 30–35 Ma cycle, but Legett *et al.* (1981) found no comparable periodicity on the Cambrian–Silurian records they examined.

The cause of the expansion of the oxygen-minimum zone and its specific relation to eustatic rise of sea level remain poorly understood. Moreover, there is no clear evidence how these cycles may relate to a suggested 26 Ma

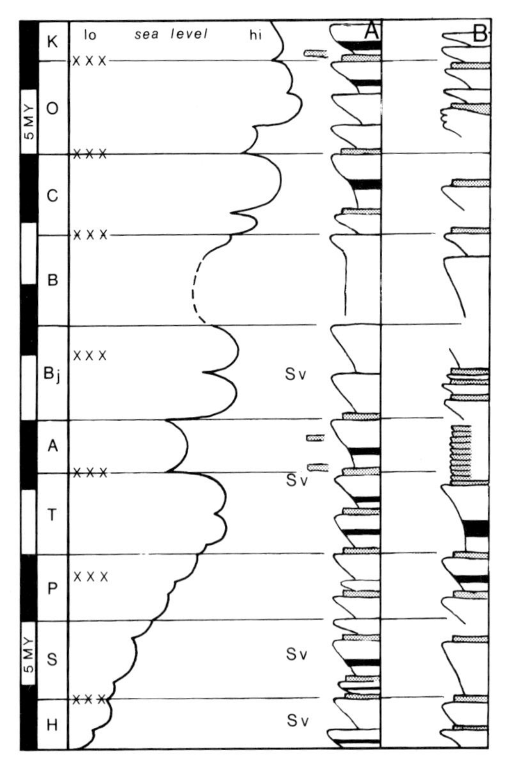


FIG. 4. Generalized successions of Jurassic oolitic ironstones (dotted) in Great Britain (A) and France (B), with associated organic-rich shales (black). Major unconformities (× × ×) after Vail and Todd (1981). Sea Level curve after Hallam (1984). Sv—oolitic ironstones in Sverdrup Basin (6 in Table 1). H—Hettangian, S—Sinemurian, P—Pliensbachian, T—Toarcian, A—Aalenian, Bj—Bajocian, B—Bathonian, C—Callovian, O—Oxfordian, K—Kimmeridgian.

pattern of extinction (Raup & Sepkowski 1986), or to periodic impact cratering (Rampino & Stothers 1984; Alvarez & Muller 1984), or whether they are actually artefacts of analysis (Lutz 1985).

Within this pattern of hypothetical cycling differentiation of its two phases varied considerably. For example, most of the Jurassic Period (about 55 Ma) was predominantly polytaxic (Hallam 1982), at least in its climate and oceanic circulation. Similarly, Late Cambrian through Middle Ordovician time (nearly 70 Ma) was marked by more or less persistent oceanic anoxia (Berry & Wilde 1978), even though there was a relatively high latitudinal climate gradient in the southern hemisphere (Cocks & Fortey 1982; Van Houten 1985). Clearly, at times some overriding control must have obscured the full expression of an oligotaxic phase.

## Discussion

### Association of oolitic ironstones and black shales

Our concern with the common association of these two distinctive lithofacies does not imply that all ironstones are laterally equivalent to or overlain by black shales, nor that all widespread black shales have correlative ironstones. Nevertheless, the general ironstone–black shale relation does provide important insights into the origin of both the ferriferous ooids and the oolitic ironstones. In this review we have focused largely on the origin of ironstones because the origin of the ferric oxide and chloritic ooids remains controversial and beyond the scope of this discussion.

Factors that have favoured production of large fluxes of iron and the accumulation of

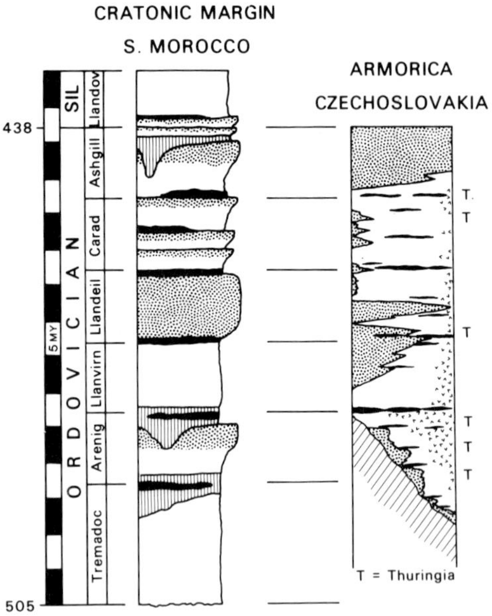

FIG. 5. Major successions of Ordovician – Early Silurian oolitic ironstones (black) in southern Morocco (9 in Table 1) after Destombes (1971), and in the Bohemian Massif (14 in Table 1) after Petranek (1964, 1974). Sandstone – dotted; mudstone – unpatterned; volcaniclastic – v; hiatus – vertical lines.

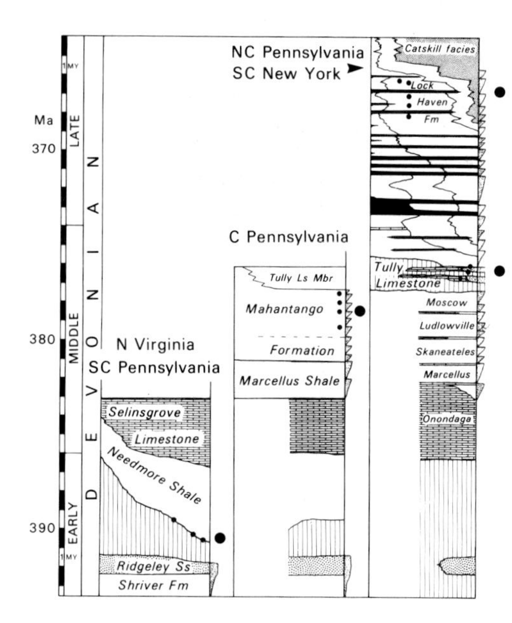


FIG. 6. Major succession of Late Early to Late Devonian oolitic ironstones (large dots) in northern Appalachian Basin. Stratigraphic framework mainly after Rickard (1975) and Berg *et al.* (1983). Minor shoaling-upward sequences are sketched along right margins. Coarse-grained sandstone – dotted pattern; fine-grained sandstone and mudstone – unpatterned; deeper-water black shale facies – stippled or black; limestone – brick pattern, hiatus – vertical lines.

ironstones (Van Houten and Bhattacharyya 1982; Van Houten and Purucker 1984) as well as preservation of organic carbon in sediments include the following: (1) dispersed continents and periods of relatively high sea level producing extensively flooded continental margins and inland seas; (2) shallow shelves with low detrital sediment supply or periodically reduced influx; (3) warm global climate and widespread deoxygenation of oceanic intermediate to deep-water masses (Borchert 1965).

According to a current hypothesis the two major episodes of Phanerozoic continental dispersal and Pangaeic assembly were accompanied by first-order sea-level changes (Vail *et al.* 1977) as well as by changes in outgassing of mantle $CO_2$, and together these activities effected systematic variation in climate (Fischer 1981, 1984). Episodes of dispersal were associated with global high sea level, higher rates of mantle outgassing and increased $P\text{co}_2$ which produced warm, equable climate. In contrast, assembly of a Pangaeic supercontinent was accompanied by lower global sea level and lower rates of outgassing and $P\text{co}_2$, producing cooler global climate, and culminating in major glaciation.

The implications of Fischer's model for oceanic and atmospheric chemistry and for the Phanerozoic record of carbonates have been discussed by Mackenzie & Piggott (1981), Worsley *et al.* (1984, 1986), Sandberg (1983), and Wilkinson *et al.* (1985). We believe that the elements of this model also bear on the relative abundance and association of ironstones and black shales during the two phases of continental dispersal (Fig. 1). These were times of exten-

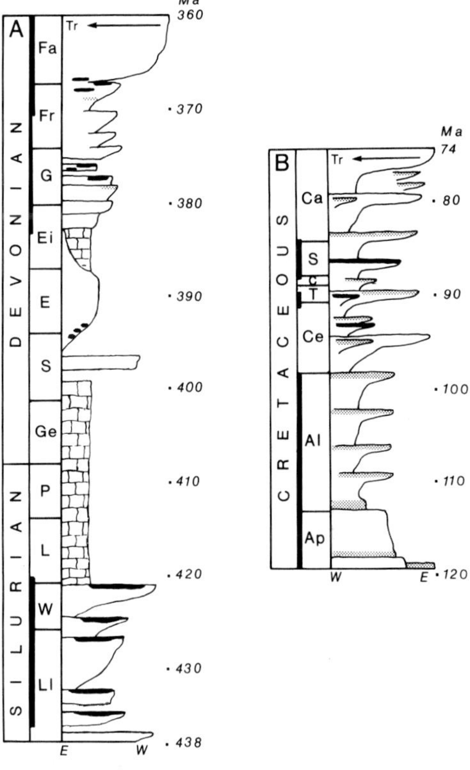


FIG. 7. Composite stratigraphic sections with prograding detrital sequences in foreland basins. (A) Early Silurian to Late Devonian deposits in northern Appalachian Basin (2 in Table 1). Ll – Llandovery, W – Wenlock, L – Ludlow, P – Pridoli, Ge – Gedinnian, S – Siegenian, E – Emsian, Ei – Eifelian, G – Givetian, Fr – Frasnian, Fa—Famennian. (B) Middle and Late Cretaceous deposits in Laramide Basin (New Mexico to Alberta). Ap – Aptian, Al – Albian, Ce – Cenomanian, T – Turonian, C – Coniancian, S – Santonian, Ca – Campanian. Oolitic ironstones – black; glauconitic sandstones – dotted; Limestone – brick pattern. Tr – transgression; E – east, W – west. Thick lines along left margins are intervals of black shale as in Fig. 1.

TABLE 2. *Oolitic ironstones and anoxic events in hypothetical 32 Ma cycles*

| POLYTAXIC PHASE | | | OLIGOTAXIC PHASE | | |
|---|---|---|---|---|---|
| Peak | Oolitic Ironstones | Anoxic Events | Anoxic Events | Oolitic Ironstones | Peak |
| Middle Miocene | O | x | x | O | Middle Pliocene |
| Middle Eocene | OO | x | | O | Oligocene |
| Latest Cretaceous | OOO | XXXX | | O | Paleocene |
| Middle Cretaceous | OOO | XXXX | | OOO | Early Late Cretaceous |
| Earliest Cretaceous | OO | | | OOO | Early Cretaceous |
| Middle Jurassic | OOO | XX | XX | OOO | Late Jurassic |
| Latest Triassic - Earliest Jurrassic | O | x | XXXX | OOOOO | Early Jurassic |
| Early Trassic | | | | | Middle Triassic |
| Middle Permian | O | x | x | | Late Permian |
| Late Carboniferous | O | x | | | Earliest Permian |
| L. Early Carboniferous | | x | | | Middle Carboniferous |
| Late Devonian | OOO | XXXX | x | | Earliest Carboniferous |
| Early Devonian | OOO | | x | OOO | Middle Devonian |
| Early Silurian | OO | XXXX | | O | Late Silurian |
| Middle Ordovician | OOOOO | XXXX | x | O | Late Ordovician |
| Early Ordovician | OOO | XXXX | x | OOO | L. Early Ordovician |
| Middle Cambrian | | x | XXXX | O | Late Cambrian |

O - oolitic ironstones; qualitative estimate ranging from poorly developed (O) to very well developed (OOOOO).
X - black shales widespread, x - black shales local or poorly developed.

sively flooded cratonic margins and inland seas (Worsley *et al.* 1984) and relatively low relief and reduced detrital supply (Hay 1981; Mackenzie & Pigott 1981), two of the important factors favouring accumulation of ironstones. In addition, the extensional tectonism which accompanied dispersal may have influenced the rate of supply of iron to the oceans from both potential riverine and sea-floor hydrothermal sources. The increased length of mid-ocean ridges and proportion of young oceanic crust accompanying the breakup of a supercontinent (Worsley *et al.* 1984) suggest increased mid-ocean ridge hydrothermal activity and flux of iron to seawater through alteration of basalt (Holland 1984, p 208, 385). However, such a hydrothermal source probably was less important than the riverine supply.

Times of increased $Pco_2$ during episodes of continental dispersal probably induced warmer global climate and higher average precipitation, as suggested by global climate simulations (Barron & Washington 1985). This condition increased the rate of chemical weathering on land (Lasaga *et al.* 1985), thus supplying more dissolved and colloidal chemical constituents, including iron and clay minerals, to the oceans. In particular, the increased rate of chemical weathering across a broad latitudinal range (Arthur & Jenkyns 1981) produced abundant kaolinite that apparently was the precursor of the ferriferous clay mineral berthierine in ironstones throughout an extensive paleo-latitudal distribution (Fig. 3). The specific geochemical mechanism involved in this transformation (Bhattacharyya 1983; Maynard 1983, pp 44–47) presumably was the addition of reduced iron (and minor magnesium) to kaolinite which contained the necessary silica and aluminium.

Reconstruction of this sort requires that kaolinite and iron accumulated while detritus was excluded. Because iron supplied by rivers is generally associated with detrital influx a direct input of iron to a basin by a large river is unlikely. However, iron may have been supplied to a basin by transportation of iron as coatings on clay minerals (Carroll 1958) that were carried into marginally reducing (suboxic) environments. There the iron could have been reduced and then reacted with kaolinitic ooids to form berthierine. From their place of origin many of the ferric oxide and berthierine ooids presumably were reworked and concentrated in ironstones (Brookfield 1971; Knox 1971).

Abundant reduced iron may also have been supplied to berthierine-forming environments through mixing of dysaerobic and suboxic shallow–intermediate water masses with marginally oxic shelf water (Borchert 1965). These weakly deoxygenated water masses could have contained ample dissolved iron derived from active weathering or sea-floor fluxes, except where dissolved sulphide was abundant enough to scavenge the dissolved iron, as in the anoxic modern Black Sea (Maynard 1983; Holland 1984). Even in this euxinic water body, however, there is a high concentration of reactive iron near the oxic–anoxic interface.

Within this general scenario an oolitic ironstone normally developed above a shoaling-upward detrital sequence (Fig. 2) that commonly had a correlative offshore black shale facies which was spread shoreward with renewed transgression produced by increased subsidence or sea-level rise. During the initial stage of the transgression detritus was trapped inshore so that both the inner and outer shelves were starved of sediment. At the same time the upper part of an intensified oxygen-minimum zone or anoxic deeper water mass spread onto the outer shelf where the low dissolved oxygen concentration favoured preservation of organic material, with or without high biologic activity (Fischer & Arthur 1977). In addition, low-oxygen iron-enriched waters spread onto the inner shelf where some of the iron was oxidised and deposited along with clay minerals. Depending on the intensity of mixing, either iron oxide or an iron-rich clay mineral was produced. During the early transgressive phase these ooids were then concentrated into ironstones by shelf currents. The effectiveness of transgressive anoxic events is illustrated by pyritization of the upper part of a thick ironstone below black shale in the Early Ordovician Wabana succession in eastern Newfoundland (Maynard 1983). Although Hallam (1975) and Hallam & Bradshaw (1979) have assigned the development of ironstones to the final stage of preceding regression, we interpret it as generally occurring during the succeeding hiatus and the intitial stage of renewed transgression. Our interpretation emphasizes the role of prolonged cessation of normal sedimentation and of the associated shoreward spread of offshore anoxic conditions.

Accumulation of black shale has commonly been considered the product of warm global climate and extensive flooding of the continents (Fischer & Arthur 1977; Wilde & Berry 1984; Arthur *et al.* 1985). Those conditions presumably led to more sluggish circulation resulting from haline-dominated, warm deep-water circulation as well as from lower initial oxygen solubility in the warm surface waters that sank to become deeper water. In this way warmer climate associated with phases of continental dispersal tended to develop oxygen-deficient oceans and accumulation of black shale. These considerations suggest that the relation between ironstones and black shale is not fortuitous. Nevertheless, detailed sedimentologic, mineralogic and geochemical analysis of relevant stratigraphic sections is required to demonstrate that this possibility is more than speculation.

### 32 Ma pattern of cycles

Although both ironstones and black shales accumulated relatively abundantly during the two long Phanerozoic episodes (Fig. 1) their association is too general to provide rigorous constraints for speculation about a possible genetic correlation. Comparison of the record of ironstones with the pattern of postulated polytaxic intervals, on the other hand, does suggest common favourable conditions (Table 2) fluctuating every few tens of million years. Moreover, in the long-enduring, composite successions (Fig. 1) the non-productive intervals correlate closely with hypothetical oligotaxic phases.

Several contradictory correlations are also introduced by this comparison, however. Among these, and particularly pertinent for the present review, some important episodes of ironstone accumulation occurred during oligotaxic phases, emphasizing the role of conditions independent of oligotaxy. For example, the thick mid-Pliocene Kerch ironstone in southwestern Russia accumulated during the latest (current) oligotaxic phase, but it did develop in a local Pontian euxinic basin. In addition, both of the principal episodes of ironstones, during Ordovician and Jurassic times, persisted through a hypothetical oligotaxic phase, suggesting that a shallow oxygen-minimum zone persisted on flooded shelves, thus producing essentially a prolonged polytaxic episode in both climate and oceanic circulation. Any speculation about the similar records of Ordovician and Jurassic ironstones with respect to the 32 Ma pattern must acknowledge the markedly different palaeogeographic condtions that prevailed during these major episodes.

## Contrasting Ordovician and Jurassic palaeogeography

### Ordovician Period

(1) Laurasian continents were widely dispersed in low latitudes while western Gondwana was essentially intact in high southern latitudes. Most of the ironstones accumulated around the northwestern margin of Gondwana and on associated Armorican blocks (Fig. 3).

(2) A generally mild climate dominated low latitudes whereas a high climate gradient prevailed in higher southern latitudes. Moreover, a decrease in temperature accompanied by Gondwanan glaciation marked the end of the period.
(3) A major high stand of sea level and a stagnant ocean with a shallow oxygen-minimum zone persisted during much of early Palaeozoic time (Berry & Wilde 1978; Wilde 1987), but abundant ironstones failed to develop until several tens of million years after the beginning of the extensive transgression in Cambrian time.
(4) The widespread shallow seas that flooded broad cratonic margins and inland areas induced a dominantly maritime climate.
(5) Detrital sedimentation predominated in higher southern latitudes whereas carbonate accumulation was widespread on broad shelves in low latitudes (Cocks & Fortey 1982).

### Jurassic Period

(1) Most of the ironstones accumulated in northern latitudes on assembled Laurasia (Fig. 3).
(2) A generally warm and equable climate prevailed throughout the period without evident glaciation.
(3) Ironstones began to accumulate during a low stand of sea level several tens of million years before the major Mesozoic rise occurred. Moreover, development of ironstones diminished somewhat as sea level rose and carbonate sedimentation became more widespread in late Jurassic time.
(4) In early Jurassic time very narrow flooded cratonic margins (Wyatt 1987) and diminished inland seas were bordered by extensively exposed coastal plains.
(5) Humid conditions prevailed and detrital sedimentation predominated in higher latitudes whereas arid conditions and carbonates and evaporites were widespread in lower latitudes (Hallam 1982).

## Major successions of oolitic ironstones

In order to evaluate more specific relations between ironstones and black shales we review a few well constrained stratigraphic reconstructions in which some of the ironstones developed in small-scale termporal patterns tentatively attributed to the 400,000 year Milankovitch climate cycle (Van Houten 1986). These examples also direct attention to the question of why ironstones failed to develop in many places during these and other generally congenial times.

### Jurassic oolitic ironstones, northwestern Europe

In the rigorously constrained record of Jurassic shoaling-upward sequences in northwestern Europe (Fig. 4) each oolitic ironstone normally is the 'roof bed' of a sequence, lying below a succeeding transgressive black shale (Hallam & Bradshaw 1979; McGhee & Bayer 1985; Bayer *et al.* 1985). Locally, accumulation of ironstones followed coal measures formed during the preceding regression. Sequence associations of this sort were repeated every few milion years. The importance of a prolonged hiatus in fostering the development of ooids before renewed transgressive muddy sedimentation is documented by the Lorraine ironstones of eastern France and adjacent Luxembourg (Mouterde & Tintant 1980; Teyssen 1984).

The regional framework of northwestern Europe (Sellwood & Jenkyns 1975; Ziegler 1982) was characterized by an extensional regime of sea-floor swells and basins influenced by repeated differential subsidence (Vail & Todd 1981). Sedimentation was able to maintain a generally level sea floor, however. Under the influence of eustatic rise and fall of sea level and intermittant subsidence (Hallam & Sellwood 1976) the local swells determined where the regional successions ended with a condensed deposit of oolitic ironstone. Additional evidence of the important role of eustasy is provided by the close correlation between Jurassic ironstones in the Sverdrup Basin (Arctic Canada; Fig. 4) and those in northwestern Europe (Embry 1982), as well as by a detailed analysis of the Jurassic record in the northern North Sea Basin (Vail & Todd 1981).

### Devonian record, northern Appalachian basin

*Black shales:* A remarkable repetition of black shales (Broadhead *et al.* 1982; Roen 1984), together with several scattered oolitic ironstones, constitutes part of the well known Middle and Late Devonian Catskill deltaic complex (Fig. 6) with its dominant motif of cyclic sequences (Rickard 1975; McGhee & Bayer 1985; Baird & Brett 1986; Brett & Baird 1986). Although this succession developed during the Acadian Orogeny it accumulated under relatively subdued tectonic conditions (Faill 1983) in a low-latitude inland sea fed by numerous small contiguous deltas (Sevon 1981).

During Middle Devonian (Givetian) time several extensive thin carbonate deposits and tongues of organic-rich mud from the oldest deep-basin facies spread shoreward across the delta platform. Then, after a significant latest Givetian stillstand (Tully Formation), increased rates of tectonism and subsidence introduced renewed detritial influx and spread on anoxic facies with repeated transgressive black shale tongues landward. This episode persisted through much of Late Devonian (Frasnian–early Famennian) time (Fig. 6). Each transgressive tongue was succeeded by a shoaling-upward sequence commonly capped by a disconformity or a coquinite rarely containing scattered pellets of glauconite (Ehrets 1981; Baird & Brett 1986; Selleck & Linsley 1984; Woodrow & Isley 1983). In this setting deep-basin anoxia apparently intensified every few million years (McGhee & Bayer 1985). A similar temporal pattern is also recorded in some late Palaeozoic transgressive-regressive sequences (Ramsbottom 1979; Ross & Ross 1985).

Current speculation about the cause of shoreline advances and retreats on this scale ranges from rapidly varying tectonically controlled subsidence (Ehrets 1981; Ettensohn 1983; Faill 1985) to eustatic rise and fall of sea level (Dennison & Head, 1975; McGhee & Bayer, 1985). Smaller-scale shoaling-upward sequences with correlative offshore black shale facies, nested within the thicker units, were repeated every 250 to 500 thousands years (Brett & Baird 1986) and apparently were caused by a short-term nonglacial forcing function such as the 400 thousand-year Milankovitch climate cycle. Short sequences of this sort have commonly been considered autocyclic (Glaeser 1979), but similar ones in Carboniferous deposits in the United States and northwestern Europe have been attributed to 400–500 thousand-year eustatic fluctuations of sea level (Busch & Rollins 1984; Heckel 1977, 1986; Ramsbottom 1979).

*Oolitic ironstones:* Several poorly-developed ironstones are associated with hiatuses or cap shoaling-upward sequences in these Middle and Late Devonian deposits. The oldest, occurring above a significant hiatus, is in the basal black shale (Beaverdam Member) of the latest Early to early Middle Devonian Needmore Shale (Fig. 6; Berg *et al.* 1983) from south-central Pennsylvania to northern Virginia (Hunter 1960; Inners 1979; Lesure 1957). This muddy formation records the initial effect of the Acadian Orogeny and grades into a western black shale facies. The overlying Marcellus (Millboro) black shale is the oldest widespread Devonian anoxic deposit in the Appalachian Basin.

The several thin late Middle Devonian upper Mahantango ironstones in central Pennsylvania (Fig. 6) cap repeated short shoaling-upward sequences of the sort that dominated Middle Devonian detrital shelf sedimentation (Faill *et al.* 1978; Sarwar 1984, personal communication; Brett & Baird 1986). Black shale is not directly associated with these ironstones, but correlative offshore sequences grade into a deep-basin facies. Moreover, the Mahantango Formation interfingers with the Marcellus (Millboro) black shale to the southwest (Hasson & Dennison 1979). Poorly developed ironstones are also present in the Tully Formation (Heckel 1973) which accumulated during the culmination of waning detrital sedimentation near the end of Middle Devonian time (Fig. 6). With renewal of detrital influx in latest Middle Devonian time a major episode of deep-basin anoxia and transgressive black shales developed, but during most of the time no ironstones accumulated even though the pattern of short sequences persisted (Fig 6; Ehrets 1981). Then, several ironstones, preserved in the Late Devonian Lock Haven Formation (Fig. 6) in north-central Pennsylvania (Luce 1981), accumulated during waning of the deep-basin stage and the beginning of the subsequent basin-filling stage. Tongues of black shale are present in this section but their relation to the ironstones has not been established. Successive shoaling-upward sequences in correlative formations to the east (Woodrow 1968), like those in the Mahantango Formation, apparently accumulated on a short time scale of several hundred thousand years.

Comparison of the relatively well constrained Silurian–Devonian succession in the northern Appalachian Basin (Rickard 1975) with the Middle and Late Cretaceous record in the Western Interior Seaway (Cobban & Hook 1984; Peterson & Ryder 1975; Stott 1984) reveals a rather similar pattern of major transgressions and regressions every few to several million years (Fig. 7) in these foreland basins. Following the accumulation of several successive ironstones during an Early to Middle Silurian anoxic event the generally low-energy Appalachian detrital regime was interrupted by a long Late Silurian – Early Devonian episode of carbonate accumulation. Return of predominantly detrital influx during the Acadian orogeny (Faill 1983) led to the local accumulation of several thin ironstones. None developed, however, during the latest Givetian–Frasnian increased influx that fed extensive tongues of black shale, or during the final Famennian westward progradation of the Catskill deltaic complex.

In the Western Interior Seaway active detrital influx created a regime of repeated prograding shorelines commonly with shallow shelf sand bodies in which glauconite accumulated. As an exception, a few minor ironstones developed during a late Cenomanian to Santonian episode marked by extensive transgressions and oceanic anoxia.

### Devonian record, northwestern Europe

Successive Middle and Late Devonian ironstones in northwestern Europe (5 in Table 1 and Fig. 3; Dreesen 1982, 1989) are associated regionally with widespread black shales. During this episode repeated transgressions (Krebs 1979; McGhee & Bayer 1985) of an anoxic facies spread from the Cornwall–Rhenish Basin (Bless *et al.* 1980) across the Ardennes shelf which had been disrupted by an extensional tectonic regime (Ziegler 1982). Then these mutually congenial conditions were eliminated by regional regression that culminated in late Famennian time, except for a brief return in latest (Strunian) Devonian time.

## Summary

Phanerozoic oolitic ironstones and black shales developed most readily during an Ordovician–Devonian and a Jurassic–Palaeogene, 150–170 Ma plate-tectonically controlled interval (Fig. 1) characterized by relatively rapid sea-floor spreading and global high stand of sea level. These conditions presumably led to intensified global chemical activity such as the $CO_2$ content of the atmosphere, producing warmer climate, higher precipitation, and increased rate of chemical weathering. Within these generally favourable episodes the remarkable record of the two distinctive facies falls along a maximum of the first-order eustatic sea level curve in the Ordovician Period whereas that in the Jurassic Period falls along the minimum of the curve (Fig. 1).

During the two long episodes development of ironstones and black shales fluctuated on similar timescales ranging from several hundred thousand to a few tens of million years. Their recurring stratigraphic association suggests that each was a specific response to mutually favourable conditions commonly accompanying a submergent mode of cratonic behaviour (Sloss 1972). Among these conditions the following predominated: (a) continental breakup and dispersal, or subdued orogeny; (b) weak oceanic deep circulation and widespread oceanic anoxia; (c) moderate detrital influx that built shoaling-upward regressive sequences followed by relatively rapid transgression associated with a hiatus and unconformity; (d) accumulation in inland seas or on continental margins where the width of shallow shelves was rapidly modified by transgression and regression. The importance of additional, more specific controlling factors, especially in producing oolitic ironstones, is emphasized by their local, lenticular development.

Both ironstones and black shales may have been favoured by polytaxic phases of a hypothetical $\approx 32$ Ma secular cycle. The record of repeated widespread oceanic anoxia correlates quite closely with this pattern. Some ironstones, however, developed during apparently subdued oligotaxic phases, notably in parts of the Ordovician, Jurassic, and Cretaceous periods. Development of ironstones during hypothetical oligotaxic phases, and the contrasting regional framework of the Ordovician and Jurassic ironstone episodes, suggest that differences in conditions favouring and those inhibiting the accumulation of ooids were not marked, and that productive conditions could be achieved by varying combinations of local favourable factors.

During productive episodes both facies were commonly associated with prograding punctuated successions in which regionally extensive black shales recurred every few million years. An especially good record of ironstones alternating rather regularly with black shales in this pattern is preserved in Jurassic sequences of northwestern Europe and in the Devonian deposits of Belgium. In contrast, ironstones are rare and very local in black shale-bearing sequences in the Middle and Late Devonian deposits in the northern Appalachian Basin and in Late Cretaceous successions in the Western interior Seaway, perhaps in part because the influx of detrital sediments associated with orogeny was too great. In such settings repeated tongues of black shale commonly accompanied renewed regional transgression, regardless of whether a hiatus had occurred. Development of an ironstone, on the other hand, depended in large part on a hiatus, both local and regional, intervening in the succession of shoaling-upward sequences and implying much reduced detrital influx. After accumulation of a single ironstone normal sedimentation was renewed, frequently under transgressive anoxic conditions. When a hiatus prevailed for several million years, as in the late Liassic Lorraine (Teyssen 1984) and Late Devonian Libyan deposits (Van Houten & Karasek 1981), repeated ironstones succeeded one another directly, on the same temporal scale as the associated detrital sequences.

The several hundred thousand-year and the several million-year time scales reviewed here suggest different modes of control, and introduce the basic problem of identifying allocyclic and autocyclic factors on both scales. The relatively well constrained long timescale records involving repeated transgressions are usually attributed to either a global eustatic or a regional tectonic control. Short timescale records can seldom be correlated reliably enough beyond a local area to establish a regional control. Nevertheless, some successions of shoaling-upward sequences containing widespread tongues of black shale point to allocyclic sea level fluctuations in a cadence of about 400 000 years (Brett & Baird 1986). Many of the ironstones were associated with similar sequences, and some recurred on a similar time scale (Van Houten 1986). The more sporadic development of ironstones, in contrast to the record of black shales, points to a role of local, autocyclic conditions controlling the production of ooids.

ACKNOWLEDGEMENTS: This review has relied heavily on the many published papers we have cited. In addition, U. Bayer, C. Brett, R. Dreesen, A. Embry, S. Guerrak, A. Hallam, P. Luce, B. Selleck, T. Teyssen and D. Woodrow have supplied useful information. We also acknowledge the helpful comments of two anonymous reviewers.

## References

ALVAREZ, W. C. & MULLER, R. A. 1984. Evidence from crater ages for periodic impacts on the Earth. *Nature,* **308,** 718–20.

ANDERSON, T. F., ARTHUR, M. A. & HOLSER, W. T. 1982. Organic carbon and sulphur in black shales and secular variation in $\delta^{13}C$ and $\delta^{34}S$. *Geological Society of America Programs with Abstracts,* **14,** 433.

ARTHUR, M.A. 1979. Paleoceanographic events—recognition, resolution, and reconsideration. *Review of Geophysics and Space Physics,* **17,** 1474–94.

—— 1982. The carbon cycle: controls on atmospheric $CO_2$ and climate in the geologic past. *In: Studies in Geophysics, Climate in Earth History.* National Academy, Washington, D.C., 55–67.

—— & JENKYNS, H. C. 1981. Phosphorite and Paleoceanography: *Oceanologica Acta,* **4,** 83–96.

——, DEAN, W. E. & STOW, D. A. V. 1984. Models for the deposition of Mesozoic-Cenozoic fine-grained organic-carbon-rich sediment in the deep sea. *In:* STOW, D. A. V. & PIPER, D. J. W. (eds), *Fine-grained sediments: deep-water processes and facies,* Special Publication of the Geological Society, London, **15,** 527–560.

——, DEAN, W. E. & SCHLANGER, S. O. 1985. Variations in the global carbon cycle during the Cretaceous related to climate, volcanism and changes in atmospheric $CO_2$. *In:* SUNDQUIST, E. G. & BROECKER, W. S. (eds), *The Carbon Cycle and Atmospheric $CO_2$: Natural Variations Archean to Present, Geophysical Monography Series,* **32,** American Geophysical Union, Washington, D. C., 397–411.

BAIRD, G. C. & BRETT, C. E. 1986. Erosion on an anaerobic seafloor: significance of reworked pyrite deposits from the Devonian of New York State. *Palaeogeography, Palaeoclimatology, Palaeoecology,* **57,** 157–193.

BARRON, E. J. & WASHINGTON, W. M. 1985. Warm Cretaceous climates: high atmospheric $CO_2$ as a plausible mechanism. *In:* SUNDQUIST, E. T. & BROECKER, W. S. (eds) *The Carbon Cycle and Atmospheric $CO_2$: Natural Variations Archean to Present, Geophysical Monography Series,* **32,** American Geophysical Union, Washington, D. C., 546–553.

BAYER, U., ALTHEIMER, E. & DEUTSCHLE, W. 1985. Environmental evolution in shallow epicontinental seas: sedimentary cycles and bed formation. *In:* BAYER, U. & SEILACHER, A. (eds), *Sedimentary and Evolutionary Cycles.* Springer, New York, 347–81.

—— & MCGHEE, G. R. 1986. Cyclic patterns in the Paleozoic and Mesozoic: implications for time scale calibrations. *Paleoceanography,* **1,** 383–402.

BERG, T. M., MCINERNEY, M. K., WAY, J. H. & MACLACHLAN, D. B. 1983. Stratigraphic correlation chart of Pennsylvania. *Bureau of Topographic and Geologic Survey, General Geology Report* **75.**

BERNER, R. A. 1987. Models for carbon and sulfur cycles and atmospheric oxygen: application to Paleozoic geologic history. *American Journal of Science,* **287,** 177–196.

—— & RAISWELL, R. 1983. Burial of organic carbon and pyrite sulfur in sediments over Phanerozoic time: a new theory. *Geochimica et Cosmochimica Acta,* **47,** 855–62.

BERRY, W. B. N. & WILDE, P. 1978. Progressive ventilation of the oceans—an explanation for the distribution of Lower Paleozoic black shales. *American Journal of Science,* **278,** 257–75.

BHATTACHARYYA, D. P. 1983. Origin of berthierine in ironstones. *Clays and Clay Minerals,* **31,** 172–82.

—— 1989. Concentrated and lean oolites: examples from the Nubia Formation at Aswan, Egypt and significance of the oolite types in ironstone genesis. *In:* YOUNG, T. P. & TAYLOR, W. E. G. (eds) *Phanerozoic Ironstones,* Geological Society, London, Special Publication **46,** 93–104

BLESS, M. J. M. *et al.* 1980. Pre-Permian depositional environments around the Brabant Massif. *Sedimentary Geology,* **27,** 1–81.

BORCHERT, H. 1965. Formation of marine sedimentary iron ores. *In:* RILEY, J. P. & SKIRROW, G. (eds.) *Chemical Oceanography.* Academic Press, London, **2,** 159–204.

BRETT, C. E. & BAIRD, G. C. 1986. Symmetrical and upward shallowing cycles in the Middle Devonian of New York State and their implications for the punctuated aggradational cycle hypothesis. *Paleoceanography,* **1,** 431–45.

BROADHEAD, R. F., KEPFERLE, R. C. & POTTER, P. E. 1982. Stratigraphic and sedimentologic controls of gas in shale—examples from Upper Devonian of northern Ohio. *Bulletin of the American Association of Petroleum Geologists,* **66,** 10–27.

BROOKFIELD, M. E. 1971. An alternative to the 'clastic trap' interpretation of oolitic ironstone facies. *Geological Magazine,* **103,** 137–143.

BUSCH, R.M. & ROLLINS, H.B. 1984. Correlation of Carboniferous strata using a hierarchy of transgressive-regressive units. *Geology,* **12,** 47–74.

——& WEST, R. R. 1987. Hierarchal genetic stratigraphy: a framework for paleoceanography. *Paleoceanography,* **2,** 141–164.

CARROLL, D. 1958. Role of clay minerals in the transportation of iron. *Geochimica et Cosmochimica Acta,* **14,** 1–27.

COBBAN, W. A. & HOOK, S. C. 1984. Mid-Cretaceous molluscan biostratigraphy and paleogeography of southwestern part of Western Interior, United States. *Geological Association of Canada Special Paper,* **27,** 257–71.

COCKS, L. R. M. & FORTEY, R. A. 1982. Faunal evidence for oceanic separations in the Palaeozoic of Britain. *Journal of the Geological Society, London,* **139,** 465–478.

COOK, P. J. & MCELHINNEY, M. W. 1979. A reevaluation of the spatial and temporal distribution of sedimentary phosphate deposits in the light of plate tectonics. *Economic Geology,* **74,** 315–30.

DENNISON, J. M & HEAD, J. W. 1975. Sea level variations interpreted from the Appalachian basin Silurian and Devonian. *American Journal of Science,* **275,** 1089–120.

DESTOMBES, J. 1971. L'Ordovician au Maroc, Essai de synthese stratigraphiques. *Bureau de Recherches Geologiques et Minieres Memoires,* **73,** 237–63.

DREESEN, R. 1982. Storm-generated oolitic ironstones of the Famennian (Falb–Fa2a) in the Vesdre and Dinant synclinoria (Upper Devonian, Belgium). *Annales de la Societe Geologique de Belgique,* **105,** 105–29.

——1989. Oolite ironstones as event-stratigraphical marker beds within the Upper Devonian of the Ardenno–Rhenish Massif. *In:* YOUNG, T. P. & TAYLOR, W. E. G. (eds) *Phanerozoic Ironstones,* Geological Society, London, Special Publication **46,** 65–78.

EHRETS, J. R. 1981. The West Falls Group (Upper Devonian) Catskill delta complex: stratigraphy, environments and sedimentation. *Guidebook for field trips in south-central New York;* 53rd Annual Meeting, New York State Geological Association, 3–30.

EMBRY, A. F. 1982. The Upper Triassic—Lower Jurassic Heiberg Deltaic Complex of the Sverdrup Basin. *Canadian Society of Petroleum Geologists Memoir,* **8,** 189–217.

ETTENSOHN, F. R. 1985. Controls on development of Catskill Delta complex basin-facies. *Geological Society of America Special Paper,* **201,** 65–77.

FAILL, R. T. 1983. Tectonic framework of the Catskill delta. *Geological Society of America Programs with Abstracts,* **15,** 132.

——1985. The Acadian orogeny and the Catskill delta. *Geological Society of America Special Paper,* **201,** 15–38.

——, HOSKINS, D. M. & WELLS, R. B. 1978. Middle Devonian stratigraphy in central Pennsylvania: a revision. *Pennsylvania Geological Survey, 4th Series, Bulletin,* G-70, 1–28.

FISCHER, A. G. 1981. Climate oscillations in the biosphere. *In:* NITECKI, M. H. (ed.), *Biotic Crises in Ecological and Evolutionary Time,* Academic Press, New York, 103–131.

——1984. The two Phanerozoic supercycles. *In:* BERGGREN, W. A. & VAN COUVERING, J. (eds.), *Catastrophes in Earth History,* Princeton University Press, Princeton, N. J., 129–150.

——& ARTHUR, M. A. 1977. Secular variations in the pelagic realm. *Society of Economic Paleontologists and Mineralogists Special Publication,* **25,** 19–50.

GARRELS, R. M. & LERMAN, A. 1981. Phanerozoic cycles of sedimentry carbon and sulphur. *National Academy of Sciences Proceedings,* **78,** 4652–56.

——, & LERMAN, A. 1984. Coupling of the sedimentary sulphur and carbon cycles—an improved model. *American Journal of Science,* **284,** 989–1007.

GLAESER, J. D. 1979. Catskill delta slope deposits in the central Appalachian Basin: source and reservoir deposits. *Society of Economic Paleontologists and Mineralogists Special Publication,* **27,** 343–57.

HALLAM, A. 1975. *Jurassic Environments,* Cambridge University Press, U.K., 261 pp.

——1981. A revised sea-level curve for the early Jurassic. *Journal of the Geological Society, London,* **138,** 735–48.

——1982. The Jurassic climate. *In: Studies in Geophysics, Climate in Earth History,* National Academy, Washington, D.C., 159–63.

——1984. Pre-Quaternary sea-level changes. *Annual Reviews of Earth and Planetary Sciences,* **12,** 205–43.

——& BRADSHAW, M. J. 1979. Bituminous shales and oolitic ironstones as indicators of transgressions and regressions. *Journal of the Geological Society, London,* **136,** 157–64.

——& SELLWOOD, B. W. 1976. Middle Mesozoic sedimentation in relation to tectonics in the British area. *Journal of Geology,* **84,** 302–21.

HASSON, K. O. & DENNISON, J. M. 1979. Devonian shale stratigraphy between Perry Bay and the Fulton Lobe, south-central Pennsylvania and Maryland. *Guidebook, 44th Annual Field Conference of Pennsylvania Geologists,* 1–17.

HAY, W. W. 1981. Sedimentological and geochemical trends resulting from the breakup of Pangaea. *Oceanologica Acta,* **4,** 135–147.

HECKEL, P. H. 1973. Nature, origin, and significance of the Tully limestone; an anomalous unit in the Catskill Delta, Devonian of New York. *Geological Society of America Special Paper,* **138,** 1–244.

——1977. Origin of phosphatic black shale facies in Pennsylvania cyclothems of mid-continent North America. *Bulletin of the American Association of Petroleum Geologists,* **61,** 1045–68.

——1986. Sea-level curve for Pennsylvanian eustatic marine transgressive-regressive depositional cycles along midcontinent outcrop belt, North America. *Geology,* **14,** 330–334.

HOLLAND, H. D. 1984. *The chemical evolution of the atmosphere and oceans,* Princeton University Press, Princeton, N. J.

HOLSER, W. T. 1977. Catastrophic chemical events in the history of the Ocean. *Nature,* **267,** 403–8.

——, 1984. Gradual and abrupt shifts in ocean chemistry during Phanerozoic time. *In:* HOLLAND, H. D. & TRENDALL, A. F. (eds.), *Patterns of Change in Earth Evolution,* Springer-Verlag, Dahlem Konference, Berlin, 123–143.

HUNTER, R. E. 1960. *Iron Sedimentation in the Clinton Group of the Central Appalachian Basin.* Ph.D. Thesis, The Johns Hopkins University, Baltimore, Maryland.

INNERS, J. D. 1979. The Onesquethaw Stage in south-central Pennsylvania and nearby areas. *44th Annual Field Conference of Pennsylvania Geologists Guidebook,* 38–55.

JAANUSSON, V. 1982. Introduction to the Ordovician of Sweden, 1–10; The Siljan District, 15–42; Ordovician of Vastergotland, 164–83. *Paleontological Contributions from the University of Oslo,* 279.

JENKYNS, H. C. 1980. Cretaceous anoxic events: from continents to oceans. *Journal of the Geological Society, London,* **139,** 171–88.

——1985. The early Toarcian and Cenomanian-Turonain anoxic events in Europe: comparisons and contrasts. *Geologisches Rundschau,* **74,** 505–18.

KAUFFMAN, E. G. 1978. Benthic environments and paleoecology of the Posidonienschiefer (Toarcian). *Neues Jahrbuch fur Geologie und Palaeontologie Abhandlungen,* **157,** 18–36.

KNOX, R.W. 1970. Chamosite ooliths from the Winter Gill Ironstone of Yorkshire, England. *Journal of Sedimentary Petrology,* **40,** 1216–1225.

KREBS, W. 1979. Devonian basinal facies. *Paleontology Special Paper,* **23,** 125–39.

LASAGA, A. C., BERNER, R. A. & GARRELS, R. M. 1985. An improved model for atmospheric $CO_2$ fluctuations over the past 100 million years. *In:* SUNDUIST, E. T. & BROECKER, W. S. (eds.), *The Carbon cycle and Atmospheric $CO_2$: Natural variations Archean to Present, Geophysical Monograph Series* **32,** American Geophysical Union, Washington, D. C., 397–411.

LEGGETT, J. K. 1980. British lower Palaeozoic black shales and their palaeo-oceanographic significance. *Journal of the Geological Society, London,* **137,** 167–176.

——, MCKERROW, W. S., COCKS, L. R. M. & RICKARDS, R. B. 1981. Periodicity in the early Palaeozoic marine realm. *Journal of the Geological Society, London,* **138,** 167–176.

LESURE, F. G. 1957. Geology of the Clifton Forge iron district, Virginia. *Virginia Polytechnical Institute Bulletin,* **118,** 1–130.

LIAO, S. F. 1958. The iron ore deposits of Tungyuen, Kweichow. *Acta Geologica Sinica,* **38,** 462–72.

——1964. A study of the lithofacies and paleogeography and metallogenesis of the Ninghsiang type of iron ores. *Acta Geologica Sinica,* 44, 68–80.

LUCE, P. B. (discussant) 1981. Mansfield ore bed. *46th Annual Field Conference of Pennsylvania Geologists Guidebook,* 146–47.

LUTZ, T. 1985. The magnetic reversal record is not periodic. *Nature,* **317,** 404–407.

MACKENZIE, F. T. & PIGOTT, J. D. 1981. Tectonic controls of Phanerozoic sedimentary rock cycling. *Journal of the Geological Society, London,* **138,** 183–96.

MAYNARD, J. B. 1983. *Geochemistry of Sedimentary Ore Deposits.* Springer-Verlag, New York.

——1986. Geochemistry of oolitic iron ores, an electron microprobe study. *Economic Geology,* **81,** 1473–83.

MCGHEE, G. R. & BAYER, U. 1985. The local signature of sea-level changes. *In:* BAYER, U. & SEILACHER, A. (eds.) *Sedimentary and Evolutionary Cycles.* Springer, New York, 98–112.

MORRIS, K. A. 1980. Comparison of major sequences of organic-rich mud deposition in the British Jurassic. *Journal of the Geological Society, London,* **137,** 157–70.

MOUTERDE, R. & TINTANT, H. (coords.) 1980. Lias. *Bureau de Recherches Geologiques et Minieres Memoires,* **101,** 75–123.

PALMER, A. R. 1983. The Decade of North American Geology 1983, Geologic Time Scale, *Geology,* **11** 503–504.

PETERSON, F. & RYDER, R. T. 1975. Cretaceous rocks of the Henry Mountains region, Utah and their relation to neighboring regions. *Four Corners Geological Society, 8th Field Conference Guidebook,* 167–89.

PETRANEK, J. 1964. Gemeinsame merkmale der Eisenlager in bohmischen und thuringischen Ordovicicum. *Abhandlungen der Deutschen Akademie der Wissenschaften zu Berlin,* **2,** 79–95.

——1974. Sedimentari zelezne rudy v ordoviku kursne hory. *Sbornik geologie ved, Loziskova geologie mineralogie,* **16,** 165–98.

RAMPINO, M. R. & STOTHERS, R. B. 1984. Geological rhythms and cometary impacts. *Science,* **250,** 68.
RAMSBOTTOM, W. H. C. 1979. Rates of transgression and regression in the Carboniferous of NW Europe. *Journal of the Geological Society, London,* **136,** 147–53.
RAUP, D. M. & SEPKOSKI, J. J. 1986. Periodic extinctions of families and genera. *Science,* **231,** 833–836.
RICKARD, L. V. 1975. Correlation of the Silurian and Devonian rocks in New York State. *New York State Museum Science Service Map and Chart Series,* **24,** 1–16.
ROEN, J. B. 1984. Geology of the Devonian black shales of the Appalachian Basin. *Organic Geochemistry,* **5,** 241–54.
ROSS, C. A. & ROSS, J. R. P. 1985. Late Paleozoic depositional sequences are synchronous and worldwide. *Geology,* **13,** 194–97.
RYAN, W. B. F. & CITA, M. B. 1977. Ignorance concerning episodes of oceanwide stagnation. *Marine Geology,* **23,** 197–215.
SANDBERG, P. A. 1983. An oscillating trend in Phanerozoic non-skeletal carbonate mineralogy. *Nature,* **305,** 19–22.
SCHLANGER, S. O. & JENKYNS, H. C. 1976. Cretaceous oceanic anoxic sediments: causes and consequences. *Geologie en Mijnbouw,* **55,** 179–84.
SCHOLLE, P. A. & ARTHUR, M. A. 1980. Carbon isotopic fluctuations in pelagic limestones: potential stratigraphic and petroleum exploration tool. *Bulletin of the American Association of Petroleum Geologists,* **64,** 67–89.
SELLECK, B. W. & LINSLEY, R. M. 1984. Sedimentology and faunal assemblages in the Hamilton Group of Central New York. *56th Annual Meeting of the New York State Geological Association Field Trip Guidebook,* 221–229.
SELLWOOD, B. W. & JENKYNS, H. C. 1975. Basins and swells and the evolution of an epeiric sea (Pliensbachian-Bajocian) of Great Britain. *Journal of the Geological Society, London,* **131,** 373–388.
SEVON, W. D. 1981. The Middle and Upper Devonian clastic wedge in northeastern Pennsylvania. *53rd Annual Meeting of the New York State Geological Association Field Trip Guidebook,* 31–53.
SLOSS, L. L. 1972. Synchrony of Phanerozoic sedimentary-tectonic events of the North American craton and the Russian Platform. *24th International Geologic Congress Montreal,* **Section 6,** 24–32.
STOTT, D. F. 1984. Cretaceous sequences of the foothills of the Canadian Rocky Mountains. *Canadian Society of Petroleum Geologists Memoir,* **9,** 85–107.
TEYSSEN, T. A. L. 1984. Sedimentology of the Minette oolitic ironstones of Luxembourg and Lorraine: a Jurassic subtidal sandwave complex. *Sedimentology,* **31,** 195–211.
VAIL, P. R., MITCHUM, R. M. & THOMPSON, S. 1977. Seismic stratigraphy and global changes in sea level, part 4. *American Association of Petroleum Geologists Memoir,* **26,** 83–97.
——& TODD, R. G. 1981. Northern North Sea Jurassic unconformities, chronostratigraphy and sea-level changes from seismic stratigraphy. *In:* ILLING, L. & HOBSON, G. (eds.) *Petroleum Geology of the Continental Shelf of North-West Europe,* Heyden, London, 216–35.
VAN DER VOO, R. 1982. Pre-Mesozoic paleomagnetism and plate tectonics. *Annual Reviews of Earth and Planetary Sciences,* **10,** 191–220.
VAN HOUTEN, F. B. 1985. Oolitic ironstones and contrasting Ordovician and Jurassic paleogeography. *Geology,* **13,** 722–24.
——1986. Search for Milankovitch patterns among oolitic ironstones. *Paleoceanography,* **1,** 459–66.
——& BHATTACHARYYA, D. P. 1982. Phanerozoic oolitic ironstones—geologic record and facies model. *Annual Review of Earth and Planetary Sciences,* **10,** 441–457.
——& KARASEK, R., 1981. Sedimentologic framework of the Late Devonian oolitic iron formation, Shatti Valley, west-central Libya. *Journal of Sedimentary Petrology,* **51,** 415–27.
——& PURUCKER, M. E. 1984. Glauconitite peloids and chamositic ooids—favourable factors, constraints, and problems. *Earth Science Reviews,* **20,** 211–243.
VEIZER, J. 1985. Carbonates and ancient oceans: isotopic and chemical record on time scales of $10^7$ and $10^9$ years. *In:* SUNDQUIST, E. T. & BROECKER, W. S. (eds.), *The Carbon Cycle and Atmospheric $CO_2$: Natural Variations Archean to Present, Geophysical Monograph Series,* **32,** American Geophysical Union, Washington, D. C., 595–601.
——, HOLSER, W. T. & WILGUS, C. K. 1980. Correlation of $^{13}C/^{12}C$ and $^{34}S/^{32}S$ secular variations. *Geochimica Cosmochimica Acta,* **44,** 579–587.
WAPLES, D. W. 1983. Reappraisal of anoxia and organic richness, with emphasis on Cretaceous of North America, *Bulletin of the American Association of Petroleum Geologists,* **67,** 963–78.
WILDE, P. 1987. Model of progressive ventilation of the Late Precambrian—Early Paleozoic ocean. *American Journal of Science,* **287,** 442–459.
——& BERRY, W. B. N. 1984. Destablilization of the oceanic density, structure and its significance to marine "extinction"events. *Palaeogeography, Palaeoclimatology, Palaeoecology,* **48,** 143–62.
WILKINSON, B. H., OWEN, R. M. & CARROLL, A. R. 1985. Submarine hydrothermal weathering, global eustacy and carbonate polymorphism in Phanerozoic marine oolites. *Journal of Sedimentary Petrology,* **55,** 171–183.
WOODROW, D. L. 1968. Stratigraphy, structure and sedimentary patterns in the Upper Devonian of Bradford County, Pennsylvania, *Pennsylvania Geological Survey, 4th Series, Bulletin,* **G-54,** 1–78.
——& ISLEY, A. M. 1983. Facies, topography, and sedimentary processes in the Catskill sea

(Devonian), New York and Pennsylvania. *Geological Society of America Bulletin,* **94,** 459–470.

WORSLEY, T. R., NANCE, R. D. & MOODY, J. B. 1984. Global tectonics and eustacy for the past 2 billion years. *Marine Geology,* **58,** 373–400.

——, NANCE, R. D. & MOODY, J. B. 1986. Tectonic cycles and the history of the earth's biocheochemical and paleoceanographic record. *Paleoceanography,* **1,** 233–263.

WYATT, A. R. 1987. Shallow water areas in space and time. *Journal of the Geological Society, London,* **144,** 111–20.

ZIEGLER, P. A. 1982. *Geological Atlas of Western and Central Europe.* Shell International, The Hague.

F. B. VAN HOUTEN, Department of Geological and Geophysical Sciences, Princeton University, Princeton, NJ 08544, USA.

M. A. ARTHUR, Graduate School of Oceanography, University of Rhode Island, Narragansett, RI 02882, USA.

# Eustatically controlled ooidal ironstone deposition: facies relationships of the Ordovician open-shelf ironstones of Western Europe

T.P. Young

SUMMARY: Ooidal ironstones are a characteristic feature of the Ordovician successions of south-west Europe and adjacent areas. They occur in shallow marine clastic sequences deposited on a broad low-topography shelf. Three stratigraphic intervals have a particularly widespread development of ironstones: the early Llanvirn, the early Caradoc and the early Ashgill. These intervals, corresponding to periods of major global eustatic sea-level rise, are marked by the development of thin, but persistent, ooidal ironstones even in relatively offshore areas. The ironstones produced by these three events have been studied in the previously contiguous areas of central Portugal and Armorica.

The development of the ironstones, and the nature of the successions within which they occur, show remarkable similarities among the three horizons; this recurrence of similar sedimentary facies defines a cyclicity. In each of these 'first-order' cycles the upper part is characterised by the occurrence of a second order of cyclicity, represented by asymmetric coarsening— and thickening—upwards sequences of siltstone and sandstone tempestites, commonly capped by phosphatic conglomerates. The latest 'second-order' cycle in each case is characterized by a more distal sequence of dark mudstones with graptolites. The chamositic ooidal ironstones lie disconformably above these upper mudstone-dominated cycles, and are interpreted as forming during 'second-order' transgressions late in major global transgressive events. Both 'first-' and 'second-order' sedimentary cyclicity is attributed to eustatic variation. The 'first-order' eustatic cycles have a period of 10–15 Ma, and largely correspond to the international stratigraphic series, while the period of the 'second-order' cycles is estimated as 0.2–0.5 Ma.

The Ordovician was one of the periods of greatest production of ooidal ironstones in geological history (Petranek 1964a, b; Van Houten & Purucker 1984; Van Houten 1985). These ironstones are mainly to be found in N. Africa and on the fragments of the Avalonian and Armorican terranes which are now largely incorporated into Europe (Van Houten 1985). A palaeogeographical reconstruction for the Ordovician (Fig. 1) demonstrates that these ironstones formed at mid- to high-latitudes, predominantly, but not exclusively, on the margins of Gondwana.

A review of the occurrences of Ordovician ooidal ironstones (to be published in full elsewhere) shows three major peaks of ironstone production (Fig. 2). These peaks occur in the early Llanvirn, the early Caradoc and the early Ashgill. They correspond closely to the phases of major global eustatic sea-level rise proposed from analyses of sedimentary facies relationships of areas on previously independent continental blocks, particularly N. America, Britain, Baltica and Australia (Fortey 1984). These peaks in the stratigraphic distribution are largely produced by thin, but laterally persistant, often 'lean' ironstones (*sensu* Bhattacharyya 1989), whereas there is a more general distribution through the Ordovician of more locally-developed, often thicker, 'concentrated', ooidal ironstones.

It seems unlikely that any single facies model can account for the occurrence of the sporadic, local, and often more economic ironstones, but the temporal relationship during the Ordovician between the widespread thin ooidal ironstones and periods of sea-level rise is striking.

## Geological and sedimentary characteristics of the Ibero-Armorican area

The sequences examined in this study are found in Armorica and central Portugal. These two areas were part of a single stable shelf during the Ordovician. The western part of Brittany (particularly the Crozon peninsula) and central Portugal were in very close proximity in the Ordovician, but became separated, probably in mid-Devonian times, by dextral transcurrent faulting on the South Armorican and Cordoba/Tomar shear zones (Fig. 3). This study focuses

*From* YOUNG, T. P. & TAYLOR, W. E. G. (eds), 1989, *Phanerozoic Ironstones*
Geological Society Special Publication No. 46, pp. 51-63

FIG. 1. Palaeogeographic reconstruction of the southern hemisphere in the late Ordovician. Areas in which ironstones were developed during the Ordovician are stippled, with the location of the study area indicated by the arrow.

on central Portugal, where it is possible to trace the individual horizons over large areas; this is supplemented with some information from the Armorican Massif.

The lithostratigraphic nomenclature for the area studied is complicated by the separate schemes applied to the different areas of Ordovician outcrop. Despite this complexity of nomenclature a strikingly widespread distribution of many sedimentary facies can be recognized across the study area (Fig. 4).

The Ordovician sequences in Armorica and the Luso-Alcudian Zone of central Iberia (Lotze 1945) are characterized by their relatively thin development. In this region the Ordovician deposits rest unconformably on a late Pre-Cambrian sedimentary sequence, although in some areas (e.g., parts of Normandy) fossiliferous Cambrian may occur between. The lower Ordovician is usually developed in sandstone facies, the 'Armorican Quartzite' of Arenig age, which locally lies on red-beds of possible Tremadoc age. The Armorican Quartzite is dominantly shallow sub-tidal in origin, and in some areas it contains ooidal ironstones (Chauvel 1971). The sequence of Llanvirn to basal Ashgill age consists of shallow marine clastic sediments, which show strong storm influence. In most areas the Llanvirn and Llandeilo are in a mudstone facies, commonly punctuated by several packets of storm-generated sandstones. The Caradoc sequences are more generally sandy, but reflect similar environmental conditions. The sediments of Tremadoc to Caradoc age were affected by belts of differential subsidence, but the topography seems to have remained very slight throughout. A period of greater activity of these minor basins occurred in the early to middle Ashgill, with volcanicity in some areas of previously high subsidence.

The middle to upper Ordovician shallow shelf clastics record the accumulation of sediments on a low topography shelf, with very widespread distribution of individual units. The alternating packets of storm-generated sandstones (tempestites) and mudstones appear to record changes in sediment supply to the shelf, and in the ability of storm activity to move the sediment on the shelf, both of which are strongly dependent on relative sea level stand. Within the packets of storm generated sandstones coarsening- and thickening-up sequences occur, most markedly in the upper Llandeilo and upper Caradoc sequences.

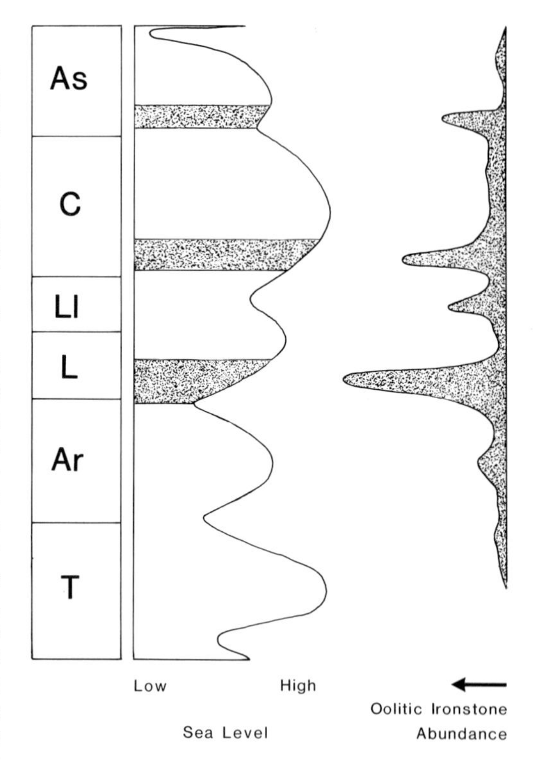


FIG. 2. The abundance of ooidal ironstones during the Ordovician, plotted against the eustatic curve of Fortey (1984). T—Tremadoc, Ar—Arenig, L—Llanvirn, Ll—Llandeilo, C—Caradoc, As—Ashgill.

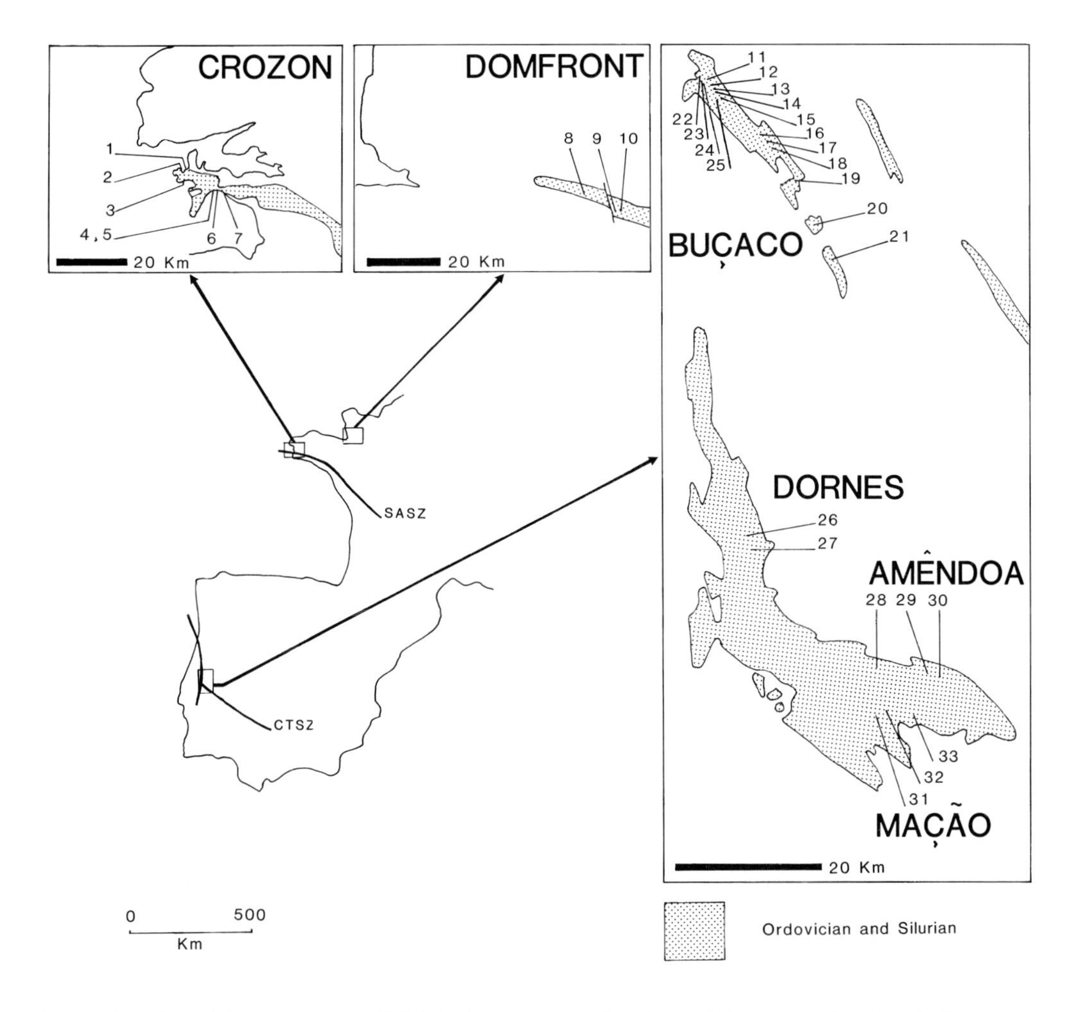


FIG. 3. Location of the sequences studied. Numbers correspond to those of the measured sections in Figs. 5–7:
1, La Mort-Anglaise; 2, Corrigou; 3, Le Veryac'h; 4, Postolonnec; 5, Postolonnec; 6, Raguenez; 7, Kerglintin; 8, La Fosse Arthour; 9, Domfront, west of river; 10, Domfront, road section east of river; 11, Favaçal; 12, Louredo; 13, Vale Saido; 14, Cacemes; 15, Zuvinhal; 16, Riba de Baixo; 17, Belfeiro; 18, Riba da Cima; 19, Poiares; 20, Rio Ceira; 21, Albegaria; 22, Porto de Santa Anna; 23, Leira Mà; 24, Louredo; 25, Zuvinhal; 26, Monte de Carvalhal; 27, Lameiros; 28, Amêndoa; 29, Ribeira de Laje; 30, Balancho; 31, Aboboreira; 32, Carregueira; 33, Pereiro. SASZ, South Armorican Shear Zone; CTSZ, Cordoba/Tomar Shear Zone.

Within the study area ooidal ironstones are developed at three horizons (Fig. 4). The lowest (♂1) is close to the base of the Brejo Fundeiro Fm. (Portugal), Postolonnec Fm. (W. Brittany) and Pissot Fm. (S. Normandy), just above the top of the Armorican Quartzite. The middle (♂2) is at the base of the Louredo Fm. (Portugal) and the base of the Kermeur Fm. (W. Brittany), while the uppermost (♂3) is at the base of the Porto de Santa Anna Fm. (Portugal) and the base of the Rosan Fm. (W. Brittany). These three horizons are of early Llanvirn, early Caradoc and early Ashgill ages respectively, precisely those times of most widespread ironstone production regionally (Fig. 2).

## Descriptions of the three main ironstone-bearing intervals

The description of the sequence bearing the ironstones is based on the study of 31 sections,

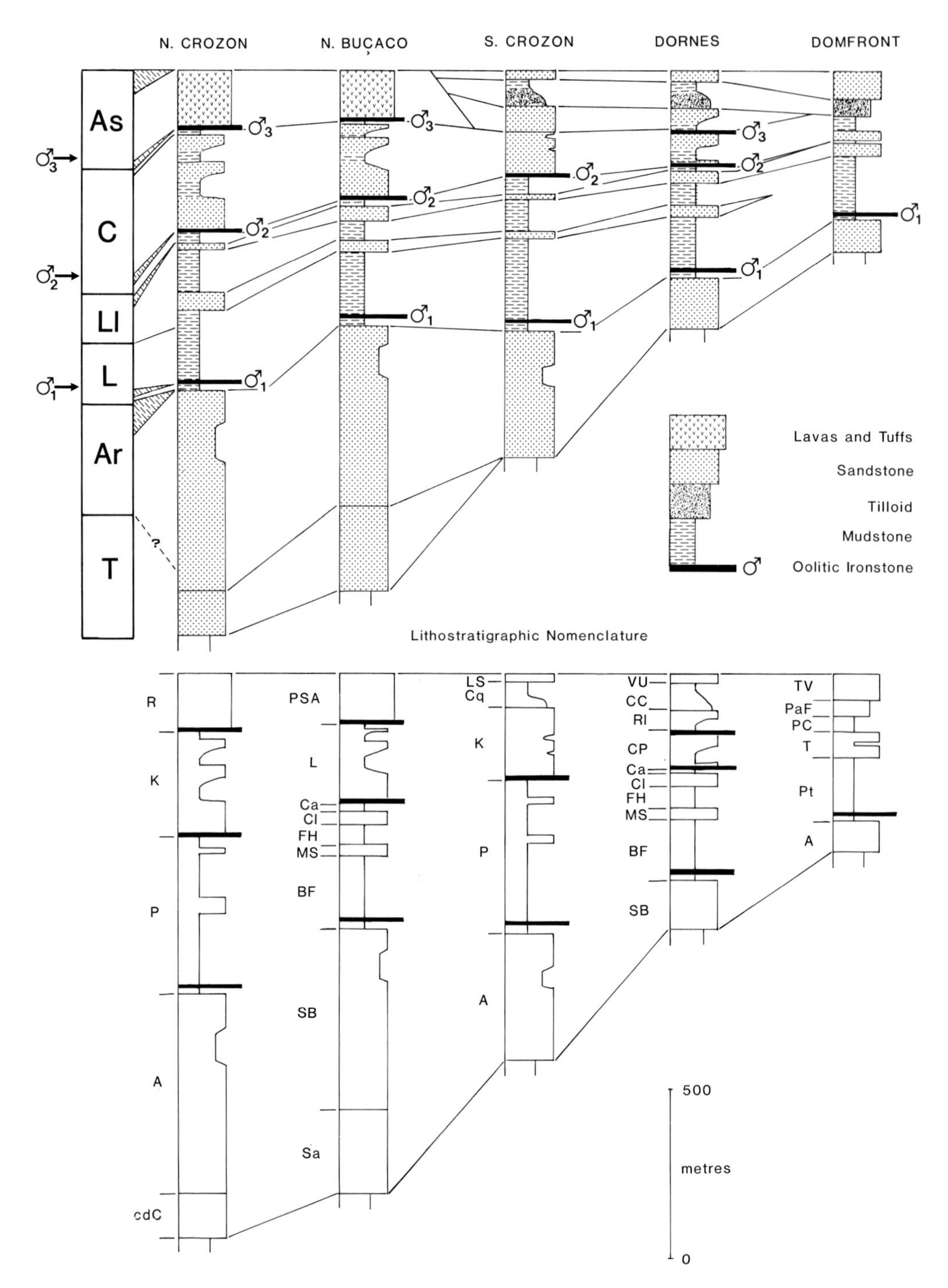


FIG. 4. Stratigraphy of the Ordovician in the study area. Correlation of sequences (upper) and lithostratigraphic nomenclature (lower). A—Armorican Quartzite Fm., BF—Brejo Fundeiro Fm., cdC—cap de la Chêvre Fm., Ca—Carregueira Fm., CC—Casal Carvalhal Fm., Cl—Cabril Fm., CP—Cabeço do Pão Fm., Cq—Cosquer Fm., FH—Fonte de Horta Fm., K—Kermeur Fm., L—Louredo Fm., LS—Grès de Lamm Saoz, MS—Monte da Sombadeira Fm., P—Postolonnec Fm., PaF—Pélites à Fragments Fm., R—Rosan Formation, Rl—Ribeira da Laje Fm., Sa—Sarnelha Fm., SB—Serra de Brejo Fm., T—Tertre Chapon Fm., TV—Tertre de la Violère Fm, VU—Vale da Ursa Fm.

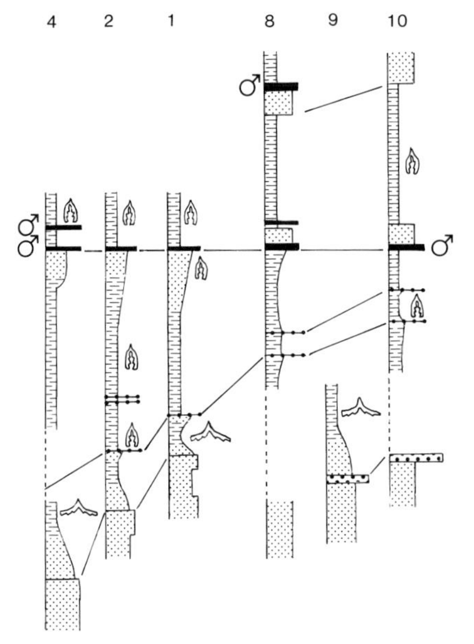


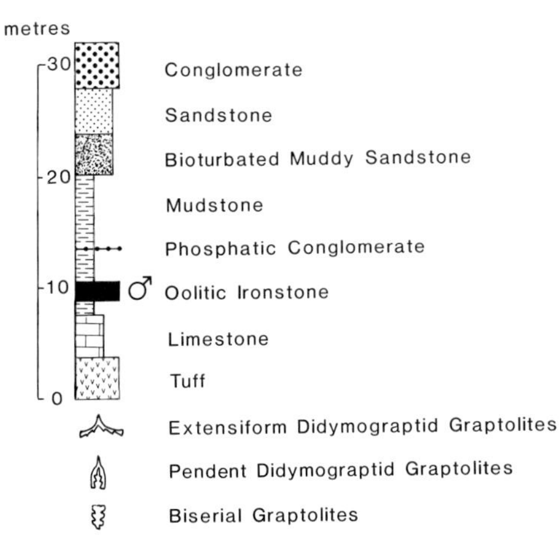


FIG. 5. Measured sections through the lower Llanvirn ironstones, showing increasing influence of the Mancellia 'rise' towards the right. Datum is the base of chitinozoan zone 7. Tie-lines indicate sedimentological marker horizons. Locations 1, 2, & 4 are in the Crozon area, locations 8, 9, & 10 are in the Domfront area.

with two additional sections discussed on the basis of published work by other authors. Biostratigraphical correlation of the sections is largely based on the study of the chitinozoan biostratigraphy by Paris (1979, 1981), but also includes correlations based on unpublished analysis of the macrofauna (Young 1985).

## The early Llanvirn

The early Llanvirn transgression was a major event throughout the European and north African areas. In much of the region it is marked by the appearance of dark mudstones (the lower part of the Calymene or Tristani shales of old authors) above the 'Armorican Quartzite' facies. In the present work this interval has been studied in the Crozon Peninsula, Brittany, and the Domfront Syncline, Normandy (Fig. 5). The nature of the transition from the Armorican Quartzite to the overlying mudstones (e.g., the Postolonnec Fm. of Crozon, the Traveusot Fm. south of Rennes, the Andouillé Fm. of the Armorican Median Syncline, the Pissot Fm. of Domfront, the Urville Fm. of the Caen area and the Brejo Fundeiro Fm. of central Portugal) had not been described in detail until recently. The biostratigraphy in particular had been hampered by poor stratigraphic localization of important material, particularly some of the graptolite collections. Recent studies of the chitinozoans from Armorica and Portugal (Paris 1981), the macrofauna in central Portugal (Romano *et al.* 1986), graptolites from western Brittany (Paris & Skevington 1979) and further detailed lithostratigraphic studies in Normandy (Robardet 1981) have begun to present a more detailed account of the biostratigraphy.

### *Lithostratigraphy.*

The middle Arenig arenaceous facies in the upper part of the Armorican Quartzite is overlain, apparently conformably, by a finer-grained facies in many areas. These mudstones, siltstones and sandstones show storm-generated features, and are closely related in facies to the underlying sandstones. They contain faunas dated as middle Arenig (by extensiform didymograptid graptolites; Paris & Skevington 1979) or as late Arenig (zones 3 and 4a of the chitinozoan biostratigraphy; Paris 1981). They are usually capped by a thin coarsening-up sequence, overlain by a phosphatic conglomerate bed, a few centimetres thick. Immediately above the conglomerate, dark mudstones yield shelly and graptolitic faunas, as well as chitinozoans, indicating an early Llanvirn age. A point of confusion in Crozon is that the well known 'lingulid beds' (e.g., in the Corrigou section) are not at the break between the Arenig and Llanvirn strata, but lie at a lower horizon, within the Armorican Quartzite Formation. In many areas additional disconformities may occur both below the Arenig graptolite faunas, and above the first Llanvirn faunas (e.g., Fig. 5, locality 10). The stratigraphy of the deposits between the Armorican Quartzite facies and the ironstone is thus rather complicated, but includes at least one major disconformity. The presence of further thin phosphatic conglomerates suggests additional disconformities, but it is not yet possible to resolve whether these may be correlated on a regional scale.

The ironstone at the base of chitinozoan zone 7 (the datum level on Fig. 5) lies at about 20m above the phosphatic conglomerate at the base of the Llanvirn in Crozon. This interval becomes thinner in the areas of more condensed deposits, until in parts of Normandy the ironstone rests directly on the disconformity. The mudstones below the ironstone are capped by a coarsening-up sequence of siltstones and fine sandstones, displaying such storm-generated features as wave ripples, fining-up sharp-based beds with planar to low-angle cross-stratification and gutters. The ooidal ironstone rests on a siderite/phosphate hardground (the siderite now pseudomorphed by chlorite). The ooidal ironstone is predominantly chamosite, with evidence of the previous presence of siderite. It contains phosphatized intraclasts, and locally laminae which may be of stromatolitic origin. The higher ironstone at Postolonnec (Fig. 5, locality 4) is of a similar nature, but the ironstones higher in zone 7 at Domfront (Fig. 5, localities 8 and 10) are mixed with ferruginous sandstone.

In the Caen area the ironstones are well developed (Joseph 1982), and range well up into the Urville Fm., the local equivalent of the Pissot Fm. Although these thicker, more arenaceous, ironstones are not the subject of this paper, they are related to the ironstones of the more argillaceous sequences described here, and the origin of the two facies must be considered together. Joseph interpreted these ironstones as being parallel giant sandwaves, and they may represent the reworking of ooids and sand in shallow-water environments.

### *Biostratigraphy*

The ironstones occur well above the first pendent didymograptid graptolites, and are therefore firmly within the Llanvirn. Biostratigraphic subdivision of the Llanvirn sediments is difficult. At the localities (Fig. 5, localities 4, 9, 10) examined by Paris (1981) the oldest Llanvirn sediments are within his biozone 4b, and lie on the phosphatic conglomerate above the beds yielding the Arenig graptolites. The top of biozone 4b coincides with an additional phosphatic pebble-bed at Domfront, and a similar bed occurs in Crozon, but it is uncertain if these reflect a regional hiatus. Above the sediments of biozone 4b age, those of biozone 5 age include the mudstones capped by the coarsening-up sequence of siltstones and fine sandstones. Paris records that the ironstone, and the sequence immediately above it, are of biozone 7 age, but that phosphatized clasts in the ironstone are of biozone 6. Paris indicates that biozone 7 is still within the lower Llanvirn.

At Domfront an ironstone occurs higher within the zone 7 deposits. Paris suggests that the ironstone at the base of the Urville Fm. in the May Syncline, which lies directly on Cambrian sediments where the Armorican Quartzite is absent, also is of zone 7 age.

## The early Caradoc

The early Caradoc ironstone of the Crozon Peninsula has been well studied because of its rich microflora, which was used to illustrate the close similarity in development of the Ordovician between the Crozon and Buçaco areas (Henry *et al.* 1974).

### *Lithostratigraphy*

The sequence containing the ironstone commences with a shallow-water storm-generated sandstone packet of Llandeilo age (the Cabril Formation of central Portugal). This packet consists of a series of coarsening-up cycles, each capped by a phosphatic conlomerate (Fig. 6). The highest of these conglomerates in Portugal lies at the contact of the Cabril Fm. with the overlying mudstones of the Carregueira Fm. In western Brittany the highest of these coarsening-up cycles contains much more finer-grained sediment than the lower cycles; the episodic reworking associated with storm-activity is represented by coquinas and by winnowing and reworking of early diagenetic nodules, rather than by the influx of coarse clastic material, except at its very top (e.g., Plage de Postolonnec). This last cycle contains macrofauna of the *Marrolithus bureaui* Biozone, commonly taken as belonging to the latest part of the Llandeilo (Henry 1980). This markedly finer-grained sedimentation suggests that this cycle was more transgressive than the earlier ones, although it was far less so than the following cycle.

The deposits of the *M. bureaui* Biozone are overlain by dark mudstones: the 'Schistes de Veryac'h' (Crozon) and the Carregueira Formation (central Portugal). These mudstones are noticeably transgressive over the earlier deposits, and their base is locally marked by the

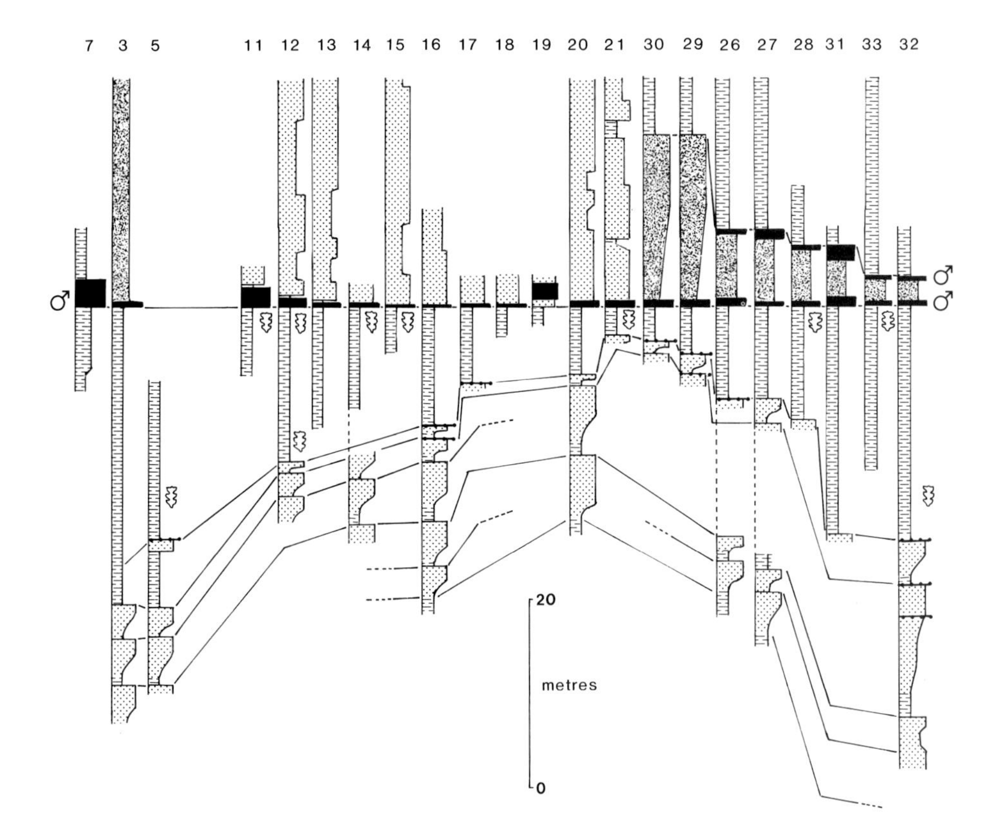


FIG. 6. Measured sections through the Lower Caradoc ironstones arranged from left to right in a north/south section across the Dornes/Amêndoa 'rise'. For key see Fig. 5. Datum is the base of chitinozoan zone 14. Tie-lines indicate sedimentological marker horizons. Locations 3, 5 & 7 are in the Crozon area, locations 11–21 are in the Buçaco syncline, locations 28, 29 & 30 are in the Amêndoa area, locations 26 & 27 are in the Dornes area and locations 31, 32 & 33 are in the Mação area.

development of phosphatic conglomerates, coquinas and reworked concretions. Where the sequence appears more continuous (e.g Le Veryac'h; Fig. 6, locality 3) the macrofauna shows a change from the assemblages of *Neseuretus, Crozonaspis, Marrolithus, Heterorthina, Apollonorthis* and the various bivalves of the *M. bureaui* Biozone, through assemblages rich in *Heterorthina* and *Placoparia,* to those characterized by gastropods and crinoids. Previous studies of the palaeoecology of Llandeilo faunas in the region (Romano 1982; Young 1985) have demonstrated that faunas with abundant *Placoparia* are generally of slightly deeper-water origin than those with *Neseuretus* and *Crozonaspis.* This faunal change is therefore interpreted as recording a relative deepening of the sea.

The dark mudstones below the ironstone are of variable thickness, as they are below the ironstones of the other two horizons. They thin from nearly 30m in S. Crozon to only 2m on the Dornes/Amêndoa 'rise', and then thicken to 25m in the Mação area. Sedimentary structures are difficult to detect in the black mudstones, but there is some field evidence for the increased abundance of silty horizons in the upper few metres. Calculation of the abundance of minerals by comparison of XRD and XRF data also supports a significant upward increase of quartz content in the upper few metres of the unit. This suggests that the mudstone unit forms a further coarsening-up sequence.

The ironstone horizon rests disconformably on the mudstone unit, with the lower part of the ironstone containing numerous phosphatized clasts derived from the mudstone. The top of the underlying sediments is marked by a concretionary layer composed of pyrite, chamosite (probably largely replacing siderite) and phosphates. The upper surface of this concretionary horizon can be seen to be erosional at some localities (e.g., Le Veryac'h; Fig. 6, locality 3). The phosphatized intraclasts in the ironstone are concentrated in the small erosional hollows of this probable hardground surface.

The ironstone is thoroughly bioturbated, with lenses of ooidal grain-ironstone remaining amongst the predominant ooidal mud-ironstone/wacke-ironstone. The original sedimentary structure of the ironstone has been largely destroyed by the bioturbation, and it is therefore uncertain to what extent the preserved textures represent biological mixing of originally discrete mud-ironstone and grain-ironstone laminae. The ooidal ironstone is generally less than 40cm thick, but in the areas of greatest subsidence (the southernmost exposures of Crozon and the northernmost localities in Buçaco; Fig. 6, localities 7 and 12) it is locally up to 3m thick. The upper part of the thin ironstone bed grades up into the sandstone tempestite/mudstone alternation, with much biogenic intermixing, except in the central part of the Buçaco syncline (Fig. 6, localities 15–18) where scouring below the sandstone tempestites has locally removed part or all of the ironstone.

*Biostratigraphy*

The chitinozoan evidence produced by Paris (1981) together with the earlier acritarch studies, provide quite detailed biostratigraphic information. The tempestite sequence of the Cabril Formation, dated by macrofauna as Llandeilo, falls within chitinozoan zone 11. In Brittany (Fig. 6, localities 3 and 5), the highest cycle below the dark mudstones yields faunas referred to the *Marrolithus bureaui* Biozone, commonly taken as belonging to the latest part of the Llandeilo (Henry 1980). The few samples examined by Paris for chitinozoans from the *Marrolithus bureaui* Biozone apparently did not permit the erection of a zonal scheme for this interval.

The dark mudstones above the *Marrolithus bureaui* Biozone contain several new faunal elements, including the first appearance in the area of the trilobite subgenus *Dalmanitina (Dalmanitina)*. This has been taken by many biostratigraphers as marking the base of the Caradoc. Further biostratigraphic evidence is provided by the occurrence of the graptolite *Climacograptus bekkeri* in the mudstones, which is known from the Kukruse Stage in Estonia (= upper part of *N. gracilis* Biozone—probably the Caradoc part of this zone). Paris (1981) indicates that the dark mudstones of the Carregueira Formation lie within chitinozoan zone 12, with zone 13 identified at the very top of the formation at some localities.

The matrix of the ironstone and the overlying sandstones, yielded faunas of chitinozoan zone 14, while the phosphatized intraclasts yielded zone 13. Zone 13 was closely dated by Paris as late Costonian/early Harnagian, while zone 14 is less closely controlled, but appears to be of early Caradoc (?Harnagian-Longvillian) age. The macrofauna of the ooidal ironstone includes several forms known from the lower Caradoc elsewhere in the Mediterranean region.

**The Lower Ashgill**

The ooidal ironstone (Fig. 3, ♂3) at the base of the Rosan Formation in Crozon and of the Porto de Santa Anna Fm. in Portugal shows facies relationships similar to those of the older ironstones (Fig. 7).

*Lithostratigraphy*

The ironstone overlies sequences with storm-generated sandstones, which, in Crozon, form a thickening and coarsening-up sequence of discrete sandstones displaying hummocky cross-stratification (HCS), whereas in Portugal the equivalent interval comprises a coarsening-up sequence capped by amalgamated sandstones with HCS. These deposits are overlain by dark mudstones, with a sparse fauna including graptolites. The contact between the top of the amalgamated sandstones with the overlying mudstone (the Galhano Member) in north-central Portugal (Fig. 7, localities 22–25) is marked by a thin ferruginous muddy sandstone with phosphatized intraclasts. In Brittany the base of the equivalent mudstones is marked by a bed of reworked, phosphatized nodules (Fig. 7, locality 6). As with the older two ironstones the mudstone unit is rather variable in thickness, ranging from 10m in S. Crozon to about 3m in the Buçaco syncline. It is not known south of the Buçaco syncline.

The 20–40 cm thick ironstone contains both chamositic mudstones and chamositic ooidal grainstones, with common intermixing of the two. Phosphatized mudstone intraclasts are abundant at the base of the bed, which rests on a concretionary layer at the top of the underlying mudstones. The ironstone is overlain by ferruginous and calcareous sandstones, and impure limestones.

*Biostratigraphy*

The biostratigraphy of this interval is not as well controlled as that of the older ironstones. According to Paris (1979, 1981) the sandstones of the well developed coarsening-up cycle,

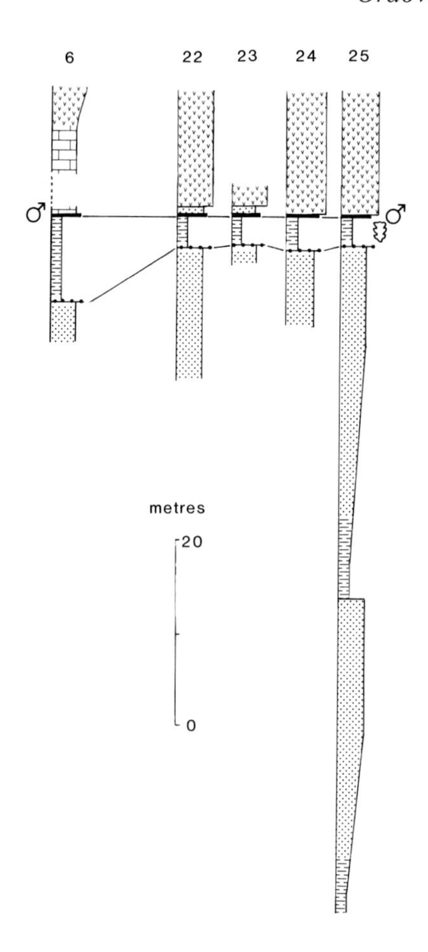


FIG. 7. Measured sections through the lower Ashgill ironstones, showing increasing influence of the Dornes/Amêndoa 'rise' towards the right. For key see Fig. 5. Tie-lines indicate sedimentological marker horizons. Datum is the base of chitinozoan zone 18. Location 6 is in the Crozon area, locations 22–25 are in the Buçaco syncline.

together with the dark mudstone unit lie within chitinozoan biozone 17. The microflora of the phosphatized intraclasts within the ironstone are also referred to biozone 17, whereas the matrix of the ironstone contains a microflora of biozone 18. Paris suggest that biozone 17 should be referred to the latest Caradoc (Actonian/Onnian) and biozone 18 to the early Ashgill (Pusgillian). Rare graptolites occur in the Galhano Member and in the Leira Ma Member in central Portugal, but so far these have not proved diagnostic. The abundant shelly faunas of the ironstone (chiefly brachiopods and bryozoans), and of the overlying sediments appear to confirm an age close to the Caradoc/Ashgill boundary, but the largely endemic nature of the trilobites and brachiopods presently precludes closer dating.

## General characteristics of the ironstone-bearing sequences

The vertical and lateral distribution of sedimentary facies of the three sequences described above is very similar. In the younger two examples the deposits below the ooidal ironstones show a marked cyclicity of coarsening- and thickening-up cycles. These progradational cycles of mudstones and storm generated sandstones are commonly capped by phosphatic conglomerates. In each case these sandstone-dominated cycles are overlain by a unit of dark mudstone, on which the ironstones lie. The dark mudstone units are interpreted as further, more distal, progradational cycles. In the more expanded sequences (e.g., the lower Caradoc in Brittany) the mudstones commence with deposits from the deepening phase of the cycle. This mudstone-dominated cycle is, in each case, very variable in thickness, with up to 30m of mudstones in the 'basinal' areas, but with the mudstone absent in the more condensed sequences. The progradational part of these mudstone-dominated cycles is developed in a distal facies, and none of the examples exhibits HCS in this part of the sequence, in contrast to the common occurrence of this style of bedding in the earlier sandstone-dominated cycles.

The importance of the hiatus below the ironstones seems to decrease basinward, and in one example (the lower Caradoc of Kerglintin, Crozon; Fig. 6, locality 7) the succession may be continuous. The time represented by the sub-ironstone disconformity is difficult to judge, but in each case the chitinozoans from the ironstone matrix were ascribed by Paris (1981) to a different zone to those in the phosphatized intraclasts. This suggests that the time involved may be significant. Similar microfloral changes were observed by Paris across some of the phosphatic conglomerate horizons. The hiatus below each of the ironstones enabled phosphatization of the concretionary layer at the top of the existing sediments, and of the intraclasts lying on the sediment surface, before the formation of the ooidal ironstone.

It is very difficult to establish any change in water depth between the sediments above and below the ironstones. The Llanvirn examples return to relatively undisturbed mudstones above the ironstone compared with the storm-generated sequences below. The Caradoc and Ashgill examples are both marked by only slight coarsening-up sequences below the ironstone, but sandstone deposition above. In these case, however, the only discernable faunal changes seem to be substrate related, and some faunal elements persist through the ironstone. Above the ironstones there is little evidence for the cyclicity which marks the sequences below them, although in some areas further cycles do occur, marked by the local development of additional ironstones. These occurrences (Fig. 5, localities 8 and 10; Fig. 6, localities 26–28, and 31–33) appear to be restricted to the areas with particularly low subsidence rates.

These generalizations allow the construction of an idealized cycle. The ironstone occurrences delimit a first order of cyclicity, corresponding to the eustatic cycles propounded by Fortey (1984). Only part of each 'first-order' cycle is characterized by coarsening-up sequences, the asymmetric expression of the 'second-order' cycles. The ironstones occur at, or near, the levels where the 'second-order' cyclicity ceases to be marked.

The petrology of the ironstones supports the idea that the ooids were developed more or less *in situ* on the open shelf, although some minor, local reworking has undoubtably taken place in some instances. The ooids in the more sandy developments (e.g., the Llanvirn ironstones of Normandy) commonly have nuclei of quartz. This is rarely, if ever, seen in the more 'offshore' developments, such as the Lower Caradoc ironstone of central Portugal. This suggests that the offshore ironstones are not merely deposits of ooids transported from other areas across the disconformity surface. Further evidence that the ooids were not transported to their present position is provided by their complete absence from the sediments immediately overlying the ironstones. If the ooids had been reworked from a distant source they should also occur in the storm deposits immediately above the ironstones. The development of sideritic concretionary layers, probably hardgrounds in some cases, also indicates that *in situ* iron enrichment of the existing sediment was possible during breaks in deposition preceding the ooid development.

## The nature of the cyclicity

The sequences studied record the effects of the 'first-order' global sea-level rises which occurred during their deposition. These cycles of 10–15 Ma period have been documented by Fortey (1984). Direct reference of these cycles to the terminology of global eustatic cyclicity described by Vail *et al.* (1977) is not possible, because there is no published detailed analysis of Palaeozoic sea-level changes undertaken with their methodology. Published eustatic curves for the Ordovician produced by other means (e.g., Fortey 1984) bear little resemblance to the curve of Vail *et al.* (1977). The period of the 'first-order' cycles described here is, however, within the variation of the Mesozoic third-order cycles of Vail *et al.* (1977).

Superimposed on this overall sea-level rise is a marked 'second-order' cyclicity of much shorter period. A rough calculation based on an overall duration for the Llandeilo of 3–7 Ma (e.g., Ross & Naeser 1984) and assuming constant rates of deposition might suggest a period of 0.2–0.5Ma for this 'second-order' cyclicity. This is of a similar magnitude to cyclicity seen in other ironstone-bearing shelf sequences (cf. the Middle Lias ironstones of the UK; Howard 1985), and has tentatively been ascribed to a 400 Ka Milankovitch cycle by Van Houten (1986).

The widespread nature of the storm-generated sandstone packets in Iberia is well documented (e.g., Brenchley *et al.* 1986), making an autocyclic explanation for the 'second-order' cycles improbable, and minor local tectonism seems unlikely to provide an explanation for the cyclicity. Small-scale sea-level variation seems the most likely explanation. The driving mechanism for this is unknown; there is no evidence for a pre-Ashgill icecap during the Ordovician, so ice volume fluctuations seem unlikely. The lack of good chronostratigraphic control makes comparison with Milankovitch cycles difficult.

The restriction of 'second-order' cycles to a limited part of each 'first-order' cycle can be interpreted in terms of the models described by McGhee & Bayer (1985). They suggest that the development of 'second-order' cycles in this manner may indicate a period of stasis of the 'first-order' process; in this case a 'first-order' eustatic lowstand. They also comment that such a punctuated sequence may be formed when progradation of the sediments has reached a point close to its stability limit, so that minor fluctuations of water depth may produce either deposition or non-depositon/erosion. The occurrences of 'second-order' cycles in the sequence studied here could either represent 'first-order' eustatic minima, or periods of proximity to the limit of stable progradation of the sedimentary system, which might themselves be due to eustatic lowstand.

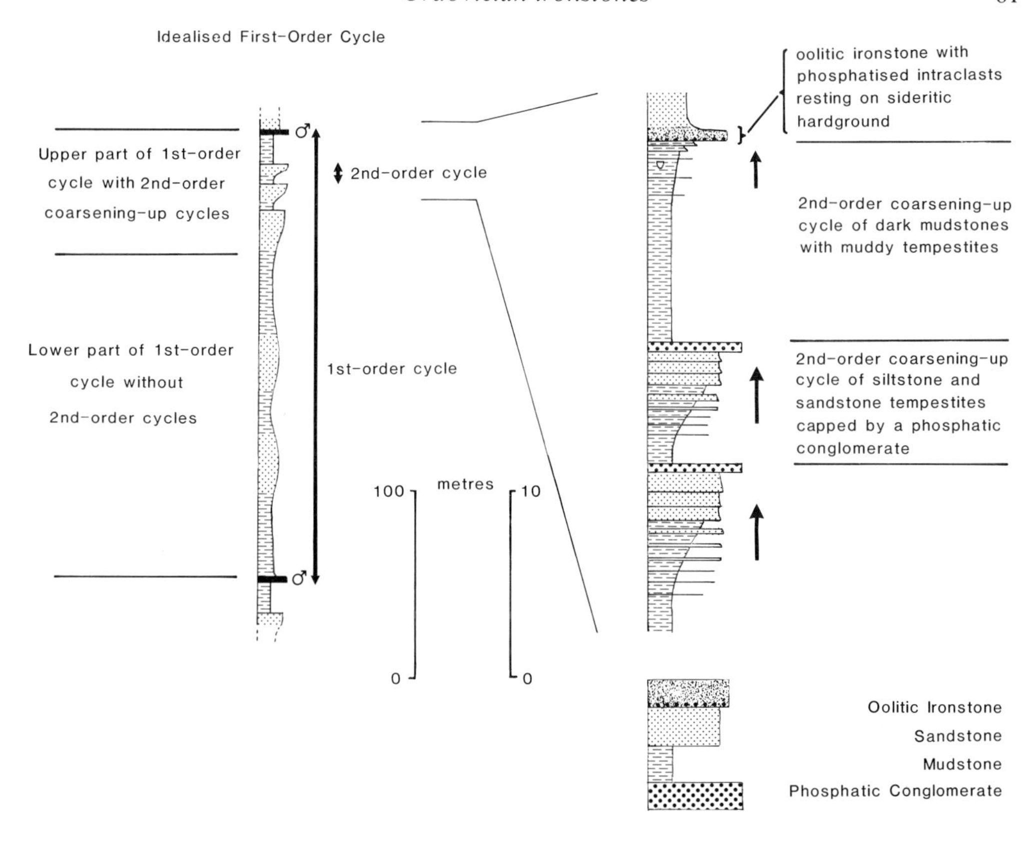


FIG. 8. Schematic illustration of the nature of the sedimentary cyclicity, showing relationship between 'first-' and 'second-order' cycles.

It is significant that the disconformities are covered by ironstones only during the transgressive phase of the cycle, and generally only late in that transgression. In the younger two examples ironstone development is generally restricted to a single horizon, although in one area (southern central Portugal; Fig. 6, localities 26–28, and 31–33) there is a second lower Caradoc ironstone. The Llanvirn transgression more commonly produced multiple ironstones (Fig. 5, localities 4, 8 and 10).

## Summary

Although the general conditions for the development of ironstones were apparently present throughout most of the Ordovician, specific environmental control limited their occurrence. One significant way in which those specific controls could be met was during minor transgressive events during periods of overall eustatic sea-level rise; periods of transgression of both the 'first-' and 'second-order' cyclicity.

It is not yet possible to determine directly what those controls were, but a break in clastic sediment input was important. The iron content of the 'normal' Ordovician mudstones in the area is very high, and given the reasonably long time interval probably involved in the formation of the ironstone, the important factor may have been the decoupling of the clastic sediment and iron supplies, and not an absolute enrichment of the iron. The transgression produced by the eustatic rise may have contributed to this, not only by interrupting sediment supply, but also by reworking terrestrial weathering products and hence supplying significant amounts of iron to the shelf seas. The ironstones may actually represent a net drop in the rate of iron incorporation into the marine sediments.

In this model the 'first-order' cycles may be largely responsible for providing a suitable source of iron, and in producing a low sediment input. The 'second-order' cycles punctuate the progradation of the shelf deposits and provide opportunity for periods of very low rates of sediment accumulation. Iron enrichment of the sediment began during the period of non-deposition, with the formation of sideritic hardgrounds. When limited amounts of sediment began to accumulate on the scoured surface the ooidal ironstone formed. The prolonged period of time during which this thin veneer of sediment formed the surface layer may have allowed sufficient degradation of organic matter to take place in the oxic and sub-oxic zones (Froelich *et al.* 1979) to allow the formation of authigenic iron silicates and inhibit subsequent sulphate reduction.

The formation of significant ooidal ironstone deposits around the fringes of Gondwana poses an important question; for although the continent may have provided a low-lying landmass which could easily have been reworked during transgressive events, the potential source areas lay over, or close to, the south pole. Some authors have postulated volcanic sources for the iron in similar non-tropical ironstones elsewhere (e.g., Dreesen 1989) but there is no evidence to link the peri-Gondwanan Ordovician ironstones with volcanicity. The majority of the ironstones occur in sequences with no evidence of volcanicity, and in those areas which do contain evidence of volcanic activity (e.g. North Wales) there is no correlation with the timing of ironstone formation (Trythall *et al.* 1987). The nature of Ordovician terrestrial processes remains unknown, but some mechanism must have existed for producing weathering products similar to those of the well vegetated humid environments usually cited as the source for the iron of Mesozoic ooidal ironstones.

ACKNOWLEDGEMENTS: The assistance of Drs. M Melou (Université de Bretagne Occidentale, Brest, France). J. L. Henry and F. Paris (Université de Rennes, France) in introducing me to the Armorican localities is gratefully acknowledged, as is the cooperation of the Serviços Geológicos de Portugal. Dr. M. Romano kindly suggested improvements to the initial manuscript. I would also like to thank Mr. M. Cooper for redrafting some of the diagrams. The research upon which this paper is based was undertaken during the tenure of an NERC Research Fellowship.

## References

BHATTACHARYYA, D.B. 1989. Concentrated and lean oolites: examples from the Nubia Formation at Aswan, Egypt, and significance of the oolite types in ironstone genesis. *In:* YOUNG, T.P. & TAYLOR, W.E.G. (eds) *Phanerozoic Ironstones,* Geological Society, London, Special Publication, **46,** 93–104.

BRENCHLEY, P.J., ROMANO, M. & GUTIÉRREZ-MARCO, J.C. 1986. Proximal and disital hummocky cross-stratified facies on a wide Ordovician shelf in Iberia. *In:* KNIGHT, R.J. & MCLEAN, J.R. (eds.), *Shelf sands and sandstones,* Canadian Society of Petroleum Geologists, Memoir **11,** 241–255.

CHAUVEL, J.J., 1971. Contribution à l'étude des minerais de fer de l'Ordovicien inférieur de Bretagne. *Mémoires de la Société géologique et minéralogique de Bretagne,* **16.**

DREESEN, R. 1989. Oolitic ironstones as event-stratigraphical marker beds within the Upper Devonian of the Ardenno-Rhenish Massif. *In:* YOUNG, T.P. & TAYLOR, W.E.G. (eds) *Phanerozoic Ironstones,* Geological Society, London, Special Publication, **46,** 65–78.

FORTEY, R.A. 1984. Global earlier Ordovician transgressions and regressions and their biological implications. pp. 37–50, *In:* BRUTON, D.L. (ed.) *Aspects of the Ordovician System,* Palaeontological Contributions from the University of Oslo, 295, Universitetsforlaget.

FROELICH, P.M., KLINKHAMMER, G.P., BENDER, M.L., LUEDTKE, N.A., HEATH, G.R., CULLEN, D., DAUPHIN, P., HAMMOND, D, HARTMAN, B., & MAYNARD, V. Early oxidation of organic matter in pelagic sediments of the eastern equatorial Atlantic: suboxic diagenesis. *Geochemica et Cosmochimica Acta,* **43,** 1075–1090.

HENRY, J-L., 1980. Trilobites ordoviciens du Massif Armoricain. *Mémoires de la Société géologique et minéralogique de Bretagne,* **22.**

——, NION, J-L., PARIS, F. & THADEU, D. 1974. Chitinozoaires, ostracodes et trilobites de l'Ordovicien du Portugal (serra de Buçaco) et du massif Armoricain: essai de comparaison et signification paléogéographique. *Communicaçòes dos Serviços Geológicos de Portugal,* **57,** 303–345, 10 plates (for 1973–4).

HOWARD, A.S. 1985. Lithostratigraphy of the Staithes Sandstone and Cleveland Ironstone formations (Lower Jurassic) of north-east Yorkshire. *Proceedings of the Yorkshire Geological Society,* **45,** 261–275.

JOSEPH, P. 1982. *Le minerai de fer oolithique ordovicien du Massif Armoricain: sedimentologie et paléogéographie.* Thèse de Docteur Ingénieur, Paris.

LOTZE, F. 1945. Zur Gliederung de Variszden der Iberischen Meseta. *Geotektoniker Forschunden Stuttgart,* **6,** 78–92.

MCGHEE, G. R. & BAYER, U. 1985. The local signature of sea-level changes. pp. 98–112, *In:* BAYER, U. & SEILACHER, A. (eds.) *Sedimenatry and evolutionary cycles,* Lecture notes in earth sciences, **1,** Springer-Verlag.

PARIS, F. 1979. Les chitinozoaires de la Formation de Louredo, Ordovicien Supérieur du synclinal de Buçaco (Portugal). *Palaeontographica A,* **164,** 24–51.

——1981. Les chitinozoaires dans le Paléozoique du sud-ouest de l'Europe. *Mémoire de la Société géologique et minéralogique de Bretagne,* **26,** 412 pp.

——& SKEVINGTON, D. 1979. Prescence de graptolites de l'Arenig Moyen à la base de la Formation de Postolonnec (Massif Armoricain); conséquences stratigraphiques et paléogéographiques. *Géobios,* **12,** 907–911.

PETRANEK, J. 1964a. Shallow water origin of early paleozoic oolitic iron ores. pp. 319–322, *In:* Van Straaten, L.M.J.U. (ed.), *Deltaic and shallow marine deposits,* Developments in Sedimentology, **1.**

——1964b. Ordovician—a major epoch in ironstone deposition. *International Geological Congress, 22nd, Delhi, Report* **15,** 51–57.

ROBARDET, M. 1981. Evolution géodynamique du nord-est du Massif Armoricain au Paléozoique. *Mémoires de la Société géologique et minéralogique de Bretagne,* **20,** 342 pp.

ROMANO, M. 1982. The Ordovician biostratigraphy of Portugal—A review with new data and re-appraisal. *Geological Journal,* **17,** 89–110.

——, BRENCHLEY, P.J. & MCDOUGALL, N.D. 1986. New information concerning the age of the beds immediately overlying the Armorican Quartzite in central Portugal. *Géobios,* **19,** 421–433.

ROSS, R.J. & NAESER, C.W. 1984. The Ordovician times scale—new refinements, pp. 5–10, *In:* BRUTON, D.L. (ed.) *Aspects of the Ordovician System,* Palaeontological Contributions from the University of Oslo, **295,** Universitetsforlaget.

TRYTHALL, R.J.B., ECCLES, C., MOLYNEAUX, S.G. & TAYLOR, W.E.G. 1987. Age and controls of ironstone deposition (Ordovician) North Wales. *Geological Journal,* **22,** 31–43.

VAIL, P.R., MITCHUM, R.M. & THOMPSON, S. 1977. Seismic stratigraphy and global changes of sea level. Part four: Global cycles of relative changes of sea level. pp. 83–97, *In:* PEYTON, C.E. (ed.) *Seismic Stratigraphy—applications to hydrocarbon exploration.* American Association of Petroleum Geologists, Memoir 26.

VAN HOUTEN, F.B. 1985. Oolitic ironstones and contrasting Ordovician and Jurassic palaeogeography. *Geology,* **13,** 722–724.

——1986. Search for Milankovitch patterns among oolitic ironstones. *Paleoceanography,* **1,** 459–466.

——& PURUCKER, M.E. 1984. Glauconitic peloids and chamositic ooids—favourable factors, constraints and problems. *Earth Science Reviews,* **20,** 211–243.

YOUNG, T.P. 1985. *The stratigraphy of the upper Ordovician of central Portugal.* PhD thesis, University of Sheffield.

T.P. YOUNG, Department of Geology, UNCC, PO Box 914, Cardiff, CF1 3YE, UK.

# Oolitic ironstones as event-stratigraphical marker beds within the Upper Devonian of the Ardenno-Rhenish Massif

**Roland Dreesen**

SUMMARY: Seven stratigraphically distinct, Clinton-type oolitic ironstones occur in the Upper Devonian of the Ardenne shelf, south of the London–Brabant Massif. They occur at the lithofacies boundary of succeeding lithological units and represent condensed deposits associated with hardgrounds and shoaling upward sequences. Microfacies analysis points to storm-generated concentration, removal and subsequent transport of the ferruginized allochems. The ironstones represent excellent marker beds which can be traced over several tens of kilometres on the palaeoshelf. Conodonts provide further evidence for the allochtonous character of the allochems and have enabled precise dating and correlation of each ironstone level with synsedimentary volcanic and tectonic events in the Rhenish Massif. Although the source of the iron is still a matter of discussion, lateritic weathering is not accepted here; instead, volcanism and evapotranspiration processes might represent potential sources for the iron in the studied ironstones.

Clinton-type oolitic ironstones occur at distinct stratigraphical levels within shelf siliciclastics, south and south-east of the London–Brabant Massif. Among these levels only the basal Famennian oolitic ironstone has been mined as an iron ore recently (1860–1946), but the lowermost level, at the base of the Frasnian, has been locally mined from at least Roman times. All of the other oolitic ironstone levels represent thin centrimetric to decimetric beds, without any economic significance, but with an important lateral extension, and hence of great stratigraphic value for basin analysis.

The Upper Devonian oolitic ironstones have been studied in detail in more than 50 outcrops and cored boreholes from several tectonic units directly south of the London–Brabant Massif (southern and south-eastern Belgium): the Namur, the Dinant and the Verviers Synclinoria (see Fig. 1). These ironstones consist of thin ferruginous ooid-bearing limestones (bioclastic wacke/packstones and grainstones) which are interbedded in either nodular shales (outer shelf) or micaceous silt- and sandstones (inner shelf). Locally (e.g., in the westernmost part of the Verviers Synclinorium and along the northern borders of the Dinant Synclinorium) some of the ironstones are associated with evaporitic deposits (calcite or dolomite pseudomorphs after anhydrite; primary dolomites).

A red staining of the host sediment is very common. Moreover, red-stained shales (e.g., the purple Upper Famenne Shales in the southern Dinant Synclinorium) or red-stained nodular limestones (e.g., the griotte-type *Cheiloceras* Limestone of the Aachen area, easternmost part of the Verviers Synclinorium) represent the most offshore facies of particular Famennian oolitic ironstone levels (Dreesen 1982b; Dreesen *et al.* 1985).

Although very similar in outcrop or in hand specimen, the Belgian Upper Devonian oolitic ironstones can readily be differentiated on the basis of the ferruginized allochems (see below), but conodont-based biostratigraphy provides the best tool to distinguish them.

## Stratigraphy

Seven distinct oolitic ironstone levels have been recognized in the Belgian Upper Devonian (Ardenne shelf). With the exception of the lowermost level (ironstone O at the base of the Frasnian) and the uppermost one (level V within the Strunian), the oolitic ironstones are concentrated within the lower half of the Famennian. It is striking that this interval, with the highest concentration of ironstones in Belgium, corresponds to an interval in the hemipelagic to pelagic settings of the Rheinisches Schiefergebirge (Western Germany) with red and green basinal shales enclosing numerous sandstone turbidites and metabentonites (volcanic tuffs) (see Fig. 2) (Dreesen 1982a).

Furthermore, level O can be biostratigraphically correlated with an important exhalative iron ore deposit (the so-called Roteisengrenzlager (Dreesen *et al.* 1985). The youngest oolitic ironstone (level V) has been correlated by means of spores and conodonts with a conspicuous volcanic bomb level (the so-called Bombenschalstein) of the Dill area in the Rheinisches Schiefergebirge (Dreesen & Streel

*From* YOUNG, T. P. & TAYLOR, W. E. G. (eds), 1989, *Phanerozoic Ironstones*
Geological Society Special Publication No. 46, pp. 65-78

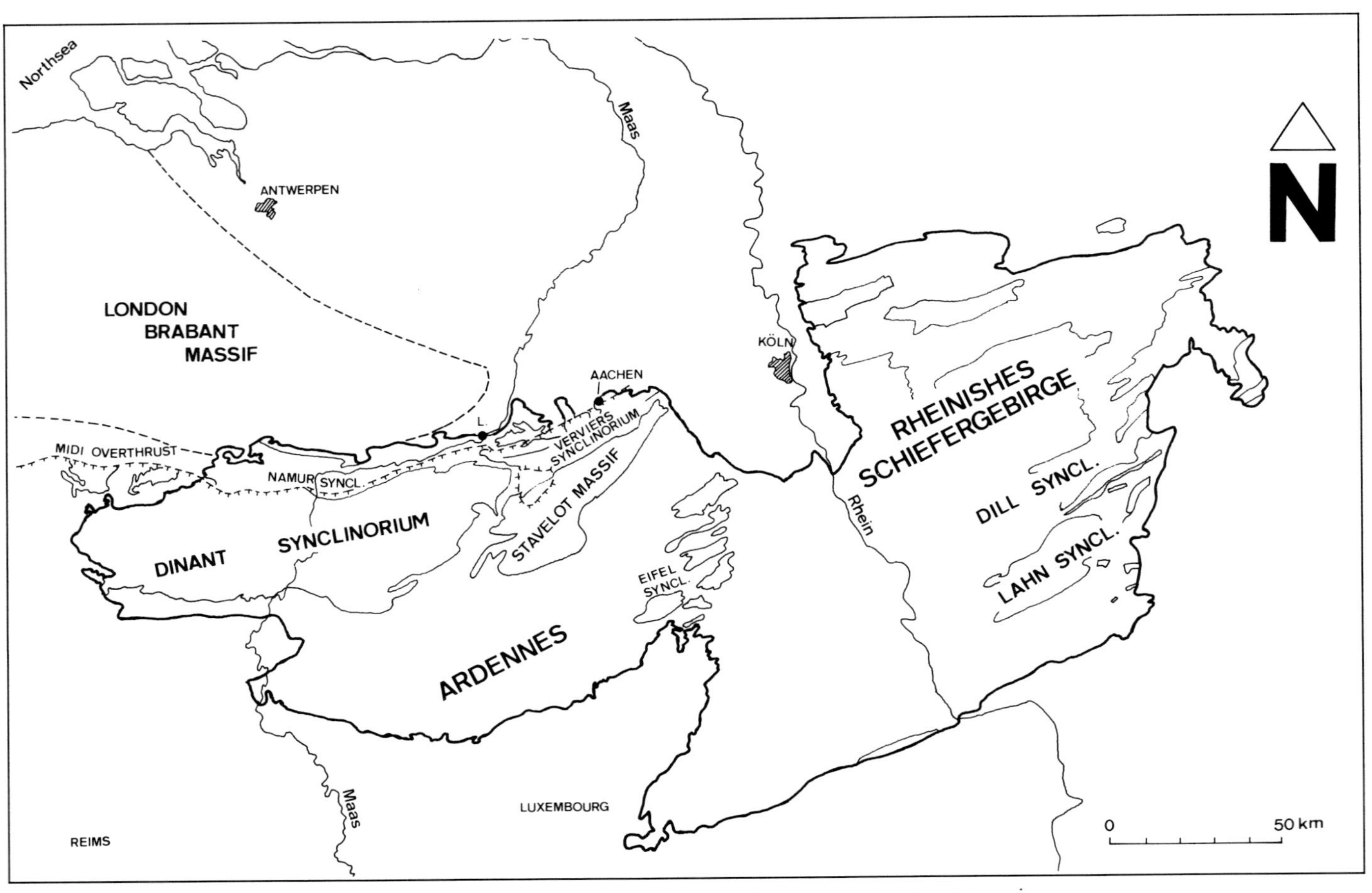


FIG. 1. Index map of the Ardenno – Rhenish Massif with the location of the tectonic units mentioned in this study.

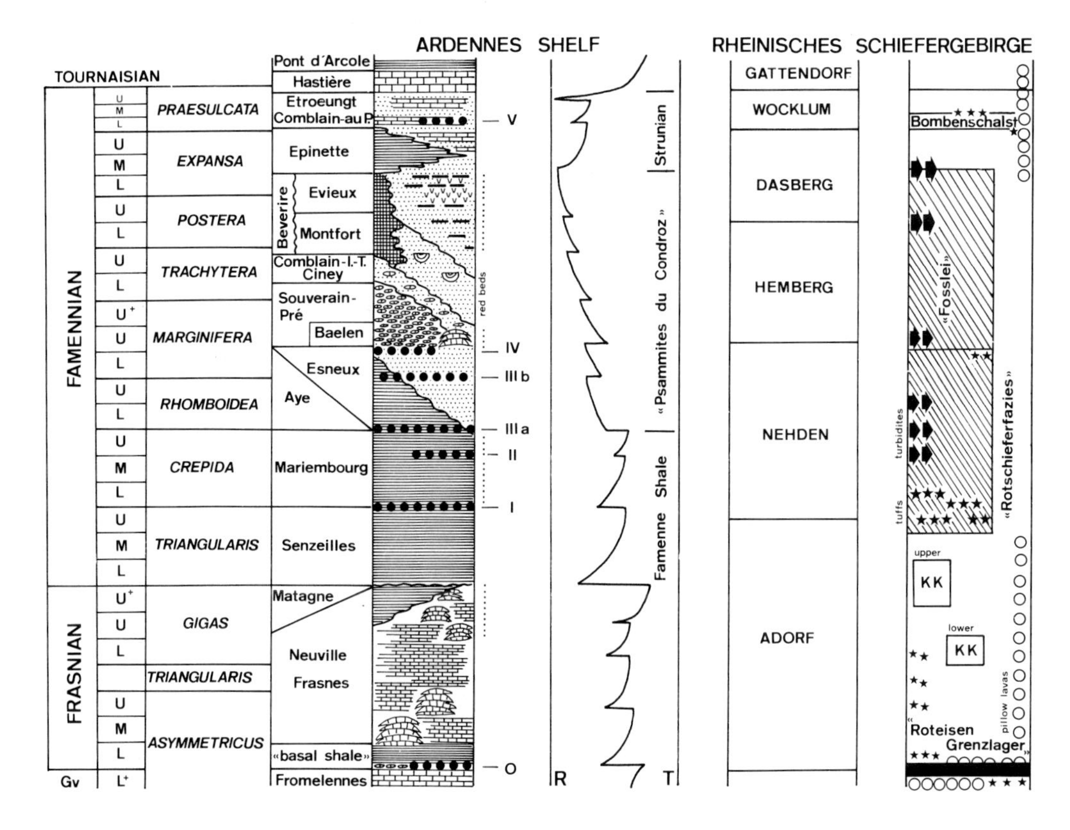


FIG. 2. Stratigraphic correlation scheme of Upper Devonian episodic events in the Ardenno–Rhenish Massif. At the left: Belgian stages and standard conodont zonation; in the centre: lithologic columnar section for the Ardenne shelf, flanked on the left by informal and formal lithostratigraphical names and on the right by the designations for the seven ironstone levels. Just to the right is a suggested sea-level curve. On the extreme right is a graphic depiction of the event deposits of the Rheinisches Schiefergebirge in terms of six German Stufen. Pillow lavas are shown by circles, metabentonites by stars, turbidites by short black arrows, and basinal red shales by diagonal shading. KK refers to particular anoxic event deposits ('Kellwasserkalke').

1985). Volcanic glass fragments, chloritized subangular extraclasts of supposed volcanic origin, and idiomorphic zircon crystals have been frequently observed in the calcareous host sediment of the Famennian oolitic ironstones.

It is striking that no ironstones have been recorded from the stratigraphic interval between oolitic ironstone levels O and I, or between ironstone levels IV and V. The first stratigraphic interval corresponds to a period of major reef development in the Frasnian of the Ardenne shelf, the second interval represents the peak of a regressive megasequence corresponding to the Condroz Sandstones (Thorez & Dreesen 1986). It is noteworthy that the near-mutual exclusion of the Upper Devonian reefs and iron oolites in the Belgian Ardenne fits the 'inverse' relationship of reefs and calcareous ooids in the Palaeozoic, as recently proposed by Eliuk (1987). The highest concentration of oolitic ironstones in the lower half of the Famennian might indicate the onset of a worldwide regression, in the mid-Famennian, possibly of glacio-eustatic origin (Caputo 1985; Streel 1986). The progradation of siliciclastics produced by this event would have prohibited any oolite formation during the Upper Famennian on the paleoshelf. Stromatoporoid reef facies reappear from the base of the Strunian on, and calcareous ooids do not occur before the Lower Carboniferous (Hastière Limestone, basal Tournaisian).

Besides a precise dating of each oolitic ironstone level, conodonts have also provided good evidence for condensation within each ironstone level, corresponding to a timespan of less than one conodont zone or less than a half-million years (Dreesen 1984). This timespan marks a period of slow deposition or even an

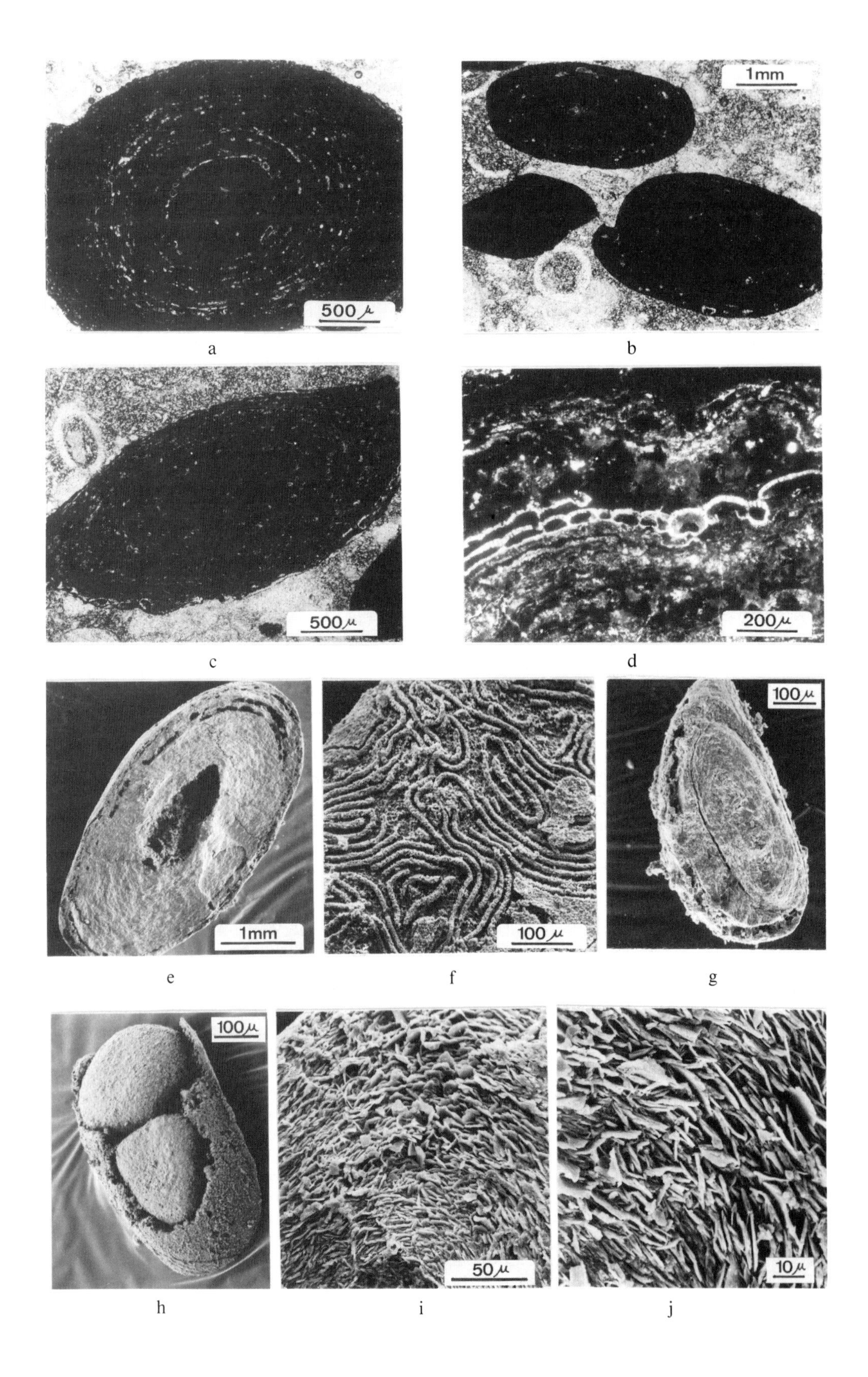

a
b
c
d
e
f
g
h
i
j

omission event, during which ferruginous-coated grains were formed in nearshore environments, preceding their rapid, high-energy transport, supposedly during severe storms, onto the open shelf. Furthermore, conodont biofacies analysis has provided additional evidence for the allochtonous character of the ironstones. For example, five of the seven studied levels contain mixed nearshore and pelagic conodont biofacies, most a mixed icriodid–polygnathid–palmatolepid biofacies (Sandberg & Dreesen 1984; Dreesen *et al.* 1986).

Although the ironstone beds are relatively thin (a few cm to a few tens of cm) they can easily be traced over tens of kilometres on the palaeoshelf. One of the best examples is oolitic ironstone level IIIa, which has been observed from Aachen (W. Germany) to Sains-du-Nord (N. France), for over more than 150 km (Dreesen 1982b)! The lateral extent differs for each ironstone level and from one tectonic unit to another. The presence of an important thrust-fault between the Namur and Dinant Synclinoria makes it difficult to appreciate the exact extent of each of the ironstone levels. In the Namur Synclinorium, oolitic ironstone level I is apparently restricted to a few tens of kilometres, especially along its southern border. It is suggested that this important accumulation of ferruginous ooids (the only level which has been industrially mined—maximum thickness of about 150 cm) resulted from trapping behind a temporary shoal or barrier of tectonic origin, the so-called Condroz Ridge, which prevented the ferruginized allochems from being removed and transported over greater distances onto the open shelf (see Fig. 8, after Thorez & Dreesen 1986). However, the same level is more widespread in the Verviers Synclinorium where it clearly indicates west-to-east directed transport.

## Microfacies

The Upper Devonian oolitic ironstones of the Ardenne display different 'ore' facies: either flattened 'flax seed'-type ferruginous ooids, highly fossiliferous 'fossil ore' facies or less ferruginous 'transgressive lag'-type deposits. 'Flax seed' and 'fossil ores' generally grade into each other and appear to be often mixed. However, the former tends to represent a more proximal facies, whereas the latter is a rather distal shelf facies, with respect to the palaeocoastline. In the 'transgressive lag'-type deposits, rounded bioclasts are only slightly impregnated by ferric oxide and Fe-rich chlorite.

The 'flax seed' ore types (for instance oolitic ironstone level) contain abundant characteristic 'Osagia'-type algal oncoids and pisoncoids (oncolites) with problematic encrusing tubular organisms (see Figs. 3a–f). Some 'fossil ore' type ironstone facies commonly contain charophytes, either related to *Sycidium* (ironstone IIIa) or *Umbella* (ironstone levels IV and V). Typical ferric oxide or chloritic moulds of those charophycean oogonia (resembling spinous berries or ampullae) have been obtained through formic acid-etching of the limestone host sediment and wet sieving of the dissolution residues (see Figs. 4d,g,h,i).

Microfacies analysis of the ferruginized allochems points to the allochtonous character of the ironstones: heterogenous ferruginized coated grains are mixed with calcitic bioclasts and fossils to form bioclastic wacke-, pack-, and grain-ironstones. The hematitic or chloritic allochems comprise ooids, superficial ooids, multiple ooids, algal oncoids, algal pisoncoids, cortoids, slightly coated rounded bioclasts and algal-encrusted intraclasts. Spastolithic oncoids exceptionally occur within ironstone level O. The more pseudo-oolitic ironstones ('fossil ore' and 'transgressive lag' ironstone facies) are essentially composed of moderately to strongly Fe-impregnated and only slightly Fe-coated, bored bioclasts such as crinoid ossicles, bryozoans, brachiopod and ostracode shells (ironstone levels O, III, IV and V) (Figs. 4a, b, e).

Each ironstone level consists of at least one concentration of ferruginized allochems or of a ferruginized algal-encrusted (microstromatolitic) hardground only, but it may also comprise several sublevels (for example

FIG. 3. (a, b, c) Thin sections of proximal 'flax seed' type oolitic ironstone level I showing *Osagia*-type hematitic algal oncoids with interlayered sparite-walled encrusting organisms. (d) Detail of the outer laminae of a hematitic algal oncoid with chaplet-like colonies of a sparite-walled encrusting organism. Note the fine pores of the calcitic walls. (e, f) SEM photographs of acid-etched hematitic algal oncoids from ironstone level I; (e) An equatorial section showing tangentially oriented strings of hemispheroidal voids after dissolution of the calcitic walls. (f) Detail of meandering hematitic molds of problematic tubular encrusting, originally sparite-walled organisms, after the removal of the exterior-most hematitic scale. (g) SEM photograph of an assymmetric, discoidal chloritic ooid from distal ironstone level IIIa. Note presence of larger chlorite platelets in the core. (h) SEM photograph of a hematitic multiple ooid from proximal ironstone level IIIa. (i, j) SEM photographs : details of the tangentially-concentric arrangement of hematite flakes within ferruginized algal oncoids.

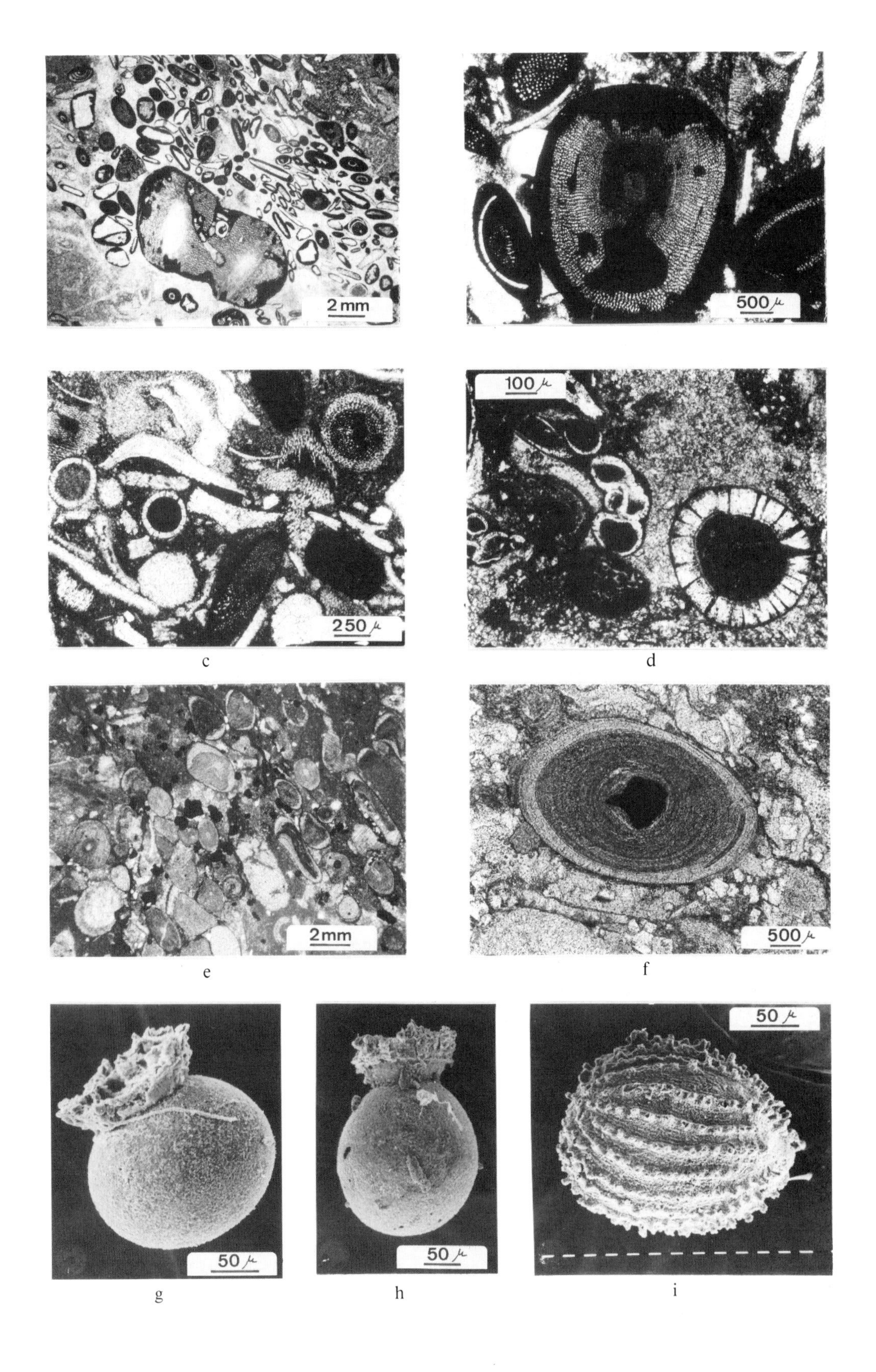
2 mm
500 µ
250 µ
100 µ
2mm
500 µ
50 µ
50 µ
50 µ
c
d
e
f
g
h
i

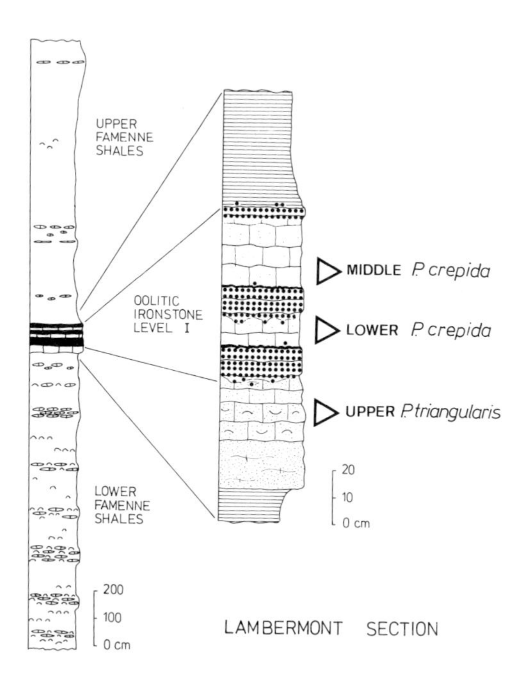


FIG. 5. Representative columnar section of oolite ironstone level I at Lambermont, near Verviers (Verviers Synclinorium, eastern Belgium). Note the shallowing upward sequence (shale, calcareous siltstone, sandy coquina, ferruginous ooids, microstromatolitic hardground), the stratigraphic condensation (in terms of conodont zones), and the three ironstone sublevels, two of which are capped by hardgrounds.

ironstone level I in the Verviers Synclinorium; see Fig. 5).

Each (sub)level surmounts a basal erosional unconformity and it is commonly capped by or encloses a ferruginized algal-encrusted (microstromatolitic) hardground. This would mean that each ironstone level does not necessarily represent one single event, but that it may consist of different superimposed events, including erosional, condensation, non-deposition and transport events. The hardgrounds originated at the end of a shoaling upward sequence (e.g., Fig. 5: Lambermont section; level I). The lithification itself occurred during reduced sedimentation, or non-deposition conditions. The calcitic allochems of the host sediment often contain cephalopods (orthoceratids, goniatites), and exceptionally even rugose corals (e.g., level IIIa) (Dreesen 1982b).

The presence of a basal erosional unconformity, the allochtonous character of the ferruginized allochems and their density stratification, the lateral decrease of ironstone bed thickness and that of the total amount of hematitic allochems, have led to the assumption that the Upper Devonian oolitic ironstones of the Belgian Ardennes represent storm deposits or tempestites (Fig. 6) (Dreesen 1982).

Ideally, proximal and distal tempestites can be differentiated on the basis of their ferruginized allochem content (see Fig. 7).

The most distal facies of some ironstones is characterized by a lack of ferruginized allochems and by a conspicuous red staining (requiring a ferric oxide pigment content of at least 1.5%). The original iron (hydr)oxide being the material with the smallest grain size, would be the first to be whirled-up and the last to settle during resedimentation; it would, therefore travel the greatest distance (e.g., the purple Mariembourg Shales as coeval deposits of ironstone levels I and II; the red Cheiloceras Limestone as coeval deposit of ironstone level IIIa (Dreesen 1982b; Dreesen *et al.* 1985).

From the above observations a working model for the origin and distribution of the oolitic ironstones is here proposed: during a temporary decrease of the siliciclastic influx calcareous coated grains originated in the shallow, protected shelf and restricted marine environments of a near-coastal, broadly embayed area (see upper half of Fig. 6). Unusually strong storms may be invoked as a cause for the winnowing of the enveloping muds leading from mudstone to packstone concentration of the originally calcareous allochems. These coated grains became piled into shoals, probably proximal to lagoons, where they could become periodically subaerially exposed. During the probably short emersion periods, vadose or phreatic weathering promoted the replacement of Mg by Fe in the original high-Mg calcite allochems (e.g., the

FIG. 4. (a–d) Thin sections of 'fossil ore' type ironstones with numerous superficial bioclastic ooids and some larger, bored crinoid ossicles. (b) Detail of an intensively bored crinoid ossicle. (c, d) *Sycidium* oogonia as nuclei of superficial ooids within a bioclastic wackestone/packstone. Note presence of pores (see Fig. 10). (e) Packstone concentration of chlorite-impregnated and chlorite-coated superficial ooids and rounded bioclasts (mainly crinoid ossicles). Black polygons represent dispersed sulphides (distal ironstone level IIIb). (f) Detail of a discoidal chloritic ooid with conspicuous concentric layering (distal ironstone level IIIa). (g, h) SEM photograph of typical ampullae-like infillings (molds) of *Umbella* (outer sparite wall removed). Ironstone level IV. (i) SEM photograph of a berry-like hematitic mold of a *Sycidium* superficial ooid. Vertical spine rows represent the hematite-infilled pores of the original sparitic wall.

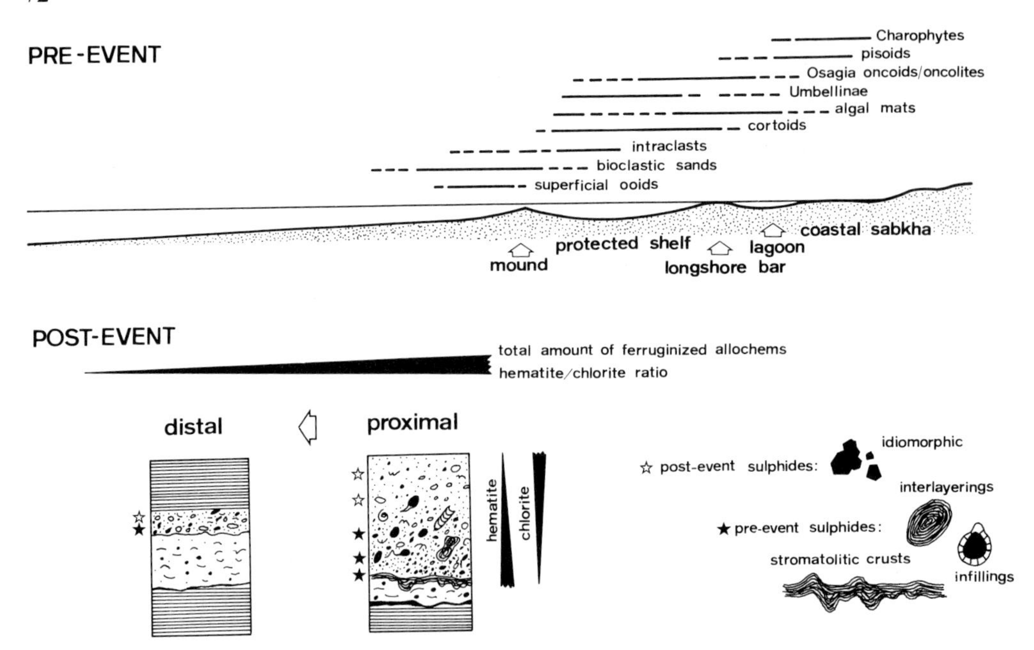


FIG. 6. Diagram showing the idealized distribution of the original carbonate coated grains prior to their ferruginization. The lateral sequence of depositional environments is taken from Thorez & Dreesen, 1986. The lower part of the figure depicts an idealized proximal/distal differentiation within a Famennian oolitic ironstone, after removal and the storm-induced transport of the ferruginized allochems.

crinoid ossicles, ooids and algal oncoids). Subsequently the ferroan calcite was oxidized to form the goethitic and later hematitic pseudomorphs. Alternatively, the more flattened, non-bioclastic coated grains (the 'flax seed' type ooids) could have originated as berthierine ooids in adjacent, low-energy muddy (lagoonal pond) areas and were subsequently altered into hematite. Relict (non-oxidized) chloritic ooids and pseudo-ooids are still present in the Upper Devonian oolitic ironstones, but they are less frequent than the hematitic ooids and they are commonly distorted by post-burial compaction.

Violent storm surges (tropical hurricanes ?) and even high-energy waves (tsunamis ?) could be evoked to explain the removal of the ferruginized coated grains and their transport onto the shelf. Finally, a strong longshore current may account for their redistribution over large shelf areas in the Ardenne (see further).

## Palaeogeography and event-stratigraphy

The source of the iron is still a matter of speculation, but the classical concept of weathering of lateritic soils, as for the Minette (Siehl & Thein 1978), cannot be accepted here, because of the lack of evidence for true humid tropical or subtropical conditions. The frequent occurrence of redbeds, evaporites and pedogenic carbonates (caliche, calcretes) in the Belgian Upper Devonian is more symptomatic of prevailing semi-arid climatic conditions of a tropical trade wind belt (Paproth *et al.* 1986). Furthermore, the presence of a high fresh feldspar content within the micaceous Condroz Sandstones would have been practically impossible to meet under humid tropical conditions (Thorez & Dreesen 1986).

Instead, the iron required probably more than one source (see Fig. 8): the iron could have been derived from the surface waters of nearby density-stratified evaporitic pans (Sonnenfeld *et al.* 1977); on the other hand the iron-rich chlorites could have originated from the diagenesis and halmyrolysis of volcanic ashes: chloritized angular bentonite relics, vermicular chlorite extraclasts, weathered fragments of volcanic glass and idiomorphic zircon crystals are frequently observed in the calcareous host sediment of the Upper Devonian oolitic ironstones.

XRD-analysis of hand-picked chloritic ooids and oncoids from Lower Famennian oolitic ironstones, revealed a IIb- chlorite polytype as

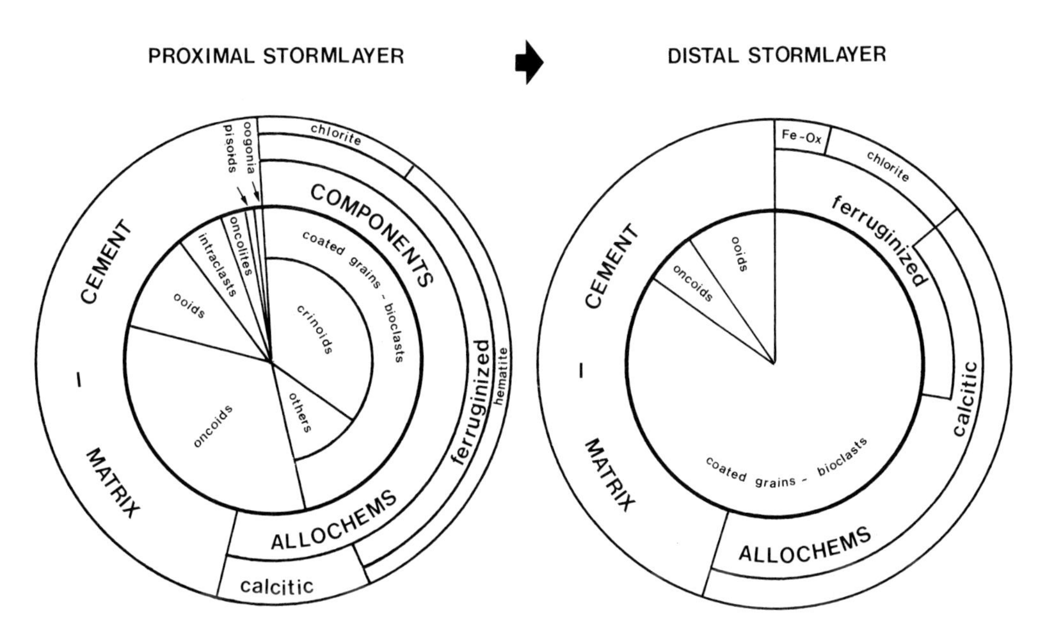


FIG. 7. Circle graphs representing semiquantitative microfacies analysis data and the ideal lateral evolution of some Lower Famennian oolitic ironstones.

the main constituent with illite, smectite and kaolinite, calcite, hematite and quartz as minor components. This chlorite is very similar to the chlorites observed in Belgian Lower Carboniferous metabentonites (Thorez, personal communication).

Microscopic analysis of polished sections revealed small inclusions and interlayerings of Fe–Cu–Zn–sulphides, within the chloritic ooids and oncoids. This base metal sulphide mineralization could have occurred under reducing conditions (generated by decaying algae) in the bottom ooze of the proposed density- stratified evaporitic lagoons (Dreesen & Thorez 1982).

SEM analysis of ferric-oxide algal oncoids shows a distinct tangential-concentric orientation of the hematite platelets (Figs. 3i,j), and the presence of intercortical hemispherical voids, related to encrusting tubular calcitic organisms. The alternation of zones of tangentially constructed laminae and zones with organisms indicates interruptions in the normal process of (original carbonate?) oncoid accretion. The persistance of sparitic walls after ferruginization might indicate the preferential replacement of micritic laminae by iron minerals. Within those walls, pores only have been infilled by hematite.

The relative scarcity of 'berthierine/chamosite' ooids (Figs. 3g, 4f, g) may result from their replacement by ferric oxide: indeed, hematitic ooids and algal oncoids commonly display a discoidal shape ('flax seed') suggesting compaction of a chloritic precursor.

During late Devonian times the Ardenne shelf corresponded to a relative shallow epicontinental sea bordering the southern, windward side of the Old Red Continent (Paproth *et al.* 1986). In the then southern hemisphere the easterly tropical trade winds drove a warm, equatorial sea current westward between the equator and the 20°S palaeolatitude (Heckel & Witzke 1979). One branch of this warm, south-equatorial boundary current passed through the Polish Basin, and was deflected towards the Ardenno-Rhenish Basin, encroaching the northern shelf area. Here, the current was forced to pass through narrow seaways, in between the Old Red Continent and episodically emergant shallow-water Ardenno–Rhenish Shoals (Fig. 9). The current was sufficiently strong to redistribute alluvial discharges of deltas prograding from the northern hinterland. The shallow Condroz shelf area and its lateral equivalents (Namur S., northern borders of the Dinant S., western Verviers S.) formed the recipient of the coarser siliciclastics (the Condroz Sandstones) whereas the finer siliciclastics (silts and pelites) were deposited in the more offshore settings of the shelf (southern Dinant S.). The more pelagic to hemipelagic settings of the Cornwall–Rhenish Basin were less contaminated by this

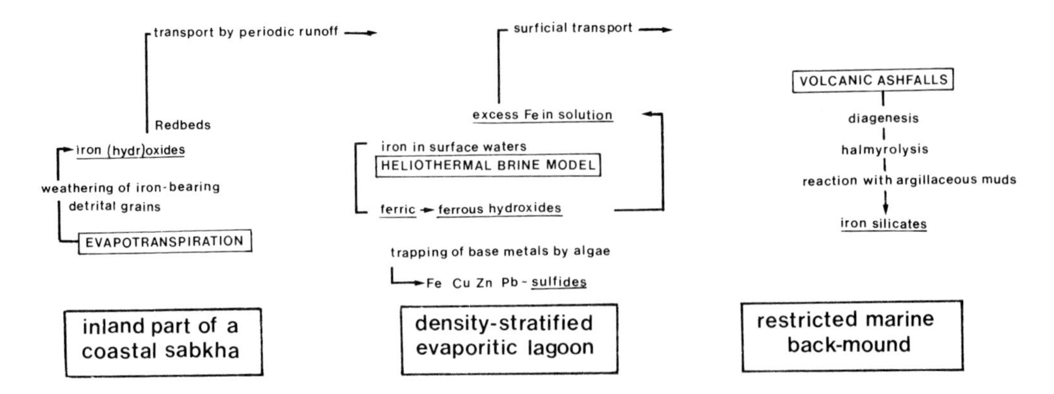


FIG. 8. Possible and potential sources for the iron in the Upper Devonian ironstones in Belgium.

detrital influx: they are characterized by silty shales, entomozoan shales and nodular shales, with locally condensed cephalopod limestones on submarine shoals. The episodic reddening of those muds and carbonates in the more basinal areas (Rheinisches Schiefergebirge) matches the episodic occurrence of oolitic ironstones on the Condroz shelf and the submarine volcanism in the Lahn–Dill area.

The original setting of the ironstones which formed during a temporary emersion of protected nearshore shelf areas is indicated by heavy black dots in Fig. 9 (reconstructed Famennian palaeogeography of NW Europe, taken from Paproth *et al.* (1986)). Tropical storms regularly affected the shallow shelf and near coastal areas, as evidenced by the presence of numerous tempestitic coquinas and of hummocky stratification within the Condroz Sandstones. Even more powerful storm waves might have affected the Condroz shelf (tsunamis) : these extremely violent processes could have done a great deal of the sedimentary work, but we have not yet learned to recognize their results. Indirect evidence pointing to a possible trigger mechanism for tsunamis are found for instance in the numerous ball-and-pillow levels of the Upper Famennian: these deformation structures have been related to the passage of seismic waves (Hempton & Dewey 1983).

Volcanic activity is known to have occurred at the time of formation of the Upper Devonian ironstones. All of these episodic events strongly suggest synsedimentary tectonic activity.

Indeed, the rhythmical progradation, during the Famennian, of fluviolagoonal, coastal barrier and tidal flat complexes, S and SE of the London–Brabant High (Thorez & Dreesen 1986) is associated with the episodic reactivation of small tectonic blocks (see Fig. 10), which produced differential subsidence and the persistence of non-deposition areas. At about the same time, inversion structures became reactivated in the Rheinisches Schiefergebirge, producing and triggering turbidites towards depocenters in the immediate surroundings (Paproth *et al.* 1986).

These tectonic blocks of the Ardenne, as well as the inversion structures in the Rhenish Massif, are bordered by NNW–SSE and WSW–ENE directed block faults, which are supposed to have been inherited from the basement. Rifting, block-faulting or tensional movements along these deep-seated faults could have easily provoked sea-level fluctuations on a shallow epicontinental shelf, bordering a relatively flat, deeply eroded 'continent'. Further withdrawal of the sea, combined with a reduced influx of siliciclastics, due to a temporary tectonic blocking of the longshore current, allowed the formation of shoaling upward cycles, of condensed sequences and of hardgrounds on the shelf. Under particular environmental conditions (diagenetic weathering and evapotranspiration processes on the inland edge of coastal sabkhas) ferruginous ooids originated, most probably through coating, impregnation and diagenetic replacement of former carbonate coated grains on the various emerging coastal and shallow marine subenvironments.

Highly explosive volcanic activity, both submarine and continental, would have produced considerable amounts of volcanic ash, which may have easily reached the Ardenne shelf by aeolian transport (note wind direction and possible continental volcanic eruption centre in Fig 9). After being dropped into the sea these ashes could have been a potential source for

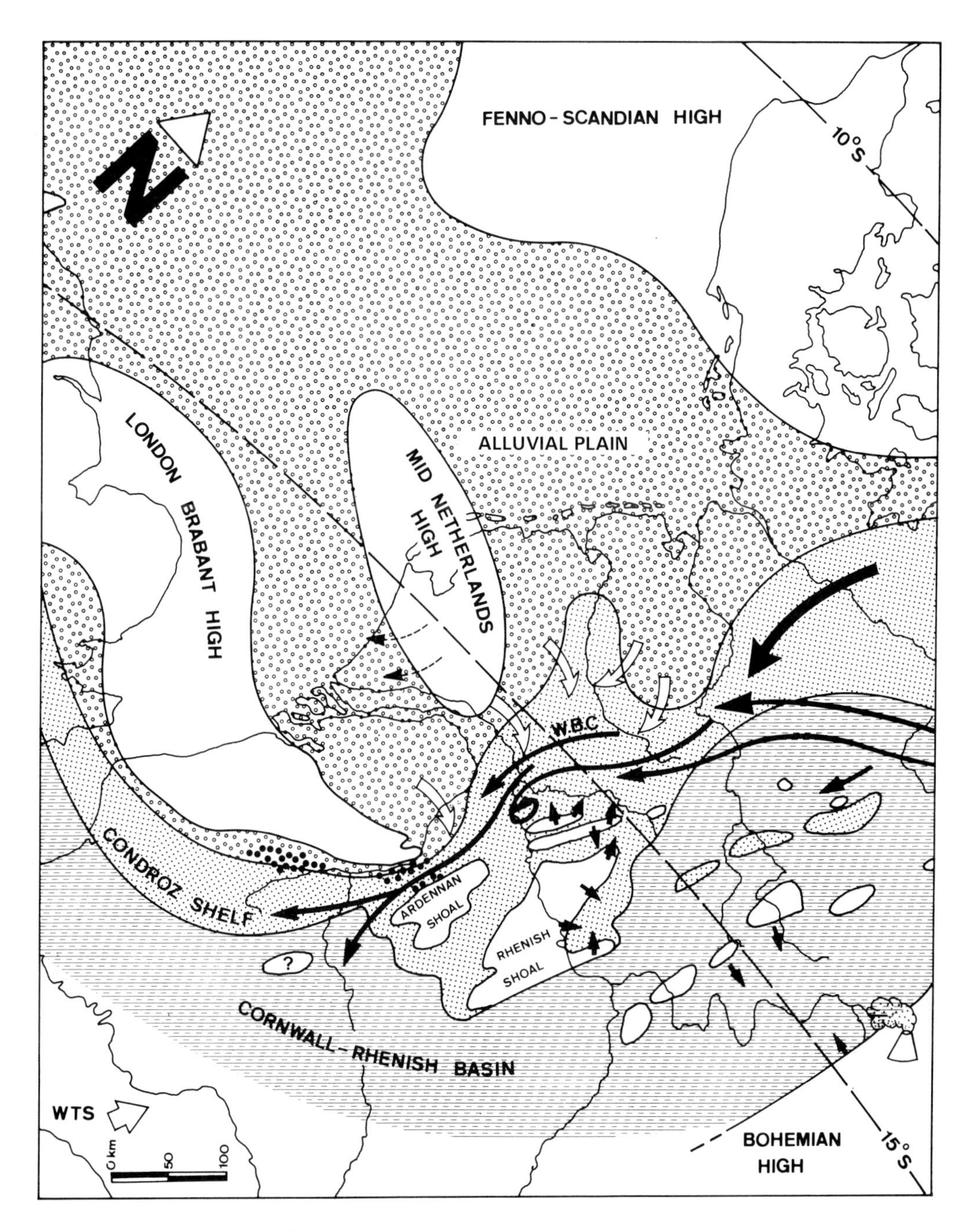


FIG. 9. Reconstructed Famennian palaeogeography for part of NW Europe (modified after Paproth *et al.* 1986.) WTS = westerly tropical storms; WBC = western boundary current; long black arrows indicate current directions; short black arrows represent detrital shedding; white arrows represent deltaic discharges; dotted arrows indicate eolian sediment transport; heavy black dots represent the original setting and supposedly storm-generated accumulation of ferruginous ooids, before their removal and high-energy wave transport to the SW and to the E. Suggested presence of continental volcanic eruption centres on the Bohemian High. Submarine volcanic eruption centres not indicated but supposedly located N and E of the Rhenish shoals.

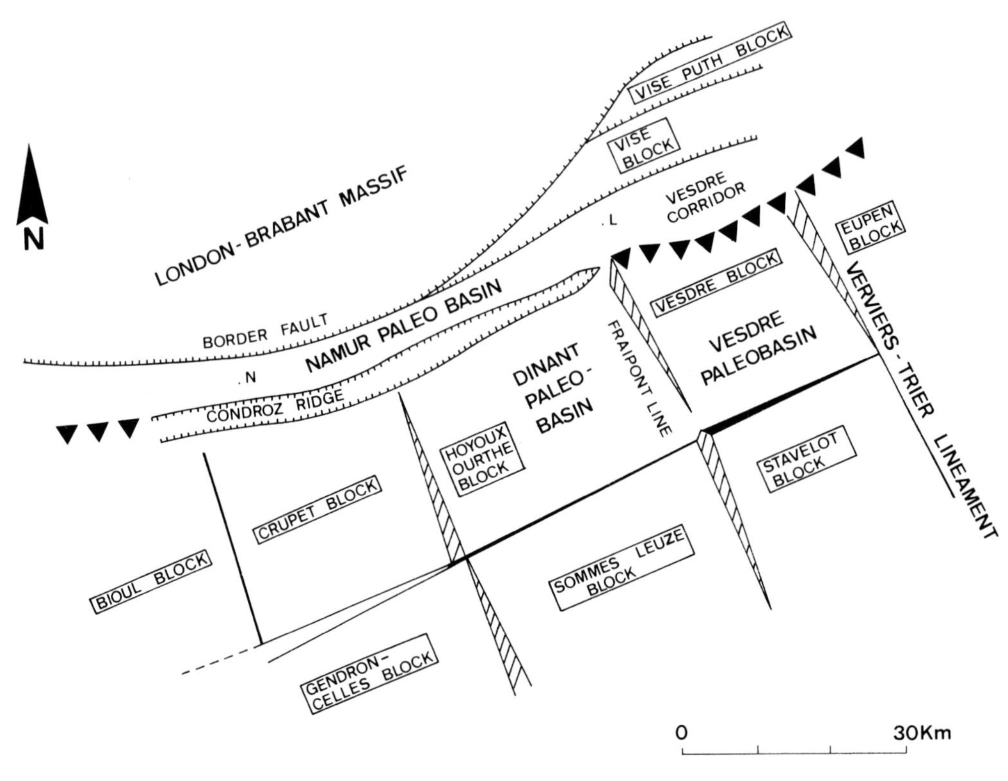


FIG. 10. Intersecting fault pattern and nine named fault blocks, the sudden movements of which produced seismic shocks that might have triggered high-energy waves (tsunamis ?) and caused downslope transport of ferruginous allochems to form the oolitic ironstones on the Ardenne palaeoshelf (after Thorez & Dreesen, 1986). Black triangles indicate the future location of the Midi Overthrust, L - Liège, N = Namur.

chlorites, through chemical reaction with marine argillaceous muds. Furthermore, explosive volcanism and seismic shocks could have triggered diverse turbulent events, such as turbidity currents, severe storms and even tsunamis, which, in their turn, would have been responsible for the transport of both the ferruginized coated grains into the open shelf (Ardenne) and the shelf sands into the basinal areas (Rhenish Massif).

The 'geophantasmogram' of Fig. 11 shows the suggested interdependence of the tectono-sedimentary processes mentioned above and the origin of episodic event surfaces and deposits. Oolitic ironstones and coeval sandstone turbidites or volcanic tuffs not only represent excellent marker beds, but they are also the mute evidence of episodic events which may have been related to pre-orogenic, intra-cratonic synsedimentary tectonics.

Although the Late Devonian generally has been considered as a relatively 'quiet' geotectonic period in the 'geosynclinal' development of the Ardenno-Rhenish Massif (Franke *et al.* 1978), tectonic movements occurred episodically and strongly influenced the depositional history of the sedimentary basin S and SE of the London–Brabant Massif.

## Concluding remarks

The Upper Devonian oolitic ironstones of the Belgian Ardenne shelf display characteristics both in favour of, and contradictory to, the model of Van Houten & Bhattacharyya (1982) for Phanerozoic oolitic ironstones.

Characteristics in favour of their model are the lithofacies discordance between successive units, the general shoaling-upward sequences and the abrupt waning of sediment supply associated with ironstone formation. Elements in contradiction to the model are the remarkable synchroneity of the Belgian ironstones with synsedimentary tectonic activity, and the general semi-arid climatologic conditions hampering a

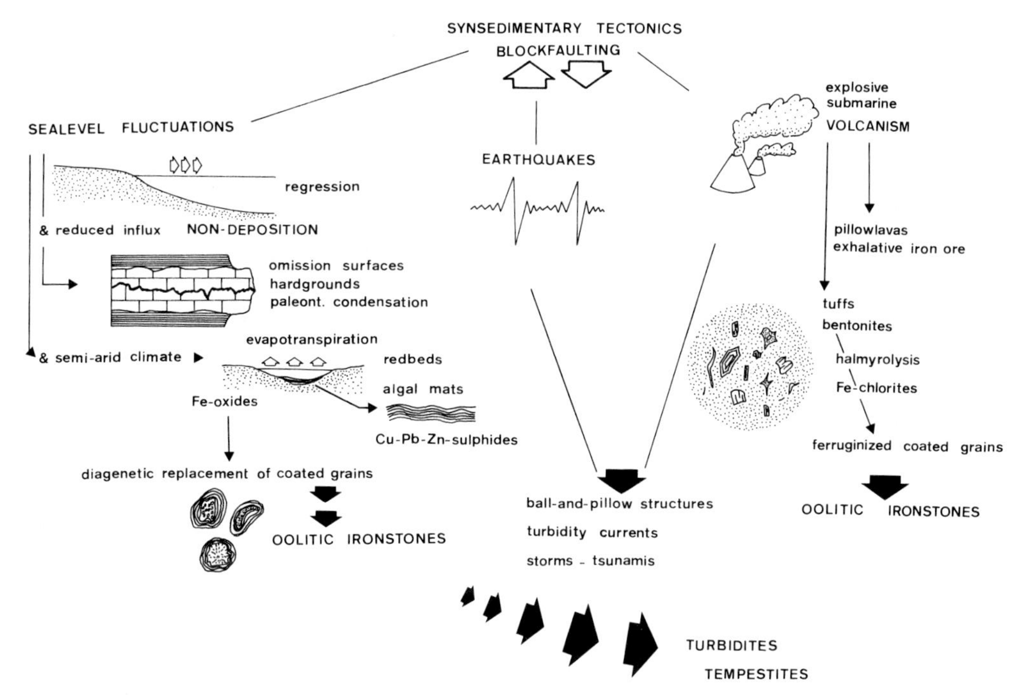


FIG. 11. 'Geophantasmogram' of strongly interdependent tectono-sedimentary processes which provided the mechanism for the formation and transport of ferruginized coated grains. The turbulent events that triggered the downslope transport were seismic shocks, which are evidenced by conspicuous ball-and-pillow levels that crosscut different depositonal environments (Thorez & Dreesen 1986). The seismic shocks, which probably fostered turbidity currents and possibly tsunamis, were produced by the episodic reactivation of block faults. The faults also may have provided feeders through which magmas fed submarine volcanic exhalations, which together with ashfalls from volcanoes, provided chemical enrichment for formation of the next cycle of ferruginization.

deep weathering of the hinterland and hence excluding a latosol-derived source for the iron.

Moreover, the presence of non-ferruginized, calcitic encrusting tubular organisms, interlayered with zones of tangentially constructed laminae (now hematite flakes) in hematitic algal oncoids and pisoncoids, points in favour of a selective replacement of particular carbonate coated grains.

ACKNOWLEDGEMENTS: The manuscript benefited from unpublished clay-mineralogical data by J Thorez (Liège, Belgium) and from comments by C.A. Sandberg (Denver, USA) and W. Ziegler (Frankfurt, W. Germany).

## References


CAPUTO, M.V. 1985. Late Devonian glaciation in South America. *Paleogeography, Paleoclimatology, Paleoecology,* **51,** 291–317.

DREESEN, R. 1982a. A propos des niveaux d'oolithes ferrugineuses de l'Ardenne et du volcanisme sysédimentaire dans le Massif Ardenno-Rhénan au Dévonien Supérieur—Essai de corrélation stratigraphique. *Neues Jahrbuch für Geologie und Paläontologie Monatshefte,* **1,** 1–11.

——1982b Storm-generated oolitic ironstones of the Famennian (Fab-Fa2a) in the Vesdre and Dinant Synclinoria (Upper Devonian, Belgium). *Annales de la Socité Géologie de Belgique,* **105,** 105–129.

——1984 Stratigraphic correlation of Famennian oolitic ironstones in the Havelange (Dinant Basin) and Verviers boreholes (Vesdre Massif) (Upper Devonian, Belgium). *Bulletin de la Société Belge de Géologie,* **93,** 197–211.

——,KASIG, W., PAPROTH, E & WILDER H. 1985. Recent investigations within the Devonian and Carboniferous North and South of the Stavelot-Venn Massif. *Neues Jahrbuch für Geologie und Paläontologie Abhandlungen,* **171,** 237–265.

——,SANDBERG, C.A. & ZIEGLER, W. 1986. Review of

late Devonian and early Carboniferous conodont biostratigraphy and biofacies models as applied to the Ardenne shelf. *Annales de la Société Géologique de Belgique,* **109**, 175–186.

—— & STREEL, M. 1985. A new event-stratigraphical marker bed in the uppermost Devonian of the Ardenno-Rhenish Massif, *In:* Meuse-Rhine Euregion Geologists Meeting in Liège, May 24, 1985, abstracts. *Annales de la Société Géologie de Belgique,* **108**, 412.

——& THOREZ, J. 1982. Upper Devonian sediments in the Ardenno-Rhenish area; sedimentology and geochemistry. *Publicaties van het Natuurhistorisch Genootschap Limburg,* **32**, 9–15.

ELIUK, L.S. 1987. Submarine hydrothermal weathering, global eustacy, and carbonate polymorphism in Phanerozoic marine oolites. Discussion. *Journal of Sedimentary Petrology,* **57**, 184–185.

FRANKE, W., EDER, W., ENGEL, W & LANGENSTRASSEN, F. 1978. Main aspects of geosynclinal sedimentation in the Rhenohercynian Zone. *Zeitschrift der deutsche Geologische Gesellschaft,* **129**, 210–216.

HECKEL, P.H. & WITZKE, B.J. 1979. Devonian world paleogeography determined from the distribution of carbonates and related lithic paleoclimatic indicators, *In:* HOUSE, M.R., SCRUTTON, C. & BASSET, M.C. (eds). *The Devonian System. Special Papers in Palaeontology,* **23**, 99–124.

HEMPTON, M.R. & DEWEY, J.F. 1983. Earthquake-induced deformational structures in young lacustrine sediments, East Anatolian Fault, Southern Turkey. *Tectonophysics,* **98**, 7–14.

PAPROTH, E., DREESEN R. & THOREZ, J. 1986. Famennian paleogeography and event stratigraphy of Northwestern Europe. *Annales de la Société Géologique de Belgique,* **109**, 175–186.

SANDBERG, C.A. & DREESEN, R. 1984. Late Devonian icriodontid biofacies models and alternate shallow-water conodont zonation, *In:* CLARK, D.L. (eds) Conodont biofacies and provincialism. *Geological Society of America Special Paper* **196**, 143–178.

SIEHL, A. & THEIN, J. 1978. Geochemische Trends in der Minette (Jura, Luxemburg/Lotheringen). *Geologische Rundschau,* **67**, 1052–1077.

SONNENFELD, P., HUDEL, P.P., TUREK, A. & BOON, A. 1977. Base-metal concentration in a density stratified evaporitic pan, *In:* Reefs and Evaporites—concepts and depositional models. *American Association of Petroleum Geologists, Studies in Geology,* **5**, 181–187.

STREEL, M. 1986. Miospores contribution to the Upper Famennian—Strunian event stratigraphy. *Annales de la Société Géologie de Belgique,* **109**, 75–92.

THOREZ, J. & DREESEN, R. 1986. A model of a regressive depositional system around the Old Red Continent as exemplified by a field trip in the Upper Famennian "Psammites du Condroz" in Belgium, *In:* BLESS, M.J.M. & STREEL, M. (eds.). Late Devonian Events Around the Old Red Continent. *Annales de la Societété Géologique de Belgique,* **109**, 285–322.

VAN HOUTEN, F.B. & BHATTACHARRYA, D.P. 1982. Phanerozoic oolitic ironstones—geological record and facies model. *Annual Review of Earth and Planetary Sciences,* **10**, 441–457.

R. DREESEN, Institut National des Industries Extractives, Rue de Chéra, 200, B-4000 Liege, Belgium.

# A depositional model for the Liassic Minette ironstones (Luxemburg and France), in comparison with other Phanerozoic oolitic ironstones

T. Teyssen

SUMMARY: Oolitic ironstones generally mark the top of a regressive, coarsening- and shallowing-upward depositional megasequence. The iron ooids themselves are preferentially formed during sea-level lowstands or in condensed sections. This general model is valid for the Minette iron formation which was deposited in Toarcian and Aalenian times at the north-eastern margin of the Liassic Paris Basin in a near-shore, shallow marine, mostly high-energy environment. The top of the regressive sequence shows intraformational reworking. The Minette, however, is a remarkably thick unit (up to 60 metres), also internally composed of coarsening-up sequences. In many places these sequences represent large-scale subtidal sand waves, elsewhere large subtidal shoals advancing over their distal finer grained facies. Strong tidal currents building up large sand waves are preferentially active during transgressions. For the Minette base level rise events (transgressions) are therefore required at the top of a base level fall sequence (regression). This apparent discrepancy can be resolved when considering the tectonic setting. Reactivation of structures of the underlying Variscian basement led to subsidence events in syndepositionally formed troughs. This model can explain a high aggradational potential and deposition of thick ironstone deposits at the top of a regressive sequence while erosion took place laterally along the same palaeocoastline.

For more than 100 years cyclic sedimentation has gained the interest of geologists and many publications and textbooks have been written on this subject (Klupfel 1916; Duff *et al.* 1967; Schwarzacher 1975; Einsele & Seilacher 1982; Bayer & Seilacher 1985). In particular, the occurrence of oolitic ironstones is often associated with cyclic deposition (Hallam 1966; Hallam & Bradshaw 1979; Van Houten & Bhattacharyya 1982; Bayer *et al.* 1985). Oolitic ironstones are furthermore very important for the interpretation of cyclic sedimentation because they are often indicative of relative sea-level lowstands (Hallam & Bradshaw 1979). In all types of cyclic sediments, however, it has generally been a problem to distinguish between allo- and autocyclicity. Allocyclicity means that an external mechanism (e.g., eustasy, climatic variation, tectonics) systematically and repeatedly changes the physical sedimentary environments which will then be reflected in the sediments. The external mechanism therefore determines what type of sediment will be deposited at the base, the middle part and at the top of each cycle. Autocyclic sediments, on the other hand, are formed by trends (fining-upward or coarsening-upward grain-size trends, compositional trends or others) which are systematically created by processes inherent to the respective depositional environment (such as progradation, migration, channel fill etc.). It is the objective of this contribution to differentiate between autocyclic and allocyclic systems in ironstone bearing sequences using the depositional model of the Minette oolitic ironstones as an example.

The Minette oolitic ironstones of Luxembourg and France are considered as the type deposit of the Phanerozoic shallow marine oolitic Minette or Clinton-type Ironstones. The Minette oolitic Ironstones were deposited in Late Liassic/Early Dogger times (Toarcian/Aalenian) at the north-eastern margin of the Liassic Paris Basin (Fig. 1) in a generally shallow marine, near-shore, high-energy environment. Many aspects of these deposits have been extensively studied (mineralogy and geochemistry: Lucius 1945; Bubenicek 1964; Siehl & Thein 1975, 1978; diagenesis: Bubenicek 1983; sedimentology: Thein 1975; Teyssen 1984; Dahanayake *et al.* 1985).

The Minette Ironstones were deposited during one of the peaks of the Phanerozoic ironstone occurrences. Main peaks are found in the Ordovician and in the Lower/Middle Jurassic (Van Houten & Bhattacharyya 1982, Van Houten 1985); minor peaks can be observed in Devonian and Cretaceous times. The global controls of ironstone formation are not yet fully clear, although the relative sea level, the assembly/dispersal of continental blocks and the global extention of continental shelves apparently played a role (Van Houten &

*From* Young, T. P. & Taylor, W. E. G. (eds), 1989, *Phanerozoic Ironstones*
Geological Society Special Publication No. 46, pp. 79-92

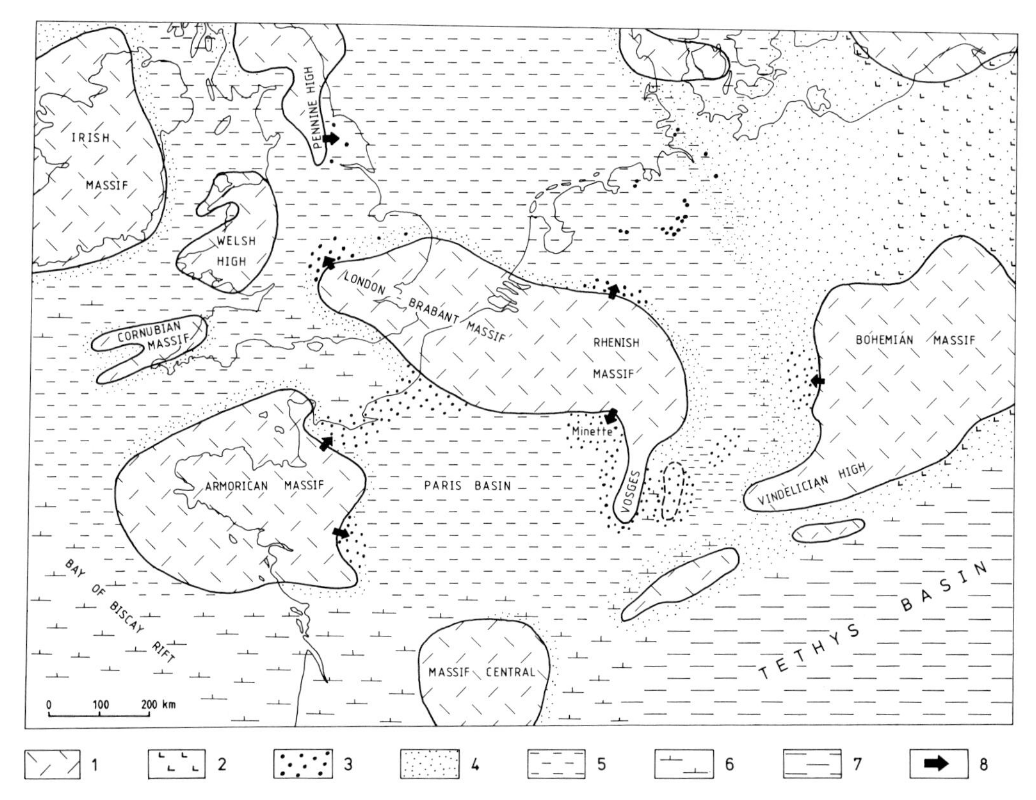


FIG. 1. Toarcian and Aalenian palaeogeography, main sources are Karrenberge (1942), Gruss & Thienhaus (1969), Hains & Horton (1969), Megnien (1980) and Ziegler (1982). The Minette Ironstones were deposited off the coastline of the Ardennes/Rhenish Massif island. The ironstones of the Upper Rhine Valley, in Central England and east of the Armorican Massif were deposited in similar geological setting. 1, area of non-deposition; 2, continental and lacustrine sedimentation; 3, oolitic ironstones; 4, deltaic, coastal and shallow marine sands; 5, shallow marine shales; 6, carbonates; 7, deeper marine shales; 8, direction of clastic influx.

Bhattacharyya 1982; Van Houten & Purucker 1984). Recently a general depositional model for oolitic ironstones of the Minette type has been established (Van Houten & Purucker 1984) which summarizes observations on ironstone bearing sequences made by many authors. This contribution will compare the depositional processes of the Minette with this general model. Such a comparison seems to be interesting because the Minette is a very thick deposit with a remarkable thickness of more than 60 metres (Teyssen 1984).

## Depositional model of oolitic ironstones

The principal depositional model outlined here has been summarized by Van Houten and his coworkers (see various references). Chamositic and glauconitic ooids and peloids as one possible type of precursor grains to limonitic and goethitic ooids are generally deposited and also primarily formed (Porrenga 1967) on shallow shelves. Accumulations of iron ooids are generally found at the top of a (regressive) coarsening-upward and shallowing-upward sequence (Fig. 2). They are marker lithologies in a (asymmetric) cyclic sedimentation (Hallam & Bradshaw 1979; Teyssen 1984). The ironstone occurs at the termination of a progradational or shoaling-upward sequence. Normally, a re-worked horizon, a hardground or an omission surface tops the ironstone indicating reduced normal sediment influx. However, in other cases the ironstones may mark the incipient transgression. Such ironstones, associated with condensed deposits, are strongly burrowed, contain fossil debris, ferruginized or phosphatized hardgrounds and mark an unconformity. Often it is not possible easily to

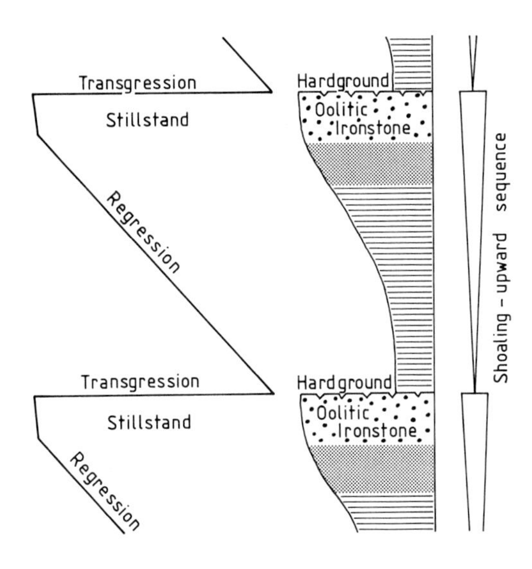


FIG. 2. Vertical sequence model of oolitic ironstone deposition, redrawn after Van Houten & Bhattacharyya (1982). Minette-type oolitic ironstones generally occur at the top of a shoaling and coarsening-upward detrital sequence deposited during a regressional trend. The top of a cycle is a submarine hardground or an omission surface, occasionally with subaerial exposure and reworking. The next sequence starts with a rapid transgression.

distinguish between these two types. Finally, many oolitic ironstones occur as local concentrations, while the other facies associated with them in the sequence have a widespread distribution. Some ironstones, however, are relatively widespread, as in the Minette.

In discussing this general model in more detail it is important to also consider the origin of the iron ooids themselves in a large-scale geological context. Sea-level lowstands or condensed sequences are probably required to produce iron ooids (possibly from chamositic and glauconitic ooids and peloids in the latter case, see above). However, sea level lowstands expose large areas of land. Lateritic weathering under warm and tropical climates may form lateritic pisolites. Most oolitic ironstones in fact were formed in tropical or subtropical settings (Van Houten & Bhattacharyya 1982; Valeton 1983). The pisolites can be eroded, transported into the sea and accumulated in near-shore environments forming oolitic ironstones of a different origin. This has been concluded from geochemical investigations (Siehl & Thein 1978; see however, Maynard 1986, for opposing interpretations) and sedimentological studies (Teyssen 1984) of the Minette. Biogenic formation of ooids in intertidal flats has been suggested for several ancient and recent deposits including the Minette by Dahanayake *et al.* (1985). Large intertidal flats as envisaged by these authors, however, are preferentially expected during high sea levels rather than during sea-level lowstands. Generally the environment of primary formation of iron ooids is different from the environment of deposition for many Minette-type deposits (Teyssen 1984; Bayer *et al.* 1985). Ooids formed during sea level lowstands in above mentioned settings will typically accumulate at the top of a regressive shoaling-upward sequence. They mark the termination of the regression.

Ooids associated with condensed sequences are formed on large, extensive shelves. Low sedimentation rates favour considerable concentrations of glauconitic pellets, peloids (as ooid precursors) and chamositic ooids (Porrenga 1967; Odin & Matter 1981; Van Houten & Purucker 1984). These are reworked and redeposited on the site of the later ironstone deposit. In this way extensive but rather thin deposits can be formed, comparable to transgressive sheet sands. In such cases, the ironstone occurs in the same position within the vertical sequence. However, it marks the onset of a transgression rather than the termination of a regression.

The first one of these two classes of ironstone formations, which is associated with a sea-level lowstand, should be more important for the genesis of larger economical and minable ironstone deposits. The sea-level lowstands are not necessarily major global lowstands. In many cases they can be local or regional relative sea-level lowstands being the result of a combination of eustacy, regional or local subsidence and of vertical tectonic movements in the hinterland. The importance of eustatic sea level lowstands for the ironstone formation using the example of the Jurassic Paris Basin will be demonstrated below.

## Liassic Minette ironstones

### Lithostratigraphy and facies

The Lower Jurassic deposits of Luxembourg and Lorraine are cyclically composed of four coarsening-upward megasequences (Fig. 3). The Minette Ironstones form the top of the uppermost cycle.

Each of the four cycles is a regressive sequence. The regressive character is more obvious in the north and north-east (towards the basin margins) than in the south-west (towards

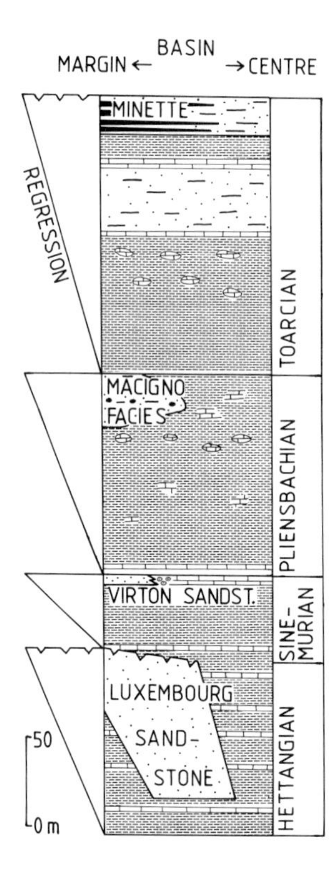


FIG. 3. Depositional cycles of the Lower Jurassic in Lorraine and Luxembourg; lithology based on Bintz *et al.* (1973). The Lower Jurassic was deposited in a series of four regressive cycles, two of which terminate in subaerial exposure (at top of Luxembourg Sandstone Formation and at top of the Minette Iron Formation). The regressive–transgressive cycles from Lorraine correspond to the Lower Jurassic coastal onlap cycles of the North Sea (cf Fig. 5).

the basin centre, Fig. 3). This is in accordance with the palaeogeographic frame (Fig. 1). Each cycle starts with a rapid deepening event. Claystones and marls were deposited in the lower part of the sequences which become more sandy towards the top. Near-shore and coastal clastics were deposited beneath the sharp top. The tops of two of the four regressive sequences were subaerially exposed (Fig. 3).

The first cycle starts with the Rhaetian Levallois Marls. These are overlain by a rhythmic mudstone sequence and bituminous shales of Hettangian age. The cycle ends with the deltaic Luxembourg Sandstone Formations which has a subaerial erosion surface at its top (Mertens *et al.* 1983).

The second cycle, Sinemurian in age, develops from the Strassen Maris and Limestones via calcareous marls to the Virton Sandstone. This sandstone is locally rich in iron ooids (near Hassel, Belgium).

The third cycle, Pliensbachian in age, begins with grey marls containing calcareous nodules and develops upward into marls and siltstones and finally to ferruginous sandstones. These high-energy, cross-bedded sandy oolitic ironstones contain iron crusts and phosphatic nodules. These deposits are the 'Macigno d'Aubange' and are well exposed at Hautcharage, Luxembourg (Bintz *et al.* 1973).

The fourth cycle is the Toarcian–Aalenian regressive sequence. This sequence consists of Toarcian organic-rich paper shales that grade upward into calcareous marls and sandstones. The Minette forms the top of this sequence and is only locally developed. Laterally along the shoreline of the Ardennes mainland a hiatus or even erosion of the uppermost Toarcian can be observed (Fig. 4; Megnien 1980: Figs 4-4, 5-9. The top of the Minette is a conglomerate (Katzenberg Conglomerate) which indicates intraformational reworking and subaerial exposure, as already noted by Klupfel (1916). The Minette is overlain by transgressive micaceous marls deposited under deeper water conditions during the Bajocian.

These four regressive cycles match reasonably well the Jurassic coastal onlap curve (Vail *et al.* 1984) developed from the North Sea (compare Figs. 3 and 5). This indicates that eustasy was generally the dominant control for facies development in the Paris Basin, as would be expected for an intracratonic basin. Paris Basin ironstones and iron rich sediments are stratigraphically equivalent to the tops of the cycles of the North Sea coastal onlap curve (Fig. 4). In the Paris Basin the ironstones are overlain by sediments deposited in deeper water environments.

The Minette Ironstones at the top of a reressive megacycle also show an internal coarsening-upward cyclic pattern (Figs. 6, 7; Lucius 1945; Bubenicek 1961; Teyssen 1984). As many as 13 coarsening-upward cycles were deposited which traditionally form the basis for the lithostratigraphic subdivision (Fig. 8). The Minette was deposited in two sub-basins (Fig. 9) which are separated by the Audun-le-Tiche Fault. These sub-basins are the Esch–Ottange Basin and the Differdange–Longwy Basin. Ironstone depostion started earlier in the Differdange–Longwy Basin but ended later in the Esch–Ottange Basin (Fig. 8).

Each of the asymmetric Minette cycles has a sharp base. The lower part is richer in clay and poorer in iron ooids (muddy facies, Fig. 6). The

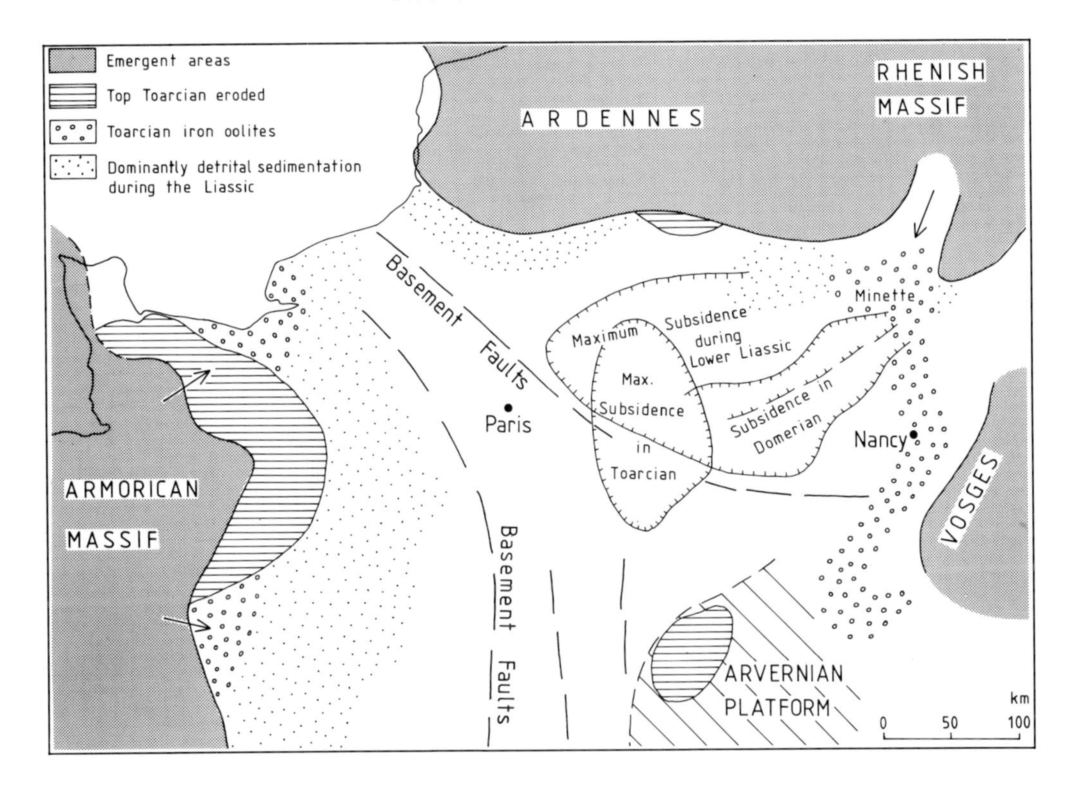


FIG. 4. Palaeogeography and subsidence of the Paris Basin controlling the deposition of the Minette Ironstones. Map based on Tintant and Lefavrais *et al.* in Megnien (1980). See text for discussion.

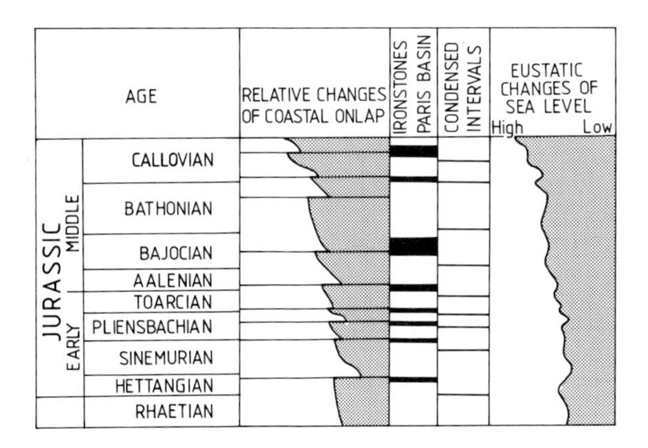


FIG. 5. Early and Middle Jurassic coastal onlap and inferred eustasy curve after Vail *et al.* (1984) and occurrences of ironstones in the Paris Basin after Megnien (1980). Horizontal lines indicate condensed horizons. The fact that ironstones in the Paris Basin are stratigraphically equivalent to the top of North Sea coastal onlap cycles indicates that eustasy was the main control of the Paris Basin regressive depositional cycles.

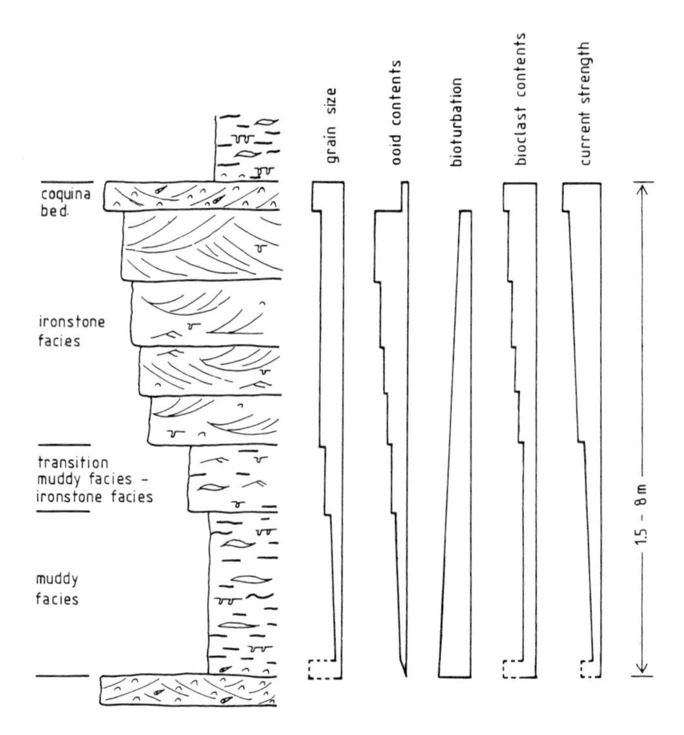


FIG. 6. Schematic sketch of a depositional cycle of the Minette deposits. The sequence leads from a muddy facies over a transitional facies to the ironstone facies with an overlying coquina bed. Grain size, content of iron ooids, bioclasts and current strength increase while the bioturbation decreases.

muddy facies is composed of linsen- and flaser-bedded silt- and mudstones, which are generally intensely bioturbated. The overlying transition facies is ripple cross-laminated and small-scale cross-bedded. The ooid content is higher than below and the bioturbation less severe. The ironstone facies contains abundant limonitic and chamositic ooids, quartz of sand and silt grain size, bioclasts and whole fossils. The quartz grains of the sand fraction are well rounded. The sediments are well sorted. Sedimentary structures are medium- to large-scale cross-bedding with bipolar current directions (Fig. 6). A coquina bed often forms the top of a cycle. From bottom to top of each cycle grain size, current strength, ooid and bioclast content increase while bioturbation decreases.

The Minette sequence is built entirely of such cycles. This internal cyclic pattern seems to be very similar to the above mentioned large-scale regressive cycles, and has indeed been interpreted in terms of allocyclic transgressive-regressive sequences (Bubenicek 1971; Thein 1975). However, the situation is much more complicated. Evidence will be presented below showing that the internal cycles are caused by a combined autocyclicity/allocyclicity (caused by migration of sand bodies/subsidence) while the regressive cycles (as the Toarcian/Aalenian cycle) are clearly allocyclic (driven by eustasy/subsidence).

## Tidal influence on deposition

The Minette ironstones were deposited under the influence of strong tidal currents. This has been reported in detail in an earlier contribution (Teyssen 1984), and will therefore only be summarized here. Cross-bedding measurements from the ironstone facies (at the upper part of each cycle) show bidirectional currents (Fig. 10). A bidirectional current distribution can also be shown in each of the cycles (Fig. 7) and throughout the extent of the whole formation (Teyssen 1984; Figs. 10 & 11). Bidirectional orientation of belemnites in the muddy facies (at the lower part of each cycle) also show bidirectional currents. The current pattern shows directions to the north and north-east and others to the west and south-west. Assuming a tidal origin of the currents, the latter has to be interpreted as ebb with respect to the palaeogeographical situation (Fig. 1); the former is supposed to represent the flood direction. Ebb directions dominate more often.

The muddy facies, however, is not a deposit of intertidal mud flats. The study of the ichnofaunal associations and of wave ripple marks indicates that the muddy facies was probably deposited in slightly deeper water than the ironstone facies. This can be confirmed by the study of sandwaves. In particular, bedding planes, and thus time lines, can generally be traced laterally from the ironstone facies in the upper part of a cycle into the muddy facies in the lower part of a cycle (Teyssen 1984; Figs 4 & 8). This indicates that the muddy facies and the ironstone facies that vertically alternate in each cycle were synchronous deposits and are lateral coexistant facies formed by the same depositional process. In many places of the outcrop zone the cycles can be interpreted as having formed by migration of large-scale tidal sand waves. The muddy facies was deposited as toesets while the cross-bedded ironstone facies was formed on the foresets of such sand waves. Migration of large tidal sand waves as currently seen on the Norfolk Banks (Stride 1985) caused the cyclic upbuilding of the Minette. In other places of the Minette outcrop zone subtidal shoals advanced over their distal muddy facies. The upbuilding of the Minette sequence is thus autocyclic in the way that the rock type found at a particular place within the cycle is determined by the depositional process and not by an external control (such as water depth).

Formation and migration of large sand wave complexes require strong tidal currents. This is the common case in transgressive situations when a region becomes submerged in water to a depth of from several metres to several tens of metres which is optimal for the development of strong tidal current patterns (cf Hammond & Heathershaw 1981). For example, most of the sand wave fields in the North Sea (Nio 1976) and in the East China Sea (Yang & Sun 1985) were active during the onset of the Holocene transgression and are drowned and inactive today. Similarly, other large sand wave fields have formed in transgressive settings (Nio 1976). To explain the Minette sequences renewed episodes of base level rises are required at the top of a

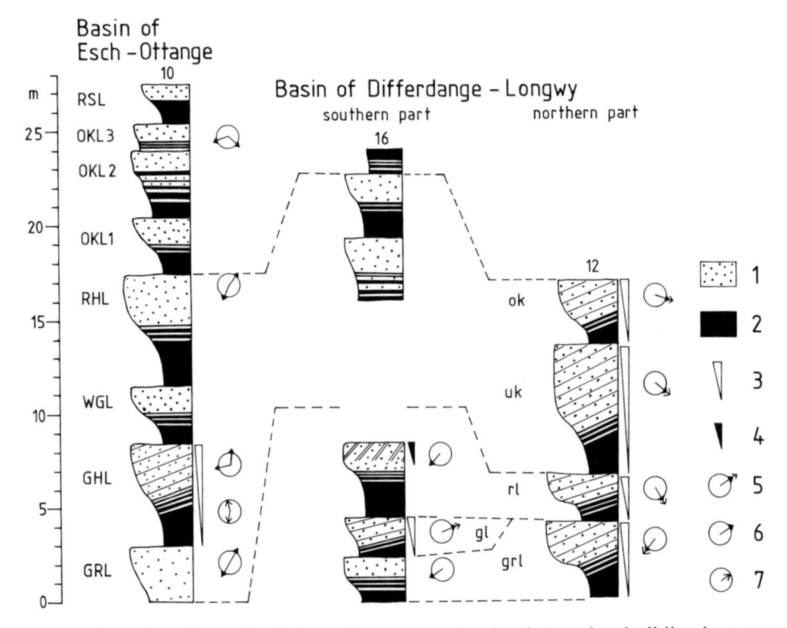


FIG. 7. Sedimentological sections from the Minette Ironstones showing internal upbuilding by coarsening-upward cycles (cf Fig. 6). See Fig. 8 for abbreviations of the cycles. Section 10 is from the Esch–Ottange Sub-basin. Sections 16 and 12 are from the Differdange–Longwy Sub-basin. The thickness of the formation decreases from south (section 16) to north (section 12) and some cycles pinch out. This is indicative of successive offlapping and a general regressive trend. Key to symbols: 1, ironstone facies; 2, muddy facies; 3, large-scale tidal sand waves building up entire cycles (formed under time-velocity symmetrical tides; see Teyssen (1984) for details); 4, small-scale tidal sand waves within the ironstone facies; 5, direction of migration of large-scale sand waves determined from dip direction of low-angle master beds; 6, tidal current directions measured within the ironstone facies; 7, tidal current directions measured within the muddy facies.

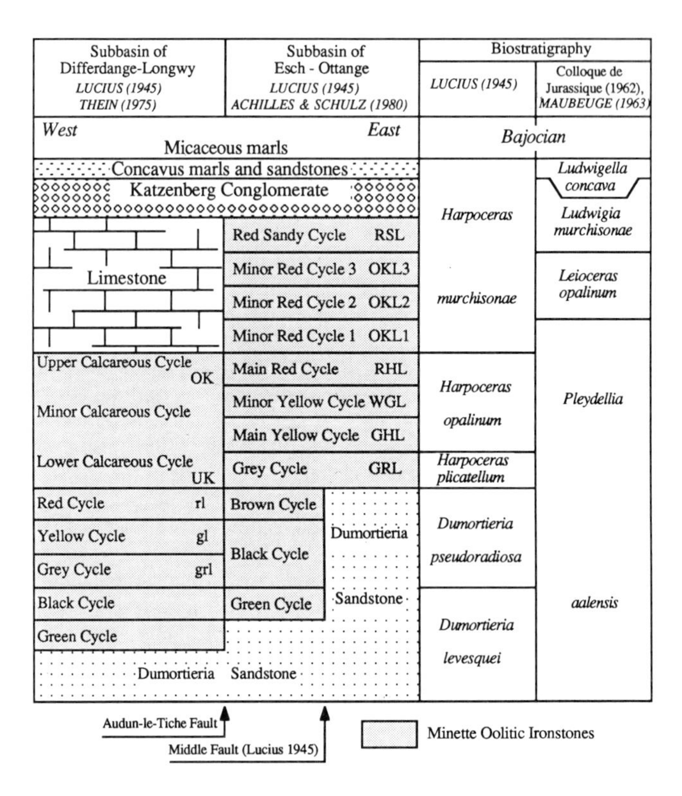


FIG. 8. Stratigraphical subdivision of the Minette Ironstones. The Green and Black Cycles from the Differdange–Longwy Subbasin and the Green, Black, and Brown Cycles from the Esch–Ottange Subbasin are very thin and badly exposed in the outcrop zone in Luxembourg. These cycles are thicker in the subsurface of France. With the cycle names abbreviations are given which are used in Fig. 7. The Toarcian–Aalenian boundary is conventionally placed at the boundary of the aalensis–opalinum zones.

regressive (base level fall) sequence. These base level rise events probably served as an (allocyclic) trigger to start the (autocyclic) deposition of sand wave complexes. The base level rises which presumably occurred during deposition of the Minette, cannot, however, be the onset of the following major deepening event leading to the micaceous marls which overlie the whole Minette Formation. They were still part of the overall regressive development because synchronous erosion and nondeposition prevailed at other parts of the basin margin along the Ardennes land (Megnien 1980; Figs. 4–4, 5–9).

### Syndepositional tectonics

The apparent discrepancy of local base level rise events at the top of an overall shoaling-upward sequence can be resolved if one considers the tectonic setting. The thickness of the Minette ironstones and the fault pattern (Fig. 9) suggest that the Minette depostional pattern is related to the tectonic elements (see also Lucius 1945; Steiner & Le Roux 1978). Furthermore, the Audun-le-Tiche Fault separates the two Minette sub-basins, in which ironstone deposition persisted for different periods (Fig. 8). The tectonic elements obviously not only influenced the thickness distribution of the whole formation but were also important for the deposition of the individual cycles. Figure 10 shows thickness and palaeocurrents for two cycles as determined in outcrop. Figure 11 shows two other cycles with their thickness distribution in the subsurface.

Some, but not all of the particular faults, however, were formed syndepositionally, as already pointed out by Lucius (1945). There are no important thickness changes across some of the faults. On the contrary, approaching the Audun–le Tiche Fault from either side the Minette shows a decrease in thickness (Lucius 1945). On the other hand, the faults have the same orientation as synclines and anticlines (Fig. 12) which were actively developing or continuing

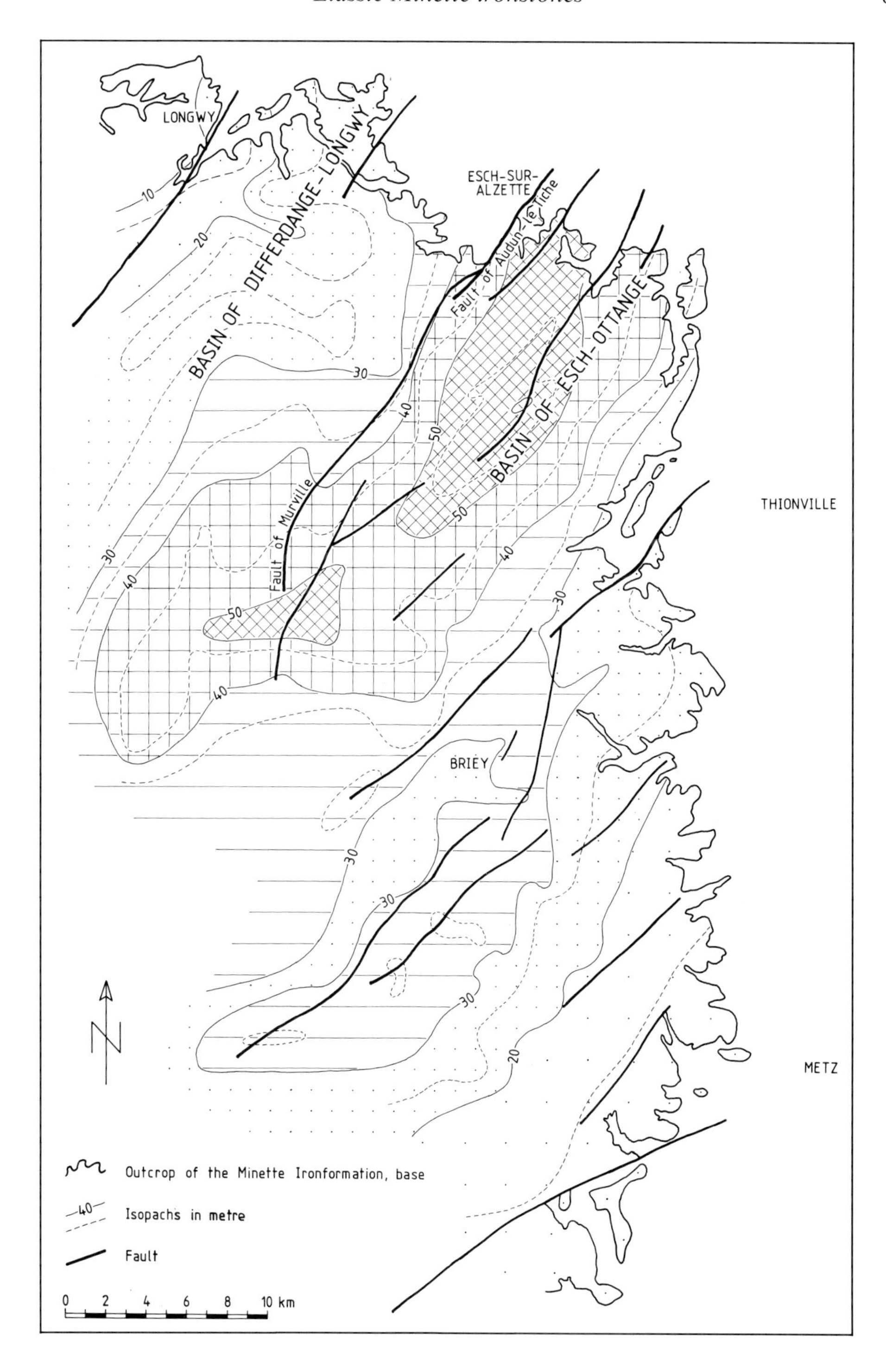


FIG. 9. Thickness of the Minette Formation after Irsid (1967), faults after Lucius (1945). The Audun–le Tiche Fault separates the Esch–Ottange Sub-basin from the Differdange–Longwy Sub-basin. Also the other faults have the same direction as the isopach lines suggesting tectonic influence on depositional patterns.

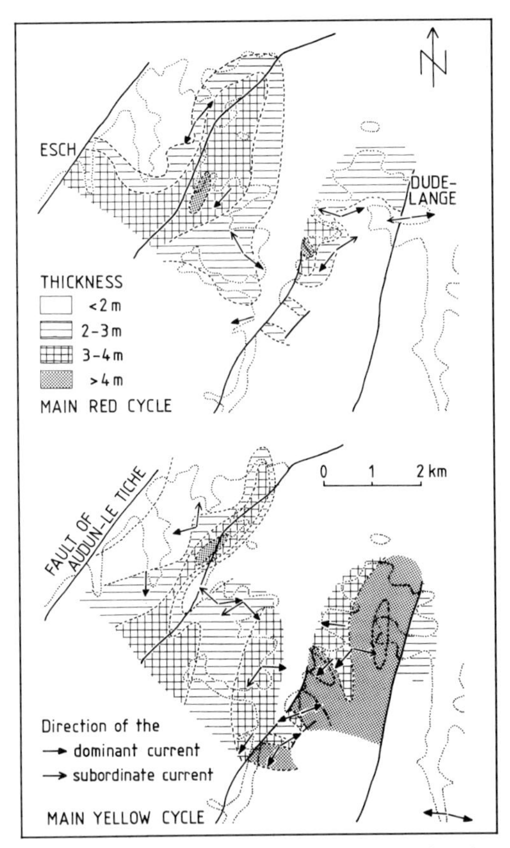


FIG. 10. Isopachs and transport directions during the depositon of the Main Yellow Cycle and of the Main Red Cycle (cf Fig. 8) of the Esch–Ottange Sub-basin. The dotted line is the outcrop of the formation. The orientation of faults and of isopachs indicates structural control on deposition also on the level of the intra-Minette cycles.

to grow during the Early and Middle Jurassic (Lucius 1945; Le Roux *et al.* 1978; Steiner & Le Roux 1978; Laugier 1971; Mertens *et al.*1983; Berners 1985). Figure 12 shows the pattern of synclines and anticlines in Luxembourg and northern Lorraine. These tectonic elements were probably formed as the response of the Mesozoic cover to underlying Variscian structures (Lucius 1945). The structures of the Mesozoic cover have the same orientation and overlie the margin of the Sarre–Lorraine Graben. This graben of Variscian origin is probably one end of a larger graben system which stretches east–west across the whole Paris Basin, and which was dissected by late Variscan wrench faults (Durandau & Koning 1985). This graben system with its lithospheric stretching is also supposed to be the prime origin for the later thermal subsidence of the entire Paris Basin. The synclines and anticlines of the Mesozoic cover overlying the margin of the Sarre–Lorraine Graben obviously influenced the subsidence pattern (Fig. 12) and thereby controlled the depositional patterns throughout the Lower Jurassic.

According to Laugier (1971) the Rhaetian (Late Triassic) Levallois Marls show greater thickness in the Luxembourg Syncline and a reduced thickness on the Moselle Anticline (see Fig. 12 for orientation). Thereafter the Hettangian Luxembourg Sandstone (cf Fig. 3) was deposited with its maximum thickness in the Weilerbach and Luxembourg Synclines (Mertens *et al.* 1983). The thickness sharply decreases beyond the Sandweiler Anticline. Sediment transport patterns also match the orientation of the structural elements (Mertens *et al.* 1983; Fig. 8). The Minette deposits are also thicker in synclines (in syndepositional troughs) and are thinner on anticlines (on syndepositional swells) as already noted by Lucius (1945). In the outcrop zone of Luxembourg the centre of the Differdange–Longwy Sub-basin is located in the Luxembourg Syncline while the centre of the Esch–Ottange Subbasin lies in the Syr Syncline. Sediment transport occurred along the axes of the synclines (Fig. 10). The two sub-basins were separated by the swell of the Sandweiler Anticline. Fracturing of the Audun–le Tiche Fault later occurred along the flank of the Sandweiler Anticline. Also the pattern of the other faults (Fig. 9) obviously was predetermined by the synclines and anticlines (Fig. 12). Subsidence events in the different synclines (i.e. syndepositional troughs) could thus account for the required base level rises (see above) and result in water depths which were optimal for development of strong tidal currents. They are therefore an allocyclic trigger for the deposition of autocyclic sequences.

### Depositional model for the Minette

The Minette ironstones were deposited at the top of a regressive sequence. This regressive sequence (Fig. 3) is one of a series of stacked regressive sequences formed as part of the transgressive–regressive cyclic fill of the intracratonic Paris Basin. This allocyclic basin fill was generally controlled by eustasy and by post-rift thermal cooling (Durandeau & Koning 1985) accounting for the subsidence. Eustatic sea level changes are supported by the match of the lithostratigraphy and occurrences of ironstones and bituminous shales in the Paris Basin with the relative coastal onlap curve (Vail *et al.* 1984) developed for the Jurassic of the North Sea.

However, the Paris Basin and the North Sea have different tectonic histories. In particular, the time of rifting and therefore cooling history differ markedly. Since the coastal onlap patterns of the Paris Basin and of the North Sea are generally correlatable, they must have been controlled by eustasy and not by flexural subsidence induced from lithospheric cooling (Watts 1972). Most of the tops of regressive sequences are marked by ironstones somewhere in the Paris Basin (Fig. 5; Megnien 1980). So far the Jurassic ironstones of the Paris Basin fit the general oolitic ironstone model described above.

The Minette ironstones in particular were deposited in high-energy near-shore conditions along one part of the palaeocoastline, while exposure and erosion of the uppermost Toarcian and non-deposition during the Aalenian can be observed at other nearby parts of the same coastline (Fig. 4; Megnien 1980: Figs, 4-4, 5-9). Strong tidal currents produced large subtidal sand wave complexes. Subsidence events in syndepositionally active troughs (todays synclines) provided the local base level rises required for the occurrence of the tidal regime. This triggered then the deposition of the autocyclic Minette ironstones. Accordingly, sedimentation patterns of the Minette (Fig. 9) portray the subsidence patterns (Figs. 4 and 12). Local subsidence is supposed to be due to reactivation of the underlying Variscian structures. Base level rise deposits at the top of a base level fall sequence can therefore be explained by interference of the basinwide regression with regional subsidence. Several subsidence events probably occurred during deposition of the Minette Formation. This is supported by the observation of lateral shifts of the depocentres from some cycles to the next ones (Figs. 8, 10, 11). It is, however, not necessary to assume an external trigger for each cycle. Some successive cycles can have been formed by migration of several large sand waves which override each other basinward and occur vertically within a sequence (Teyssen 1984). The Minette sequence is thus formed by a combination of autocyclic and allocyclic processes. The cycles are autocyclic because the rock type/facies observed at any given level within a cycle is not predetermined by an external mechanism. The sequence is allocyclic due to the subsidence triggers. At least part of the repetitive element is thus allocyclic while the internal organisation is autocyclic.

The mode of formation of the Minette ooids is not yet fully clear and will not be addressed in detail in this paper. There are, however, two recent models for the Minette. One (Siehl & Thein 1978) postulates formation of the ooids on land as lateritic pisolites; the other (Dahanayake *et al.* 1985) proposes formation by biogenic processes in extensive intertidal flats. Both theories, however, require supply of the ooids from the landward side. Ooid supply from the landward side (north-east) is also supported by the Minette palaeocurrent pattern (Teyssen 1984). Sedimentological studies (Siehl & Thein 1978; Teyssen 1984) made clear that the site of Minette deposition is not the site of ooid formation, as it is the case with most other European Liassic/Dogger Minette type oolitic ironstones (Bayer *et al.* 1985).

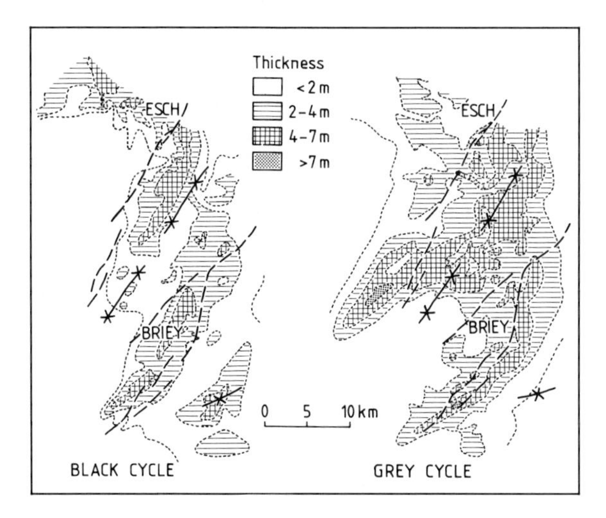


FIG. 11. Structures and thickness of the Black and Grey Cycles in the subsurface, redrawn after Le Roux *et al.* (1978).

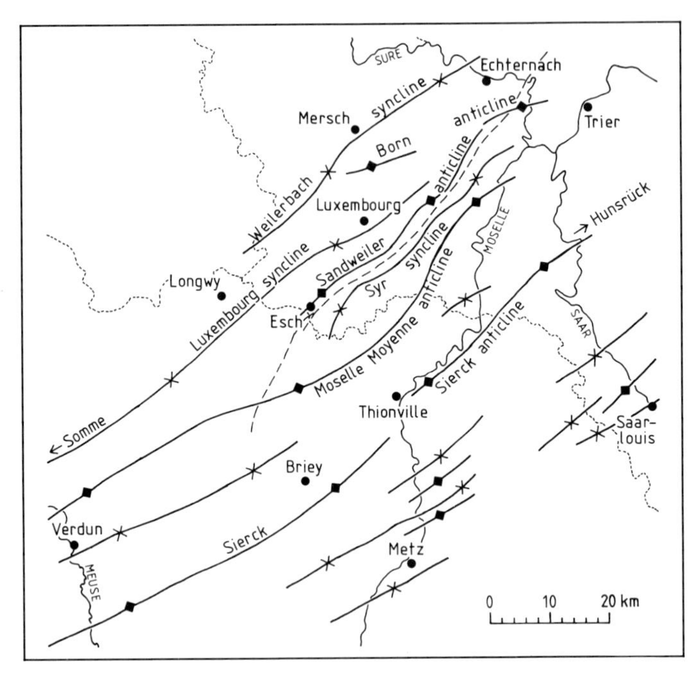


FIG. 12. Synclines and anticlines in Luxembourg and northern Lorraine, based on Bintz *et al.* (1973) and Laughier (1971). These structures were forming or continuing to grow during the Early/Middle Jurassic; see text for discussion. The Luxembourg Syncline is tentatively connected with the Somme Syncline. Field evidence is inconclusive to decide whether it was the south-western end of the Luxembourg or that of the Weilerbach Syncline where the Differdange–Longy sub-basin of the Minette was located in. Alternatively, the Weilerbach Syncline could be connected to the Somme Syncline.

## Conclusions

Oolitic ironstones generally form, often repeatedly, in allocyclic megasequences at the top of a coarsening-upward regressive sequence. Large shelves or large emergent areas are required to produce the ooids. Oolitic ironstone are often only locally developed and thin, because the aggradational potential is normally rather low at the end of the regressive part of a transgressive–regressive cycle. If this regression leads to subaerial exposure, the aggradational potential is even negative and erosion may occur.

The Minette Ironstones form a special case. Whereas erosion or non-deposition occurred along most margins of the Paris Basin, local subsidence created a high aggradational potential in northern Lorraine. Repeated subsidence events permitted the culmination of the regression to persist for a long time. The regressive development became stationary at the end and produced a very thick ironstone sequence in a combined autocyclic/allocyclic process. This combination of controls reasonably accounts for the accumulation of an unusually large volume of oolitic ironstones.

ACKNOWLEDGEMENTS: Many ideas expressed in this paper were developed in discussions during field trips led by the author to the Minette outcrops in Luxembourg. The participants of an excursion organised for the Symposium on modern and ancient tidal deposits, Utrecht, 1985, were particularly helpful. M. Epting and T. Aigner (Rijswijk) and A. Muller (Aachen) read an earlier draft of the manuscript and made many valuable suggestions for improvements.

## References

ACHILLES, H & SCHULZ, H-J. 1980. Geologische Untersuchungen in der Minette des Escher Beckens (Luxemburg). *Revue Technologie Luxembourg,* **3**, 93–141.

BAYER, U., ALTHEIMER, E. & DEUTSCHLE, W. 1985. Environmental evolution in shallow epicontinental seas: Sedimentary cycles and bed formation. *In:* BAYER, U & SEILACHER, A. (eds)

*Sedimentary and evolutionary cycles: Lecture Notes in Earth Sciences,* **1**, Springer, & Heidelberg, 347–381.

——& SEILACHER, A. (eds) 1985. *Sedimentary and evolutionary cycles: Lecture Notes in Earth Sciences,* **1**, Springer, Heidelberg.

BERNERS, H.P. 1985. *Der Einfluss der Siercker Schwelle auf die Faziesverteilung meso - kanozoischer Sedimente im NE des Pariser Beckens—Ein Sedimentationsmodell zum Luxemburger Sandstein (Lias), spezielle Aspeckte zur strukturellen Anderung der Beckenkonfiguration und zum naturramlichen Potential.* Ph.D. thesis, Technical University Aachen.

BINTZ, J., HARY, A & MULLER, A. 1973. Luxembourg. *In:* WATERLOT, G. & BEUGNIES, A. *Ardennes—Luxembourg—Guides Geologiques Regionaux,* Masson, Paris, 1–206.

BUBENICEK, L., 1961. Recherches sur la constitution et la repartition des minerais de fer dans l'Aalenien de Lorraine. *Sciences de la Terre,* **8**, 1–204.

——1964. Etude sedimentologique du minerais de fer oolitique de Lorraine — *In:* AMSTUTZ, C.G. (ed): *Sedimentology and ore genesis,* **2**, Amsterdam Elsevier (113–122).

——1971. Geologie du gisement de fer de Lorraine—*Bulletin de Centre Recherches de Pau,* 5, 223–320.

——1983. Diagenesis of iron-rich rocks—*In:* LARSEN, G. & CHILINGAR, G.V. (eds). *Diagenesis in sediments and sedimentary rocks: Developments in sedimentology,* **25b**, Elsevier, Amsterdam, 495–511.

——Colloque du Jurassique a Luxembourg 1964. Grand-Ducal, Imprime St. Paul, Luxembourg.

DAHANAYAKE, K., GERDES, G. & KRUMBEIN, W. 1985. Stromatolites, oncolites and oolites biogenically formed *in situ. Naturwissenschaften,* **72**, 513–518.

DUFF, P.McL.D., HALLAM, A. & WALTON, E.K. 1967. *Cyclic Sedimentation: Developments in Sedimentology,* **10**, Elsevier, Amsterdam.

DURANDAU, A. & KONING, A. 1985. Contribution a la connaissance de l'origin su Bassin de Paris a partir d'un graben initial. Interet economique. *Computes Rendus de l'Academic de Science Paris,* **301**, 737–742.

EINSELE, G. & SEILACHER, A. 1982. *Cyclic and event stratification.* Springer, Heidelberg—Berlin.

GRUSS, H. & THIENHAUS, R., 1969. Palaogeographie und Entstehung der Eisenerze des Oberaaleniums (Dogger β) Nordwestdeutschlands—*Beiheft geologisches Jahrbuch,* **79**, 167–172.

HAINS, B.A. & HORTON, A. 1969. *British regional geology: Central England.* Her Majesty's Stationary Office, UK.

HALLAM, A. 1966. Depositonal environments of British Liassic Ironstones considered in the context of their facies relationship. *Nature,* **209**, 1306–1307.

——& BRADSHAW, M.J. 1979. Bituminous shales and oolitic ironstones as indicators of transgressions and regressions, *Journal of the Geological Society, London,* **136**, 157–164.

HAMMOND, F.D.C. & HEATHERSHAW, A.D., 1981. A wave theory for sand waves in shelf-seas. *Nature,* **293**, 208–210.

IRSID. 1967. *Atlas geologique du gisement de fer de Lorraine.* Institut de recherches de la siderurgie, Maizieres-les-Metz.

KARRENBERG, H. 1942. Palaogeographische Uebersicht uber die Ablagerungen der Dogger B - zeit in West- und Sudwestdeutschland. *Archives Lagerstattenforschungen,* **75**, 78–79.

KLUPFEL, W. 1916. Ueber die Sedimente der Flachsee im Lothringer Jura. *Geologishes Rundschau,* **7**, 97–109.

LAUGIER, R. 1971. *Le Lias inferieur et moyen du Nord-est de la France. Sciences de la Terre,* Memoir 21.

LE ROUX, J., STEINER, P., PIRINON, B. & BELLORINI, J. P. 1978. Subsidence et sedimentation dans l'est du Bassin de Paris. Manifestations tectoniques au Jurassique Moyen dans le Synclinal de Joeuf (Moselle et Meurthe-et-Moselle)—*103e Congres national des societes savantes, Nancy,* 1978, sciences, **IV**, 363–374.

LUCIUS, M. 1945. Die luxemburger Minetteformation und die jungeren Eisenerzbildungen unseres Landes. Beitrage zur Geologie von Luxemburg. *Publique Service Carte de geologie Luxembourg,* **4**, 350 pp.

MAYNARD, J.B. 1986. Geochemistry of oolitic iron ores, an electron microprobe study, *Economic Geology,* **81**, 1473–1483.

MEGNIEN, C. (eds), 1980. Synthese geologique du Bassin de Paris—vol.I Stratigraphie et paleogeographie, *Memoir 101, Bureau de Recherches Geologie et Mineralogie.*

——Vol II Atlas. *Memoir 102 Bureau de Recherches Geologie et Mineralogie*

MERTENS, G., SPIES, E.D. & TEYSSEN, T. 1983. The Luxemburg Sandstone Formation (Lias), a tide-controlled deltaic deposit. *Annales Societe geologique de la Belgique,* **106**, 103–109.

NIO, S.D. 1976. Marine transgressions as a factor in the formation of sandwave complexes. *Geologie en Mijnbouw,* **55**, 18–40.

ODIN, G.S. & MATTER, A. 1981. De glauconiarium orgine. *Sedimentology,* **28**, 611–641.

PORRENGA, D.H. 1967. *Clay mineralogy and geochemistry of recent marine sediments in tropical areas*—Dissertation, University of Amsterdam.

SCHWARZACHER, W. 1975. *Sedimentation models and quantitive stratigraphy: Developments in Sedimentology,* **19**, Elsevier, Amsterdam.

SIEHL, A. & THEIN, J. 1975. Stratigraphische Auswertung geochemischer Daten der Minette Luxemburgs. *Zeitschrift fur geologishe Gessellschaft,* **126**, 167–181.

——1978. Geochemische Trends in derMinette (Jura, Luxemburg/Lothringen). *Geologisches Rundschau,* **67**, 1052–1077.

STEINER, P. & LE ROUX, J. 1978. Le Bajocien et le Bathonien dans L'est du Bassin Parisien. Structure—Lithostratigraphie—*103e Congres national des societes savantes, Nancy,* 1978, sciences, IV, 393–403.

STRIDE, A.H. 1985. Indications of suspension transport of sand over the Norfolk Banks—*Symposium on modern and ancient clastic tidal deposits, Abstracts,* University of Utrecht, 130.

TEYSSEN, T. 1983. *Gezeitenbeeinflusste Sedimentation und Sandwellenentwicklung in der Minette (Toarcium/Aalenium, Luxemburg/Lothrigen).* PhD.thesis, University of Bonn.

——1984. Sedimentology of the Minette oolitic ironstones of Luxembourg and Lorraine: a Jurassic subtidal sandwave complex. *Sedimentology,* **31**, 195–211.

THEIN, J. 1975. Sedimentologisch-stratigraphische Untersuchungen in der Minette des Differdinger Beckens (Luxemburg). *Publique Service Carte de geologie, Luxembourg,* **24**, 1–66.

VAIL, P.R., HARDENBOL, J. & TODD, R.G. 1984. Jurassic unconformities, chronostratigraphy, and sea-level changes from seismic stratigraphy and biostratigraphy. *In:* SCHLEE, J.S. (ed.). *Interregional unconformities and hydrocarbon accumulation. Memoir 36,* of the Association of American Petroleum Geologists, 129–144.

VALETON, I. 1983. Klimaperioden lateritischer Verwitterung und ihr Abbild in den synchronen Sedimentationsraumen. *Zeitschrift fur geologische Gesellschaft,* **134**, 413–452.

VAN HOUTEN, F.B. 1985. Oolitic ironstones and contrasting Ordovician and Jurassic paleography. *Geology,* **13**, 722–724.

——& BHATTACHARYYA, D.P. 1982. Phanerozoic oolitic ironstones—geologic record and facies model. *Annual Review of Earth & Planetary Sciences,* **10**, 441–457.

——& PURUCKER, M.E. 1984. Glauconitic peloids and chamossitic ooids—favourable factors, constraints, and problems. *Earth-Science Reviews,* **20**, 211–243.

WYATTS, A.B. 1982. Tetonic subsidence, flexure and global changes of sea level, *Nature,* **297**, 469–474.

YANG, C. & SUN, J. 1985. Tidal sand ridges on the shelf of East China Sea. *Symposium on modern and ancient clastic tidal deposits, Abstracts,* University of Utrecht. 166–168.

ZIEGLER, P.A. 1982. *Geological atlas of western and central Europe*—Elsevier, Amsterdam.

T. TEYSSEN, KSEPL/Shell Research BV, Volmerlaan 6, 2288GD Rijswijk, The Netherlands.

Present address: Petroleum Development Oman, Exploration Department, PO Box 81, Muscat, Sultanate of Oman.

# Concentrated and lean oolites: examples from the Nubia Formation at Aswan, Egypt, and significance of the oolite types in ironstone genesis

**D. P. Bhattacharyya**

SUMMARY: Two textural types of oolitic ironstones occur in the Late Cretaceous Nubia Formation at Aswan, Egypt. These are: (1) relatively thin bedded (5–20 cm) muddy or 'lean' oolites containing up to 30% ooids, proto-ooids and peloids scattered in a matrix of ferruginous mud, and (2) thicker beds (as much as 2.5 m thick) of 'rich' or 'concentrated' oolites comprising 60–80% ooids.

Sedimentologic and textural attributes of the 'concentrated' oolites suggest that these represent ooids transported and redeposited away from their site of formation. In fact, in ironstone-bearing successions elsewhere the concentrated oolites are either associated with winnowed sandstones or, in sand-poor deposits, they substitute for sandstone. Together the ooids and the detrital sand form bar and sand wave complexes in upward shoaling marginal marine sediments. In unbioturbated or less bioturbated occurrences these sandbodies of ferruginous ooids are replete with primary sedimentary structures indicative of strong current transport. Thin beds of similar ferruginous ooids in sequences of off-shore carbonate or argillaceous basinal mudstone, such as those in the Silurian Clinton Group of eastern United States, are storm deposits characterized by basal scours and hummocky cross-stratification. The muddy or 'lean' oolites, on the other hand, are commonly associated with prograding subtidal mudflat deposits. Textural characteristics of these ironstones suggest that they probably represent the primary sites of ooid formation. Different stages of ooid formation preserved in these 'lean' ironstones and their relationship with the concentrated oolites provide information which is significant for a more realistic and comprehensive genetic modelling for the Aswan ironstones.

In the model proposed here the deposition of muddy, ferruginous sediments on prograding, near-shore, subtidal mudflats begins during the waning stages of detrital sediment supply. The ooids are formed on these shallow-water mudflats by mechanical accretion of detrital kaolinitic clay and precipitated iron hydroxide minerals, mostly around faecal pellets or flocculated peloids, under the influence of gentle wave or current agitation. Periodic high-energy events, such as storm-generated or tidal currents, transport the ooids to off-shore areas to be reworked there into oolitic sand bar and shoal complexes. Following deposition, the kaolinitic ooids, peloids and matrix are transformed into berthierine under a chemically reducing diagenetic environment.

In the study of Phanerozoic oolitic ironstones most attention has been paid to the relatively thick (up to 20 m) ironstone beds with ooid concentration in excess of 60%. These commercially exploitable 'rich' or 'concentrated' oolites are rather spectacular in their host sedimentary sequences. It is, therefore, not surprising to find these at the centre of attention. On the other hand, some ironstone-bearing sequences have been reported (James & Van Houten 1979; Bhattacharyya 1980) to contain thinner (commonly between 5 and 20 cm, but may be as much as 40 cm thick) ironstone beds with less than 30% ooids set in a ferruginous, muddy matrix. Most occurrences of these 'lean' oolites have only been cursorily mentioned or escaped attention altogether. Consequently, the formulation of hypotheses about the origin of oolitic ironstones has relied heavily on the attributes of the 'concentrated' oolites, and the potential of the 'lean' oolites in answering some of the questions of berthierine-bearing oolitic ironstone genesis has not been explored.

The purpose of this paper is to document the attributes of the 'lean', muddy, ferruginous oolites of one such occurrence, the Nubia Formation at Aswan, Egypt (Fig. 1) where the 'lean' and the 'concentrated' oolites occur in close association, and to discuss the significance of various attributes of these two types of ferruginous oolites in understanding the origin of Phanerozoic oolitic ironstones.

Throughout this paper the terms 'ooids', 'oolites', 'proto-ooids', 'peloids' and 'ironstones' are used in the sense elaborated by Van Houten & Bhattacharyya (1982).

*From* YOUNG, T. P. & TAYLOR, W. E. G. (eds), 1989, *Phanerozoic Ironstones*
Geological Society Special Publication No. 46, pp. 93-103

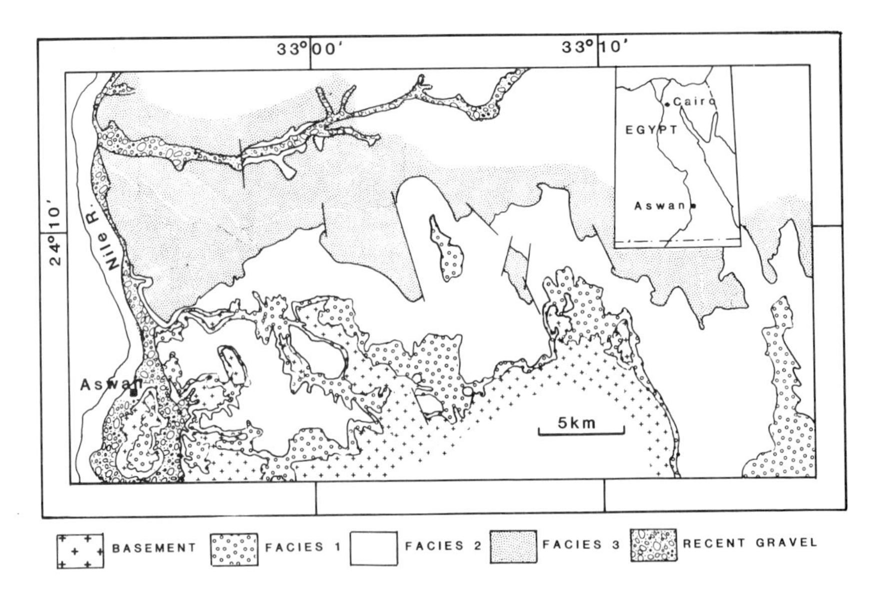


FIG. 1. Geological map of Aswan area, Egypt.

## General geology of the Aswan ironstone deposits, Egypt

Oolitic ironstones at Aswan are associated with the Late Cretaceous Nubia Formation which comprises three major units (Fig. 2). A comprehensive regional account of the stratigraphy and sequence of events during the deposition of Nubia Formation has been given by Van Houten *et al.* (1984). The lowest unit, Facies 1 (Fig. 2), is between 10 and 100 m thick, and is composed of a sequence of upward fining, lenticular, pale yellow to pale brown or grey, kaolinite-rich, coarse-grained quartzose sandstones with palaeosol horizons. These have been interpreted (Bhattacharyya 1980) as the deposits of a low-sinuosity fluvial system. The facies overlies the peneplained basement granites and gneisses of Precambrian age. The intermediate unit, Facies 2, is between 17 and 75 m thick, and is composed of winnowed quartz-sandstones, mudstones, kaolinitic claystones, and oolitic ironstones in three basinwide, shoaling upward sequences (Figs. 2 & 3).

The uppermost facies, Facies 3, comprises a 120–200 m thick succession of spectacularly cross-bedded fluvial to paralic, sandstones, interdistributary claystones and deltafront mudstones. This facies was deposited during a brief span of rapid coastal progradation following a period of reactivated basement uplifts in the south and east (Van Houten *et al.* 1984).

Since the ironstone-bearing Facies 2 is the focus of this paper, a detailed description of this unit is given below.

### Facies 2 and associated ironstones

Facies 2 of the Nubia Formation around Aswan comprises two distinct parts (Fig. 2). The lower part is composed of two subunits of tabular–planar cross-bedded sheet-like sand-bodies, each between 0.5 and 5 m thick, with a 2 to 11 m thick, laterally extensive sequence of thoroughly bioturbated silty mudstone separating the two sandstone units (Figs. 2 & 3). The upper part, on the other hand, is composed of a sequence of claystone, fine-grained sandstone and beds of oolitic ironstone in three major shoaling-upward cyclic units referred to here, from bottom to top, as A, B and C-cycles (Fig. 2). The constituent beds of each of these cycles are designated in the text by the prefixes A, B and C (e.g. A-oolite, C-clay etc.). Vestiges of a fourth cycle are also present locally, but in most areas it is absent because of an erosion surface at the base of the overlying Facies 3. The total thickness of the upper part of Facies 2 varies between 14 and 37m with an average of about 24 m.

The basal unit of each cycle is composed of a greenish grey to grey, laminated and massive claystone with a sharp, but comformable, contact with the underlying beds. The claystone grades upwards through silty claystone, small-scale tabular cross-bedded and rippled

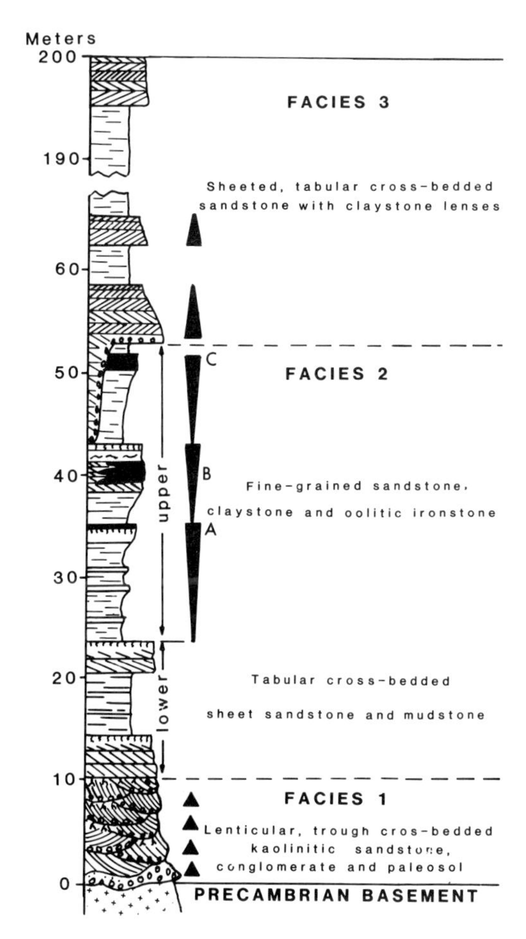


FIG. 2. Generalized facies sequence of the Nubia Formation at Aswan, Egypt. ▲ = fining-up and ▼ = coarsening-up sequences. The letters A, B and C indicate the three ironstones referred to in the text.

sandstone, into a burrowed, muddy ferruginous oolite or both (Fig. 4).

The lowest, A-cycle, is the thickest (Figs. 3 & 4), with an average thickness of about 12 m (range 2–24 m). A major part of this cycle is composed of dark greenish grey to dark grey, friable, sandy and silty claystone, commonly interbedded with lenses of intensely bioturbated sandstone, the thickest parts of which locally show small sets of tabular cross-beds along the lower halves. The uppermost unit of this cycle is a thin (commonly between 10 and 20 cm, maximum 40 cm) bed of muddy, oolitic and/or peloidal ironstone (A-oolite) that overlies a conspicuously bioturbated muddy sandstone. The A-ironstone is commonly of ooid-poor, 'lean' type containing clay-rich ooids and peloids. Locally, it also contains lenticular beds of 'concentrated' clay-rich ooids. The muddy sandstone below the ironstone bed contains at least two distinctive burrow-zones. The lower zone is dominated by *Teichichnus* and the upper zone contains an abundance of *Diplocraterion* and *Thalassinoides* with sporadic *Chondrites*. The ironstone bed is laterally persistent throughout the outcrop belt and, together with the muddy sandstones forms an extensive ledge in outcrops (Fig. 4). The A-oolite locally contains fragments of bone, fish teeth and numerous spheroidal ferruginous and siliceous nodules, 1 to 5 cm in diameter, along the uppermost bedding surface. In many places, burrows in the top few centimetres of the sandstone, immediately below the ironstone, are filled with ferruginous peloids and ooids. In the south-eastern part of the area the A-oolite is poorly developed, and is usually replaced by a thin bed of ferruginous sandstone with abundant nodules. Ferruginized moulds of the marine bivalves, *Inoceramus* and *Isocardia,* occur locally in the ferruginous sandstone and ironstone.

The B-cycle is dominated by fine-grained sandstone, and contains subordinate claystone and ironstone. It is more uniform in thickness than the A-cycle (average 7.5 m, range 5–13 m). Locally thick (up to 2 m) beds of highly oolitic ('concentrated') ironstone (B-oolite) laterally interfinger with the sandstone. The sandstones in this cycle are mostly thin-bedded (less than 20 cm), well winnowed, and either tabular-planar cross-bedded or ripple-bedded/ripple-laminated. The uppermost 10–50 cm of this cycle consist of a set of 5–10 cm beds of brown, reddish brown or ochre-yellow ferruginous intensely bioturbated sandstone with *Skolithos* and *Diplocraterion.* In outcrops this burrowed sandstone also forms a laterally extensive ledge (Fig. 4).

The lowermost 5 to 6 m of the C-cycle is composed of pale green to grey, finely laminated claystone, within which there are several lenticular, 10 to 15 cm thick beds of reddish brown, highly bioturbated ferruginous claystone. The claystone grades upwards through purplish brown, finely laminated silty claystone into a sequence of dark brown peloidal and oolitic ironstones (C-oolite) with a maximum aggregate thickness of about 2.5 m (Fig. 4). In any outcrop the C-ironstone consists of several beds (the maximum recorded is 9) of dark brown 'lean' and 'concentrated' oolites, each between 5 and 80 cm in thickness, with 1 to 5 cm thick purplish brown claystone interbeds (Fig. 5). The vertical and lateral variations of the ironstones

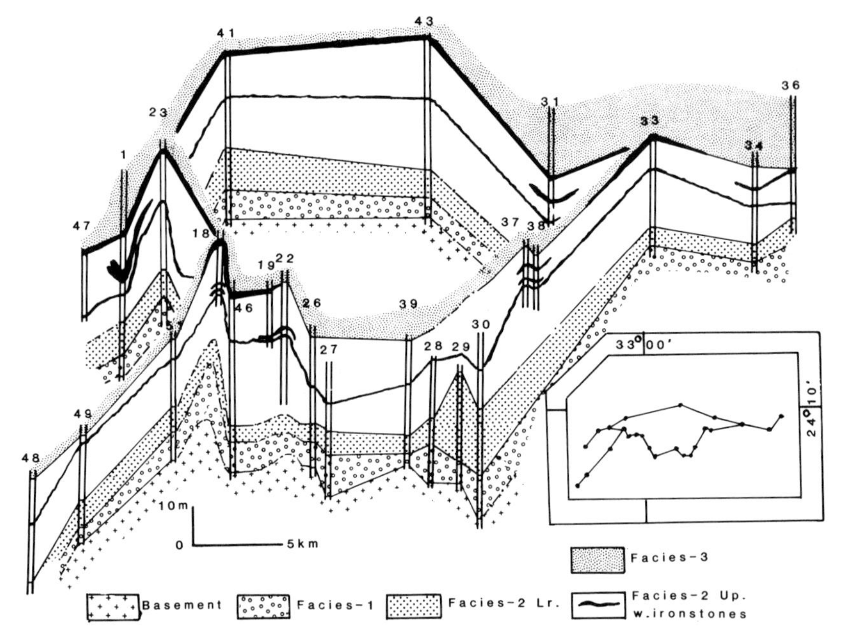


FIG. 3. Spatial distribution and interrelationship of the Nubia lithofacies around Aswan, Egypt. Thicknesses of the ironstone units are exaggerated for better definition.

are similar to those of the A-ironstone, except that some of the 'concentrated' oolites in the C-ironstone are rich in hematite ooids. Rare fragments of bone and fish teeth also occur in this ironstone unit. The C-cycle maintains a fairly uniform thickness laterally. The continuity is, nevertheless, locally interrupted by significant scour at the base of the overlying facies (Fig. 3). The C-claystone interfingers with thin lenticular, rippled or bioturbated wedges of a fine-grained sandstone similar to those found in other cycles at the eastern end of the study area (Fig. 3).

All the ironstone beds in the three cycles are commonly bioturbated and massive, but parallel lamination due to a variation in ooid/peloid

FIG. 4. Outcrop of Facies 2 along a wadi wall, 10 km east of Aswan.

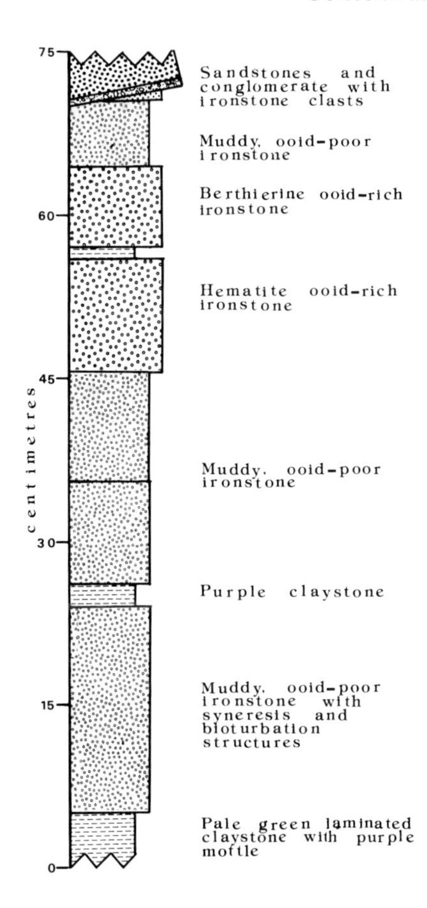


FIG. 5. Vertical profile of the C-ironstone in an outcrop 1 km northeast of Aswan. Note interbedding of 'lean' and 'concentrated' oolites and thin slivers of claystones separating the ironstones.

types is locally present (Fig. 6). Dense, ferruginous mud-rich types of A and C-oolites also show local syneresis structures (Fig. 5). Such structures, however, are more common in the C-ironstones. The claystones in the upper part of Facies 2, especially those in the C-cycle, contain abundant oxidized plant debris. Whole leaf impressions of fern and dicotyledon plants occur, rarely, in the sandstones and claystones. The claystones and the clay fractions of the muddy sandstones are dominated by kaolinite. In fact, the B- and C-claystones are almost entirely kaolinitic with only traces of illite and berthierine. The A-clay, on the other hand, contains an appreciable amount of berthierine and illite. Facies 2, in general, is sandier to the south. About 40 km southwest of Aswan the entire sequence is dominated by fine-grained sandstone with thin claystone intercalations, and is capped by a ferruginous bone-fragment conglomerate which is laterally equivalent to the C-ironstone unit. No oolitic ironstone occurs in this locality, but ferruginous claystones and sandstones occur in equivalent positons in the sequence. Vertebrate remains in the conglomerate unit include tooth plates of a freshwater lungfish *Ceratodus,* crocodile bones and fragments of turtle carapace.

## Petrography of the 'lean' and 'concentrated' oolites

The A-ironstone is commonly green-mottled, dark brown to reddish brown ferruginous mudstone with scattered ooids and peloids. Sand content varies between 10 and 60%, clayey peloids and ooids comprise up to 30% and red hematite cement and clay matrix consitute between 40 and 80% of the rock. The ooids and peloids in the A-ironstone are commmonly berthierine-rich. The uppermost part of the bed is locally ooid-rich, containing 60 to 80% berthierine ooids. The texture of the A-oolite generally grades from peloidal to oolitic from bottom to top. In the 'concentrated' parts of the

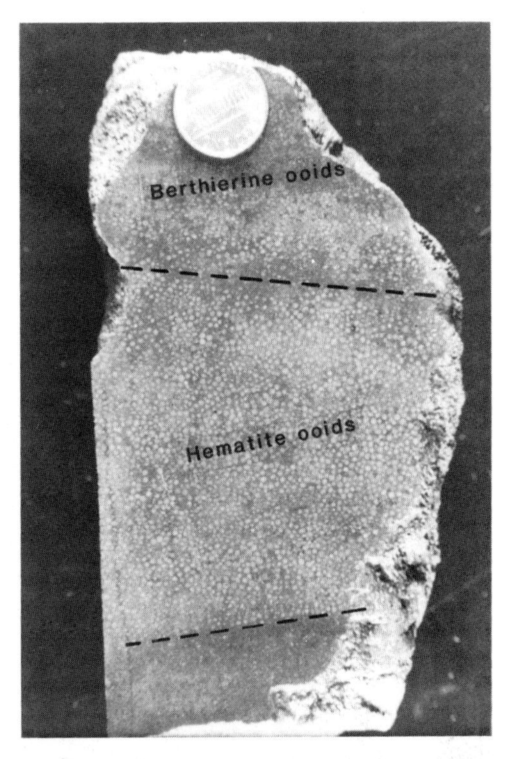


FIG. 6. 'Concentrated' C-oolite hand sample with a medial zone of hematite ooids and two border zones of berthierine ooids. The coin is 2 cm in diameter.

oolite local laminae of peloids and ooids alternate with ferruginous mud on a millimetre-scale. Many of the ooids and peloids, especially the clay-rich ones, are strongly flattened or hook-ended due to compaction-collapse, particularly in the ooid/peloid-rich portions of the rock. In contrast, the ooid and peloids 'floating' in cement or matrix are relatively undeformed. The clay minerals in the ooids and matrix are locally extensively replaced by siderite, hydroxy-apatite and calcite.

The B-ironstone is a 'concentrated' oolite with ooid content generally in excess of 60%. Detrital quartz and hematite cement constitute the remainder of the rock. The ooid population is mostly hematite-rich, and is commonly well sorted. The B-ironstone, however, locally also contains discrete laminae of berthierine-rich ooids along its upper and lower contacts. The general absence of any primary sedimentary structure, except the local parallel lamination, and an admixture of finer sand grains suggest pervasive bioturbation of the B-ironstone.

Hematite ooids in the B-ironstone are commonly spherical, oval or discoidal in shape, and some show compaction denting of one against another (Fig. 7). The latter suggests that the ooids were still in a semi-plastic state at the time of their concentration in the ironstone bed. Spastoliths are conspicuously absent, however, in the hematitic B-ironstone as well as in the hematite ooid-rich portions of the other ironstones. In contract, the hematite ooids frequently show syneresis cracks radiating from the centre. Ooids with a fragment of another ooid as a nucleus are common in the B-ironstone. Although the clay minerals are not identifiable in these hematite ooids under the light microscope, leaching of iron oxide minerals from the ooids in the laboratory and subsequent X-ray diffractometry of the residual clays reveal that an appreciable amount (10–30%) of clay (both berthierine and kaolinite) is intimately mixed with the hematite in the ooid sheaths. In the B-ironstone, hematite cement is pervasive. However, an early generation of radially disposed berthierine cement locally envelops the ooids. In such samples, the clay in the ooids is also dominated by berthierine even though the ooids are hematite-rich. The C-ironstone unit is texturally quite similar to the A-ironstone, and contains both 'lean' and 'concentrated' oolite beds (Fig. 5). Deep red or chocolate brown, muddy 'lean' types of the C-ironstone contain all types and sizes of berthierine peloids, proto-ooids and ooids, as observed in the A-ironstone. In the 'concentrated' types, however, instead of being ooid-rich the ironstone is locally peloid-rich. The upper part of any dense muddy ironstone bed in the C-unit is commonly hematite-rich (Fig. 5), and opaque to transmitted light. In addition, the C-ironstones also contain a complete spectrum of ooids, from berthierine-rich to hematite-rich. Even in a single thin section transitions from clay-rich through mixed clay-hematite to hematite-rich ooids are present.

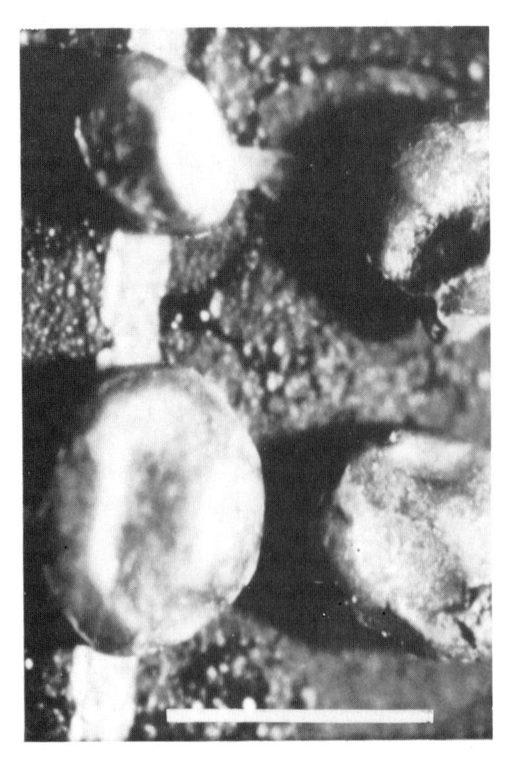

FIG. 7. Compaction dents on hematite ooids isolated from B-oolite. Scale bar = 1mm.

Alternations of clay-rich and hematite-rich ooids on a microscopic scale are locally present (Fig. 8). Some beds of 'concentrated' C-oolite contain three distinct zones: a medial zone with hematite ooids and two marginal zones with berthierine ooids (Fig. 6). Moreover, the degree of clay mineral replacement by apatite in the C-ironstone is much greater than in the A-ironstone. In common with other ironstones in the sequence, the clay-rich ooids and peloids in the C-ironstone unit are commonly strongly deformed; the hematite ooids being least deformed (Fig. 6), and the mixed ooids showing a variable degree of deformation.

In the transitional ooids with discernible alternate hematite and berthierine-rich cortical layers, two types of hematite-rich sheaths are found: (a) thicker (about 0.1 mm thick) and (b) thinner (less than 0.01 mm thick). The latter type

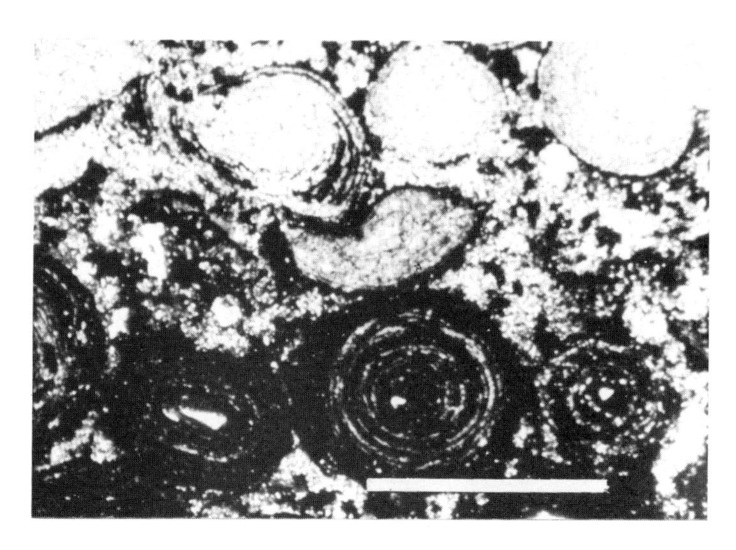

FIG. 8. Thin section-scale alternation of hematite and berthierine ooids-rich laminae, 'concentrated' C-oolite. Scale bar = 1.5 mm. Plane-polarized light.

is clearly identifiable in berthierine-rich ooids (Fig. 9). Considering the pair of hematite and berthierine-rich sheaths as a single unit, the ooids contain betwen 10 and 45 pairs depending on the diameter of the ooid. However, in any specific ooid-rich bed or lamination the number of these cortex-pairs in the ooids remains within narrow limits (Fig. 10). Locally, parts of some hematite sheaths appear abraded. Subsequent growth around those abraded parts gives rise to coalesced sheaths (Fig. 9).

The textural attributes of the lenticular, thin ferruginous claystone beds within the C-clay unit are of special interest because they shed additional light on the mode of formation of ironstones associated with them. These claystones are composed of somewhat irregular aggregates (floccules?) and peloids of hematite and goethite in a clay matrix and *vice versa*. These aggregates are subrounded to irregularly deformed in outline, and show internal fabric similar to the peloids and peloidal cores of the ooids in ironstones. Nevertheless, the textural relations between these clay and hematite aggregates suggest that these were probably deposited as a result of colloidal flocculation by a mechanism similar to the one postulated for the Eocene 'Sawdust Sand' of Tennessee and Kentucky, U.S.A. (Pryor & Vanwie 1971) rather than as faecal pellets. Clay in the matrix, peloids and floccules, is dominated by kaolinite, but also includes an appreciable amount of berthierine. This rock apparently has all the essential characteristics of a clayey ironstone except that it contains no ooids. Instead, flocculent texture of iron oxide and clay minerals is best developed in these rocks. Although these beds appear highly bioturbated, internal, somewhat diffuse lamination due to a variation in the relative proportion of iron oxide and clay mineral aggregates is locally present. Muddy ironstones beds of A- and C-oolites commonly display similar textures.

Most of the ooids in the Aswan ironstones have a peloidal nucleus. Detrital quartz grains or other clasts are clearly subordinate as ooid nuclei (less than 10% of the ooid nuclei in any specific ironstone bed). The muddy ironstones commonly show a complete range of development of oolitic texture. At one end of the spectrum there are peloids with no oolitic cortex. Ooids with a peloidal core and concentric, tangential cortices dominate the other end. In between, there are proto-ooids which have only a very thin oolitic coating. These textural stages are difficult to identify (in iron oxide-rich muddy ironstones opaque to transmitted light); they are better illustrated by the berthierine-rich ironstones.

## Discussion

Facies-2 rocks are shallow marine deposits that probably accumulated in an embayment. Body fossils are sparse, but marine trace-fossils abound in these rocks. The scarcity of body fossils suggests that, for the most part, the embayment was probably restricted rather than open marine. Southward, this restricted embay-

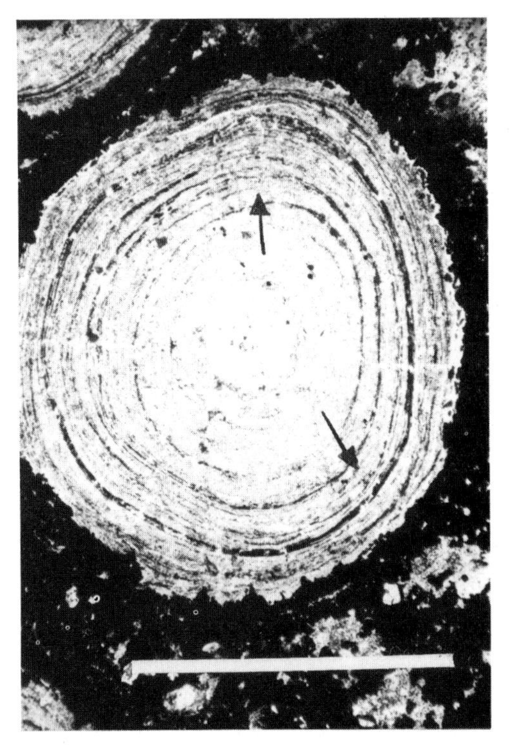

FIG. 9. Ooid with alternate berthierine and hematite-rich sheaths. Arrow points to abrasion and growth-coalescence of sheaths. Corroded outer margin of the ooid is due to replacement by hematite cement. Scale bar = 1 mm. Plane-polarized light.

ment merged, at times, with local fresh to brackish water marshes inhabited by a variety of vertebrates such as *Ceratodus,* crocodiles and turtles.

Three aggradational cycles in this sequence represent at least three minor but basinwide oscillations of relative sea level, each characterized by a coastal progradation and upward shoaling. Within this framework, the A- and C-cycles had rather similar development, culminating in an oolitic and peloidal ferruginous mudflat near the fair-weather wave base. The ferruginous sediments were deposited during the waning stages of detrital sediment supply as indicated by a dominance of iron oxide mud, lack of wave and current-generated sedimentary structures and extensive bioturbation.

Sedimentologic properties of the B-cycle, in contrast, indicate shoaling up above the wave base and generation of a series of sand waves or 'breaker-bars' (Goldring & Bridges 1973) characterized by highly winnowed, thin, cross-bedded or rippled quartz sandstones with local supply (during periodic storms?) of ferruginous oolitic sand from nearby source areas, such as the ferruginous mudflats of the A- or C- ironstones. The depositional setting of the Aswan ironstone-bearing sequence was rather similar, in general, to the modern low-tide, low-energy, prograding muddy coastline of Louisiana, U.S.A. (Beall 1968), but probably with an even lower tidal range and a much lower sediment supply.

Extensive bioturbation in the upper part of each of the cycles, concentration of iron therein and the occurrence of phosphate (e.g. bone fragments, fish teeth) and nodules in the ferruginous beds suggest that the upper muddy ironstone/ferruginous sandstone parts of the cycles are condensed like the hardgrounds in carbonate sequences (Van Houten & Bhattacharyya 1982). Carbonate hardgrounds are commonly iron encrusted. In the ironstones, however, the concentration of iron is about an order of magnitude higher than in the calcareous hardgrounds. There is, apparently, a significant inverse correlation between the rate of detrital sedimentation and iron concentration in the resultant sediments of any ironstone-bearing sequence. This can be explained as a balance between the detrital and chemical components of

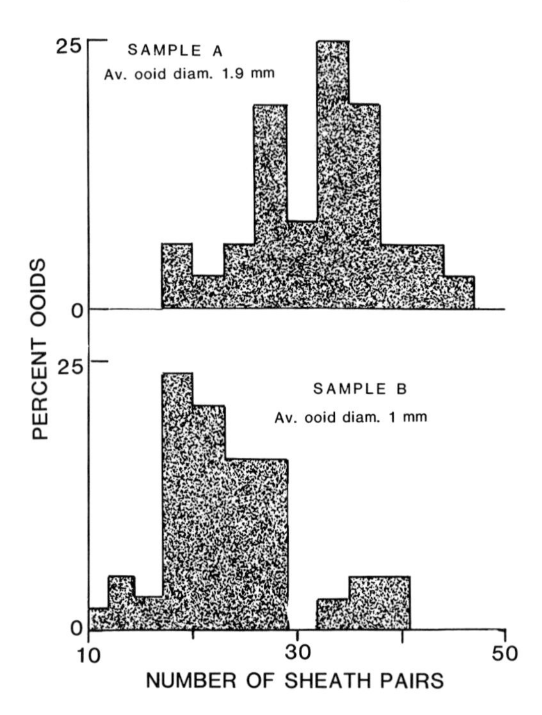


FIG. 10. Frequency histogram of the number of sheath-pairs within the ooids in two samples of 'concentrated' oolites.

such a depositional system as follows. In marginal marine environments with substantial fresh water input a significant amount of iron is brought into the system as metal-organic colloids. The majority of these colloids flocculate and precipitate upon mixing with sea water (Crerar *et al.* 1980; Sholkovitz 1978; Boyle *et al.* 1977). Such a system is also commonly dominated by suspended sediment load, which acts as a diluent for the precipitated iron. The basin gradually becomes sediment-starved, as the detrital sediment load wanes and the relative proportion of iron in the resultant deposits gradually increases. Highly ferruginous sediments are deposited only during the periods of maximum starvation of such a basin.

Textural attributes of the Aswan ironstones suggest that the 'lean', muddy ironstone beds of A- and C-cycles probably represent the principal site of ooid formation. The 'concentrated' B-oolite, on the other hand, is a product of transportation during higher-energy events like storms or regional tidal currents, and continued winnowing and concentration of ferruginous ooids within the wave breaking zones. Local accumulations of ooid or peloid-rich beds in the sequence of A- and C-ironstones, likewise, probably represent ooid/peloid bars and shoals which have been winnowed and mechanically concentrated during short-term high-energy events.

The most compelling evidence that the muddy A- and C-ironstones are primary sites of ooid formation is the occurrence in these ironstones of a variety of textural components, such as the peloids, proto-ooids and ooids, which represent various stages of ooid development. These components are easily identified in the berthierine-rich ironstones. Dense, hematite-rich parts of the ironstones probably also contain similar hematite-rich components. However, such ironstones are opaque to transmitted light, and polish poorly for reflected light microscopy due to the extremely fine grain-size of the constituent minerals. As a result, identification of different stages of ooid development in these muddy, hematite-rich ironstones is difficult.

Although the peloids are mostly faecal pellets, some, such as those in the lenticular ferruginous claystones within the C-clay of Aswan, are interpreted as inorganic floccules.

The hematite and berthierine-bearing ironstones from Aswan have certain common, as well as diverse, characteristics which bear on their mode of formation and subsequent depositional/diagenetic history. In both varieties the number of ooids with a peloidal nucleus greatly exceeds those with a detrital grain nucleus. Clay-rich ooids never have an ooid fragment as nucleus; such ooid-fragment nuclei are locally common in hematite ooids. Berthierine ooids, especially in 'concentrated' oolites, are generally strongly deformed, whereas the hematitie oolids are commonly spheroidal or flattened parallel to bedding. Radial syneresis cracks occur only in the hematite ooids.

These fundamental differences are clearly related to the physical properties of the constituent minerals, and of the ooids themselves. Clay-rich ooids would easily suffer plastic deformation under stress because of higher water content (both structural and interstitial), plasticity and cohesiveness of the constituent clays relative to the iron oxide minerals. On the other hand, the iron oxide minerals, especially hematite, would dewater quickly in the process of recrystallization from amorphous hydroxide precursors which would result in syneresis cracking and fragmentation. Moreover, crystalline iron oxide mineral aggregates are brittle compared with those of clay minerals. Consequently, ooids rich in iron oxide minerals would undergo brittle deformation, and those with syneresis cracks would fragment easily under stress. An ooid fragment formed in this way might subsequently form the nucleus of another ooid. In the 'concentrated' oolite beds the formation of laminae of the different kinds of ooids is probably controlled by the nature of ooids available in the source beds, and the intrinsic difference in hydrodynamic properties between the ooid types during their transportation and concentration in oolite shoals.

The sedimentological and textural characteristics of the Aswan ironstones provide an increased understanding of the processes responsible for the formation of these rocks, not only in Aswan but also elsewhere in the Phanerozoic sedimentary record. Apparently two conditions must be met before any ironstone can be deposited in a marginal marine basin. These are: (a) sufficient input of fresh water carrying iron and some kaolinitic clay, and (b) gradual starvation of the basin. Under these conditions the muddy ironstones form first on the prograding subtidal mudflats from precipitated ferruginous mud, detrital clay floccules and faecal pellets under the lowest-energy hydraulic regime, probably below the fair weather wave base. With gradual shoaling of the mudflat and concomitant increase in the wave and current energies the peloids and floccules mechanically accrete detrital kaolinitic clay and precipitated iron oxide particles forming ooids and proto-ooids. Kaolinite is common in coastal mixing zones because of its greater flocculation capability (Whitehouse *et al.* 1960; Doyle & Sparks 1980). The relative amount of detrital clay input in the system determines the formation of ooid types, either clay-rich or iron oxide-rich. But, commonly, gradual waning of the detrital clay input induces gradual evolution of the ooid types, from initial clay-rich ooids to final iron oxide ooids. Mixed ooids of alternating sheath composition probably form in response to seasonal variation in relative proportion of detrital clay supply. Additional evidence for such a mode of ooid formation from the resultant internal fabric of the ooids is given by Bhattacharyya & Kakimoto (1982).

After their formation on prograding, near-shore, subtidal ferruginous mudflats the ooids may be reworked to form thicker deposits of 'concentrated' oolites.

Concentrated oolitic ironstones have been documented elsewhere in the Phanerozoic record, for example the Devonian ironstones of Libya (Van Houten & Karasek 1981), the Jurassic ironstones of Luxembourg (Teyssen 1984) and the Frodingham ironstone of England (T.P. Fletcher 1987, personal communication), where winnowing into oolite shoals and bars occured under wave or current dominated conditions.

Locally the ooids may be transported into deeper parts of the basin during storm events, and be deposited there as a thin sheet-like bed (tempestite) such as those in the Clinton Group of eastern USA. Such tempestites are characterized by a basal scour, super- and sub-jacent basinal sediments, admixtures of bioclastic and/or siliciclastic debris, and hummocky cross-stratification.

Following the deposition of these ferruginous sediments early diagenetic remobilization of iron, probably by microbial reduction, organic decay or even by the introduction of transgressive anoxia from deeper oxygen minimum zone, initiates the process of *berthierinization* of the detrital kaolinite in the ooids, peloids and matrix by substitution of $Fe^{2+}$ for $Al^{3+}$ (Bhattacharyya 1983). Released $Al^{3+}$ and $Si^{4+}$ along with $Fe^{2+}$ in the pore solution locally precipitates berthierine cement. Excess silica in the pore solution probably precipitates opal as reported in some ironstones (Rohrlich 1974). Progressive diagenesis may also locally produce siderite and pyrite.

ACKNOWLEDGEMENTS: The study was supported by NSF Grants EAR77-06007 and 0IP75-07943, the latter administered through the University of South Carolina. An initial draft was critically reviewed by F.B. Van Houten and an anonymous reviewer to whom the author is greatly indebted.

## References

BEALL, A.O. Jr. 1968. Sedimentary processes operating along western Louisiana shoreline. *Journal of Sedimentary Petrology,* **38,** 869–877.

BHATTACHARYYA, D.P. 1980. *Sedimentology of the Late Cretaceous Nubia Formation at Aswan, southeast Egypt, and the origin of the associated ironstone.* Ph.D. thesis, Princeton University.

—— 1983. Origin of berthierine in ironstones. *Clays and Clay minerals,* **31,** 173–182.

—— & KAKIMOTO, P.K. 1982. Origin of ferriferous ooids: An SEM study of ironstone ooids and bauxite pisoids. *Journal of Sedimentary Petrology,* **52,** 849–857.

BOYLE, E.M., EDMONDS, J.M. & SHOLKOVITZ, E.R. 1977. The mechanism of iron removal in estuaries. *Geochimica et Cosmochimica Acta,* **41,** 1313—1324.

CRERAR, D.A., MEANS, J.L., YURETICH, R., BOROSIK, M., AMSTER, J., HASTINGS, D., KNOX, G., LYON, K. & QUIETT, R. 1980. Hydrogeochemistry of the New Jersey coastal plains II: Transport and deposition iron, carbon and selected elements. *Chemical Geology,* **33,** 23–44.

DOYLE, L.J. & SPARKS, T.N. 1980. Sediments of the Mississippi, Alabama and Florida (MAFLA) continental shelf. *Journal of Sedimentary Petrology,* **50,** 905–916.

GOLDRING, R. & BRIDGES, P. 1973. Sub-littoral sheet sandstones. *Journal of Sedimentary Petrology,* **43,** 736–747.

JAMES, H.E. & VAN HOUTEN, F.B. 1979. Miocene goethitic and chamositic oolites, northern Colombia. *Sedimentology,* **26,** 125–133.

PRYOR, W.A. & VANWIE, W.A. 1971. The 'Sawdust Sand'—An Eocene sediment of floccule origin. *Journal of Seimentary Petrology,* **41,** 763–769.

RHOHRLICH, V. 1974. Microstructures and microchemistry of iron oolites. *Mineralium Deposita,* **9,** 133–142.

SHOLKOVITZ, E.R. 1978. The flocculation of dissolved Fe, Mn, Al, Cu, Ni, Co and Cd during estuarine mixing. *Earth and Planetary Science Letters,* **41,** 77–86.

TEYSSEN, T.A.L. 1984. Sedimentology of the Minette oolitic ironstones of Luxembourg and Lorraine: a Jurassic subtidal sand wave complex. *Sedimentology,* **31,** 195–211.

VAN HOUTEN, F.B. & BHATTACHARYYA, D.P. 1982. Phanerozoic oolitc ironstones—geologic records and facies model. *Annual Review of Earth and Planetary Sciences,* **10,** 441–457.

——, —— & MANSOUR, S.E.I. 1984. Cretaceous Nubia Formation and correlative deposits, eastern Egypt: Major regressive-transgressive complex. *Geological Society of America Bulletin,* **95,** 397–405.
——& KARASEK, R.M. 1981. Sedimentologic framework of Late Devonian and oolitic iron formation, Shatti valley, west-central Libya. *Journal of Sedimentary Petrology,* **51,** 415–427.
WHITEHOUSE, R., JEFFREYS, L.M. & DEBBRECHT, J.D. 1960. Differential settling tendencies of clay minerals in saline water. *Proceedings of 7th Conference on Clays and Clay Minerals,* 1–79.

D.P. BHATTACHARYYA, Earth & Planetary Sciences Department, Washington University, St. Louis, MO 63130, USA.
Present address: 1518 Central Street, #2F Evanston, Ill 60201, USA.

# Stratigraphic and environmental patterns of ironstone deposits

**Ulf Bayer**

SUMMARY: Evaluation of the accumulation of Minette- and Clinton-type ironstones in terms of stratigraphic successions and regional facies patterns provides a sequence of types from condensed 'iron shot' muddy carbonates over barrier bars to distal tempestite deposits. A lagoonal origin of iron-oolites is derived and discussed. Biogenetic carbonate production and a biogenic contribution to iron-oolite formation, as well as replacement patterns of carbonate by iron-minerals indicate that (1) ironstone formation is related to environmental changes in the course of transgressive – regressive cycles and (2) formation of iron-ooids occured in environments with short term fluctuations between oxidizing and reducing conditions. Adopting this dynamic viewpoint allows dissolution and precipitation of iron 'side-by-side' without contradicting classical models.

Although Phanerozoic ironstones are of low economical value today, they are still fascinating objects for geochemical, petrographical, sedimentological and stratigraphical work. Easily recognizable by their commonly bright red colour in the field, they provide remarkable stratigraphic markers and frequently yield a rich fauna for the palaeontologist and stratigrapher in otherwise monotonous sequences. In addition, they generally mark a drastic temporal change in the sedimentological regime, and they are frequently indicative of a local diastem or, at least, a condensed sequence (Bitterli 1979; Gygi 1981; Embry 1982; Van Houten & Purucker 1984; Bayer *et al.* 1985). In palaeontological terms, faunal breaks and replacements occur sometimes in association with ironstones (Berman 1963; Bayer & McGhee 1984, 1985). In addition, the well recognized position of many ironstones at the top of coarsening upward cycles provides an interesting relationship to sea-level fluctuations (Aldinger 1957; Van Houten & Karasek 1981; Van Houten & Purucker 1984; Bayer & McGhee 1984, 1985).

Although ironstones are a common facies type in certain intervals of earth history (Van Houten & Bhattacharyya 1982), the genesis of these rocks is still poorly understood. In many ways ironstones show similarities with carbonate deposits, both on a microscopic and on a macroscopic scale. Ferruginization models of carbonates, therefore, have a long tradition for the explanation of ironstone formation. Replacement models have been referred either to early diagenesis (Alling 1947; Bayer *et al.* 1985) or to late diagenetic ferruginization of carbonate ooliths (Kimberley 1979). The arguments, however, are not necessarily conclusive, as the discussion of Kimberley's model showed (e.g., Bradshaw *et al.* 1980). Nevertheless, close relationships between carbonate deposits and ironstones sometimes occur on a basinwide scale (Hunter 1970). In addition, many of the iron-oolite bearing beds are calcarenites by definition. The similarity between ironstones and carbonates have been recognized in various regional studies, and the formation area of iron-oolites has been reconstructed in terms of lagoonal environments which closely resemble the environments of carbonate-oolite formation (Sheldon 1965, 1970; Dimroth, 1975, 1979; Bayer & McGhee, 1985).

The 'lagoonal model' is of special interest because it integrates a wide variety of observations made on ironstones. The main purpose, therefore, will be to test this model in terms of accumulation patterns, the relation of ironstones to diastems, and their relation to carbonates. This will be carried out not by means of a detailed local study, but by considering the temporal and regional accumulation patterns of ironstones on larger regional scales. Finally, an attempt is made to explain the value of a 'lagoonal formation centre'.

## The Clinton group – a general model of ironstone occurrence?

The Silurian Clinton Group certainly provides a 'model study' of ironstone occurrences because of the extensive research which has been carried out since the middle of the last century. Although the lithologically very similar Minette may be more famous, it is only the Clinton Group

*From* YOUNG, T. P. & TAYLOR, W. E. G. (eds), 1989, *Phanerozoic Ironstones*
Geological Society Special Publication No. 46, pp. 105-117

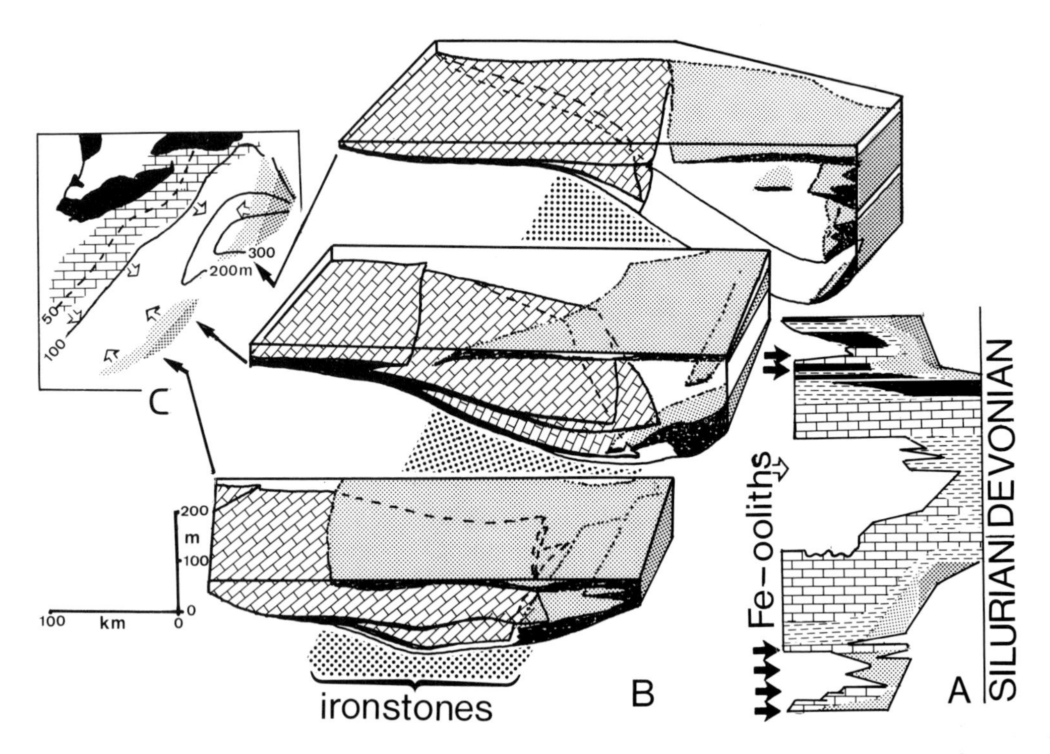


FIG. 1: Ironstone occurrence in the Middle Appalachian basin. A: Chronostratigraphic chart (Silurian-Middle Devonian) based on Rickard (1975). Ironstone occurrences indicated by arrows; symbols, as usually understood, black shales = black. B: Three-dimensional exploded view of sediment accumulation during the Clinton Group; data from Hunter (1975, 1970). The ironstones predominantly accumulated in the temporal interaction zone of carbonate tongues and sandstone tongues, as indicated. Arrows indicate the approximate regional location of cross-sections in Fig. 1C. C: Major facies distribution during Clinton-time modified after Hunter (1960, 1970). The Great Lakes (black) are given for scale. Open arrows indicate the general direction of sediment transport, as discussed by Hunter (1970). Isolines refer to sediment thickness.

where studies on a basin-wide basis are readily available in monographic papers like Gillette (1947) and Hunter (1960, 1970). An excellent summary of the stratigraphy of the New York state area has been given by Rickard (1975), and there are numerous studies on geochemical, petrographical, sedimentological and stratigraphical problems. It will not be the purpose here to discuss the literature as has been excellently done by Hunter (1970). Instead, the Clinton will serve to formulate some concepts by use of a subjective, but intentional selection from the available literature.

The general patterns of sediment accumulation during Clinton time, as it has been worked out by Hunter (1960, 1970) and Rickard (1975) are summarized as follows (see Fig. 1).

(1) On a temporal Scale (Fig. 1A), the ironstones are not randomly distributed, but they are situated within diastems which are in part erosional and in other parts non-depositional. This is not only true for the Silurian; the rare ironstone occurrences of the Devonian seem to occupy similar positions. Chamosite-oolites are known from the Mahantango Formation of central Pennsylvania, topping coarsening upward cycles (e.g., Ellison 1974). Significantly the diastems associated with the ironstones provide faunal breaks, indirectly evident from Rickard's correlation chart, and definitely stated by Berman (1963).

(2) The ironstones are related to prograding shallow water sediments (Fig. 1B). Hunter's analysis shows that the ironstones accumulated along the 'hinge lines' of the basin, as indicated. Sheldon (1965, 1970) proposed a model for the Birmingham (Alabama) ironstones in which they originated in a lagoonal environment. Folk (1960) and Gleich & Warhauser (1981) concluded that the Keefer Formation and other prograda-

tional sandstones represent barrier bar sequences with associated lagoonal interbeds. The ironstones, therefore, may have been related to an extremely flat relief of the basin. Nickelsen & Cotter (1983, p.24) disagree with the barrier bar model, however, they described in much detail a profile at Danville (Pennsylvania) and pointed out that the iron-bearing beds are storm generated tempestites (Nickelsen & Cotter 1983, pp. 99–109). In addition, Nickelsen & Cotter (1983, p. 16) proposed an environmental scenario for the Middle Silurian with oolite banks at the western marginal carbonate facies and megarippled sand shoals at the eastern marginal sand facies which provides sufficient similarities with Folk's (1960) concept. Chowns & McKinney (1980) and Bearce (1973) pointed out that the Ordovician and Silurian 'Birmingham' ironstone units commonly contain intraformational ironstone clasts which apparently were derived in place by reworking of the ironstones. Such reworked material can be found elsewhere in the Clinton and indicates that, at least, most ironstones are secondary deposits.

(3) The general setting (Fig. 1C) shows an eastern source of clastic sediments and a western carbonate environment separated by a zone of distal clay deposition. The ironstones cannot be related to either of these major facies regimes, but occur, as Hunter points out, in the interaction zone between them. However, eastward and westward derived ooliths can be distinguished to some extent by their cores, which are either predominant quartz grains or biogenic fragments (Lindgren 1933). The oolitic ironstones are closely associated with calcarenites, even in the sandstone facies (Hunter 1970). Lindgren (1933) and Alling (1947) emphasized a relationship between iron-oolites and associated fossil fragments. Alling (1947) even concluded that the 'ores were originally limestones which were replaced by iron compounds' and that 'the "fossil" ores are the result of the partial replacement of the carbonate of fossils and the subsequent oolitic coatings'. At least, there is wide agreement that the iron-oolites formed in a shallow marine environment and that oolitic hematite and chamosite (or their precursors) are primary minerals forming near the site of deposition (Hudson & Maynard 1986).

The Clinton ironstones, thus, provides us with a set of observations and concepts which may be provocative in parts, due to the choice of references and the way they have been put together in a context. However, using this scenario, it could be of interest to test the extent to which these concepts may be valid. As James (1966) stated: 'The lack of modern-day examples of iron-stones and iron-formation deposition makes the subject of origin a fertile one for speculation'. Although it is difficult to go the consequent 'mathematical' way of axioms and conclusions when all detailed local observations provide some differences and alternative interpretations, on a sufficiently general level one may test models, which here are given in terms of the special example.

## Stratigraphic patterns

The frequent association of ironstones with shoaling upward sequences has long been recognized (e.g. Hemingway 1951; Aldinger 1957, 1965; Van Houten & Bhattacharyya 1982) and has been reviewed by Van Houten & Purucker (1984). Transgressive–regressive cycles associated with prograding shorelines may locally induce coarsening upward sequences. In such situations the ironstones most commonly are located at the top of the coarsening upward cycle. Examples have been provided by Van Houten & Karasek (1980) from the Devonian of Lybia, Embry (1982) from the Liassic of the Sverdrup Basin (Canadian Arctic Archipelago), Bayer & McGhee (1985) and Bayer *et al.* (1985) from the South German Aalenian. There is, however, some difference in the interpretation of the stratigraphic position of ironstones. Embry placed the ironstones in the basal transgressive portion of each cycle, while Bayer *et al.* (1985) considered them the top of the shoaling cycle, following the argument of Hallam & Bradshaw (1979) that the ironstones represent the highest energy level. Ironstones are usually associated with erosional events, indicated by reworked pebbles and erosional (omission) surfaces, but in addition, they may be underlain and/or overlain by a diastem, allowing different interpretations.

On the other hand, starvation of sediments is another important property of many ironstones. The associated fossil debris frequently shows intensive destruction by microboring organisms (Schloz 1972; Bayer *et al.* 1985), and hardgrounds associated with ironstones gave rise to the build-up of foraminiferal–serpulid bioherms, perhaps in interaction with blue-green algae (Gatrall *et al.* 1972; Bitterli 1979; Bayer *et al.* 1985). Development of iron-oolites in connection with condensed sequences and hardgrounds has been considered by Geyer & Hinkelbein (1971), Bitterli (1979), and Fürsich (1979). These patterns apply especially to thin, but regionally widespread ironstones (McGhee & Bayer 1985) which, in some cases, may be related to the condensed sections of the seismic stratigraphic chart proposed by Vail *et al.* (1984). In addition, ironstones are not obviously causally related to major movements of the strand-line (Berman 1963), although they occur at the top of shoaling upward sequences. During diastems they spread widely over basin areas and then may even affect the ecological system within the basin (Berman 1963; Bayer & McGhee 1984).

On a regional level, the depositional environments of ironstones may appear more complicated. From a detailed study of the South German Aalenian deposits it is possible to trace the ironstone facies from a condensed and repeatedly reworked nearshore environment over a zone of iron-oolite accumulation into a distal area where the ironstones were sedimented as tempestites (Bayer & McGhee 1985; Bayer *et al.* 1985). This facies sequence has been found in a very similar way by Bitterli (1979) studying Oxfordian strata and by Dreesen (1982) in the Famennian of Belgium. These observations indicate that winnowing is an essential process in iron-oolite formation, rather than starvation alone. The regional pattern, in general, is associated with a size decrease in the iron-ooids providing larger sizes up to 'oncolites' nearshore and smaller ooid sizes in the more distal deposits. Especially the large shallow water ooids commonly show repeated overgrowth by sessile formaminifera, indicating alternating periods of reworking and aggregation of oolitic layers and of stable periods which allowed the foraminifera to settle and to grow.

Considering starvation as an interval of reduced detrital input, iron may be enriched by precipitation from the sea-water or the porewater or may be dissolved and redistributed from detrital grains. Winnowing allows alternating periods of mud covering and reworking and, therefore, provides potentially a regime of alternating reducing and oxidizing conditions which has been postulated for ironstones containing ooids with alternating layers of berthierine/chamosite and goethite or for deposits with alternating horizons of chamositic and goethitic oolites (e.g., Finkenwirth & Simon 1969). The two processes, however, are not exclusive; they may occur in a temporal succession at the same locality or at the same time at different regions within a basin, as proposed by models of Bitterli (1979) and Bayer & McGhee (1985) which propose shallow water reworking of iron-ooids and distal redeposition as tempestites.

Some information about the relationship between argillaceous sediments and ironstones can be gathered from the Oxfordian occurrences of Europe. Figure 2 illustrates the occurrence of an ironstone at the transition from the Oxford Clay to the carbonate oolites of the Coral Rag in northern France. The ooids of the basal ironstone are limonitic, however, they show repeated foraminiferal overgrowth and ooid patterns identical to those found in the calcereous Coral Rag. An alternating sequence of clay and interbedded carbonate layers is developed above the ironstone. Within these layers the ferruginization decreases upward whereby reworked pebbles show the more intensive ferruginization. Transforming this vertical sequence into a lateral facies pattern would provide us with a model with ooids derived from a carbonate platform. However, the shoals were deposited in a muddy environment, where they may have been covered by clay, and where they may have been reworked repeatedly as is indicated by the intraformational pebbles. The more distal the deposits are, the more frequently they have been redeposited. In this process, ferruginization increases, indicating an early diagenetic replacement of aragonite as postulated by Alling (1947) and considered possible by Bayer *et al.* (1985).

Ironstones in the North German Corallian (Oxfordian) are also associated with carbonate ooids (Simon 1969) and they accumulated at transitions between the carbonates and clastic sequences, both stratigraphically and regionally. In addition, the iron oolites frequently are related to unconformities, as indicated in Fig.3 (Simon 1969). These unconformities sometimes show traces of lateritic weathering. Laterites, therefore, have been considered the source for the iron oolites. However, at least for the profiles in Fig. 3 it is obvious, that the ironstones are closely related to clastic tongues invading the carbonate regime. The ironstones' relationship to laterites, therefore, is at least obscured. Simon (1969) pointed out that the ironstone bearing sequences of the Corallian are cyclic, revealing a pattern as in Fig. 2, i.e. a sequence: clastic - iron oolite — carbonate oolite. Even for this carbonate facies, some relation between ironooliths and clastic/argillaceous input seems to be suggestive.

Although these examples from the Corallian may not be general ones, they are not unique. At least on a regional scale a similar situation holds for the Northampton ironstone which accumulated between a carbonate platform in the south-

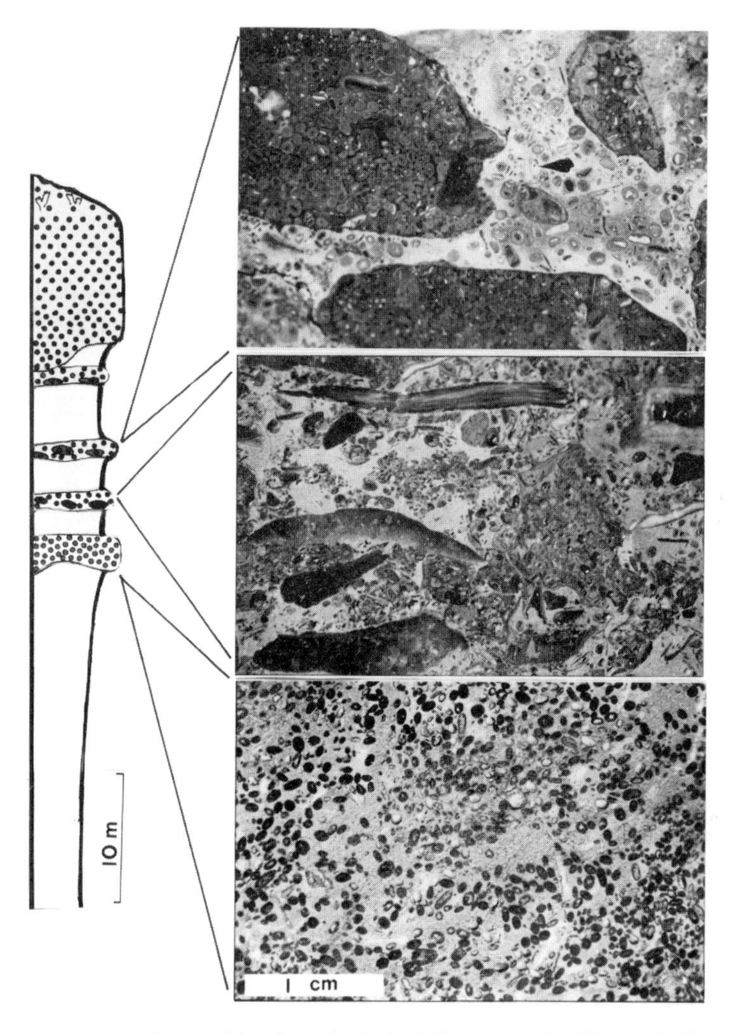


FIG. 2: Ironstone occurrence at the transition from the Oxford Clay to the Corallian at the Vaches-Noires (Villers-sur-Mer, northern France). The first calcareous bed provides an ironstone, while ferruginization decreases in the succeeding 'event' beds. However, reworked pebbles in these beds still show ferruginization of different degrees.

west (Cotswolds) and clastic sedimentation in the north (Yorkshire) revealing the regional pattern of the Appalachian Clinton deposits to some extent.

## Regional patterns

In middle Europe, five temporal intervals of major iron-ore accumulations occur in the Jurassic (McGhee & Bayer 1985): in the Sinemurian, Pliensbachian, Aalenian, Callovian and Oxfordian. The Oxfordian ones have already been discussed, including regional information. The regional patterns of iron-oolite accumulation in the Sinemurian and Aalenian are illustrated in Fig. 4. The facies distribution in these examples is almost identical, although they are temporally separated by 30 million years. The Aalenian includes the famous Minette type in Lorraine and the large iron-ore deposits reveal the lithology of the Minette type, although the Minette is distinguished slightly by its geographical position and its age. The Minette is located at an embayment and accumulated from the Late Toarcian to the Middle Aalenian.

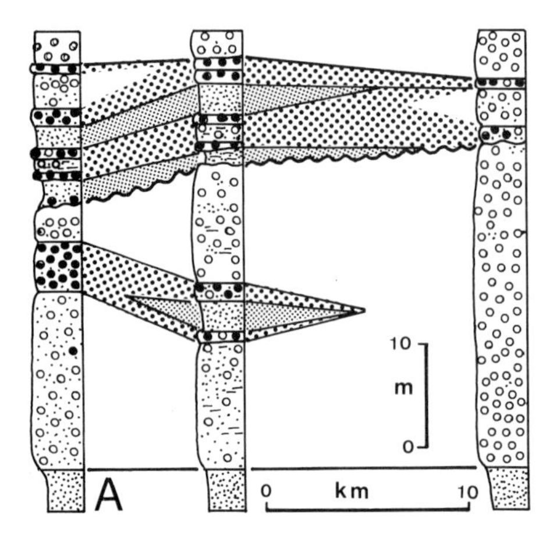


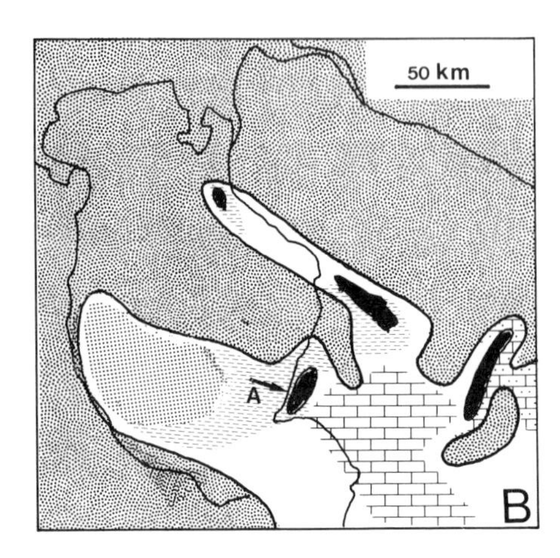


FIG. 3: Stratigraphic (left) and regional (right) occurrence of ironstones (black circles and patches) in the oolitic carbonates of the North German Corallian (modified after Simon, 1969). In the map the location of the stratigraphic profiles is indicated by 'A'. The iron-oolites are stratigraphically related to erosional surfaces (wavy line) and to clastic tongues extending into the carbonate ooliths (white circles). Regionally, the ironstones occur at the transition of argillaceous sediments and carbonates within a large embayment.

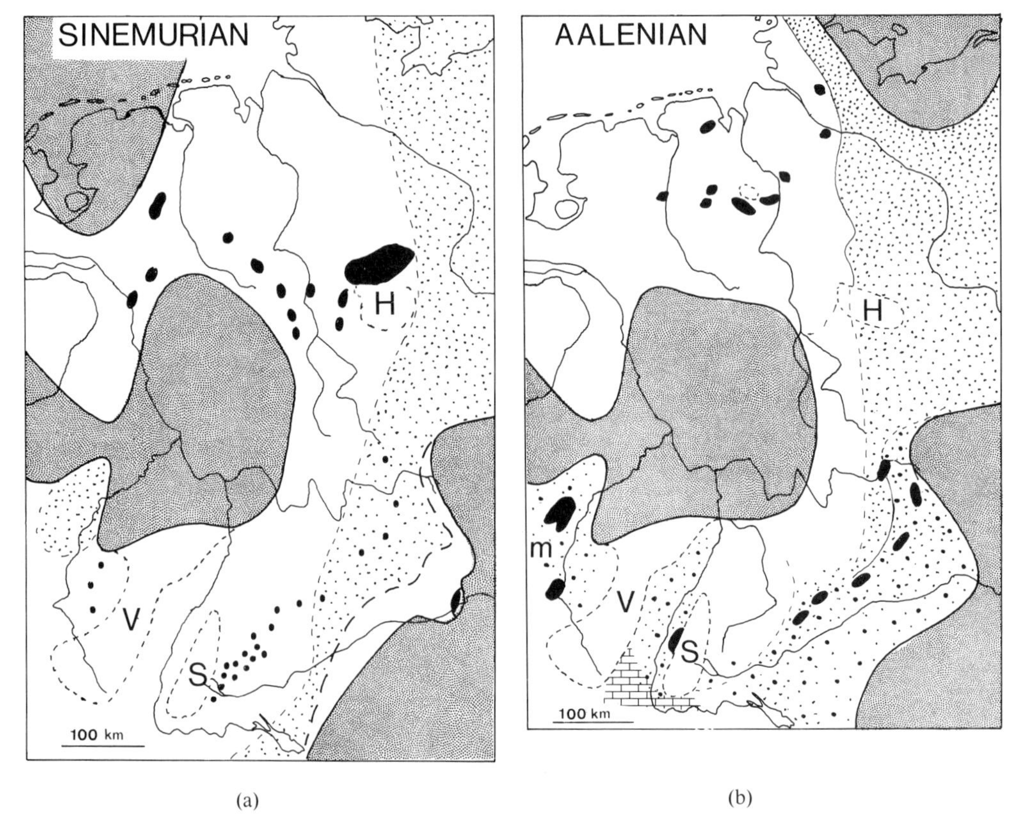


FIG. 4: Paleogeographic occurrence of ironstones in the Sinemurian and Aalenian of Middle Europe (ironstones = black patches; symbols, as usually understood). Land areas are shadowed, palaeohighs are indicated, H: Harz, S: Black Forest, V: Vosges. m: Minette. Maps after Aldinger (1968), Bayer & McGhee (1985), Hoffman (1969), Ziegler (1978) and other sources.

The Sinemurian and Aalenian paleogeographic pattern is different from the Upper Jurassic ones (Figs 2 and 3). The sediments are almost exclusively argillaceous, grading eastward and close to islands into sandy deposits. The iron ores, however, are not randomly distributed. In the Sinemurian they mainly accumulated along the northern flank of the Rhenic Island where several of these occurrences accumulated in local traps which were provided by active faults (Thienhaus 1969). Exceptions are the iron ores which accumulated near the Harz which formed a local paleohigh, perhaps an island without an associated larger landmass. These occurrences will be discussed below in more detail.

Iron ores did not accumulate in South Germany in Sinemurian time. However, iron-shot (iron-oolite bearing) condensed carbonates developed which regionally are widely distributed and which provide an example for a major break in the faunal sequence (Bayer & McGhee 1985). This South German iron-shot bed developed at the temporal transition from a clastic sequence into carbonates. The ironstones did not accumulate in the eastern sandy facies, but in the argillaceous facies, as in the Clinton pattern.

In the Aalenian, the pattern changes slightly. In South Germany, the iron ores accumulated in bars paralleling the coast line. The detailed structure of these bars has been studied by Wild (1951) while there wre still active workings. Additional studies (Werner 1959) indicate a longshore transport of these bars, and both studies showed that they are not related to special traps, but can be considered as true bar deposits. Lucius (1949) and Teyssen (1983) showed that the Minette at its type locality also provides an example of such a bar deposit.

Again, there are exceptions. The North German iron ores are not related to a landmass. They surround local palaeohighs produced by active salt domes at that time (Gruss & Thienhaus 1969). The accumulations are of bar type; however, they were favoured by the synsedimentary tectonics which also may have provided traps as in the Sinemurian example.

Of special interest are the 'isolated' occurrences of iron-oolites which are well separated from a larger landmass which could supply them with iron-clasts of lateritic origin as has been proposed by Siehl & Thein (1978, 1986). The Liassic occurrences near the Harz have been studied in much detail (e.g., Hoffmann 1969). Fig. 5 summarizes the patterns in the vicinity of this palaeohigh which in the Sinemurian and the Pliesnbachian was surrounded by argillaceous and clastic sediments. Close to the island, we find an area with a condensed sequence consisting mainly of iron-oolitic marls and limestones. The major iron ores reaching several metres in thickness, however, accumulated at the margin of this local platform, either in syntectonical traps or in bars. In the Pliensbachian we even find a lateral transition into bioclastic carbonates. The platform itself provides an example of starved sedimentation, there was little clastic input and sediment build-up was dominated by biogenetic production and chemical precipitation; however, a carbonate platform did not evolve. Ironshot limestones in a similar environment were described by Gatrall *et al.* (1972). They studied the distribution of ferruginized 'snuff boxes' and associated iron-ooids which occur on a local high in Dorset. The *discites* time-interval in particular provides an example of oolites distributed from a local 'snuff box' ridge. The 'snuff boxes' are nodules with concentric layers which have been built-up by blue-green algae, sessile foraminifera and serpulids. The 'snuff boxes', therefore, resemble

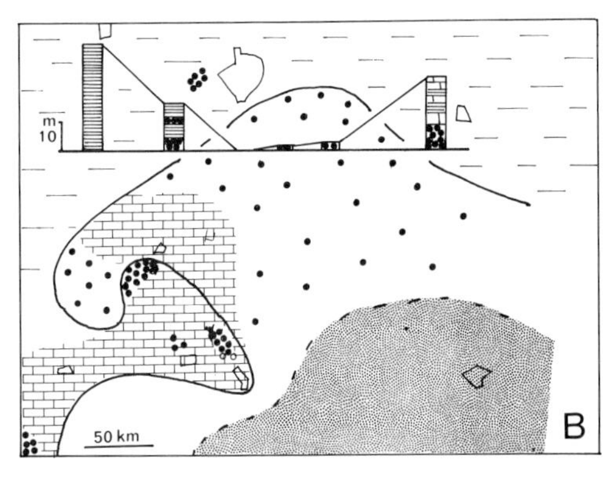


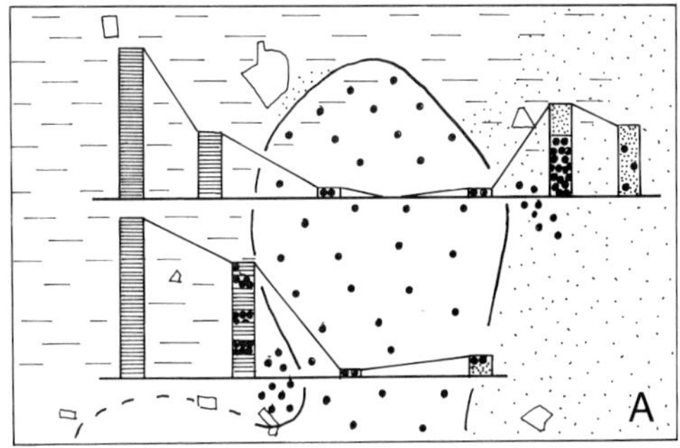


FIG. 5: Palaeogeographic setting near the Harz in Sinemurian (a) and Pliensbachian (B) times. The profiles provide approximate cross-sections over the paleohigh. Iron-oolites: black circles; land area (?) shadowed, other symbols, as usually understood. Modified after Hoffman (1969).

the biogenetic crusts observed by Bitterli (1979) in association with the erosional facies of ironstones. Similar incrustations and nodules have been found by Bayer *et al.* (1985) in the Aalenian ironstone facies and in the carbonate facies of the lower Muschelkalk, and they interpreted these biogenetic structures as evidence for lagoonal environments, mainly with regard to the probably associated blue-green algae. A rather similar lithology occurs in the Middle Jurassic Bajeux facies and a variety of Jurassic hardgrounds containing iron-ooids (Fürsich 1979).

The 'lagoonal model' proposed by Sheldon (1965, 1970) for the Clinton, thus, may well be extended to Jurassic ironstone accumulations, assuming a lagoonal environment located on the palaeohighs or along the coast lines, and secondary accumulation of ooids in barrier bars from which they have been distributed into deeper parts of the basin by storms. The type locality of the Minette may even serve as the prototype of such a model. Teyssen (1983, 1984) clearly illustrated the bar-like nature of the ironstones, paleogeographically, those are located in front of a large embayment from which all the sediments have been eroded today (Bayer & McGhee 1985). Certainly the Oxfordian example from North Germany (Fig.3) provides a typical 'lagoonal' setting, and it indicates that the relation of ironstones to a particular lithology — calcerous or sandy — may be less important than regionally or temporally varying depositional conditions.

## Biogenic contribution to oolite formation

Iron oolites of some major deposits like the Aalenian ones are known frequently to have quartz grains as cores. However, as soon as biogenetic detritus becomes a major constitution of the sediments, the cores of the ooids will be shell debris. Ferruginization of these particles will first occur in vacuoles, as they are primarily available in bryozoa and echinoderms (Alling 1947). Cavities may also be provided by microboring organisms and then will ferruginize. The iron then may increasingly replace the carbonate and coat the grain (Fig. 6a). Such grains can form an essential compound of ironstones and are indicative of slow sedimentation rates, perhaps under repeated reworking.

Laminar thickening of the coating produces ooids which in some environments may grow to oncolite size and shape, even composite iron-oolites are common in shallow water environments. Early diagenetic shrinkage and breaking of limonitic oolites has been observed (Bayer *et al.* 1985), and the fragments may serve as cores after reworking. Larger ooids frequently have repeated overgrowth by sessile foraminifera (Fig. 6b) showing that these ooids were formed under marine or, at least, brackish conditions. The foraminifera also contribute to the growth of the ooid. The calcareous cores and the enclosed foraminiferal overgrowth are frequently ferruginized. The original extent of the shell fragment, as well as the foraminifera, is only observable in polarized incident (reflected) light (Fig. 6b). Comparison with the picture taken under transmitted light shows the extent of ferruginization. Careful analysis of the incident light picture reveals a lighter zone at the margin which resembles less ferruginized carbonate oolites (cf. Fig. 2).

An example of alternation of berthierine and true carbonate layers including sessile foraminifera is shown in Fig. 7a. Berthierine, however, may also replace carbonate (Fig. 7b). These oolites are certainly indicative for changing environmental conditions, but an early diagenetic origin of the berthierine cannot be excluded.

The examples given in Figs 6 and 7 demonstrate that biogenetic activity was to some extent involved in the formation process of oncolite type iron-ooids. In an analogy with the previously discussed 'snuff boxes', it seems reasonable to consider that blue-green algae could have been involved; alternatively, bacteria could have contributed to the growth of calcareous and ferruginous layers (Chowns 1986). These possibilities should be explored by more detailed analyses. The classical iron-oolites, however, are not suited to such studies because they are secondary accumulations and usually well sorted by winnowing. Studies on the origin of iron-oolites should be carried out at the place of their genesis, which here is considered at the condensed proximal deposits yielding the largest, oncolite size, ooids.

## Conclusions

Evidence for a 'lagoonal origin' of iron-oolites has been derived from well known examples. There seems to be little doubt that many iron-oolites accumulated in bars and their regional position suggests these are barrier bars, although the proximal facies commonly is unknown. Some examples, however, indicate that this is typically a condensed iron-shot mudrock which

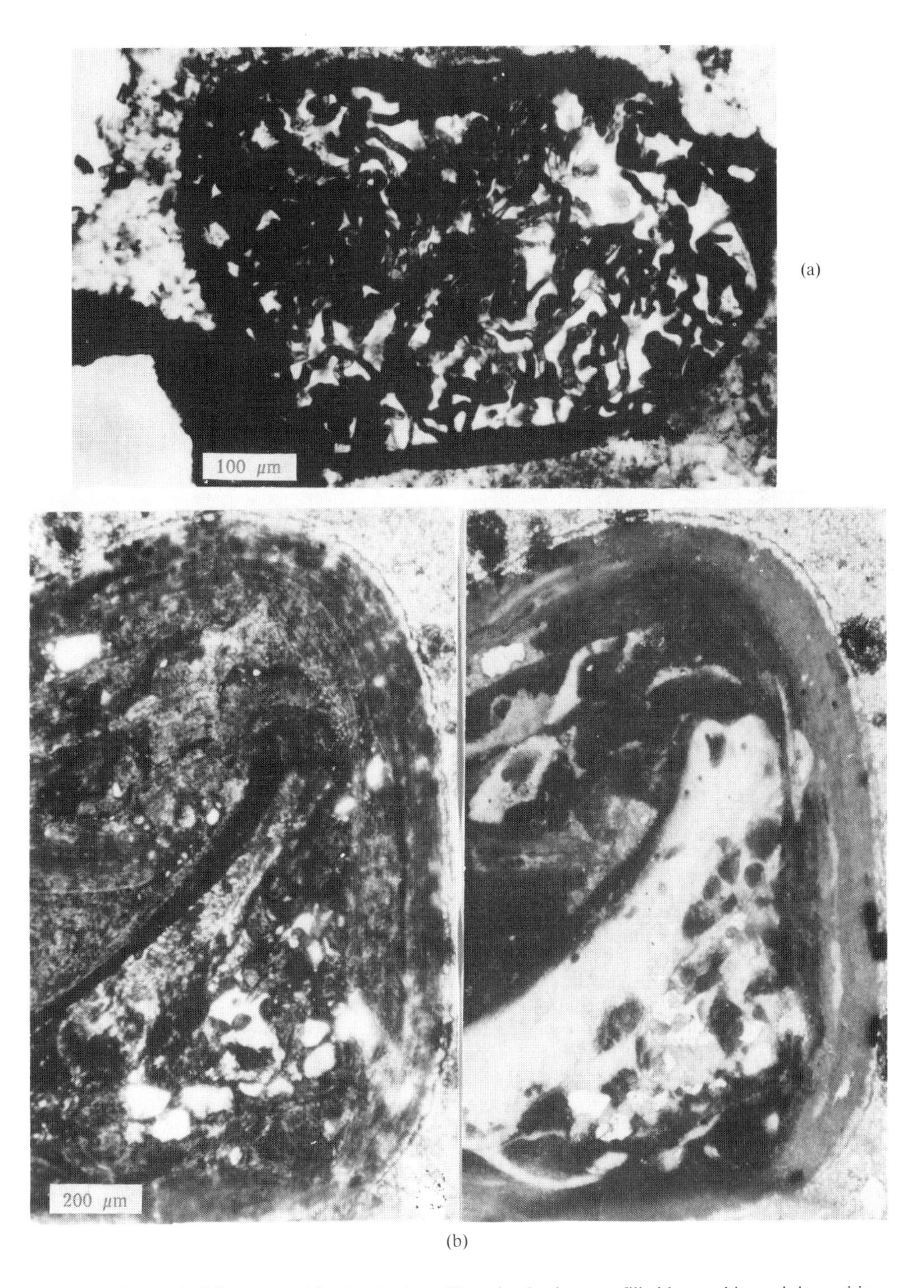


FIG. 6: (a) Coated shell fragment with microborings. The microborings are filled by goethite and the cavities partially have been extended during ferruginization. Transmitted light. Aalenian (*concava*-Zone), Achdorf, South Germany. (b) Ferruginization of a shell fragment and of overgrown foraminifera in an ooid from the ironstone at the Vaches-Noires (cf. Fig. 2). Left: transmitted light, right: incident light.

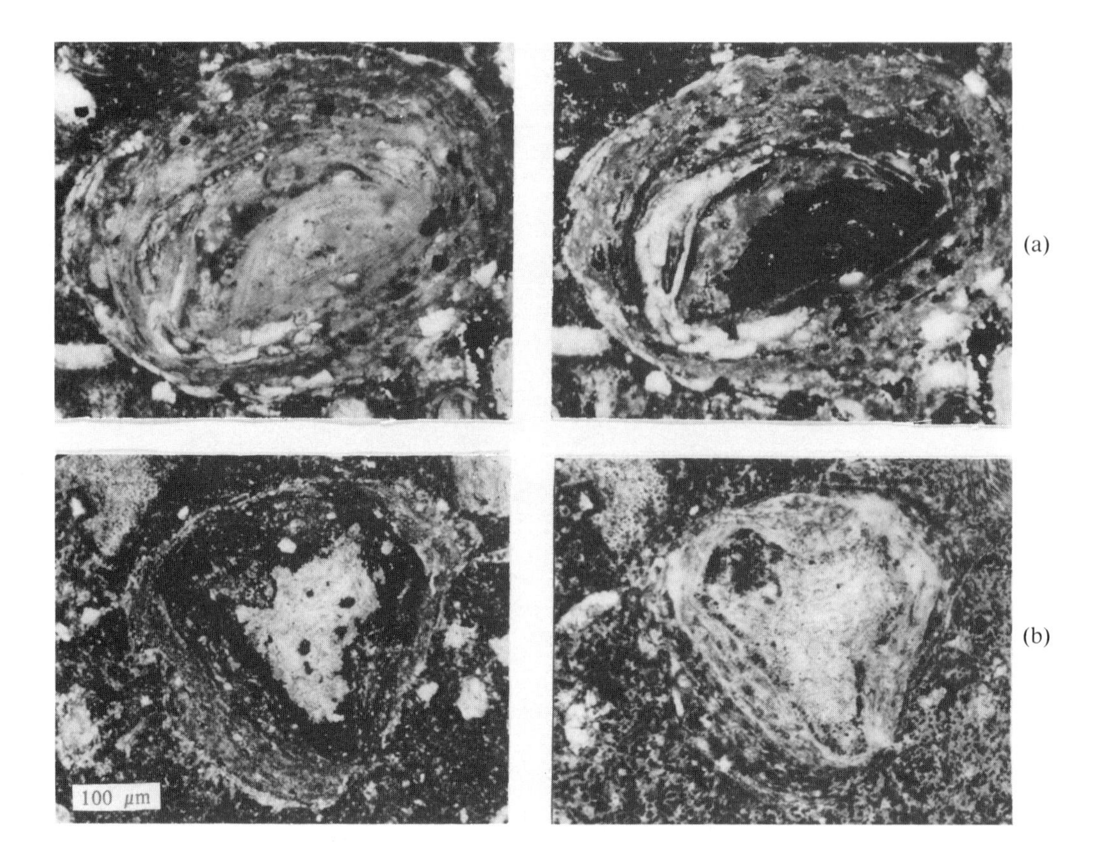


FIG. 7: (a) Ooid with alternating layers of carbonate and berthierine (black under crossed nicols, right). Transmitted light, right: crossed nicols. Aalenian (*murchinsonae*-Zone), Achdorf, South Germany. (b) Replacement of carbonate by berthierine in an ooid. Transmitted light, right: crossed nicols. Aalenian (*murchinsonae*-Zone), Achdorf, South Germany.

frequently has been reworked, and which may yield biogenetic structures like 'snuff boxes' or incrustations. Sedimentation of ironstones distally of the bars can usually be attributed to storm-induced event beds. In this environmental sense, the Clinton may serve as a general model for a variety of marine ironstones. There are, of course, other deposits like 'Trümmererze' and lateritic ores and there are many ironstones which have mostly pellet cored ooids and very little fossil debris (Van Houten, personal communication). These ironstones, therefore, may be well distinguished from the Clinton type ironstone with its many coated grains, abundant fossil 'hash', and quartz cored ooids. However, the discussed Jurassic iron-oolites feature these properties.

The lagoonal model is in good agreement with the observations made by Van Houten (1986) and Van Houten & Bhattacharyya (1982). They found that ironstones predominantly accumulated at stable cratons through times of global sea-level highstand and globally warm and humid climate. This framework provides low reliefs and widely extended shallow epeirogenic seas which favour the development of extended shallow bays and lagoonal environments.

The lagoonal model, as stated by Sheldon (1970), even permits integration of rather different concepts of iron-oolite formation. Within a short geological time-interval such a lagoonal environment may vary from saline to brackish conditions, it may switch between oxidizing and reducing conditions, it may be a 'carbonate environment' for some time after the mud has been swept out, and the carbonate material may repeatedly be covered by a mud blanket. Even periods of exposure and weathering are possible; a flat relief requires only a small sea-level fall for wide areas to become emergent. Oxidizing and reducing conditions are easily exchanged, temporally and regionally, and the barrier bars provide the possibility for excellent grain-size sorting, as found in many iron-oolites. The iron may well derive from weathering on land, but we also can explain the accumulation of ironstones near small palaeohighs and islands. Borchert's (1960) well known facies model still is applicable, not on a regional scale but on a temporal scale of successive states the lagoonal environment may assume. As the relation of ironstones to diastems implies, a dynamical view of this facies type may be more valuable than the classical static view of previous facies models.

ACKNOWLEDGEMENTS: The studies on ironstones originated at, and were financially supported by, the SFB 53 at Tubingen. A Heisenberg grant from the DFG then allowed continue studies as research scientist at the Universities of Leicester, Birmingham, Rutgers (New Brunswick) and Princeton. I thank J. Hudson, A. Hallam, G.R. McGhee and F.B. Van Houten for these opportunities and for valuable discussions. The manuscript benefited from the valuable criticism of unknown reviewers and extensive comments and corrections by F.B. Van Houten.

## References

ALDINGER, H. 1957. Eisenoolithbildung und rhythmische Schichtung im suddeutschen Jura. *Geologisches Jahrbuch,* **74,** 87–96.

——1965. Uber den Einfluss bon Meeresspiegel-schwankungen auf Flachwassersedimente im Schwabischen Jura. *Tschermaks mineralogische und petrographische Mitteilungen 10,* **61,** 167–182.

ALLING, H.L. 1947. diagenesis of the Clinton hematite ores in New York. *Bulletin of the Gelogical Society of America,* **58,** 991–1018.

BAYER, U., ALTHEINER, E. & DEUTSCHLE, W. 1985. Environmental evolution in shallow epicontinental seas: Sedimentary cycles and bed formation. *In:* BAYER, U. & SEILACHER, A. (eds) *Sedimentary and Evolutionary Cycles. Lecture Notes in Earth Sciences No. 1,* Springer, Heidelberg, 347–381.

——& MCGHEE, G.R. 1984. Iterative evolution of Middle Jurassic ammonite faunas. *Lethaia,* **17,** 1–16.

——&——1985. Evolution in marginal epicontinental basins: The role of phylogenetic and ecological factors. *In:* BAYER, U. & SEILACHER, A. (eds) *Sedimentary and Evolutionary Cycles. Lecture Notes in Earth Sciences No. 1,* Springer, Heidelberg, 164–220.

BEARCE, D.N. 1973. Biostratigraphy and facies relations in the lower Middle Ordovician of easternmost Tennessee. *American Journal of Science,* **273A,** 261–293.

BERMAN, B.L. 1963. Hematite facies in the Silurian of the Appalachian basin, (abs.) *Special Paper 73, of the Geological Society of America,* 114–115.

BITTERLI, P.H. 1979. Cyclic sedimentation in the upper Bathonian—Callovian of the Swiss Jura Mountains. *Symposium Sedimentation jurassique W. European. A.S.F. Publication,* **1,** 99–106.

BORCHERT, H. 1960. Genesis of marine sedimentary iron ore. *Bulletin of the Institution of Mining and Metallurgy,* **640,** 261–279.

BRADSHAW, M.J. *et al.* 1980. Origin of oolitic ironstones—discussion. *Journal of Sedimentary Petrology,* **50,** 295–304.

CHOWNS, T.M. 1986. Petrology of Lower Jurassic Chamosite-Siderite Ironstones from Northeast England (Abs.). *Bulletin of the American Association of Petroleum Geologists,* **70,** p. 573.

——& MCKINNEY, F.K. 1980. Depositional facies in middle-upper Ordovician and Silurian rocks of Alabama and Georgia. *Geological Society of America 1980 Annual Meeting, Atlanta, Georgia. Excursion Southeast Geology,* **2,** 323–348.

DIMROTH, E. 1975. Depositional environment of the iron-rich sedimentary rocks. *Geologische Rundschau,* **64,** 751–767.

——1979. Models of physical sedimentation of iron formations. *In*: WALKER R.G. (ed) *Facies Models.* Geoscience Canada, Reprint Series 1, 175–182

DRESSEN, R. 1982. Storm-generated oolitic ironstones of the Famennian (Falb–Fa2a) in the Vesdre and Dinant synclinoria (Upper Devonian, Belgium). *Annales de Societé geologique de la Belgique,* **105,** 105–129

ELLISON, R.L. 1974. Stratigraphy and palaeontology of the Mahantango formation in south-central. *General Geology Report 48 of the Pennsylvanian Geological Survey.*

EMBRY, A.F. 1982. The Upper Triassic-Lower Jurassic Heiberg deltaic complex of the Sverdrup basin. *In*: EMBRY A.F. & BALKWILL, H.R. (eds) *Arctic Geology and Geophysics. Memoir 8 of Canadian Society of Petroleum Geology,* 189–215.

FOLK, R.L. 1960. Petrography and origin of the Tuscorarra, Rose Hill, and Kiefer Formations, lower and middle Silurian of eastern West Virginia. *Journal of Sedimentary Petrology,* **30,** 1–58.

FINKENWIRTH, A. & SIMON, P. 1969. Das Eisenerzlager des Lias gamma der Grube Echte. *In*: *Sammelwork Deutsche Eisenerziagerstatten. Beiheft Geologiches Jahrbuch,* **79,** 59–84.

FURSICH, F. 1979. Genesis, environments, and ecology of Jurassic hardgrounds. *Neues Jahrubch fur Geologie,* **158,** 1–63.

GATRALL, M. *et al.* 1972. Limonitic concretions from the European Jurassic, with particular reference to the 'snuff boxes' of southern England. *Sedimentology,* **18,** 79–103.

GEYER, O.F. & HINKELBEIN, K. 1971. Eisenoolithische Kondesation-Horizonteim Lias der Sierra Espuna (Provinz Murcia, Spanien). *Neues Jahrubuch fur Geologie und Palaeontologie, Monatshefte,* 398–414.

GILLETTE, T. 1947. The Clinton of western and central New York. *Bulletin of the New York State Museum, 341.*

GLEICH, G. & WARHAUSER, S.H. 1981. Stratigraphy and petrography of the Silurian Keefer Sandstone in the Central Appalachians. *Abstracts of the Geological Society of America,* **13,** 134.

GRUSS & THIENHAUS 1969. Die marin-sedimentaren Eisenerze des Dogger in Nordwestdeutschland. *In*: *Sammelwerk Deutsche Eisenerzlagerstatt, Beiheft Geologisches Jahrbuch,* **79,** 121–213.

GYGI, R.A. 1981. Oolitic iron formation: marine or not marine? *Eclogae Geologiea Helvetiae,* **74,** 233–254.

HALLAM A. & BRADSHAW, M.J. 1979. Bituminous shales and oolitic ironstones as indicators of transgressions and regressions. *Journal of the Geological Society, London,* **136,** 157–164.

HEMMINGWAY, J.E. 1951 Cyclic Sedimentation and the deposition of ironstones in the Yorkshire Lias. *Proceedings of the Yorkshire Geological Society,* **28,** 67–74.

HOFFMAN, K. 1969. Palaogeographie der nordwestdeutschen Lias-Eisenerze. *In*: *Sammelwerk Deutscher Eisenerzlagerstatten. Beiheft Geologisches Jahrbuch,* **79,** 104–110.

HUDSON, T.W. & MAYNARD, J.B. 1986. Petrography and diagenesis of Sedimentary ironstones in Silurian of Appoalachian Mountains (abs). *Bulletin of the American Association of Petroleum Geologists,* **70,** 602.

HUNTER, R.E. 1960. Iron sedimentation in the Clinton Group of the Central Appalachian basin. Ph.D. dissertation, The John Hopkins University, Baltimore.

——1970. Facies of iron sedimentation in the Clinton Group. *In*: FISHER, G.W. (eds) *Studies of Appalachian geology, Central and Southern.* John Wiley & Sons, New York, 101–121.

JAMES, H.L. 1966. Chemistry of the iron-rich sedimentary rocks. *In*: FLEISCHER, M. (ed) *Data of Geochemistry, 6th ed., Professional Paper of the U.S. Geological Survey 440, W,* 1–61.

KIMBERLEY, M.M. 1979. Origin of oolitic iron formations. *Journal of Sedimentary Petrology* **49,** 111–132.

LINDGREN, W. 1933. *Mineral Deposits,* McGraw Hill Book, Co., New York 26–280.

LUCIUS, M. 1945. Die Luxemburger Minetteformation undiejungere Eisenerzbildungen unseres Landes, *Publique Service de Carte Geologique de Luxembourg V.*

MCGHEE, G. R. & BAYER, U. 1985. The local signature of sea-level changes. *In*: BAYER, U. & SEILACHER, A. (eds) *Sedimentary and evolutionary cycles Lecture Notes in Earth Sciences No. 1.* Springer, Heidelberg 98–113.

NICKELSON R.P. & COTTER E. 1983. Silurian depositional history and Alleghanian deformation in the Pennsylvania Valley and Ridge. *Guidebook 48th annual field conference of Pennsylvania Geology,* 99–109.

RICKARD, L.V. 1975. Correlation of the Silurian and devonian rocks in New York State. *New York State Museum Science Map Chart 24.* 1–16.

SCHLOZ, W. 1972. Zur Bildungsgeschichte der Oolithenbank (Hettangium) in Baden Wurttemberg. *Arbeitsgemeinschaft Institut fur Geologie und Palaeontologie von Universitat Stuttgart NF 67.* 101–212.

SHELDON, R.P. 1965. Barrier island and lagoonal iron sedimentation in the Silurian of Alabama (abs). *Special Paper 82 of the Geological Society of America* 182.

——1970. Sedimentation of iron-rich rocks of Llandovery Age (Lower Silurian) in the southern Appalachian basin. *Special Paper 102 of the Geological Society of America* 107–112.

SIEHL, A. & THEIN, J. 1978. Geochemische Trends in der Minette (Jura/Luxemburg/Lothringen). *Geologisches Rundschau,* **67,** 1052–1077.

——1986. Origin of Minette-Type Ironstones (abs.). *Bulletin of the American Association of Petroleum Geologists* **70,** 648.

SIMON P., 1969. Die Lias-Eisenserze der Grube Friederike. *In*: *Sammelwerk Deutscher Eisenerzlagerstatten, Beiheft Geologisches Jahrbuch,* **79,** 40–58.

TEYSSEN, T. 1983. Tide-controlled sedimentation and sand wave-evolution in the Liassic minette iron formation of Luxemburg and NE France (abs). *Fourth Annual Meeting of the International Association of Sedimentologists.*

——1984. Sedimentology of the Minette oolitic ironstones of Luxembourg and Lorraine: a Jurassic subtidal sandwave complex. *Sedimentology,* **31,** 195–211.

THIENHAUS, R. 1969. Ubersicht uber die Dogger-Eisenerze Norwestdeutschlands. *In*: *Sammelwerk Deutscher Eisenerzlagerstatten. Beiheft Geologisches Jahrbuch,* **79,** 124.

VAIL R.P. *et al.* 1984. Jurassic unconformities, chronostritigraphy and sea-level changes from seismic stratigraphy and biostratigraphy. *Memoir 36 of the American Association of Petroleum Geologists,* 129–144.

VAN HOUTEN, F.B. 1985. Oolitic ironstones and constrasting Ordovician and Jurassic Palaeogeography. *Geology,* **13,** 722–624.

——& BHATTACHARYYA, D.P. 1982. Phanerozoic oolitic ironstones — geologic record and facies model. *Annual Reviews of Earth Planetary Sciences,* **10,** 441–457.

——& KARASE, R. 1981. Sedimentology framework of the Late Devonian oolitic iron formation, Shatti Valley, west-central Libya. *Journal Sedimentary Petrology,* **51,** 415–427.

——& PURUCKER, M.E. 1984. Glauconitic Peloids and Chamositic Ooids—Favourable factors, Constraints, and Problems. *Earth Science Reviews,* **20,** 211–243.

WERNER, F. 1959. Zur Kenntnis der Eisenoolithfacies des Braunjura beta von Ostwurttemberg. *Arbeitgemeinschaft fur Geologie und Palaeontologie von Institute Technical hoch-*

schule Stuttgart NF23, 169pp.
WILD, H. 1951. Zur Bildungsheschichte der Braunjura-beta-Floze und ihrer Begleitgesteine in NO-Wurttemberg. *Geologisches Jahrbuch,* **65**, 271–298.
ZIEGLER, P.A. 1978. North-Western Europe: Tectonics and Basin Development. *Geologic en Mijnbouw,* **57**, 589–626.

U. BAYER, Institute für Geologie und Paläontologie, Universitat Tübingen, Sigwartstr. 10, D-7400 Tübingen.
Present address: Institut für Erdöl und organische Chemie, ICH5, Kernforschungsanlage Jülich, D-5170 Jülich.

# Fabrics

# The application of analytical transmission electron microscopy to the study of oolitic ironstones: a preliminary study

**C.R. Hughes**

SUMMARY: Analytical transmission electron microscopy (ATEM) has been used by the author to obtain textural, crystallographic and quantitative chemical information from examples of berthierine-rich and chamosite-rich oolitic ironstones. Individual phases were often found to be too small to have been readily analysed by other techniques. Berthierine laths (of the order of 20 × 500 to 1000 nm) and goethite grains (of the order of 25 nm in diameter) were found to exhibit unorientated microtextures, which were independent of both depositional and oolitic fabrics. Chemical analyses of berthierines revealed fairly uniform compositions regardless of textural association. Chemical data from the chamosite-rich samples appeared analogous to that of the berthierine-rich material, although average crystal sizes were found to be much larger (chamosite laths varied from 0.1 × 1.5 μm to 0.5 x 2 mμ). The compositional data suggest that the chamosite-rich oolitic ironstones are simply the late stage diagenetic (or low grade metamorphic) recrystallized equivalents of berthierine-rich precursors. No direct evidence indicative of the mode of origin for berthierine and goethite ooliths was found. It is suggested that ancient ironstone ooliths were probably formed via similar mechanisms to those which generate modern and ancient carbonate ooliths, although obviously the original sediment and interstitial fluids would have been very different.

Sorby (1857) was the first to examine an oolitic ironstone under the petrological microscope, when he looked at material from the Cleveland Hill Main Seam of North Yorkshire. Since that time the textural features of oolitic ironstones, which are so beautifully amenable to light microscopy, have been reported extensively in the literature. One problem however with the petrographic examination of this kind of material, is the extremely fine grain size of many minerals present. Poorly characterized phases are frequently intergrown in a similar way to those in many mudrocks.

X-ray diffractometry (XRD) has been used since the early 1930s to identify and characterize the very fine grained phases present in oolitic ironstones (Jung 1931). Although a very powerful tool, XRD provides no direct textural information, and as lattice parameters are averaged for a large number of crystals the technique will not always resolve the presence of minor components.

With the development of transmission electron microscopy (TEM) during the 1950s and '60s it became possible to accumulate a range of textural and crystallographic information from single clay-size crystals. This included lattice and ordering data not previously resolved by XRD. Together XRD and TEM proved a very useful complementary pair of techniques, although no record could be traced of any TEM studies of oolitic ironstone mineralogy prior to this decade.

Over the past couple of decades petrographic studies have been considerably enhanced with the emergence of scanning electron microscopy (SEM). For example, Bhattacharyya and Kakimoto (1982) used SEM to compare the fine structure of ironstone ooliths with that of bauxite pisoliths. More recently the usefulness of SEM has been boosted with the introduction of high-resolution computeried backscattered electron imagery and energy dispersive X-ray spectroscopy (EDS), as demonstrated by Kearsley (1989).

Mineral compositional data are 'traditionally' obtained by mechanically disaggregating the rock and attempting to separate the various minerals for wet chemical or and/or spectroscopic analysis. Of course with very fine grained closely related phases, total separation is almost impossible. Even with the emergence of widespread combined electron beam/X-ray spectroscopic instrumentation the separation problem has been only partially resolved. Further, X-ray spectroscopy can distinguish neither the oxidation state of iron nor detect the presence of water. Nevertheless the electron microprobe (EMPA) and ATEM have proved particularly useful in the study of phyllosilicate chemistry. Qualitative (and with care at least semi-quantitative) chemical analyses are also possible by SEM EDS. In all three techniques, chemical

*From* YOUNG, T. P. & TAYLOR, W. E. G. (eds), 1989, *Phanerozoic Ironstones*
Geological Society Special Publication No. 46, pp. 121-131

analysis (via X-ray spectroscopy) of very fine minerals can be related directly to thin section petrography, an obvious advantage over conventional geochemical techniques. Velde (1985) has summarized published EMPA data from ironstones and new data is continually appearing (e.g., Maynard 1986). ATEM is not a widespread facility and very little ATEM data from ironstones has been published (Curtis *et al.* 1985).

Statistically quantitative chemical analyses by the wavelength dispersive X-ray spectrometry (WDS) of EMPA can be an order of magnitude more precise than the energy dispersive X-ray spectrometry (EDS) of ATEM. This is because in WDS absolute X-ray counts are evaluated sequentially for each element, whilst the much lower overall counts of ATEM EDS have to be normalized. The precision electron optics of ATEM however, permit the resolution of textures much more finely than by EMPA. This facilitates the avoidance of contamination in single-crystal analyses from adjacent phases, which cannot readily be done in the microprobe. ATEM is particularly useful because chemical, textural and crystallographic information can be obtained from grains of the order of only tens of nanometres in size, too small to isolate by other methods. However, even with ATEM problems arise especially if chemical analyses are required from phases smaller than 50nm in diameter.

## Methods and problems

Ultra-thin rock sections were prepared for ATEM examination by ion-beam thinning, following the techniques of Phakey *et al.* (1972). Data were collected using a Philips 400T TEM/STEM system with EDAX energy-dispersive X-ray analysis. X-ray emission spectra were processed according to the ratio method of Cliff & Lorimer (1975). The method assumes that below a certain critical thickness, sample matrix effects (primarily X-ray absorption and flourescence) are sufficiently close to zero that characteristic X-ray intensities are almost directly proportional to element concentrations. The ATEM system at Sheffield has been calibrated for the analysis of phyllosilicates by determining proportionality constants relative to silicon from sections of known standard micas. Unfortunately the system has not been calibrated for the analysis of oxides and carbonates, so compositional data collected from these phases can only be qualitative.

The fundamental limits of the Cliff & Lorimer ratio method are defined by the relatively poor counting statistics of thin film ATEM EDS, particularly with respect to elements present in low concentrations. If thin film criteria are to be satisfied, then the necessarily tiny mass excited by the electron beam can only generate X-rays accordingly.

A thorough examination of the sources of experimental error which are particularly appropriate in the ATEM examination of phyllosilicates has been carried out at Sheffield (Hughes *et al.* 1987; Hughes 1987). From this work it was concluded that during analysis of single crystals great care must be taken to avoid adjacent phases. The operator should also maintain minimum X-ray absorption pathways between the irradiated volume and the EDS detector. Additionally the physical and chemical stability of each phase under examination should be established under a range of operating conditions, to ensure that significant specimen damage does not occur during analysis. This usually entails analysing phases using the lowest beam current, largest beam diameter configuration giving statistically valid X-ray count rates.

## Sample selection

In this preliminary study four berthierine-rich and three chamosite-rich ironstone samples were selected.

A. Berthierine-rich ironstones

(1) Main Seam, Kettleness Member, Cleveland Ironstone Formation, England, Lower Jurassic.
(2) Frodingham Ironstone (×2), England, Lower Jurassic.
(3) Wasia Formation ironstone, Arabia, Upper Cretaceous.

B. Chamosite-rich ironstones

(1) Tremadoc ironstone, Wales, Ordovician.
(2) Betws Garmon Ironstone, Wales, Ordovician.
(3) Ordovician ironstone, Portugal.

The results and conclusions may not be widely applicable, but they certainly indicate the potential of ATEM in the detailed mineralogical examination of ironstones.

## ATEM data from berthierine-rich oolitic ironstones

### Textures

The samples studied all contained the iron-rich phases berthierine, goethite and siderite. At the very high magnifications (usually greater than 25 000×) needed to resolve individual crystals of berthierine and goethite all four samples revealed similar textures (see Figs. 1 & 2). Individual goethite particles were commonly sub-equigranular, less than 25 nm in diameter ranging up to around 50 nm, occurring as diffuse

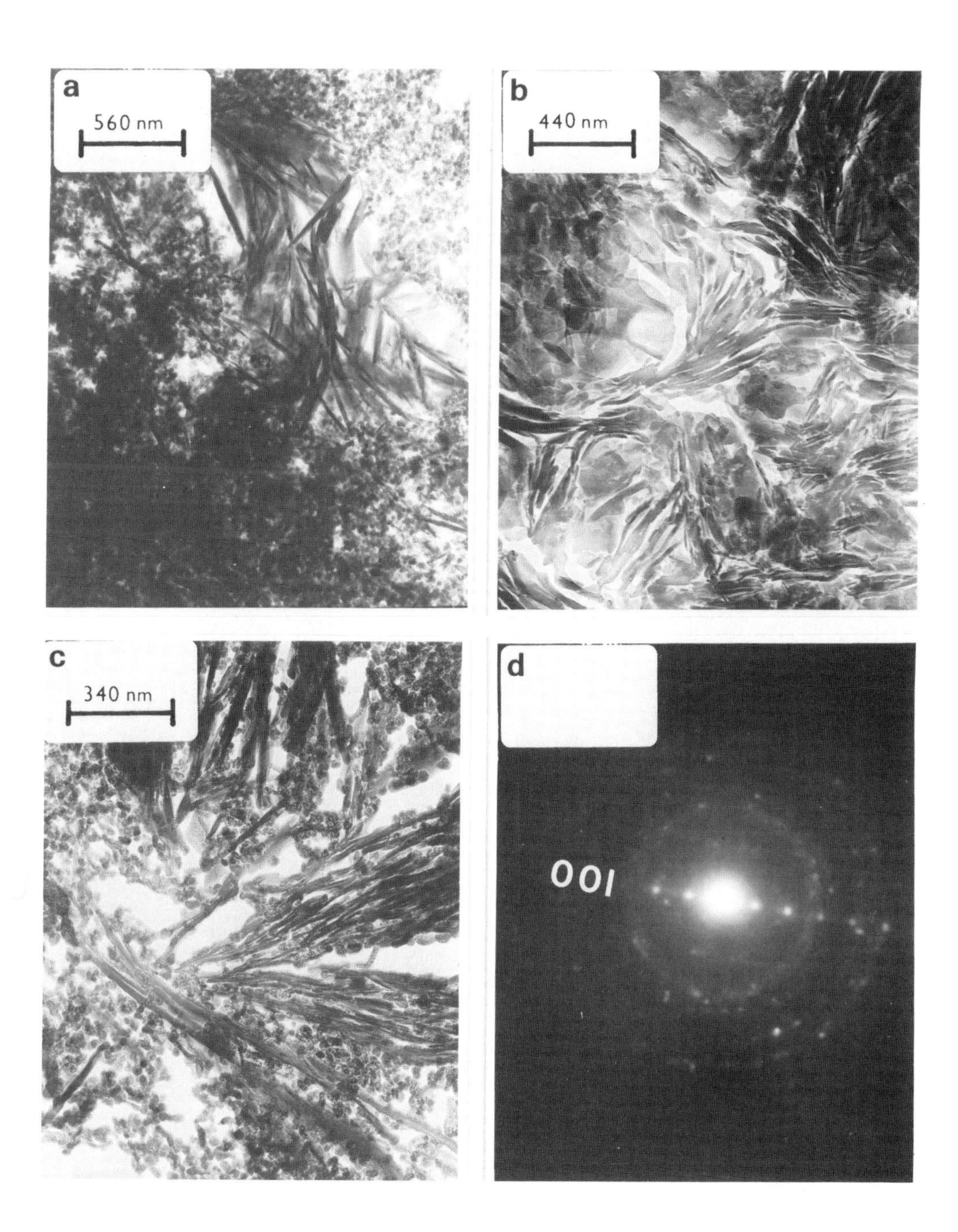


FIG. 1. Transmission electron micrographs and an electron diffraction pattern from sample of oolitic bertherine-rich Frodingham Ironstone. (a) A clump of berthierine blades surrounded by goethite within an oolith cortex. (b) Intimately admixed berthierine and goethite with diffuse amorphous meterial (possibly organic?) within oolith cortex. (c) Intimately admixed blades of berthierine and particles of goethite from groundmass between ooliths. (d) Electron diffraction pattern *d*(001) = 0.705 nm for berthierine plus diffuse ring pattern from ultra-fine goethites.

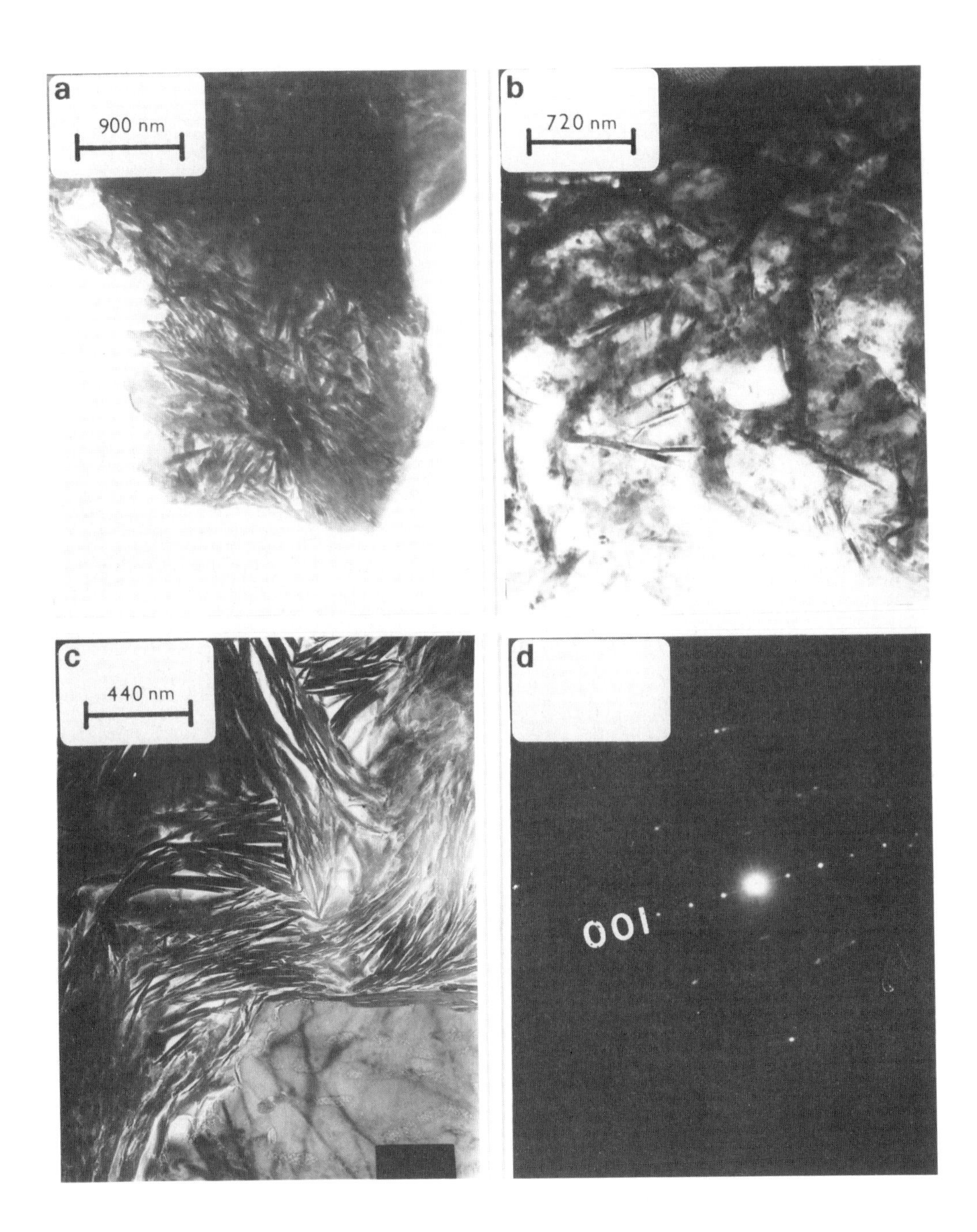


FIG 2. Transmission electron micrographs and an electron diffraction pattern from berthierine-rich oolitic ironstones. (a) An array of berthierine blades within an oolith core (Cleveland Ironstone). (b) Blades of berthierine suspended in goethitic matrix, from groundmass between ooliths (Wasia ironstone). (c) Bundles of berthierine within groundmass with a carbonate crystal showing signs of beam damage (Cleveland Ironstone). (d) Electron diffraction pattern $d(001) = 0.705$ nm for berthierine (Frodingham Ironstone).

masses variably admixed with fibrous berthierine. Individual berthierine crystals appeared to be of the order 20 x 500 to 1,000 nm, and were either randomly distributed among goethite grains or concentrated in matted unorientated clusters. These morphologies were found to be similar in all four samples. Further, on the basis of these ultra-textures alone, it was impossible to distinguish oolith cortex areas from core areas, or intra-oolith from inter-oolith areas. This was because of the uniformity of individual mineral morphologies. The only textural differences were caused by variable concentrations of each phase, and these were not specific to particular areas.

No preferred orientation of berthierine was observed within individual oolith laminae, which seemed to be defined only by their bounding surfaces and/or by changes in mineral proportions. Between individual mineral grains in some ooliths, a low electron-contrast amorphous phase was detected which could not be resolved by electron diffraction nor by EDS. Organic material often displays such characteristics, and is common in carbonate ooliths.

The replacive siderite crystals identified under the light microscope were generally found to be too large and not sufficiently thinned for examination by ATEM. Carbonates in general do not respond well to ion-beam thinning perhaps due to their brittle crystal cleavage, and chemical data would be more readily obtained by EMPA.

Mineral identification was confirmed by electron diffractometry (see Figs. 1–3).

**Chemical composition**

Experience has shown that the smallest useful analytical spot size for ATEM is about 50 nm, and the best quality data are usually collected using spot sizes at least ten times greater. Clearly both the goethite and berthierine crystals were too small for optimum analysis, even by ATEM standards.

In the very few cases where the edges of siderite grains had been sufficiently thinned for analysis, they were found to disintegrate under the beam.

Although the individual berthierine crystals were found to be too small to isolate for analysis, it was found that good quality data could be obtained by collecting spectra simultaneously from several apparently identical crystals in close proximity. Table 1 lists the calculated means of the most precise ATEM data recorded from each of four samples. Weight per cent oxide ratios have been recalculated on the basis of nine

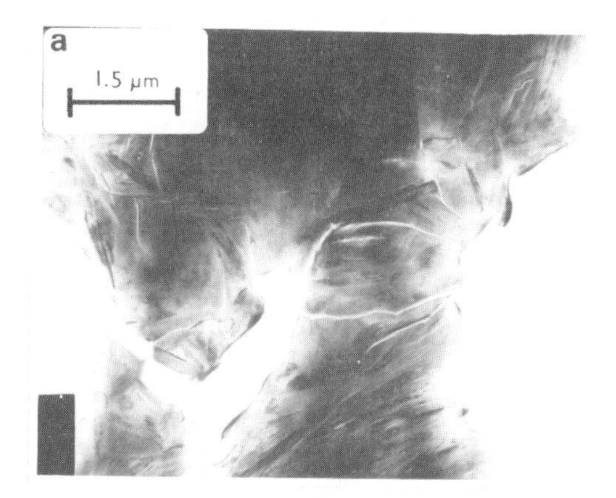

FIG. 3. Transmission electron micrograph and diffraction patterns from Betws Garmon Ironstone chamosite. (a) Interlocking chamosite lath texture. (b) Lattice image from single lath showing predominant 1.4 nm spacing of lattic fringes with marked disconformities probably due to stacking disorder. (c) Electron diffraction pattern *d*(001) = 1.43 nm for chlorite.

TABLE 1. *ATEM mean berthierine data*

| | CISN07 | $\sigma^{n-1}$ | F83N12 | $\sigma^{n-1}$ | F85N20 | $\sigma^{n-1}$ | WASN12 | $\sigma^{n-1}$ |
|---|---|---|---|---|---|---|---|---|
| Si | 1.41 | 0.04 | 1.54 | 0.06 | 1.57 | 0.08 | 1.29 | 0.06 |
| Al (TET) | 0.59 | 0.04 | 0.46 | 0.06 | 0.43 | 0.08 | 0.71 | 0.06 |
| Al (OCT) | 0.52 | 0.13 | 0.37 | 0.11 | 0.39 | 0.18 | 0.57 | 0.09 |
| Ti | 0.02 | 0.02 | 0.00 | 0.00 | 0.00 | 0.00 | 0.00 | 0.00 |
| Fe (III) | 0.21 | 0.01 | 0.22 | 0.02 | 0.21 | 0.03 | 0.11 | 0.01 |
| Fe (II) | 1.86 | 0.12 | 1.99 | 0.15 | 1.92 | 0.24 | 1.73 | 0.10 |
| Mn | 0.01 | 0.01 | 0.00 | 0.00 | 0.00 | 0.00 | 0.00 | 0.00 |
| Mg | 0.15 | 0.01 | 0.29 | 0.14 | 0.36 | 0.16 | 0.48 | 0.08 |
| Ca | 0.06 | 0.02 | 0.05 | 0.11 | 0.03 | 0.02 | 0.00 | 0.00 |
| Na | 0.00 | 0.00 | 0.00 | 0.00 | 0.00 | 0.00 | 0.00 | 0.00 |
| K | 0.15 | 0.01 | 0.03 | 0.10 | 0.01 | 0.03 | 0.00 | 0.00 |
| OCT | 2.77 | 0.04 | 2.87 | 0.08 | 2.88 | 0.10 | 2.97 | 0.04 |
| INT | 0.21 | 0.03 | 0.08 | 0.10 | 0.04 | 0.05 | 0.00 | 0.00 |
| OH | 4.00 | | 4.00 | | 4.00 | | 4.00 | |
| *n* | 7 | | 12 | | 20 | | 12 | |

(Re-calculated on the basis of five oxygens and four hydroxyls, Fe(II): Fe(III) ratio has been arbitrarily chosen at 9:1.) CISN07: Main Seam, Cleveland Ironstone Formation. F83N12: Frodingham Ironstone. F85N20: Frodingham Ironstone. WASN12: Wasia ironstone

oxygens, for direct comparison with the published formula of Bailey (1980):

$$(Fe^{2+}, Mn^{2+}, Mg)_{3-x}(Fe^{3+}, Al)_x(Si_{2-x}Al)_x O_5(OH)_4$$

The ATEM data fall well within the range of Bailey's formula. Additionally the Frodingham and Cleveland data are comparable with EMPA berthierine data from the same deposits, published by Maynard (1986).

Although the statistics are very poor, the presence of alkali elements is indicated in the Cleveland and Frodingham berthierines. Maynard (1986) has already noted the occurrence of potassium in Frodingham berthierine which he suggests may indicate a transitional phase between berthierine and glauconite. From Table 1 it is interesting to note that the berthierines with the lowest octahedral totals appear to contain the most potassium. Whether present within the structure (perhaps going some way towards balancing the octahedral charge imbalance), or as some sort of contaminant due to juxtaposition of alkali-metal bearing phases which were not identified, has not been determined with certainty. However apart from carbonate no other alkali-bearing phases were identified in ATEM, which would suggest that the potassium at least may be sited within the berthierine structure.

In order to compare the above ATEM data with published analyses the compositional field of 29 published analyses were plotted on triangular diagrams (Fig. 4). From these the ATEM data appear to fall within the compositional field indicated by published analyses. If it is assumed that ATEM is probably the only direct method of ensuring a reasonably uncontaminated analysis due to the minute size of the berthierines, then on the basis of the ATEM data it might be suggested that the compositional range of natural berthierines may be more limited than suggested in the literature. The spread of published data in Fig. 4., especially with regard to silica and aluminium may be partly due to contamination by quartz and kaolinitic phases which are commonly found in ironstones.

## ATEM data from chamosite-rich oolitic ironstones

### Textures

It is generally accepted that the chamosite-rich oolitic ironstones are recrystallized late stage diagenetic and greenschist metamorphic products, after berthierine-rich ironstones. A consequence of the recrystallization is a marked increase in crystal size, and a less intimate admixture of the component phases. In the three samples selected, the iron-oxide components (now hematite or magnetite) and siderites are too coarsely crystalline for much textural infor-

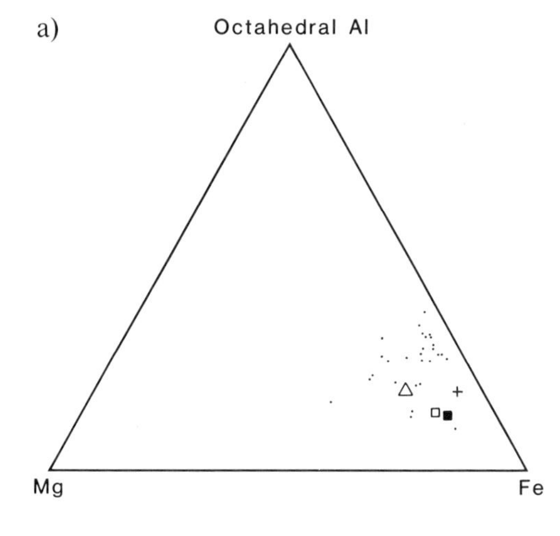


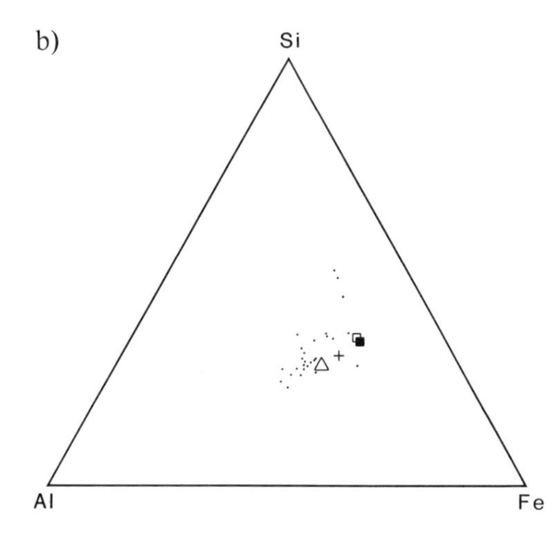


FIG. 4. Triangular plots illustrating the compositional field of berthierine as defined by published analyses, with mean TEM EDS data. Open triangle, Wasia ironstone; squares, Frodingham ironstone; cross, Cleveland Ironstone; others not differentiated, data included from Velde *et al.* (1974), Brindley (1982), Ijima and Matsumoto (1982), and Marshall (1983).

mation to be gleaned by ATEM. Examination was therefore largely confined to the chamosite.

Well crystalline chlorites, with a distinct interlocking bladed configuration were found in all three samples (for examples see Fig. 3). The size of individual crystals varied within each sample from 0.1 x 1.5 μm to 0.5 x 2 μm and greater (getting within the range of normal optics). No preferred orientation was noted although individual crystals did not appear to interlock with crystals of adjacent oolite rims. The much larger crystals facilitated better chemical analysis, although distinct evidence of weathering was found in the form of poorly ordered kaolinites.

### Chemical composition

When calculated to chlorite molecular formulae on the basis of 20 oxygens and 16 hydroxyls the ATEM EDS compositions come out close to ideal iron-rich chamosites (Table 2).

The compositional range of the common trioctahedral chlorites can be usefully depicted as in Fig. 5, after Hayes (1970). The ironstone chlorites plot clearly at the iron-rich end of the chamosite compositional field. The major element composition of berthierines is often very close to that of chamosite, so berthierine data can readily be recalculated to the chlorite molecular formula. This has been done so that the compositional field of berthierine can be compared directly with that of chamosite. Most berthierines plot within or close to the field of chamosite.

## Discussion and preliminary conclusions

This limited study has demonstrated the potential of ATEM when applied to the examination of oolitic ironstones. The overall procedures are difficult and time consuming; however the rewards in terms of textural crystallographic and chemical information are substantial.

### Textures

From the berthierine-rich oolitic ironstone samples examined to date it would appear that individual berthierine and goethite crystals are so small that they can only readily be imaged using TEM. The fact that berthierine and goethite morphologies are fairly uniform both within and between ooliths would suggest that they formed after burial. This is supported by the observation that crystal orientations appear to be independent of depositional fabrics.

The apparent lack of crystal orientation relative to oolith laminae would suggest that the oolith-forming processes did not directly influence crystal orientation. The concentric structure of the ooliths would appear to be the product of successive spherical to ovoid bounding surfaces only. The exact nature of these surfaces was not determined; in some cases

TABLE 2. *ATEM mean data Ordovician ironstone chamosites*

| | PCN06 | $\sigma^{n-1}$ | BCN06 | $\sigma^{n-1}$ | TCN08 | $\sigma^{n-1}$ |
|---|---|---|---|---|---|---|
| Si | 5.17 | 0.11 | 5.36 | 0.13 | 5.31 | 0.21 |
| Al (TET) | 2.83 | 0.11 | 2.64 | 0.13 | 2.69 | 0.21 |
| Al (OCT) | 2.87 | 0.21 | 1.89 | 0.15 | 1.70 | 0.49 |
| Ti | 0.00 | 0.00 | 0.01 | 0.01 | 0.00 | 0.00 |
| Fe (III) | 0.69 | 0.03 | 0.87 | 0.01 | 0.96 | 0.07 |
| Fe (II) | 6.22 | 0.30 | 7.80 | 0.13 | 8.62 | 0.68 |
| Mn | 0.04 | 0.02 | 0.01 | 0.011 | 0.002 | 0.007 |
| Mg | 1.81 | 0.06 | 1.27 | 0.05 | 0.68 | 0.05 |
| Ca | 0.0 | 0.00 | 0.03 | 0.03 | 0.01 | 0.02 |
| Na | 0.00 | 0.00 | 0.03 | 0.06 | 0.09 | 0.11 |
| K | 0.00 | 0.00 | 0.05 | 0.01 | 0.00 | 0.00 |
| OCT | 11.63 | 0.13 | 11.86 | 0.06 | 11.97 | 0.28 |
| INT | 0.00 | 0.00 | 0.11 | 0.08 | 0.09 | 0.12 |
| OH | 16 | | 6 | | 8 | |
| *n* | 6 | | 6 | | 8 | |

(re-calculated on the basis 20 oxygens and 16 hydroxyls, Fe(II): Fe(III) ratio has been arbitrarily chosen at 9:1.) PCN06: Ordovician ironstone, Portugal. BCN06: Betws Garmon Ironstone, Snowdonia. TCN08: Tremadoc ironstone, Snowdonia.

there appeared to be a textural break between laminae, in others simply a marked difference in berthierine to goethite ratios. A low electron-contrast intercrystalline amorphous material may represent a significant organic component; however, confirmation would require careful investigation using a light element EDS detector.

Chamosite-rich ironstones are much more coarsely crystalline than berthierine-rich ironstones. This would concur with the model that chamosite-rich oolitic ironstones represent the recrystallized products after berthierine-rich ironstones. It is interesting to note that even after such a radical change in crystal size there remains a lack of preferred crystal orientation, yet the concentric laminar structure remains. This suggests that the oolith bounding laminar surfaces have not been fully bridged during recrystallization.

It is perhaps worth noting that on the sub-millimetre scale, there are several orders of magnitude differences in size between various structures as illustrated in Fig. 6. The berthierine and goethite crystals imaged by ATEM are too fine to be resolved with any clarity by SEM. If such minute crystal sizes are typical then it is likely that the tangential and radial crystal fabrics recorded in the literature (e.g. Bhattacharyya and Kakimoto 1982; Nahon *et al.* 1980; Kearsley 1989) may be aggregates or packets of unresolved minute crystals rather than single large crystals. One feature of ironstone ooliths which is readily imaged by SEM, but not by TEM, is the very porous nature of cortex and core areas. This micro-porosity is a very significant feature which seems to have been ignored by many workers. Primary ironstone ooliths appear to be very open structures with enormous internal surface areas on the micron and sub-micron scale. During oolith formation and very early diagenesis, not only would the mineral fabric be ideally suited to fully react with pore fluids, but there would be more than adequate room for microbial activity (many sediment dwelling prokaryotes are of the order 0.5μm in their largest dimension).

### Chemistry

In the material examined so far, both berthierine and goethite were found to be too small for single-crystal chemical analysis by ATEM under ideal operating conditions. However, with great care, resonable berthierine data were collected from clumps of crystals. This proved to be fairly consistent within each sample with no obvious chemical distinction between berthierines, whether from oolith cores or cortices, or from grain coating cement or matrix. A similar uniformity was identified in the chamosites of the chamositic ironstones examined.

Overall the ATEM data appears to show less intersample variation for berthierine than is evident in the literature. Coupled with the grain size information, this would suggest that many published berthierine analyses may represent

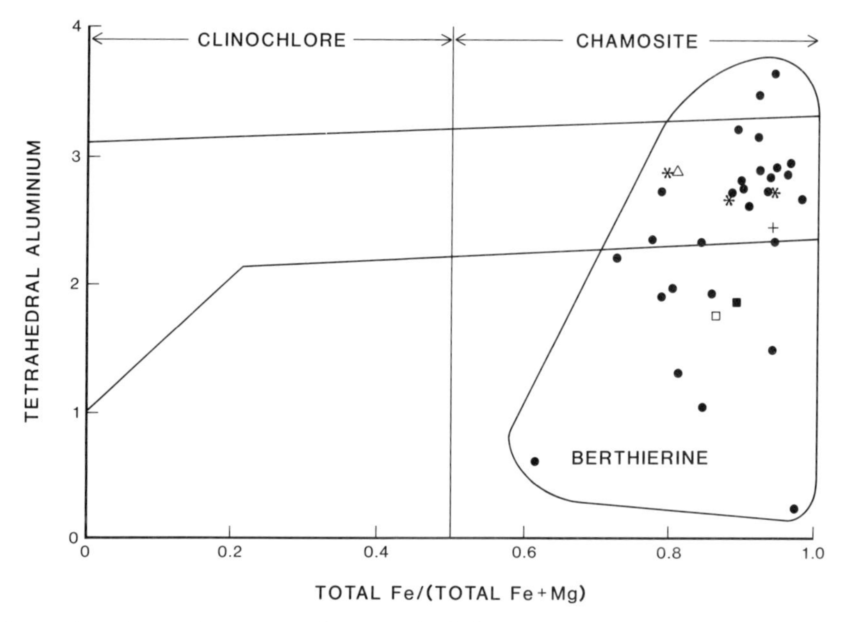


FIG. 5. Compositional range for common trioctahedral chlorites (straight edged box) after Hayes (1970), modified after Curtis *et al.* (1986). Ironstone chamosites indicated as stars. Also included for comparison is the compositional field for berthierine (curved edged box), using the data plotted in Fig.1 (recalculated to chlorite molecular formulae).

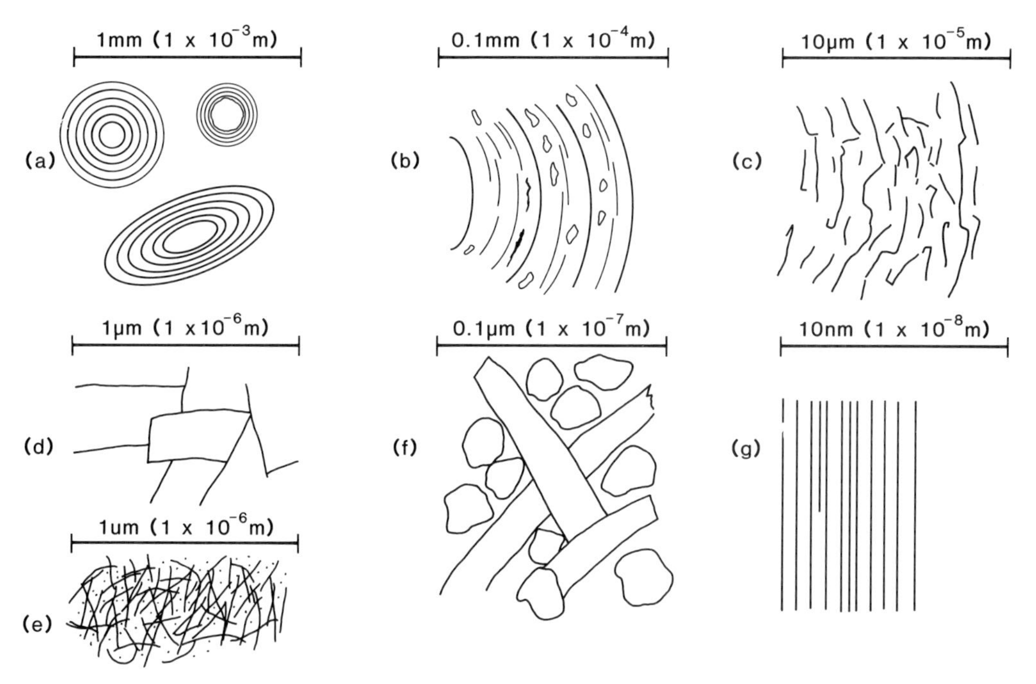


FIG. 6. Size scale of components within ironstone ooliths. (a) Individual ooliths resolved as concentrically coated grains. (b) Individual sheaths resolved to reveal both continuous and discontinuous laminae. (c) Laminar surfaces resolved as bounding porous areas of flakey crystal aggregates. (d) Individual chamosite blades resolved as interlocking network. (e) At same magnification berthierine resolved as matter mass. (f) Individual berthierine and goethite crystals resolved. (g) When phyllosilicate structure sufficiently ordered lattic images resolved.

partial mixtures (particularly with kaolinite and quartz). Gross compositional variation between individual oolith coatings previously reported (e.g. Bhattacharyya 1983), may well reflect different concentrations of ultra-fine minerals, and not variation in individual mineral compositions as has been previously assumed.

The ATEM chemical data confirms that ironstone-berthierines and chamosites have very similar compositions corroborating a genetic link.

### Paragenesis

The ultra-fine mineral assemblage berthierine plus goethite contains both a reduced ferrous phase and an oxidized ferric phase. This assemblage could be a stable one but not in the presence of organic matter. There are two simple equally applicable explanations. The goethite could be a primary diagenetic phase after colloid dehydration within an essentially oxic zone. Subsequently this could have been partially reduced by organic matter during burial, releasing ferrous iron to react with any aluminosilicates present, precipitating secondary berthierine. Alternatively berthierine could have been the primary authigenic phase, a diagenetic product after amorphous precursors; the goethite forming later during reworking and secondary oxidation. Of course any combination of these ideas can be construed as potentially responsible for the various assemblages seen in the rocks today, often further complicated by carbonate activity and the consequent presence of diagenetic siderite.

Chemical equations have been written for these kinds of reaction (Bhattacharyya 1983; Iijima & Matsumoto 1982; Curtis 1985) and Maynard (1986) has constructed simple phase diagrams from thermodynamic models; however it must be stressed that these can only be approximations. Certainly the earliest actual reaction pathways would have been an intimate part of oolith formation, and would have been very much controlled by the kinetics of individual competing reactions and possibly influenced by microbial activity.

The mode of origin for the oolith structure is perhaps the one feature of these rocks about which there is widespread disagreement, as is evident from many of the papers in this volume. Most workers appear to be happy in agreeing to disagree with each other over the merits of the various models.

The textural evidence revealed by ATEM so far suggests that neither tangential nor radial crystal orientations are particularly evident within the oolith structure. This lack of concordance between the mineral micro-assemblages and their bounding laminae suggests that they may have been generated before, during or after oolith formation. Clearly the microtextures give little indication of the actual oolith generating mechanism.

Perhaps the simplest hypothesis is that each oolith lamina represents a preserved successive grain coating event, punctuated by a change of conditions. Whether the controlling mechanism was predominantly physical or chemical cannot easily be determined. A mechanical component is evident from the reworked nature of many ironstones. Even abraded contacts between individual successive oolith laminae are found, indicating that ooliths have been moved between grain coating events. There is however no unequivocal textural evidence of mechanical adhesion as a significant process during individual grain coating events.

Modern primary aragonite ooliths are the products of episodic chemical precipitation, both with and without significant microbial activity, and variably influenced by mechanical abrasion (summarized by Leeder (1982). Aragonite precipitation may occur during episodic burial while the growing ooliths sit stranded and colonized by blue-green algae within bedforms, the algae perhaps enhancing precipitation (Davies *et al.* 1978). Additionally, Deelman (1978) has shown that aragonite precipitation can also be stimulated at oolith surfaces by intermittent agitation. Apart from the subaqueous environment, these processes are akin to those of modern vadose carbonate pisolite generation, in that the concentric structure is built up during successive intervals of grain coating carbonate precipitation from saturated interstitial waters. In both cases the forces of crystal growth are usually quite adequate to displace adjacent grains and permit concentric rims to develop.

Modern iron-rich pisolites occur in some laterites and if these are analogous to carbonate pisolites, then uniformitarianism would suggest that ancient ironstone oolites may well have formed in the same way as carbonate oolites. Obviously the chemistry of the original sediment and interstitial fluids would have been very different.

ACKNOWLEDGEMENTS: I would like to thank C. Curtis for his support and encouragement, J. Whiteman for his advice and instruction on the technical side of ATEM, and T. Young for hours of discussion and help with sample selection.

## References


BAILEY, S.W. 1980. Structures of layer silicates. *Mineralogical Society Monograph No.5,* 1–124.

BHATTACHARYYA, D.P. 1983. Origin of berthierine in ironstones. *Clays and Clay Minerals,* **31,** 173–182.

——& KAKIMOTO, P.K. 1982. Origin of ferriferous ooids; an SEM study of ironstone ooids and bauxite pisoids. *Journal of Sedimentary Petrology,* **52,** 849–857.

BRINDLEY, G.W. 1982. Chemical compositions of berthierines — a review. *Clays and Clay Minerals,* **30,** 153–155.

CLIFF G. & LORIMER, G.W. 1975. The quantitative analysis of thin specimens. *Journal of Microscopy,* **103,** 203–207.

CURTIS C.D. 1985. Some mineral precpitation and transformation during burialdiagenesis. *Philosophical Transactions of the Royal Society of London,* **A 315,** 91–105.

——, HUGHES, C.R., WHITEMAN, J.A. & WHITTLE, C.K. 1985. Compositional variation within some chlorites and some comments on their origin. *Mineralogical Magazine,* **49,** 375–386.

DAVIES, P.J., BUBELA B. & FURGUSON J. 1978. The formation of ooids. *Sedimentology,* **25,** 703–730.

DEELMAN, J.C. 1978. Experimental ooids and grapestones: carbonate aggregates and their origin. *Journal of Sedimentary Petrology,* **48,** 503–512.

HAYES, J.B. 1970. Polytypism of chlorite in sedimentary rocks. *Clays and Clay Minerals,* **18,** 285–306.

HUGHES, C.R. 1987. The composition and origin of layer silicates in iron-formations and ironstones: a preliminary analytical transmission electron microscopial study. *PhD thesis.*

——, CURTIS, C.D., WHITEMAN, J.A. & HEPING, S. 1987. Applications of analytical transmission electron microscopy to clay mineral geochemistry studies. Published by the Clay Minerals Society, in a set of course notes, following a workshop entitled 'Electron Microscopy and Microprobe Techniques in Clay Analysis', in October 1986.

IIJIMA, A. & MATSUMOTO, R. 1982 Berthierine and Chamosite in Coal measures of Japan. *Clays and Clay Minerals, 30,* 264–274.

JUNG, H. 1931. Untersuchungen uber den Chamosit von Schmiedefeld i. *Chem. Erde, 6,* 275–306.

KEARSLEY, A.T. 1989. Iron-rich ooids, their mineralogy and microfabric: clues to their origin and evolution. *In:* YOUNG, T.P. & TAYLOR, W.E.G. (eds) *Phanerozoic Ironstones* Geological Society, London, Special Publication, **46,** 141–164

LEEDER, M.R. 1982. *Sedimentology Process and Product,* George Allen and Unwin.

MARSHALL, J.F. 1983. Geochemistry of iron-rich sediments on the outer floor off northern New South Wales. *Marine Geology,* **51,** 163–175.

MAYNARD, J.B. 1986. Geochemistry of oolitic iron ores, an electron microprobe study. *Economic Geology,* **81,** 1473–1483.

NADEAU, P.H., WILSON M.J., MCARDY W.J. & TAIT J.M. 1984. Interparticle diffraction: a new concept for interstratified clays. *Clay Minerals,* **19,** 757–769.

PHAKEY, P.P., CURTIS C.D. & OERTAL G. 1972. Transmission electron microscopy of fine-grained phyllosilicates in ultra-thin rock sections. *Clays and Clay Minerals,* **20,** 193–197.

SORBY H.C. 1857. On the origin of the Cleveland Hill ironstone. *Proceedings of the Polytechnic Society of the West Riding of Yorkshire,* **3,** 457–461.

VELDE, B. 1985. Clay Minerals — A Physico-Chemical Explanation of their occurrence. *Developments in Sedimentology 40.* Elsevier.

——, RAOULT, J.F. & LEIKINE, M. 1974. Metamorphosed berthierine pellets inmid-Cretaceous rocks from north-eastern Algeria. *Journal of Sedimentary Petrology,* **44,** 1275–1280.


C. R. HUGHES, Department of Geology, Beaumont Building, The University, Brookhill, Sheffield, S3 7HF, UK.

# The formation of goethitic ooids in condensed Jurassic deposits in northern Switzerland

A.U. Gehring

SUMMARY: Goethitic ooids from condensed sequences close to the Middle/Upper Jurassic boundary in the Swiss Jura mountains were analysed for texture and chemistry. The sequences are characterized by discontinuities of sedimentation and reworking, which indicate a marine aerobic depositional environment. The ooids consist mainly of isometric goethite particles with apatite and clay mineral flakes. The coexistence of these mineralogical phases can be explained by different adsorption processes due to electrostatic forces or ligand exchange reactions. Both types of processes are discussed and illustrated schematically.

Oolitic ironstones are widespread throughout the Jurassic stratigraphy of northwestern and central Europe. Due to the chemical complexity and variability no generally acceptable theory of their formation has been proposed.

A large number of studies reflect the multitude of different geological aspects of the formation of oolitic ironstones. Most authors have concentrated on petrography and stratigraphy. Geochemical information mainly comes from bulk analysis of oolitic ironstones (e.g. James 1966; Siehl & Thein 1978). Detailed investigations concerning sedimentological (Teyssen 1984) and microbiological aspects (Dahanayake & Krumbein 1986) have been presented. Recently the formation of ironstones has been reviewed by Van Houten and co-workers. They concluded that ironstones were formed in marine, near shore environments during episodes of mild climate and widespread transgression of continents (cf. Van Houten & Bhattacharyya 1982; Van Houten & Purucker 1984).

In contrast to the oolitic ironstones, little is known about the genesis of the ferriferous ooids which are their most important components. These spherical or ellipsoidal ooids mainly consist of berthierine (7Å trioctahedral serpentine) or of iron oxide, often goethite ($\alpha$-FeOOH) and less frequently hematite ($\alpha$-$Fe_2O_3$). The few investigations of ferriferous ooids yield information of the textures (Radszweski & Schneiderhöhn 1962; Bhattacharyya & Kakimoto 1982) and the chemical composition (e.g. Halbach 1968; Rohrlich 1974).

Up to now there have been few studies which deal both with the geological and the chemical aspects of ooid formation. The aim of this study is to review and discuss constraining geological and chemical conditions necessary for the formation of goethitic ooids. The oolitic condensed deposits of the Middle/Upper Jurassic boundary in northern Switzerland will be used as an example.

## Facies differences

Northern Switzerland (Fig.1) can be subdivided into two facies realms for the Middle and Upper Jurassic (cf. Trumpy 1980). In the northwest there was a shallow-water marine carbonate platform, whereas in the southeast there was a deeper, low-energy depositional setting with more muddy sedimentation (cf. Ziegler 1982; Débrand-Passard & Courbouleix 1984a,b). In late Callovian and earliest Oxfordian time the facies differences between the two realms disappeared and highly condensed sequences with a very rich ammonite fauna were deposited over both areas. These facies changes correlate with a worldwide transgressive event (Hallam 1984; Vail *et al.* 1985), so that the condensed deposits may have formed in slightly deeper water than the underlying formations.

Sequences at localities 1 to 3 (Fig. 1) are formations of the north-western depositional realm. The commonly iron-oolitic, condensed deposits usually overlie a crinoidal and bryozoan limestone (Dalle nacrée) and pass upwards into marls (Renggeri marls) and then into marls rich in corals (Liesberg beds). The successions at localities 4 to 6 (Fig. 1) are assigned to the south-eastern depositional realm. Here condensed, iron-oolitic beds lie between fine arenaceous limestones or marls (Varians bed) and limestones with siliceous sponges (Birmenstorf beds). A detailed description of the geological sections can be found in Bolliger & Burri (1970), Persoz & Remane (1973) and Gygi & Marchand (1982).

The condensed deposits are usually less than 10

*From* YOUNG, T. P. & TAYLOR, W. E. G. (eds), 1989, *Phanerozoic Ironstones*
Geological Society Special Publication No. 46, pp. 133-139

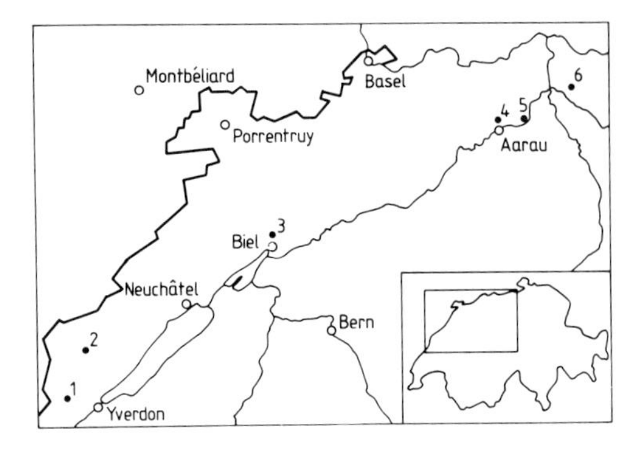


FIG. 1. Map showing the locations of the condensed oolitic sequences. Coordinates refer to the national topographic maps of Switzerland (scale 1:25000). 1. Baumes 529.700/183.375; 2. S. Sulpice 533.675/195.575; 3. Reuchenette 585.700/225.400; 4. Schellenbrucke 646.500/252.150; 5. Jakobsberg 653.900/252.400; 6. Oberehrendingen 669.075/260.100.

metre thick with abundant discontinuities. The times of interrupted sediment accumulation spanned several chronostratigraphic zones. In the north-east normal sedimentation resumed in the early Oxfordian. To the south east the sedimentary hiatuses lasted longer and continuous sedimentation began again in the middle Oxfordian (Gygi & Marchand 1982). The dating for the beginning of the hiatuses in the two realms cannot accurately be determined because of erosion and subsequent reworking of ammonite beds, but is thought to have commenced in early Callovian (Gygi 1981).

## The condensed oolitic sequence

Within the condensed oolitic sequences it is often possible to distinguish beds with greenish chamositic ooids from those containing reddish-brown goethitic ooids (Fig. 2). The chamositic ooids are contained in a marly matrix whereas the goethitic ooids are found in a calcareous matrix.

The basal layers of the oolitic sequences are often bounded by hardgrounds. Within the sequences there are horizons with ferriferous encrustations which indicate discontinuities in sedimentation in the Callovian and early Oxfordian. These encrustations show μm-sized, cauliflower-like structures which are interpreted as being of biogenic origin (Gehring 1986). The encrustations appear together with coated lithoclasts and bioclasts, relics of reworked hardgrounds, as well as goethitic ooids which may be partly broken. Extreme variation in the sedimentological features of the clasts reveals that rather complex processes of reworking, corrosion and biogenic erosion took place.

The ooids in these horizons are exclusively goethitic. Based on palaeoecological and sedimentological data, Gygi (1981) postulated an aerobic marine depositional environment. This finding is in good agreement with the mineralogical characteristics of the ooids, since ferric hydroxides form under aerobic conditions.

## Samples and methods

Samples were collected from condensed deposits at six localities (Fig. 1). They were first examined with polarized microscopy and X-ray powder diffractometry (XRD; type: Philips X-ray generator PW 1130/00/60, 40kW, 25mA; proportional detector probe 1965/20/30). On polished thin sections electron microprobe analysis (EMP; type: SEMQ ARL) was undertaken. The analyses were performed along an axis from the periphery to the centre of individual ooids (cf. Gehring 1985).

The texture of the ooids was investigated by transmission electron microscopy (TEM; type: Philips EM-300). In order to detect semi-quantitatively the chemical composition of single ooids, a scanning transmission electron microscope (STEM; type JEOL 100EX) connected to an energy dispersive spectrometer system (EDS; type: Tracor Northern 5400) was used. Numerical analysis of the data was

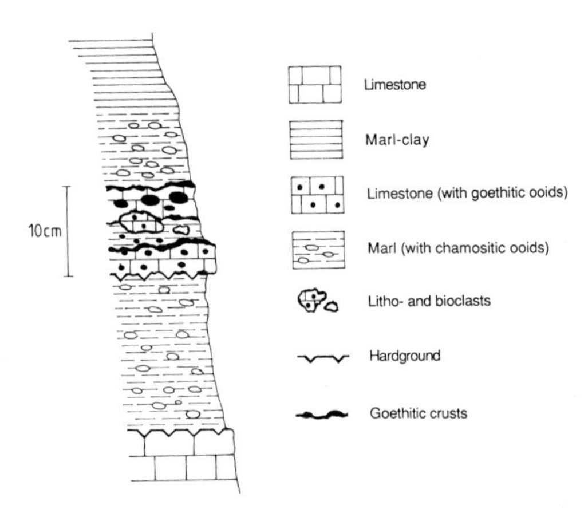


FIG. 2. Composite stratigraphic section of the condensed oolitic sequences close to the Middle/Upper Jurassic boundary in northern Switzerland.

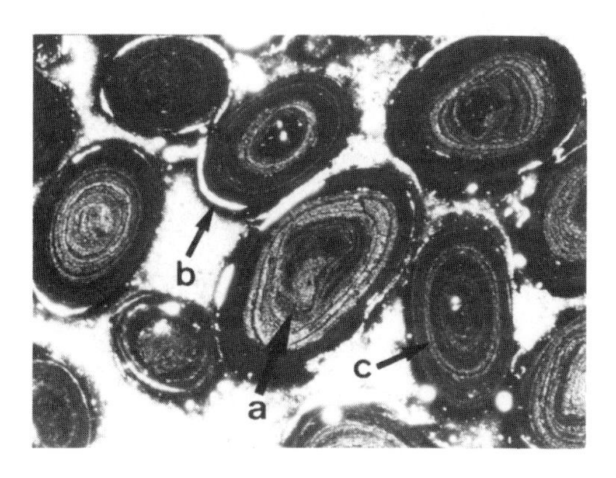


FIG. 3. Micrograph of goethitic ooids in a micritic matrix; ooid with a broken ooid as nucleus (arrowed a); shrinkage cracks filled with calcite (arrowed b); colour variations within an ooid (arrowed c). x40.

performed with SSQ-programm of Tracor Northern. For the TEM and STEM studies the oolitic samples were embedded in beeswax and polished. Prepared thin sections were heated and with fine tweezers thin sliced ooids could be removed and fixed on a TEM copper grid (diameter 2mm). Afterwards the samples were thinned out under an argon ion beam (Barber 1970) and coated with an about 150Å thick carbon film.

## Results and discussion

Under the light microscope the spherical to ellipsoidal ooids show a shell-like structure. The majority of ooids do not contain a distinct nucleus. Within the shell there are areas of variable colour intensity (Fig. 3), indicating different mineralogical compositions. XRD of the ooids showed that goethite is the most important ferriferous component. Hematite is the second and subordinate, iron oxide phase. XRD also revealed fluorapatite and sometimes illite-like impurities. The estimation of the Al-substitution in goethites from XRD data according to the method described by Schulze (1984) demonstrates that the goethites in the ooids have an Al amount of less than 2mol%. This amount of Al-substitution in the goethite under investigation is low when compared with goethites from minette-type iron ores (cf. Schellmann 1964).

The goethitic ooids suggest that changes due to diagenesis have occurred. The fissures and shrinkage of the ooids are the most prominent indication for diagenesis (Fig. 3). Shrinkage can be explained by a dehydration process during the alteration from ultrafine crystallized ferrihydrite ($5Fe_2O_3.9H_2O$) to goethite. Cavities produced through shrinkage were filled with calcite. This points to a relatively late diagenetic process in which the sediment was already lithified.

Another indication of diagenetic alteration are dark-red hematitic spots in the ooids. The hematite may have been formed as a consequence of dehydration of goethite (Berner 1969) or alternatively as a direct alteration product of ferrihydrite (Schwertmann & Murad 1983). EMP analysis of individual ooids shows that Fe, Ca and P, measured as oxides, together indicate a constant relationship between the different elements (Fig. 4). Fe correlates negatively with the two other elements. The $P_2O_5/CaO$ ratio is approximately 0.75 which indicates apatite. XRD confirms this interpretation. Apatite has often been found in other Jurassic ironstones (e.g. Maynard 1986; Halbach 1968). The inverse correlation between goethite and apatite within the ooids has already been described and explained by changes in chemical conditions during ooid formation (Gehring 1985). Beside Fe, Ca and P, the EMP analysis reveals highly variable concentration of Si and Al. The concentration and distribution of Si and Al in the ooids seems random. They may form part of clay mineral particles or other contaminants.

In the TEM goethite in the ooids shows a mosaic-like texture (Fig. 5a). Single grains have an isometric shape and an average diameter of

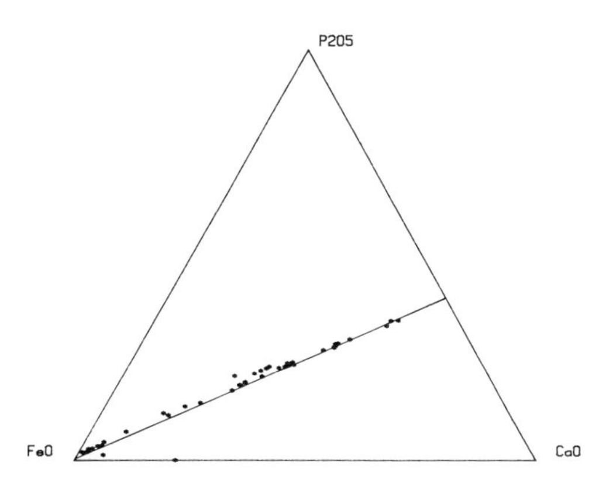


FIG. 4. Electron microprobe analyses of an individual goethitic ooid. The concentrations of the three principal elements Fe, Ca and P are determined from their oxide form of a total analysis.

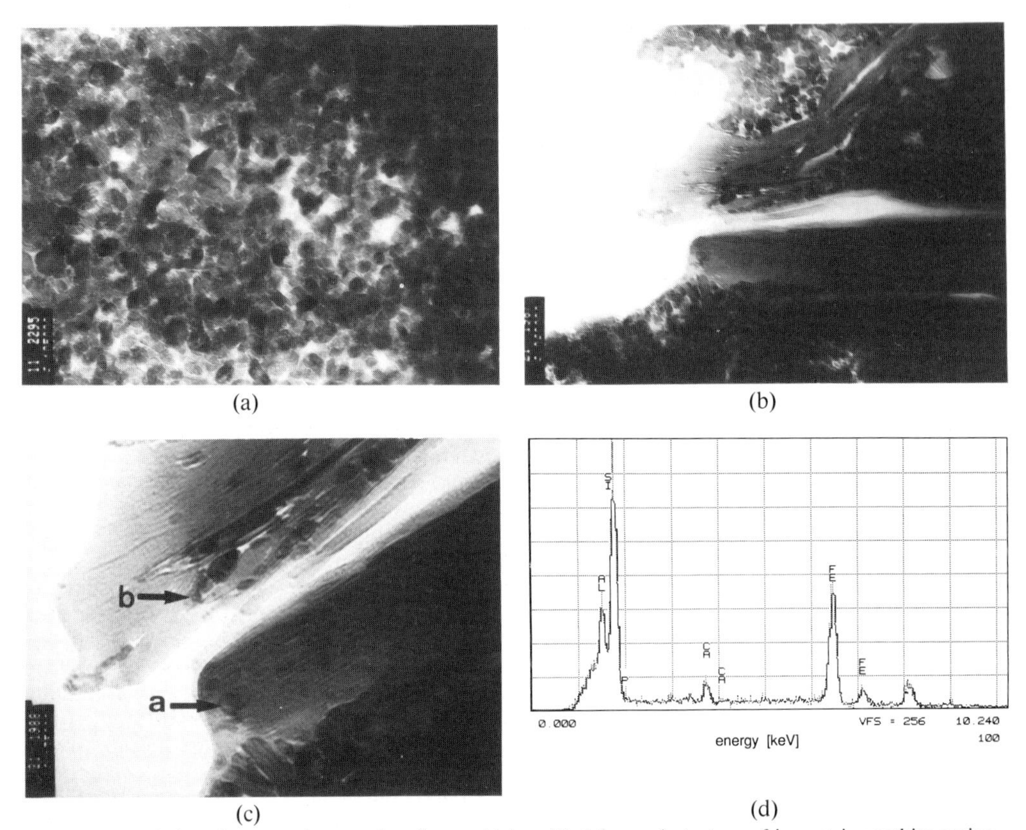


FIG. 5. Transmission electron micrographs of a goethitic ooid: (a) mosaic texture of isometric goethite grains, x80 000; (b) illite flakes associated with goethite grains, x46 000; (c) enlarged detail of Fig. 3(b), goethitic grains appear on illite surfaces (arrowed a) and between partially split layers of illite flakes, x 135 000; (d) semiquantitative analysis of the area shown in (b).

50–100nm. This shape of goethite was also described in ooids from other oolitic ironstones and lateritic pisoids (Radszewski & Schneiderhöhn 1962, Amouric *et al.* 1986). In contrast to the pisoids the goethite in the ooids does not display a needle-like morphology (Amouric *et al.* 1986) and therefore suggests that these ooids were not formed in lateritic soils. This interpretation conflicts with that of Siehl & Thein (1978), in which they suggest that oolitic ironstones form as an enrichment of lateritic ooids in a near-shore environment.

Within the mosaic textures the isometric grains sometimes appear together with other particles. The latter particles form distinct, often fillet-like flakes (Fig. 5b,c). Semiquantitative chemical analysis of areas rich in these flakes shows enrichments of Fe, Si and Al (Fig. 5d). Considering morphological data (Fig. 5b) the Fe peaks can be assigned to the isometric goethite grains. Therefore the fillet-like flakes are likely to be the structures containing Al and Si. The presence of Al and Si in flakes is strongly indicative for clay minerals. XRD data argue in favour of illite, which is by far the most common clay mineral in the Upper Jurassic sequences in northern Switzerland (Gygi & Persoz 1986).

Illite flakes show partially split layers (Fig. 5b,c). Lateral enlargement of single illite layers has been recognized previously and explained by a loss of potassium. Substantial loss of potassium can lead to a complete splitting of the layers and the formation of smectites or vermiculite layers (e.g. Srodon & Eberl 1984). The partial splitting (Fig. 5b,c) can only mean that the illite is poor in potassium and explains why a potassium peak is absent in the EDS spectra (Fig. 5d). Goethite grains are found between partially split layers of illite flakes (Fig. 5b,c). Assuming that the goethitic ooids were formed in an aerobic marine environment (Gygi 1981) in which the concentration of soluble $Fe^{3+}$ is negligible (cf. Murray & Gill 1978) it must be concluded that the iron(III) hydrous oxides in the split interfaces of illite flakes did not precipitate there; rather iron(III) hydrous oxides migrated into already split layers. The concentric array of apatite within the goethitic ooids (Gehring 1985) as well as the texture of these ooids give clues to a primary genesis. It can be excluded that the goethitic ooids were formed by replacement of a carbonate precursor since carbonate ooids are not found in condensed ammonite-rich environments. Furthermore, the formation of the examined goethitic ooids as a consequence of the oxidation of a chamositic precursor is improbable since the goethitic ooids are Al-poor. Maynard (1986) has concluded that a low Al concentration is inconsistent with the formation of goethitic ooids by oxidation of chamositic ooids.

## Adsorption processes and the formation of goethitic ooids

Adsorption processes play an important role in the mobilisation and nucleation of cations and anions in natural aquatic environments (Stumm & Morgan 1980). Metal oxides and hydroxides but also clay minerals show a strong affinity for surface chemical processes (cf. Schindler & Stumm 1987). However, geological systems do not allow us to study adsorption processes directly but the simultaneous occurrence of mineralogical phases within concentric particles suggests that surface chemical processes were also important in ancient sedimentary environments.

The chemical and mineralogical data in this study show that apart from goethite the ooids contain apatite and clay minerals. The simultaneous occurrence of these mineralogical phases gives a hint that adsorption processes could be the important cause for the formation of the goethitic ooids under investigation.

In the following paragraphs the formation of the examined goethitic ooids is described schematically from the perspective of surface chemistry (Fig. 6). Iron(III) hydrous oxides have a large specific surface which favours adsorption processes (e.g. Yates & Healy 1975). Several models attempt to explain the above processes (Stumm & Morgan 1981, and references therein). The models can be divided into two groups. The first group proposes that electrostatic forces are the major cause of adsorption (Stern 1924). The second group postulates chemical processes (e.g. ligand exchange reactions) as being the most important cause of adsorption (Breeuwsma & Lyklema 1973). However, many different parameters (e.g. pH, activity and ion exchange capacity) influence the adsorption processes of ions, chemical complexes and clay minerals (cf. Stumm & Morgan 1981).

Goethitic ooids formation could have been initiated by adsorption of particulate Fe(III) compounds (e.g. ferrihydrite or amorphous iron(III) hydrous oxides) on sediment particles (Fig. 6, 1 + 2). The source for these compounds in a depositional environment is likely to be the reducing sediment where ferrous iron is mobilized and later re-precipitated as an iron(III) hydrous oxide under aerobic

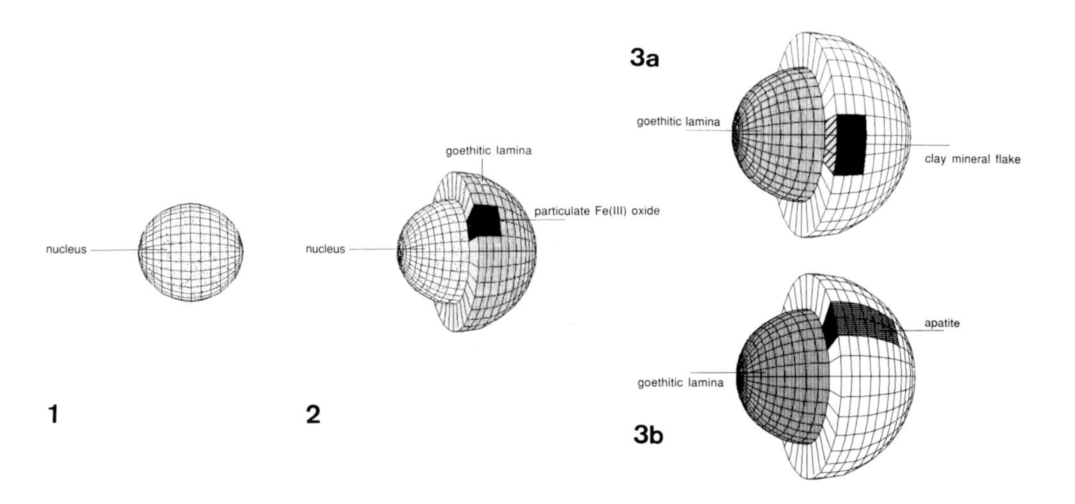


FIG. 6. Schematic interpretation of different adsorption processes during the formation of goethitic ooids (for explanation see text).

conditions at the sediment/water interface (cf. Shiller *et al.* 1985). α-FeOOH can be formed subsequently by ageing processes of the iron(III) hydrous oxide precursor (cf. Flynn 1984). Growing goethtic ooids have large specific surfaces which give the possibility for the adsorption of different ions and particles. The concentric array of apatite and the illite impurities within the goethite ooids suggest that different adsorption processes could have occurred (Fig. 6, (3a, b)). Furthermore, the mineralogical variations within the ooids can be interpreted by changes of the chemical conditions, probably due to a decrease of the pH or an increase of the phosphate concentration, which could influence the surface charge of the growing ooids (cf. Schindler & Stumm 1987). The adsorption of clay minerals onto iron(III) hydrous oxide surfaces can be explained by electrostatic interactions as postulated in the model of Stern (1924). It has been suggested that the appearance of apatite in goethitic ooids is caused by adsorption due to chemical surface processes (Gehring 1985). Stumm & Sigg (1979) suggested the following ligand exchange reaction for the adsorption of phosphate onto an iron oxidic surface (≡FeOH represents a surface group):

$$\equiv FeOH + H_2PO_4^- = \equiv FeHPO_4^- + H_2O.$$

Apatite formation on a goethitic surface could be explained by nucleation of the phosphate with Ca(II). During the formation of goethitic ooids the above described processes would be repeated many times.

## Conclusion

Based on geological information from oolitic condensed sequences close to the Middle/Upper Jurassic boundary and the above considerations concerning the chemistry and texture of goethitic ooids, it is concluded that:

(1) Goethitic ooids were formed in an aerobic depositional environment characterized by reworking during periods of discontinuous sedimentation.
(2) Ooid formation could have been caused by adsorption processes whereby particulate iron(III) hydrous oxide was adsorbed onto a nucleus site and subsequently aged to α-FeOOH.
(3) Clay mineral flakes could have been adsorbed onto goethitic laminae due to electrostatic interactions during the genesis of the ooids.
(4) The adsorption of apatite during ooid formation can be explained by a ligand exchange reaction followed by nucleation processes.
(5) The different adsorption processes stated in (2), (3) and (4) would have had to be sequentially repeated during ooid growth.

ACKNOWLEDGEMENTS: The author would like to thank D. Oigaard and C. Holderegger for carrying out the TEM and the STEM work; also C. Gehring and D. Martill for a critical reading of the manuscript.

## References

AMOURIC, M. BARONNET, A., NAHON, D. & DIDIER, P. 1986. Electron microscopic investigations of iron oxihydroxides and accompanying phases in lateritic iron crust pisoids. *Clays and Clay Minerals,* **34,** 45–52.

BARBER, D.J. 1970. Thin foils of non-metals made for electron microscopy by sputter-etching. *Journal of Materials Science,* **5,** 1–8.

BERNER, R.A. 1969. Goethite stability and the origin of red beds. *Geochimica et Cosmochimica Acta,* **33,** 605–615.

BHATTACHARYYA, D.P., & KAKIMOTO, P.K. 1982. Origin of ferriferous ooids: a SEM study of ironstone ooids and bauxite pisoids. *Journal of Sedimentary Petrology,* **52,** 849–857.

BOLLIGER, W., & BURRI, P. 1970. Sedimentologie von Schelf-Carbonaten und Beckenablagerungen im Oxfordian des zentralen Schweizer Jura. *Beiträge zur geologischen Karte der Schweiz (N.F.),* **140.**

BREEUWSMA, A., & LYKLEMA, J. 1973. Physical and chemical adsorption of ion in the electrical double layer on hematite (α-$Fe_2O_3$). *Journal of Colloidal and Interface Science,* **43,** 437–448.

DAHANAYAKE, K., & KRUMBEIN, W.E. 1986. Microbial structures in oolitic iron formations *Minerallum Deposita,* **21,** 85–94.

DÉBRAND-PASSARD, S., & COURBOULEIX, S. 1984a. Synthèse géologique de Sud-Est de la France (stratigraphie et paléogéographie). *Memoires du Bureau de Recherches géologiques et minieres,* **125,** pp. 615.

——1984b. Synthèse géologique du Sud-Est de la France (atlas). *Memoires du Bureau de Recherches géologiques et minières,* **126,** pp. 18, 61 maps.

FLYNN, C.M.jr., 1984. Hydrolysis of inorganic iron(III) salts. *Chemical Reviews,* **84,** 31–41.

GEHRING, A.U. 1985. A microchemical study of iron ooids. *Eclogae Geologicae Helvetiae,* **78,** 451–457.

——1986. Mikroorganismen in kondensierten Schichten der Dogger/Malm-Wende im Jura der Nordostschweiz. *Eclogae Geologicae Helvetiae,* **79,** 13–18.

GYGI, R.A. 1981. Oolitic iron formations: marine or not marine? *Eclogae Geologicae Helvetiae,* **74,** 233–254.

——& MARCHAND, D. 1982. Les faunes de Cardioceratinae (Ammonoidea) de Callovian terminal et de l'Oxfordian inférieur et moyen (Jurassique) de la Suisse septentrionale: stratigraphie, paléoécologie, Taxonomie préliminaire. *Geobios,* **14,** 517–571.

——& PERSOZ, F. 1986. Mineralostratigraphy, litho- and biostratigraphy combined in correlation of the Oxfordian (Late Jurassic) formations of the Swiss Jura range. *Eclogae Geologicae Helvetiae,* **79,** 385–454.

HALBACH, P. 1968. Zum Gehalt von Phosphor und anderen Spurenlementen in Brauneisenerzooiden aus dem fränkischen Dogger beta. *Contribution to Mineralogy and Petrology,* **18,** 241–251.

HALLAM, A. 1984. Pre-Quaternary sea-level changes. *Annual Review of Earth and Planetary Science,* **12,** 205–243.

JAMES, H.L. 1966. Chemistry of the iron-rich sedimentary rocks. *U.S. Geological Survey, Professonal Papers,* **400-W,** 61p.

MAYNARD, J.B. 1986. Geochemistry of oolitic iron ores, an electron microprobe study. *Economic Geology,* **81,** 1473–1483.

MURRAY, J.W., & GILL, G. 1978. The geochemistry of iron in Puget Sound. *Geochimica et Cosmochimica Acta,* **42,** 9–19.

PERSOZ, F., & REMANE, J. 1973. Evolution des milieux de dépôt au Dogger supérieur et au Malm dans le Jura neuchâtelois méridional. *Eclogae Geologicae Helvetiae,* **66,** 41–70.

RADSZEWSKI, D.E., & SCHNEIDERHÖHN, P. 1962. Elektronenmikroskopische Untersuchungen von Nadeleisenerzooiden in Pulverpraperaten und Dunnschliffen. *Beiträge zur Mineralogie und Petrographie,* **8,** 349–353.

ROHRLICH, V. 1974. Microstructure and microchemistry of iron ooliths. *Mineralium Deposita,* **9,** 133–142.

SCHELLMANN, W. 1964. Zur Rolle des Aluminiums in Nadeleisenerz Ooiden. *Neues Jahrbuch Mineralogie, Monatshefte,* 49–56.

SCHINDLER, P.W., & STUMM, W. 1987. The surface chemistry of oxides, hydroxides, and oxide minerals. In STUMM, W. (ed.) Aquatic surface chemistry. *Wiley, New York,* 83–107.

SCHULZE, D.C. 1984. The influence of aluminium on iron oxides VIII. Unit cell dimension of Al-substituted goethites and estimation of Al from them. *Clays Clay Miner.,* **32,** 36–44.

SCHWERTMANN, U., & MURAD, E. 1983. Effect of pH on the formation of goethite and hematite from ferrihydrite. *Clays and Clay Minerals,* **31,** 277–284.

SHILLER, A.M., GLESKES, J.M., & PRICE, N.B. 1985. Particulate iron and manganese in the Santa Barbara Basin, California. *Geochimica et Cosmochimica Acta,* **49,** 1239–1249.

SIEHL, A. & THEIN, J. 1978. Geochemische Trends in der Minette (Jura Luxemburg/Lothringen). *Geologische Rundschau,* **67,** 1052–1077.

SRODON J. & EBERL, D.D. 1984. Illite. *In:* BAILEY, S.W. (ed) *Micas. Reviews in Mineralogy,* **13,** 495–544.

STERN, O. 1924. Zur Theorie der elektrolytischen Doppelschicht. *Zeitschrift für Elektrochemie und angewandte physikalische Chemie,* **30,** 508–516.

STUMM, W., & MORGAN, J. 1981. *Aquatic chemistry.* Wiley-Interscience, New York.

——& SIGG, L. 1979. Kolloidchemische Grundlagen der Phosphor-Elimination in Fällung, Flocking und Filtration. *Zeitschrift für Wasser- und Abwasser-Forschung,* **12,** 73–83.

TEYSSEN, T.A.L. 1984. Sedimentology of the Minette oolitic ironstones of Luxembourg and Lorraine. *Sedimentology.* **31,** 195–211.

TRÜMPY, R. 1980. *Geology of Switzerland, Part A: An Outline of the Geology of Switzerland.* Wepf & Co, Basel, New York.

VAIL, P.R. HARDENBOL, J., & TODD, R.G. 1985. Jurassic unconformities, chronostratigraphy and sea-level changes from seismic stratigraphy and biostratigraphy. *Memoirs American Association of Petroleum Geologists,* **36,** 129–144.

VAN HOUTEN, F.B. & BHATTACHARYYA, D.P. 1982. Phanerozoic oolitic ironstones — geologic record and facies model. *Annual Review of Earth and Planetary Science,* **10,** 441–457.

——& PURUCKER, M.E. 1984. Glauconitic peloids and chamositic ooids — favorable factors, constraints, and problems. *Earth-Science Reviews,* **20,** 211–243.

YATES, D.E., & HEALY, T.W. 1975. Mechanism of anion adsorption on ferric and chromic oxide/water interfaces. *Journal of Colloidal and Interface Science,* **52,** 222–228.

ZIEGLER, P.A. 1982. *Geological atlas of western and central Europe.* — Elsevier, Amsterdam.

A.U. GEHRING, Geological Institute, Swiss Federal Institute of Technology, CH–8092 Zurich, Switzerland.

# Iron-rich ooids, their mineralogy and microfabric: clues to their origin and evolution

**A.T. Kearsley**

S U M M A R Y: The controversy surrounding oolitic ironstone origins has been prolonged by a lack of published discriminatory characters which could be applied to determine the sedimentary processes responsible for the genesis of individual iron-rich ooids. This is in part due to the limitations of optical petrography and has led to confusion between authors as to the nature of the ooids each is describing. A new and simple descriptive system which documents ooid mineralogy and fine structure has been used to distinguish ooids of different origins and a classification of iron-rich ooids into three broad mineralogical categories is proposed in this paper. These Fe-oxide/hydroxide, berthierine/chlorite and siderite dominated groups are further divided by characteristic textural criteria into 15 minor categories. The abundance of ooids which easily fit the three bulk mineralogical categories probably reflects common diagenetic modifications, rather than preservation of discrete primary phases. The wide variety of ooid microfabrics observed by transmitted light microscopy, secondary electron, backscattered electron and X-ray emission imagery reflect differing modes of ooid accretion and early diagenesis, in addition to later diagenetic and weathering processes.

Combined mineralogical and textural classification of individual ooids can indicate the relative importance of individual processes (such as lateritic ooid/pisolith growth, replacement of carbonates by berthierine or primary berthierine precipitation) in the generation of a particular ironstone.

Wherever diagnostic textural features can be seen in the matrix or cement of an oolitic ironstone it may be relatively easy to infer depositional and diagenetic processes and thereby determine the origin of both the sediment as a whole and of the ooids in particular. Unfortunately, many ironstones show very little original structure due to post-depositional modification of both texture and mineralogy and the few unambiguous clues remaining are often only to be found within ooids. In quartz arenites and calcarenites iron-rich ooids often occur as isolated and derived grains whose precise identification and interpretation can yield important information about sediment provenance, transport and diagenesis.

Iron-rich ooids display a bewildering variety of mineralogies and fine structures whose diversity is only now becoming apparent (Fig. 1 and Table 1). Several independent sedimentary processes can be responsible for the generation of these concentrically laminated grains yet bulk ooid mineralogy has proven inadequate as a reliable indicator of palaeoenvironment of origin as it is often substantially modified during diagenesis, metamorphism and subsequent erosional weathering.

Ooids are often composed of complex intergrowths of several microcrystalline mineral components which are difficult to distinguish by conventional thin section petrography. The textures and microfabrics of these fine minerals record a sequence of paragenetic events and thus provide the key to understanding ooid origin and evolution. Fine structure will therefore be of increasing importance in the recognition of sedimentary processes and palaeoenvironments as it is possible to distinguish unambiguous primary growth and secondary replacement textures and to interpret a likely sequence of diagenetic mineral transformations without having to rely upon generalized models of pore-fluid evolution (themselves based upon modified and often unreliable textures within the rock matrix or cement).

## Techniques in ooid mineralogical, textural and microfabric studies

Transmitted light microscopy of thin sections has been very widely applied in ironstone petrography and can yield rapid mineralogical and textural information on a coarse scale, but its use is severely limited by internal refraction effects at grain boundaries within a 30 micron section. This produces poor spatial resolution, and generally restricts reliable textural analysis to lamellae coarser than 10 microns in width.

*From* YOUNG, T. P. & TAYLOR, W. E. G. (eds), 1989, *Phanerozoic Ironstones*
Geological Society Special Publication No. 46, pp. 141-164

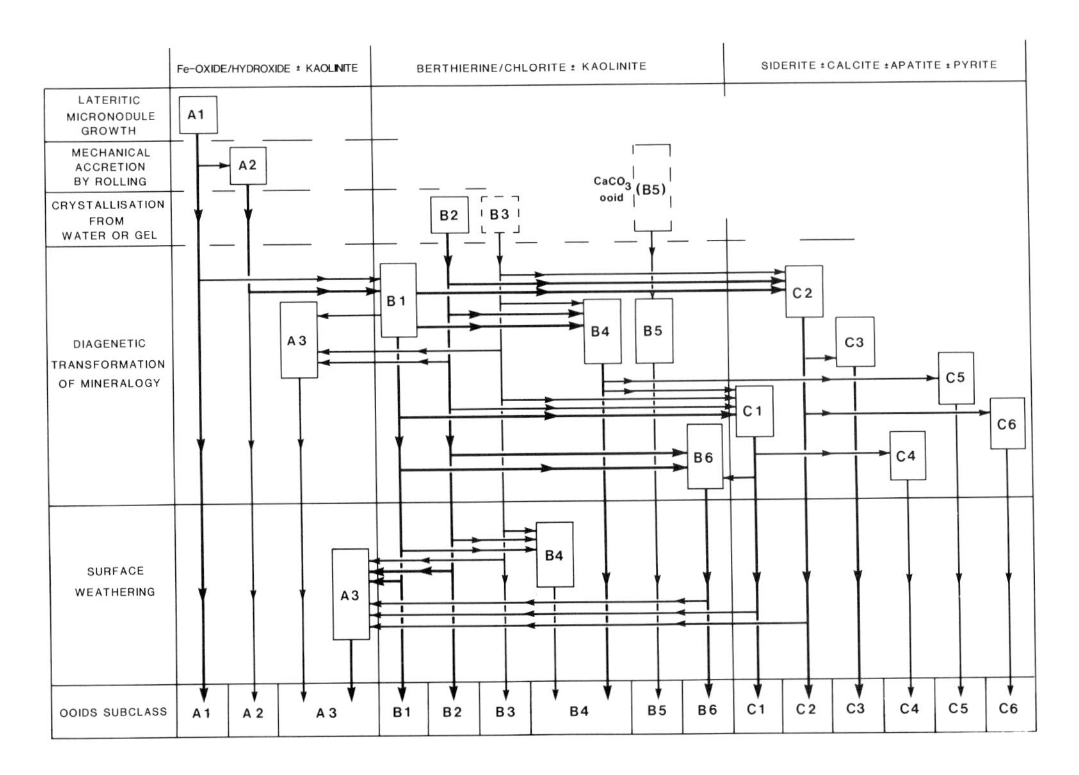


FIG. 1. The origin, evolution and classification of iron-rich ooids. Bold lines represent important processes, thin lines those of lesser importance and rarer occurrence.

Minerals with similar colours are often difficult to distinguish, and dark or opaque minerals (such as goethite and apatite) may effectively obscure any fine structure, making interpretation of their relative ages of growth impossible. These problems have considerably hampered the interpretation of ooid microstructure, but optical petrography using crossed polarizers is still probably the best method by which mineral fabrics may be seen in the extinction patterns of areas otherwise considered homogeneous and devoid of oriented microstructure.

Scanning electron microscopy (SEM) has been used to document ooid microfabrics (Bhattacharyya & Kakimoto 1982) by examination of gold-coated fracture surfaces. This secondary electron imagery (SEI) provides excellent three-dimensional information, but the sputtered gold coat and complex surface topography make X-ray microanalysis of individual particles very difficult. Identification of different minerals has generally been by recognition of characteristic crystal shapes (often giving ambiguous results) and mineral intergrowths are not easily distinguished. Transmitted light microscopy has normally been used in conjunction with SEI to provide the rather coarser scale textural information.

Backscattered electron imagery (BEI) is particularly well suited to the study of ironstones as it allows clear differentiation of a wide variety of minerals (Fig. 2) without the loss of spatial resolution encountered in transmitted light petrography. The technique has now been widely applied in both sandstone and mudstone petrography and is available on most modern scanning electron microscopes (Krinsley *et al.* 1983; White *et al.* 1984). Carbon-coated and polished sections are normally used, and particles as small as 0.25 microns can be distinguished, the commonest ooid minerals (Table 2: Fe chlorite/berthierine, kaolinite, siderite, calcite, apatite and silica) are easily recognised by their characteristic brightness in the image. Textural details normally lost in optical petrography are rapidly revealed by compositional contrast BEI. This is invaluable as a guide to reliable positioning of the electron beam for quantitative analysis of monomineralic areas. It is indeed always necessary to use X-ray elemental analysis when BEI is employed as the

TABLE 1. *Calculated electron backscattering coefficients (ETA) for common sedimentary minerals*

| *MINERAL* | *FORMULA* | *ETA* |
|---|---|---|
| Zircon | $ZrSiO_4$ | 0.2411 |
| Pyrite | $FeS_2$ | 0.2298 |
| Hematite | $Fe_2O_3$ | 0.2228 |
| Goethite | $FeOOH$ | 0.2084 |
| Pyrolusite | $MnO_2$ | 0.2049 |
| Jarosite | $KFe_3(SO_4)_2(OH)_6$ | 0.1910 |
| Rutile | $TiO_2$ | 0.1836 |
| Siderite | $FeCO_3$ | 0.1791 |
| Fe-Chlorite | $Fe_{10}Al_2[Si_6Al_2O_{20}](OH)_{16}$ | 0.1771 |
| Rhodochrosite | $MnCO_3$ | 0.1744 |
| Vivianite | $Fe_3(PO_4)_2.8H_2O$ | 0.1615 |
| Francolite | $Ca_5(PO_4)_{2.5}(CO_3)_{0.75}F$ | 0.1583 |
| Mg/Fe-Chlorite | $Fe_6Mg_4A1_2[Si_6Al_2O_{20}](OH)_{16}$ | 0.1568 |
| Biotite | $K_2Mg_2Fe_2Ti[Si_6Al_2O_{20}](OH)_4$ | 0.1536 |
| Glauconite | $K_2(Fe,A1,Fe,Mg)\ [Si_7AlO_{20}](OH)_4.H_2O$ | 0.1477 |
| Ankerite | $MgFeCa_2(CO_3)_4$ | 0.1443 |
| Calcite | $CaCO_3$ | 0.1422 |
| Gypsum | $CaSO_4.2H_2O$ | 0.1385 |
| K-Feldspar | $KA1Si_3O_8$ | 0.1370 |
| Muscovite | $K_2Al_4[Si_6Al_2O_{20}](OH)_4$ | 0.1311 |
| Illite | $K_{1.5}Al_4[Si_{6.5}A1_{1.5}O_{20}](OH)_4$ | 0.1291 |
| Silica | $SiO_2$ | 0.1253 |
| Na Feldspar | $NaA1Si_3O_8$ | 0.1243 |
| Dolomite | $CaMg(CO_3)_2$ | 0.1237 |
| Smectite | $Na_2Al_{10}Mg_2[Si_{24}O_{60}](OH)_{12}$ | 0.1225 |
| Kaolinite | $Al_4[Si_4O_{10}](OH)_8$ | 0.1185 |
| Mg Chlorite | $Mg_{10}Al_2[Si_6Al_2O_{20}](OH)_{16}$ | 0.1178 |
| Carbon | C | 0.0640 |

A high ETA value means that many electrons will be detected, a high signal output produced and a bright area seen in the image. Pyrite (bright)>hematite>geothite>siderite>francolite>berthierine>calcite>silica>kaolinite (dark).

same backscatter signal may be produced by different minerals (e.g., calcian siderite and francolite produce very similar pale grey tones). X-ray microanalysis of an apparently single point can itself produce misleading results as the electron beam may penetrate beyond the desired mineral grain and generate X-ray emission from surrounding areas. This problem may be recognized and avoided by element mapping of interesting areas, thus digital X-ray emission imagery is useful to provide both textural and analytical data (Fig. 3 i–j). The automated analysis of crystallographic orientation as applied to quartz petrofabrics is unfortunately not possible with ironstone ooids even in massive monomineralic patches unless sophisticated pseudo-Kikuchi pattern images are obtained (Dingley 1981). Identification of iron phyllosilicate structure (i.e. berthierine or chlorite) also requires electron or X-ray diffraction study (XRD), neither of which is available within the SEM. Fabric studies therefore must depend upon mineral platelet long-axis orientation measurement, subjective description or micrographs.

Thus a combination of transmitted light miscroscopy, XRD, SEI, BEI and digital X-ray emission imagery is necessary to adequately document ooid mineralogy, texture and microfabric. For very detailed examination of fine clay mineral intergrowths and their lattice orientations, high-resolution transmitted electron microscopy is essential, but the difficult preparation techniques and limited area of images combine with high cost to restrict the application of TEM in ironstone petrography (Hughes, this volume).

## Classification of iron-rich ooids

Examination of a wide variety of ironstone ooids using SEI, BEI and optical petrography has demonstrated that a great deal of mineralogical, textural and microfabric information is available for the interpretation of their origin and sub-

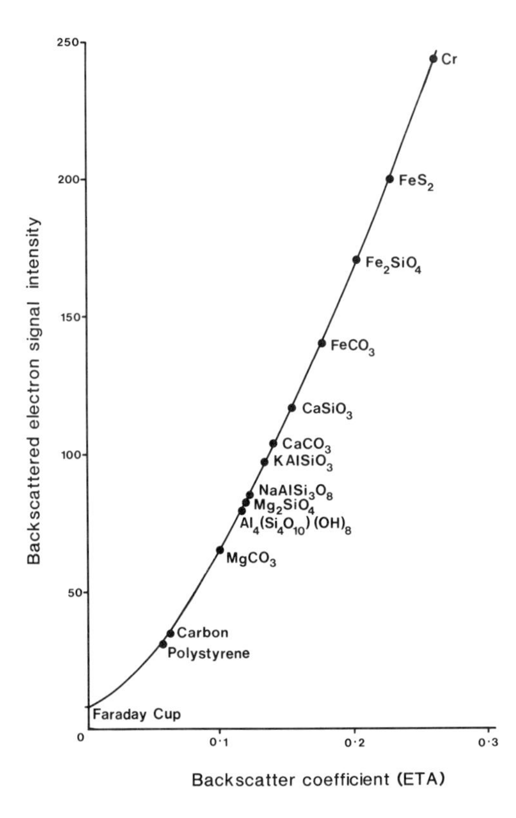


FIG. 2. Measured backscattered electron detector signal output plotted as a function of calculated backscattering coefficient for mineral standards, showing potential of BEI in differentiating ironstone constituents.

sequent evolution. There is currently no unified classification of iron-rich ooids, which has resulted in a loss of precision in published descriptions (see comments in Trendall (1983)). A preliminary scheme is proposed here (Table 2), it is not intended to be a final and immutable edifice, but merely to provide a framework and to highlight areas of microfabric investigation which may result in a better understanding of the origins of different ooid types.

Any descriptive system which is to be widely used by non-specialist workers must be founded upon a relatively simple large-scale classification. Despite the problems of convergent diagnetic development of the same minerals it is most convenient to produce a coarse classification based upon the predominant *ooid bulk mineralogy*.

Three broad classes are proposed, each defined by the predominance (>50% of ooid cortex by volume) of the appropriate minerals — it should be noted that their boundaries are necessarily subjective and rather arbitrary as subsequent modification may shift the proportions of the bulk minerals, taking the ooid from one major class to another (Fig. 1). Classification on *microfabric* alone would be equally unsatisfactory as it may often reflect either relict texture or later recrystallization rather than the characteristic growth form of the present mineralogy. Thus a descriptive scheme must include details of both ooid bulk chemistry and the texture of each mineral.

## The use of microfabric as an indicator of sedimentary process

The dependence of ironstone mineralogy upon sedimentary and diagenetic processes has been widely discussed (Bubenicek 1983; Dimroth 1979; Kimberley 1979; Maynard 1986) and can be seen within the sedimentary geochemical models of Curtis & Spears (1968). Textural and particularly microfabric evidence has, however, received less attention (Bhattacharyya & Kakimoto 1982). Differing mechanisms of ooid growth and evolution produce characteristic microfabrics which are diagnostic of the processes responsible (however, from the summary below it can be seen that one should not be too optimistic in ascribing a particular microfabric to a single growth mechanism). Nine major processes are probably responsible for the diversity of observed ooid types.

(1) Precipitation from capillary pore fluid within fine grained sediments subjected to periodic wetting and drying yields extremely fine (sub micron) crystals of goethite which are often subsequently oxidized to hematite. The oxide/oxyhydroxide films are apparently structureless, thin and irregular, both coating and penetrating individual blocks and grains of the rock (Figs. 4d, f). This process is not restricted to laterites and often occurs in intertidal sediments where early diagenetic sulphides are periodically subjected to oxidation. The microfabric is unmistakeable, although it may be subsequently modified.

(2) Precipitation may occur from open pore or surface water. This process results in relatively large (1 micron diameter), well formed tabular crystals of phyllosilicates such as berthierine or glauconite. Clusters of radiating platelets in a honeycomb or random 'house-of-cards' arrangement are common in encrusting cement (Fig. 5 c).

TABLE. 2 *Classification of Iron-Rich Ooids, their textures and probable origins*

|  | | |
|---|---|---|
| *Class A. Fe Oxide/Hydroxide ± Kaolinite Dominated* | | |
| | A1. | Goethite ± Hematite or Kaolinite (irregular)<br>Precipitation during lateritic weathering |
| | A2. | Goethite ± Kaolinite or Hematite (concentric laminae + nucleus)<br>Mechanical accretion of laminae |
| | A3. | Goethite ± Kaolinite ± Berthierine (patchy laminae)<br>Diagenetic or weathing alteration of class B ooids |
| *Class B. Berthierine/Chlorite ± Kaolinite Dominated* | | |
| | B1. | Berthierine ± Fe Oxide/Hydroxide (poorly defined laminae)<br>Transformation of A2 Fe-oxide + Kaolinite mix |
| | B2. | Berthierine + (concentric laminae)<br>Authigenic growth of ber hierine from fluid |
| | B3. | Berthierine + (concentric laminae)<br>Authigenic growth of berthierine from gel? |
| | B4. | Kaolinite ± Berthierine (concentric, spastolithic, deformed)<br>Diagenic degradation of berthierine |
| | B5. | Berthierine (poor, concentric laminae + bioclast textures)<br>Diagentic replacement of calcium carbonates |
| | B6. | Chlorite ± Silica (patchy)<br>Late diagenetic/metamorphic alteration of B1-3, C1 |
| *Class C. Siderite ± Calcite ± Phosphate ± Pyrite Dominated* | | |
| | C1. | Siderite, Berthierine, Kaolinite (displacive, disruptive)<br>Displacive growth of siderite in B4 |
| | C2. | Siderite (spongy, laminar)<br>Fine replacement of ooid laminae |
| | C3. | Calcite + Siderite (granular, laminar)<br>Fine replacement of siderite C2 |
| | C4. | Calcite ± Siderite (coarse, crystalline)<br>Coarse replacement and cement growth in ooid cores |
| | C5. | Apatite (concentric, laminar)<br>Micronodule growth or replacement of berthierine in B4 |
| | C6. | Pyrite (blocky)<br>Replacement of siderite C2 |

Unless growth has occurred within a gradually opening fracture there is rarely sign of well-developed fabric, although compositional zonation may be apparent. Fluctuations in the fluid chemistry may change the abundance of seed nuclei and the rate of crystal growth, producing alternate layers of coarse and fine platelets. On a scale of microns this can appear as concentric lamination (Fig. 5 g).

(3) Precipitation can occur within an oxide/hydroxide/hydrous-aluminosilicate gel which has concentrated the necessary elements from a dilute surrounding fluid (Harder, this volume). It is likely that this process would again generate a random orientation of platelet (001) planes. A double diffusive mechanism may be responsible for the intercalation of layers of differing mineralogy, with zones concentric to the gel surface. It is perhaps easier to preserve unstable coexisting minerals within a gel than sequentially to precipitate them from a reactive surface fluid without alteration of the underlying lamina. Biogenic mucus sheaths may also provide a site for element fixation, generating subspherical ooids (Dahanayake & Krumbein 1986).

(4) Mechanical accretion of components may occur by adhesion to the surface of a rolling grain. This process may require either electrostatic binding of new components or their capture by a sticky surface (of possible biological origin). The concentric laminae may represent separate phases of accretion, possibly linked to repeated burial and ex-

humation, and may selectively collect certain minerals from the sediment substrate. Clay mineral flakes can be either mechanically bound in random orientations, or electrostatically bound parallel to existing crystal planes. Subsequent rolling would reorientate flakes and generate a concentric tangential fabric. The incorporation of tangentially oriented detrital grains (such as heavy minerals, quartz silt or smaller ooids as reported by Chauvel & Guerrak (this volume) is perhaps the only reliable indicator that mechanical accretion has occurred; even this may only reflect a brief episode of rolling while the major mode of growth is precipitation.

(5) *In situ* transformation of minerals may occur by reaction between themselves and the pore fluid. The bulk mineralogy may change and accompanying secondary textures may either obliterate or pseudomorph the previous microfabrics. New mineral growth *can* generate randomly oriented fans of stacked crystallites which may cut across the relict fabric, but clay mineral transformations probably rarely give appreciable fabric alteration (as the oriented platelets react without movement or creation of significant new pore space). Thus the existence of a well defined microfabric within a phyllosilicate component does not necessarily imply that it is the primary mineral.

(6) Dissolution of a primary mineral, followed by growth of secondary assemblage is very common, and may proceed by mechanisms akin to the migrating solution film proposed for carbonate neomorphism (Bathurst 1976). If the rates of dissolution and precipitation are very similar a fine scale pseudomorphing can occur, with the primary fabric preserved. When dissolution proceeds more rapidly than precipitation, voids are filled by coarser and randomly oriented crystallites. Unless the migrating dissolution front is arrested and preserved (a rare event) it can be very difficult to distinguish fine microfabrics produced by this mechanism from those of processes (2) to (5). This is particularly a problem in interpretation of berthierine and kaolinite microfabrics. Whenever calcite or siderite is the secondary mineral the process is relatively easy to recognize as replacement is rarely complete, and patches of neomorphic carbonate produce a crude pseudomorphing of the fine phyllosilicate precursor fabric (Fig. 6g).

(7) Displacement of an existing microfabric by growth of a later mineral can usually be easily recognized. The secondary phases often grow upon included mineral nuclei (cf. Bhattacharyya & Kakimoto 1982), or upon more finely crystalline areas of replacement which are often at the ooid rim forming an eggshell (Fig. 6e). The growth of large euhedral displacive crystals forces the pre-existing softer microfabric aside, producing folds and indentations (Fig. 6e).

(8) Compactional reorientation may unfortunately destroy the most diagnostic features of a primary microfabric. Individual mineral platelets may be rotated into a tangential orientation to the surface of a rolled grain or a foliation parallel to the compactional horizontal may be generated during early sediment burial. Strongly cemented portions of ooids may resist compaction generating eggshell spastoliths (Fig. 6c), but many kaolinite dominated ooids show intense deformation (Fig. 6a). Crushing of ooids onto carbonate rhombs or silica sand grains may also disrupt the diagnostic microstructure.

(9) Shrinkage on dehydration or transformation can result in collapse of a highly porous fabric and the contraction of concentric laminae, producing small folds and lobes. This may superficially resemble the crenulation common in microbially bound ooids and oncoids.

## The features of the classes and subclasses of iron-rich ooids

The almost ubiquitous modification of ooid morphology, mineralogy and microfabric by post depositional pore fluid activity and compactional deformation often prevents the reliable analysis of ooid origins. Using the criteria discussed above, and based upon a

Fig. 3. a–h, PPL; i,j, digital X-ray maps; a,b ooids of subclass A1 (see Fig. 5 a,b) showing poor resolution in almost opaque Fe oxides, Lower Cretaceous, St. Chinian, France; c–g, berthierine replaced bioclasts in B5 ooids, gastropod, bivalve, ostracod, echinoid spine and crinoid brachial ossicles respectively; h, berthierine-replaced echinoderm stereom nucleus in B2 ooid, green facies of Raasay Ironstone; i,h compacted B4–C5 ooid; i, berthierine = white; ferroan calcite = blue; j, francolite = yellow; calcite = green; illite = blue, Raasay Ironstone.

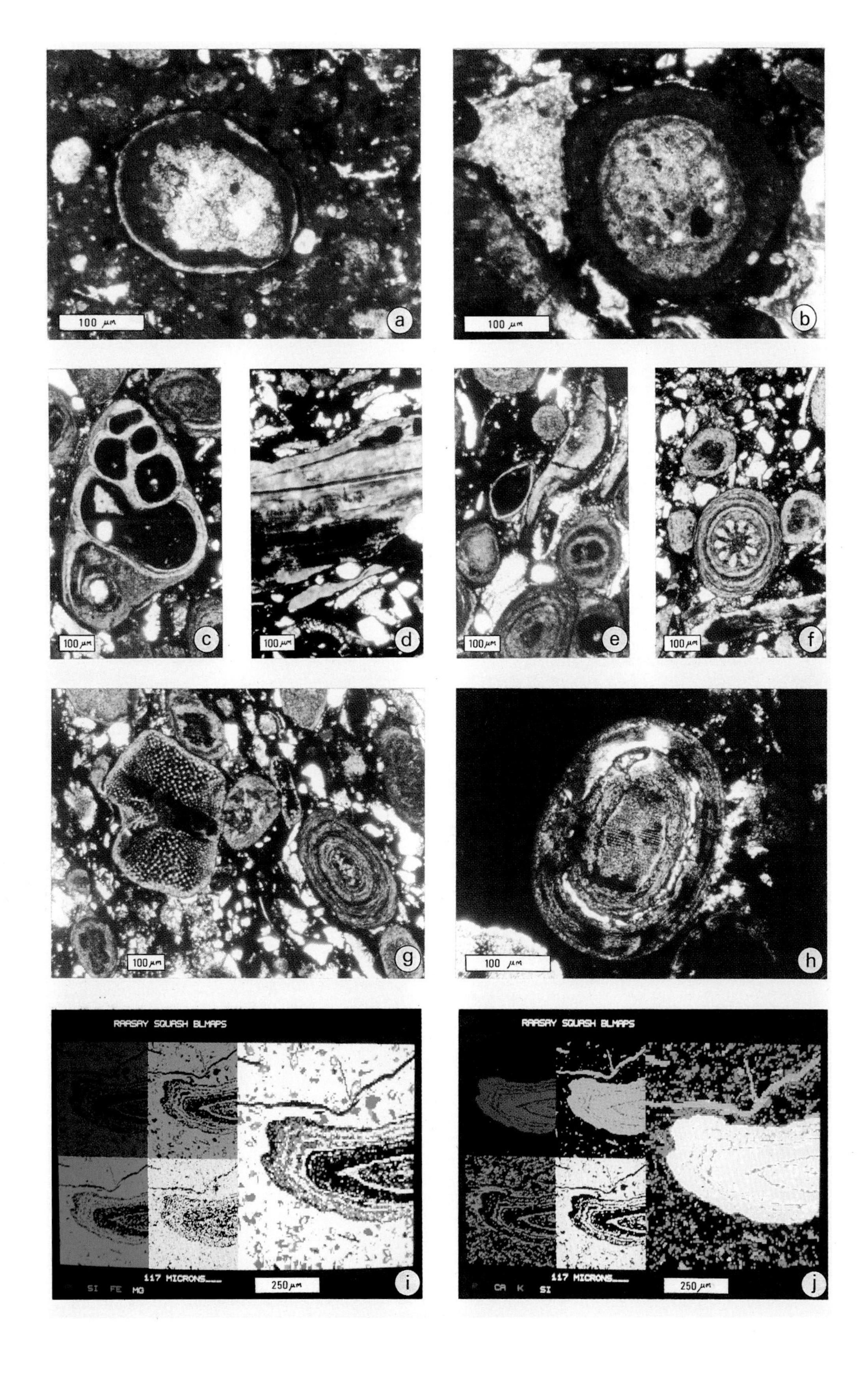
100 µm
a
100 µm
b
100µm
c
100µm
d
100µm
e
100µm
f
100µm
g
100 µm
h
RAASAY SQUASH BLMAPS
117 MICRONS
250µm
SI FE MG
i
RAASAY SQUASH BLMAPS
117 MICRONS
250µm
P CA K SI
j

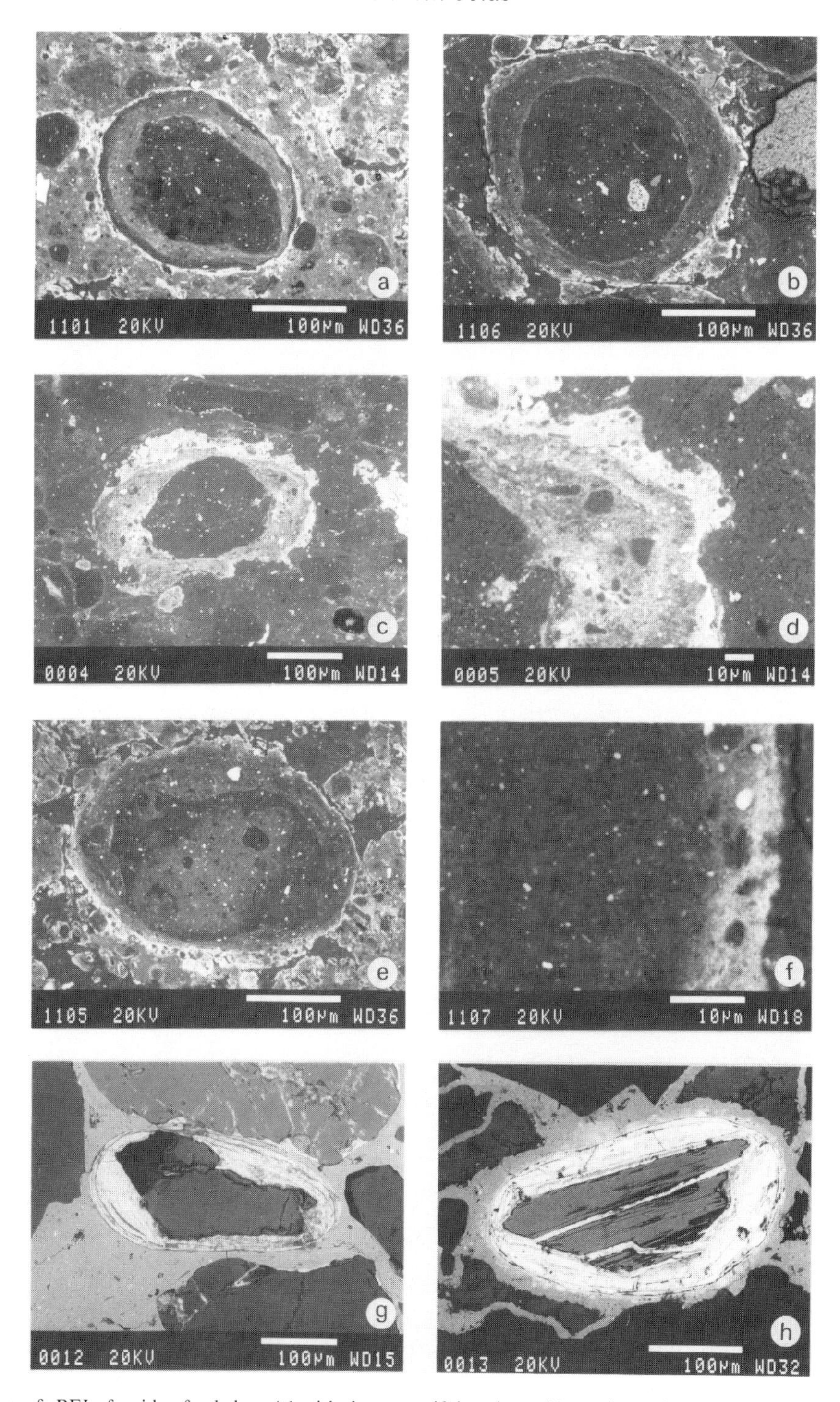


FIG. 4. a–f, BEI of ooids of subclass A1 with dense ramifying rims of hematite (pale), majority of ooid and matrix composed of kaolinite (dark) with abundant irregular inclusions. Lower Cretaceous, St. Chinian, France; g–h, BEI of ooids of subclass A2, geothite-rich cortex surrounds detrital cores of quartz in g and basaltic hornblende in h, cement around silica grains is apatite. Miocene, Suffolk, England (sample provided by P. Balson).

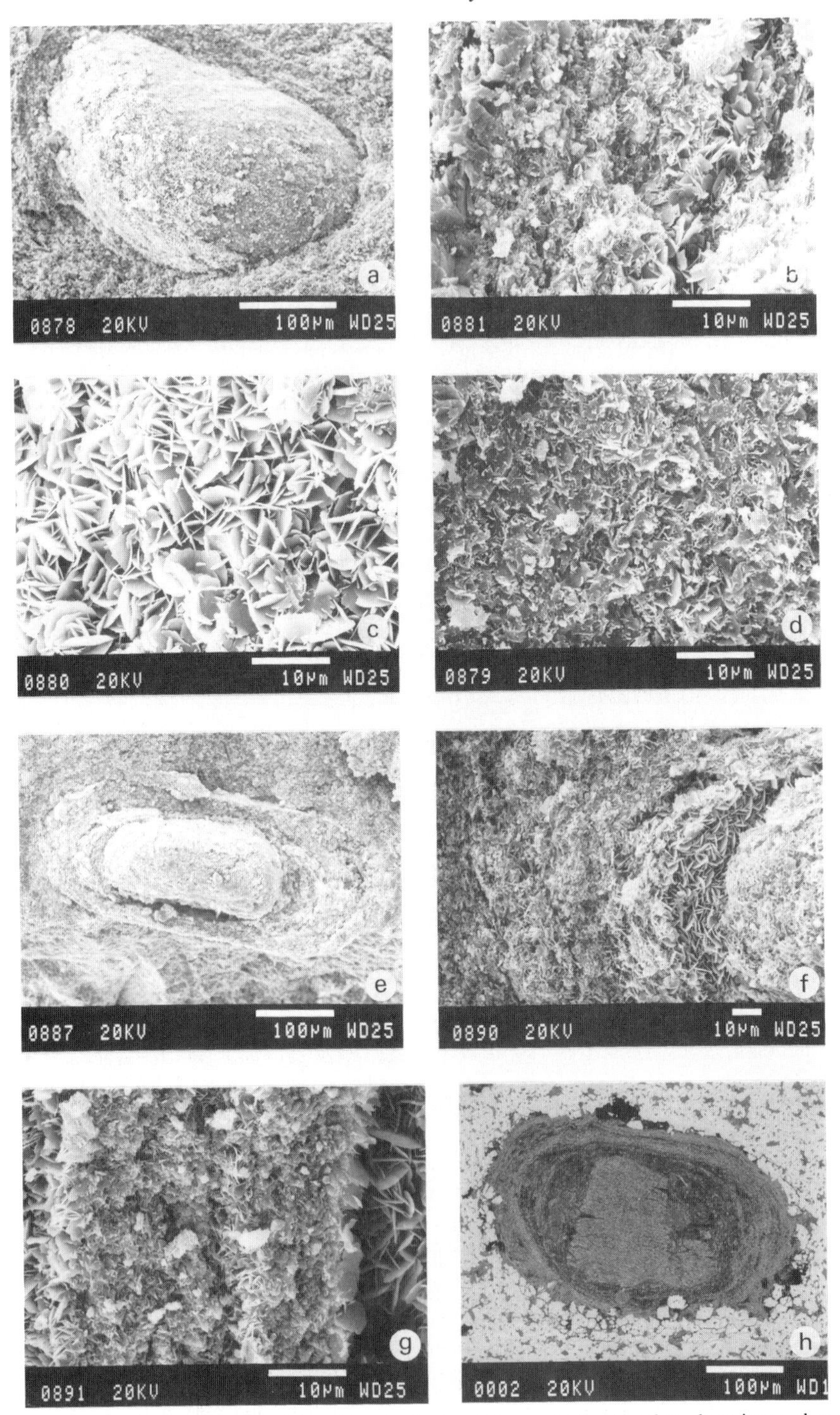


FIG. 5. a–g SEI; h, BEI of berthierine-rich ooids of sub-class B2; a ooid fractured to show internal surface of lamina in coarse radial berthierine; b, detail of a, outer lamina of fine siderite, concentric layers of differing berthierine crystal sizes; c, surface of coarse radial lamina; d, surface of lamina beneath c; e, oblate ooid; f, coarse berthierine in crescentric discontinuous lamina in e; h, ooid with laminated berthierine crust as core, surrounded by rhomic siderite in matrix. Avicula Seam, Penny Nab Member, Cleveland Ironstone Formation, Lower Jurassic, Staithes, England.

survey of available ironstones a detailed subdivision of the three mineralogical classes is proposed in Table 1.

**Class A. Fe oxide/hydroxide ± kaolinite dominated ooids**

Black, dark brown, red or orange coloured ooids are extremely common in ironstones, oolitic limestones and sandstones (examples include: Highworth Grit, U. Jurassic of Oxfordshire; Cleveland Ironstone Formation, L. Jurassic of N. Yorkshire; Rosedale Ironstone, M. Jurassic of N. Yorkshire; Miocene phosphatic sandstones of Suffolk and L. Cretaceous laterites of S. France). Their mineralogies have been reported as varying proportions of goethite, hematite and occasionally magnetite (with common berthierine/chlorite, kaolinite, silica and bauxite). The iron oxides/hydroxides may form the entire ooid coat or be present as discrete laminae between more phyllosilicate-rich layers, and may represent either the primary mineral phase (grown *in situ* or mechanically accreted) or a secondary alteration from an iron-rich phyllosilicate (or rarely carbonate) precursor. At least three subclasses can be distinguished, probably resulting from differing processes.

*Subclass A1.* Goethite and hematite ooids with irregular internal structure and complex outlines.

These ooids are predominantly a mixture of yellow-brown goethite FeOOH (often aluminous) and subordinate kaolinite. They may occasionally contain red-purple hematite $Fe_2O_3$ (with a low Al content), but rarely show any silica or carbonate minerals. Their coarse structure is normally an irregular subspherical to ovoid outline with a dense margin of high iron content passing gradually inward to a core with higher kaolinite or bauxite. The rim is often very complex with fine projections into the surrounding rock (Figs. 4b,d). Concentric lamination may be formed from several discrete zones of iron enrichment (Figs 4a, b, c). There is normally no organized microfabric, except for stacked vermicules of kaolinite (or their 'ghosts'). Irregular inclusions of high kaolinite content are common but are randomly distributed (Fig. 4d). Bhattacharyya & Kakimoto (1982) suggest that a radial crystallite fabric may be visible in some SEI micrographs, but none of the samples studied for this paper show discernable crystals.

*Interpretation*

These grains are almost certainly generated by the percolation of fluids during lateritic weathering, giving local enrichment of iron as grain coatings, matrix replacement and capillary pore space cement. Increased iron oxidation and aluminium/silicon leaching produce hematite and bauxite and remove kaolinite. The role of grain movement and exposure to larger volumes of fluid is uncertain, a stronger concentric fabric may be generated by compaction (on rolling) to produce better lamination. These stronger, better cemented and laminated structures can survive reworking and may be one source of the derived subclass A2 ooids which are discussed by Siehl & Thein (this volume). There is thus a strong preservation bias, with the more characteristic but weaker grains likely to be destroyed during movement into an environment of higher preservation potential.

A1 ooids in their site of origin are unlikely to undergo mineralogical transformation other than further oxidation or retrogressive rehydration and reduction, unless significant quantities of kaolinite survive, making later aluminosilicate generation possible.

A1 ooids are rarely seen as important ironstone constituents, although they are common in laterite pockets on carbonate palaeokarsts in southern France. The development of similar grains is described in detail by Ambrosi & Nahon (1986), Nahon *et al.* (1980), Tardy & Nahon (1985) and Parron & Nahon (1980).

*Subclass A2.* Goethite and kaolinite ooids with a detrital core and smooth outline.

The cortex of these ooids is predominantly composed of yellow-brown goethite with minor kaolinite, berthierine and rarely hematite. Silica, carbonate and phosphate minerals are often present either as small inclusions or intergrown within laminae.

The coarse structure is normally a very smooth, tabular to ovoid outline produced by laminae which thicken and thin to fill and cover any irregularities in the often angular core (Fig.4g). The very fine-scale structure cannot be resolved using conventional BEI and the iron oxyhydroxide microfabric needs to be documented using TEM. Individual laminae show variations in the proportions of goethite and kaolinite which stand out as well defined bands in BEI, although patchy intergrowths also occur within the laminae. Quantitative microanalysis is difficult due to the fine grain size and many analysis points show X-ray spectra which may be from a mixture of minerals. The laminae are

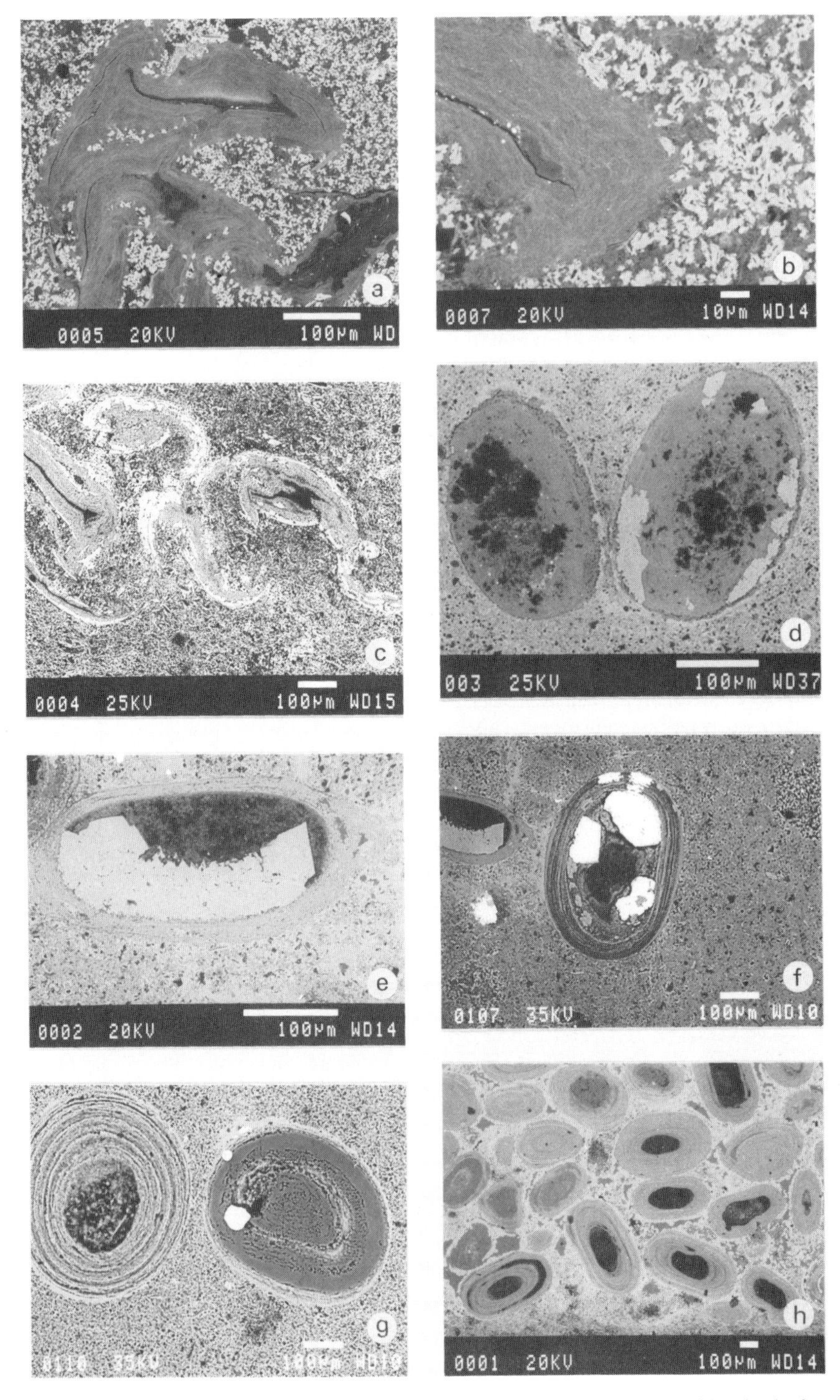


FIG. 6. a–h, BEI of ooids of classes B and C; a, b, spastolithic kaolinite-berthierine ooids of sub-class B4; c, spastolithic ooids with fractured siderite-replaced rims, sub-class C2; d, e, berthierine ooids with displacive siderite showing change from sub-class B4 to C1; f, blocky pyrite (white) replacing siderite in transition C2-C6; g, calcite (dark) replacing siderite and berthierine laminae, C3; h, coarse siderite pore-filling cement around sideritised ooids, C1 and C2. Avicula Seam.

generally between 5 and 10 microns in thickness, forming a cortex of up to 100 microns depth. The ooid nucleus is usually either a thick plane-laminar fragment of iron oxyhydroxide crust, or a sand grain. The former may be fragments of larger oncoids (cf. Siehl & Thein, this volume). Detrital cores are often silica, but are very frequently zircons, glauconites, chlorite stacks or even basaltic hornblendes, cleaved grains are often penetrated by the cortex minerals (Fig. 4h).

*Interpretation*

Despite the lack of microfabric evidence there are several clues which suggest an origin by rolling accretion. The very smooth, regular outline which covers nucleus topography is clearly not a single surface cement layer, and the laminar overlap (seen in virtually all these ooids) is indicative of accretion and polishing abrasion on a rolling grain. The preponderance of heavy minerals as nuclei in sandstones dominated by quartz also suggests that density was important in the concentration of oolitic grains by winnowing. It is most unlikely that the cortices were generated by highly oxidizing pore fluids as mineralogically unstable cores (such as the hornblende seen in Fig. 4h) are unaltered. Thus the model for generation of overlapped laminae by *in situ* iron enrichment in highly developed laterites (described by Nahon *et al.*, (1980, Fig. 8) is not applicable in this instance. The minor quantities of berthierine present may be either accreted or a product of transformation of kaolinite by iron enrichment. It is unlikely that berthierine was the major primary mineral as the ooids show considerable resistance to compaction, and the generation of kaolinite during berthierine alteration usually results in a soft plastic ooid (Fig. 6a).

An origin by *precipitation* of primary iron oxyhydroxides from surface fluid and subsequent grain modification by periodic rolling cannot be ruled out. Later mineralogical transformation by reaction of kaolinite and Fe oxyhydroxides can yield berthierine (Bhattacharyya 1983) which may inherit the platelet microfabric of any accretion-rolling-compacted kaolinite. Limited pore fluid supply of aluminium and silicon to ooids lacking kaolinite may restrict their ability to transform, and they may thus be preserved in an almost pristine state.

The experimental results of Bhattacharyya (1983) and the commonly observed association of microfabrics produced by rolling with mixed oxide and berthierine mineralogies strongly suggest that partial conversion of subclass A2 ooids is a common fate and may be important as a precursor stage for many phyllosolicate-rich ooids.

This type of ooid is probably very common, good examples occurring in Miocene phosphatic sandstones from East Anglia, UK (P. Balson, personal communication). The strength of well formed ooids is such that they can survive compaction and have high preservation potential. They are not necessarily of marine origin, and the majority of ooids described by Siehl & Thein (1978, this volume) may be of this type or modified from subclass A1.

*Subclass A3.* Goethite, kaolinite, siderite and berthierine ooids with patchy, concentric laminated cortices.

This subclass encompasses a diverse collection of ooids, the result of convergent secondary modification to produce a mixture of goethite, hematite, berthierine, chlorite, kaolinite, siderite and silica. The coarse structure is a smooth ovoid outline formed of concentric and continous laminae, the nucleus is often poorly defined and of the same mineralogy as the cortex. The fine structure is of irregular, patchy intergrowths with variation in mineral proportions between individual laminae whose boundaries are often poorly defined and overgrown.

*Interpretation*

Destruction of a poorly concentric and tangential phyllosilicate microfabric occurs due to growth of massive, blocky iron oxyhydroxides, in some examples pseudomorphing coarse siderite which itself replaces or displaces the platelet fabric. Selective transformation of individual laminae reflects differences in both the primary mineralogy and particularly the microfabric between concentric layers. More porous and coarsely crystalline phyllosilicate laminae are very susceptible to access of diagenetic and groundwater fluids and may therefore be preferentially oxidized. The strong red-brown and green zonation seen in thin section may thus reflect porosity differences between laminae of the same primary composition.

A3 ooids are generated by diagenetic or erosive weathering oxidation of largely alumino-silicate (and occasionally sideritic) ooids. They occur in reddened surface crusts of Cleveland Ironstone Formation exposures and are probably the most common type of ooids in weathered samples. When relict microfabrics are identifiable the precursor ooid type can be established.

### Class B. Berthierine/chlorite ± kaolinite dominated ooids

Pale green, grey and white coloured ooids are commonly found in fresh samples of ironstones (examples include: Cleveland Ironstone Formation; Northampton Sand Ironstone, M. Jurassic of Northamptonshire; U. Ordovician ironstones of Gwynedd, N. Wales). They normally consist of berthierine or chlorite, kaolinite, silica, minor iron oxide, calcite and francolite. The origin of these phyllosilicate rich ooids has been widely discussed (Bhattacharyya 1983; James & Van Houten 1979; Adeleye 1980; Maynard 1986) and the consensus supports two common origins: by transformation of iron oxyhydroxide/kaolinite ooids, or primary growth from gel, pore or surface fluid. The contentious model of carbonate replacement (Kimberley 1980a, b) has been dismissed by most authors (Hallam & Bradshaw 1979; Bradshaw *et al.* 1979; Maynard 1986), but may rarely occur.

The considerable range of microfabrics displayed by this class of ooids necessitates subdivision into six groups. These are not of equal importance, and the origin of one group (B3) is uncertain.

*Subclass B1.* Berthierine and iron oxyhydroxide ooids with poorly defined cortical laminae.

Ooids of this subclass possess a detrital core covered by concentric but poorly defined cortical laminae. The cortex is a compact mixture of dominant berthierine (or chlorite if sufficiently matured) and fine iron oxyhydroxides. Randomly oriented berthierine platelets are not confined by cortical lamina boundaries, and cross-cut concentric layers. There is no strong sub-tangential platelet fabric in any part of the ooid.

#### *Interpretation*

These ooids probably represent *in situ* transformation of a class A iron oxyhydroxide ooid.

The experimental transformation of kaolinite and iron hydroxide mixtures to berthierine has been documented by Bhattacharyya (1983). It seems likely that a natural admixing by accretion onto a concentrically laminated ooid (class A) would also rapidly generate iron-rich aluminosilicates. The major theoretical problem concerns the availability of the necessary aluminium. The intense compactional deformation suffered by most kaolinite ooids (unless protected by a sideritic 'egg-shell' rim (Fig. 6a, c)) suggests that grains with sufficient kaolinite to totally transform to berthierine are unlikely to survive rolling accretion. The availability of soluble aluminium to react with an iron oxyhydroxide ooid has been questioned by Maynard (1986). Despite Maynard's assertion that aluminium mobility is negligible there is no doubt that large quantities *can* be transported within porous sediments, occasionally giving berthierine replacement of carbonate grains (Fig. 7a–h). Thus the relatively strong oxyhydroxide ooids *are* potential phyllosilicate precursors.

Textural evidence poses a further problem for identification of B1 ooids, an *in situ* transformation of a pre-existing ooid should retain some indication of the primary microfabric. In the mechanically accreted oxyhydroxide ooids this would be of concentric laminae with random or at best poorly tangential internal fabric. If the ooid is of lateritic origin there should be either no organized microfabric at all or a poorly developed radial platelet orientation. The secondary phyllosilicates are thus likely to grow as randomly arranged aggregates producing an initially porous network. These crystallites probably act as nuclei for further pore-occluding growth, finally generating a solid, compact berthierine-dominated ooid. Subclass B1 ooids should therefore not be confused with the many berthierine ooids which show development of tangential crystallite orientations in some laminae and random orientation in others, suggesting phases of rolling compaction interspersed with unrestricted growth. This is clearly not compatible with an origin by total *in situ* transformation of a primary oxyhydroxide ooid and such ooids are therefore referred to subclass B2. The absence of a detrital nucleus in some B1 berthierine ooids (and concentric lamination to the core) is also incompatible with initial mechanical accretion, and a crystalline growth mechanism from fluid must then be invoked.

Thus subclass B1 ooids are difficult to reliably identify and can only be recognized by the following characterisitics: a compact mixture of dominant berthierine (or chlorite if sufficiently matured) and fine iron oxyhydroxides; a random crystallite orientation throughout the bulk of the cortex, with platelets often crossing boundaries from one lamina to another; a detrital core. All these characters (except the latter) may unfortunately be destroyed by diagenetic or metamorphic recrystallization; nevertheless these ooids probably occur wherever sufficient aluminium is available for transformation of Class A ooids.

*Subclass B2.* Pure berthierine ooids with well defined concentric cortical laminae, random and sub-tangential platelet orientation.

Finely crystalline pure berthierine ooids are common and often surprisingly well preserved despite being relatively fragile. They rarely show the extreme deformation suffered by kaolinite-rich ooids and thus may be capable of withstanding the lamellar compaction associated with tangential reorientation of crystallites during grain rolling.

The coarse structure is of well defined laminae (typically 20 microns thick) which either compose the entire ooid or cover topography by overlap onto a detrital core. Cores may be ooid fragments (Fig. 8c), quartz grains (Fig. 8e) or irregular masses of fine berthierine (Fig. 8f). In polarized light the concentric laminae show different green tones, apparently interlayers of berthierine and kaolinite. In BEI the laminae again appear as distinct bands easily mistaken for pale berthierine and relatively dark kaolinite. Both effects are actually due to differences in berthierine grain size between laminae, with coarse berthierine crystals (giving high lamina porosity) appearing paler in thin section, and the poor backscatter resolution (at low magnifications) averaging dark resin and widely spaced pale berthierine to give an overall dark tone (Fig. 8c,e,f). High-magnification BEI clearly shows that no kaolinite is present and berthierine platelets are coarse (Fig. 8g). Low-magnification BEI does not always provide simple compositional contrast!

The berthierine texture is particularly interesting, with small platelets (0.5 to 1 micron) generally showing random orientation (Fig. 8h; Fig. 5f,g). Coarser crystals (2 to 10 microns) may either show a poorly defined tangential fabric (Fig. 8g) or be present as lenticular masses with more random or even sub-radial orientation (Fig. 5g). Surface views of ooids often show a crude radial orientation (Fig. 5c) but these may be cements filling voids created by tensional opening on compaction of a coherent plastic lamina around a better cemented core (Figs. 5e,f).

*Interpretation*

Growth of sub-radial fabric suggests growth from an enveloping fluid, while the precipitated coarser platelets may be periodically reoriented to a sub-tangential fabric by grain rolling. Thus a combination of chemical precipitation and ooid tumbling is proposed. Collapse and shrinkage of coarser, more open laminae during dewatering and bulk sediment compaction often results in ooid pinching or crenulation (Fig. 8e,f), which should not be confused with a much coarser microbial oncolite fabric.

Subclass B2 is the most common berthierine ooid type found in many ironstones (examples include: Cleveland Ironstone Formation, L. Jurassic of N. Yorkshire; Millepore Bed, Lebberston Member of Cloughton Formation, M. Jurassic of N. Yorkshire; L. Cretaceous of Saudia Arabia) and has frequently been described by previous workers (Hughes, this volume). The highly porous platelet framework makes them particularly susceptible to diagenetic alteration and their common modifications are shown in Fig. 1.

*Subclass B3.* Pure berthierine ooids with well defined cortical laminae and random platelet orientation.

These ooids have a fine concentric structure of discrete laminae each built of randomly oriented berthierine crystallites of almost uniform size; a detrital nucleus may or may not be present. There is *no* subtangential fabric.

*Interpretation*

The origin of these ooids is uncertain. They have probably undergone little or no transportation during growth (as there is no tangential fabric), yet their fine concentric structure and crystallite orientation suggest growth from an enveloping medium, probably a gel.

The elements necessary for berthierine formation in surface and pore fluids are often below the concentration required to satisfy kinetic constraints. Curtis & Spears (1968), Talbot (1974) and Harder (1978, this volume) have suggested that preconcentration in gel phases may enhance reaction rate and allow generation of berthierine in otherwise unfavourable environments. The fluctuation in supply and depletion of iron, magnesium, aluminium and silicon may set up migrating zones of reaction within a gel and result in repeated events of randomly-oriented crystal growth. This might generate a concentrically zoned structure but it seems unlikely that this process generates many true ooids as their smooth subspherical shape and evidence of rolling during growth (tangential fabric) suggest a firm discrete ooid surface, rather than an easily damaged and laterally extensive gel layer. Larger pisoliths may form by this gel mechanism, with biological assistance (Dahanayake & Krumbein 1986).

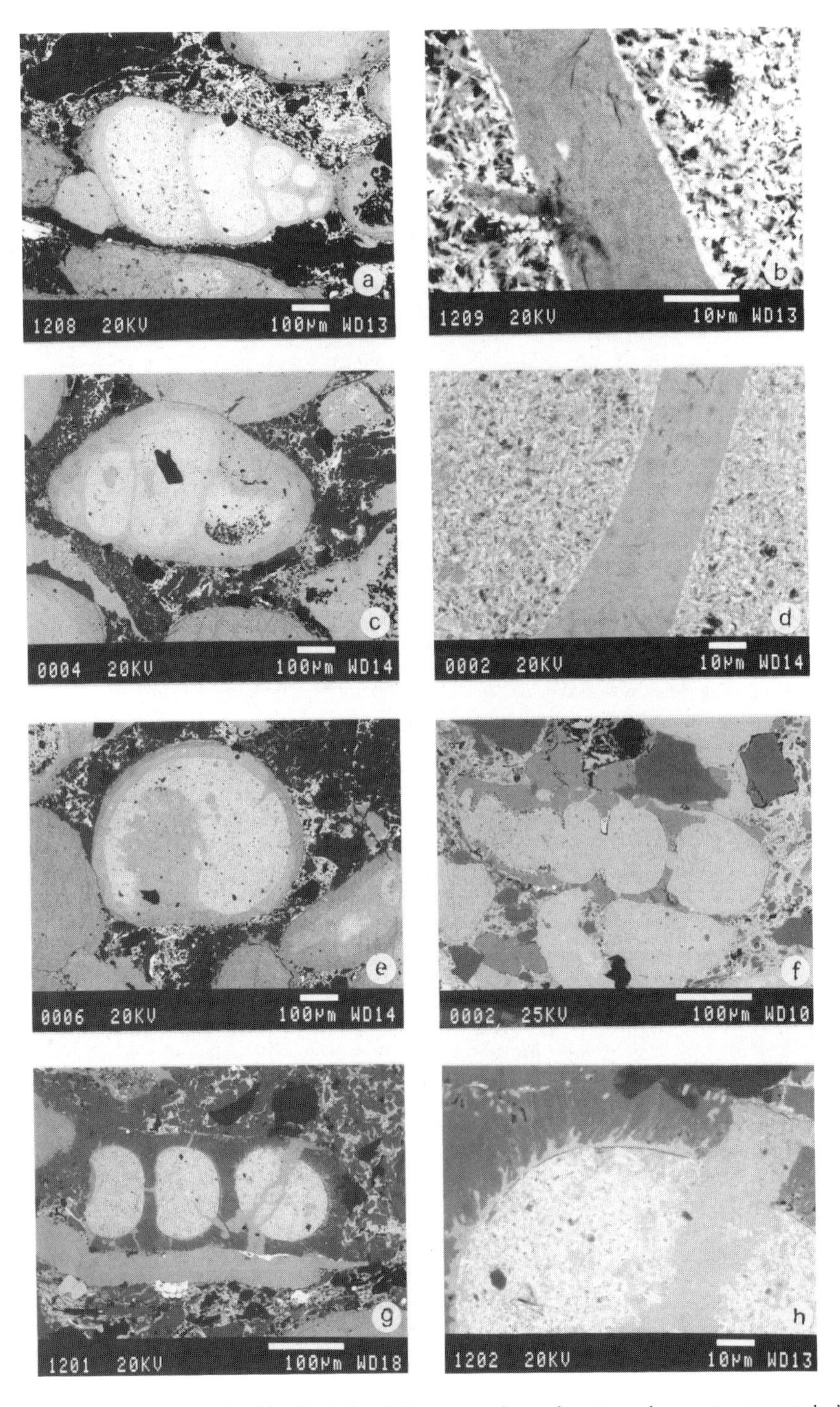


FIG. 7. a–h BEI of bioclasts and ooids of subclass B5; a–e, sections of gastropods, quartz appears dark, berthierine mid grey and hematite-rich berthierine as pale grey, shell totally replaced by fine berthierine; g, shell, matrix and cortex all fractured and veined by fine berthierine; f–h, berthierine replacement of calcite nodosariine foraminiferan tests. Ardnish Ironstone, Lower Jurassic, Isle of Raasay, Scotland.

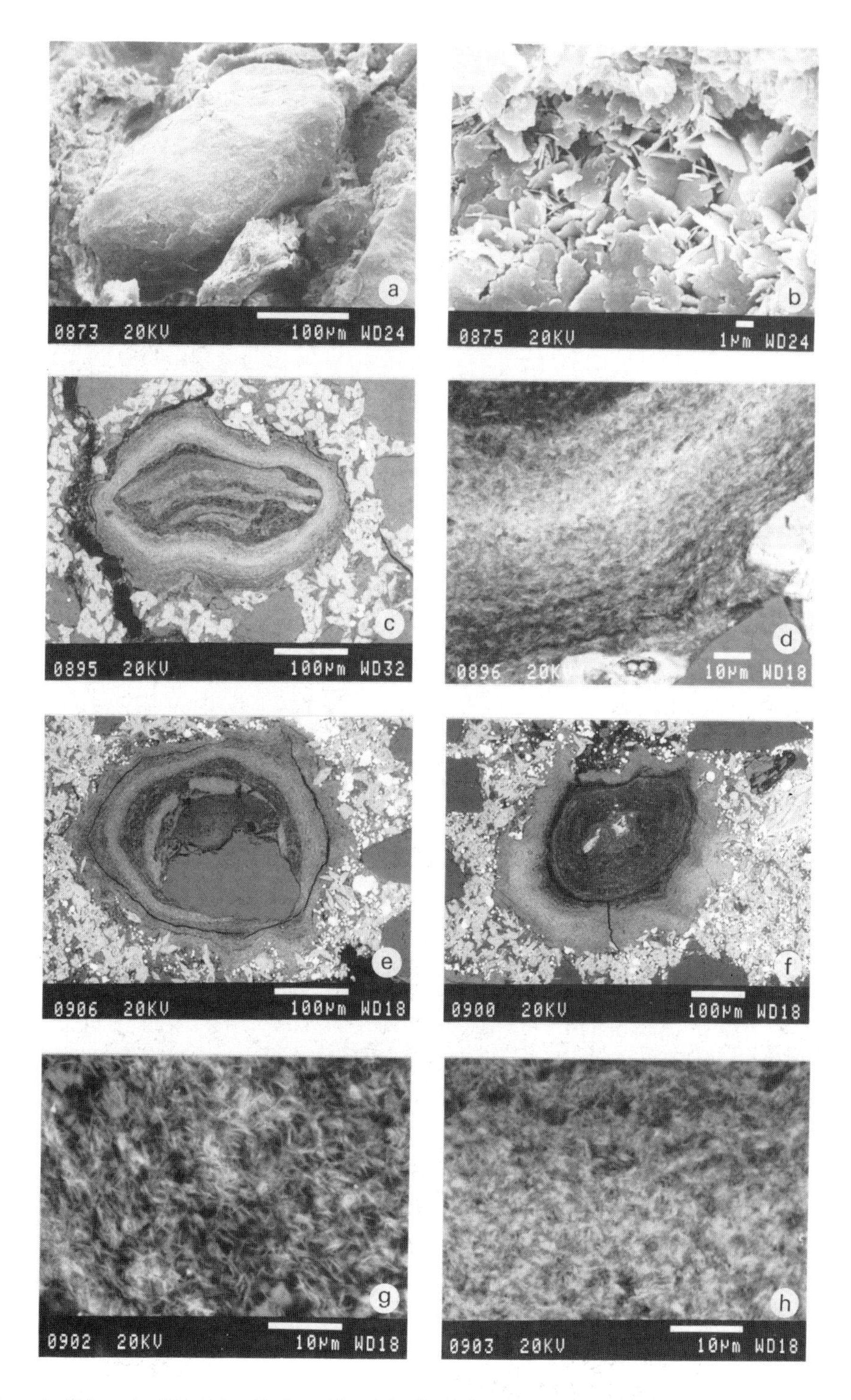


FIG. 8. a–b, SEI; c–h, BEI of berthierine-rich ooids of subclass B2; a,b, show apparent sub-tangential fabric on surface of fractured ooid; c, ooid with pisoid? fragment as core; d, sub-tangential berthierine of varying crystal size; e, quartz grain as eccentric nucleus; f, ooid without detrital core, outer laminae shrunk; g, 'dark' coarse and oriented berthierine in inner lamina of f; h, 'pale' fine and randomly oriented berthierine in outer lamina of f. Millepore Bed, Lebberston Member, Cloughton Formation, Middle Jurassic, Scarborough, England.

Ooids of this subclass are rare and difficult to distinguish from those of subclass B2; they are a relatively unimportant contributor to most ironstones, and are easily modified by diagenetic and weathering processes.

*Subclass B4.* Kaolinite-rich ooids.

White, cream or pale yellow, flattened ooids almost always occur in the same rocks as green berthierine ooids. Their predominant minerals are kaolinite and berthierine, with varying quantities of siderite, calcite, opaline silica, iron oxyhydroxide and francolite. The gross outline is very variable due to effects of compactional deformation, and ooids are frequently found to be squashed around more competent siliciclastic grains or ptygmatically folded.

Three different types of ooid structure are common.

(1) Fine concentric lamination of sub-tangentially oriented kaolinite and berthierine (Figs 6a,b).
(2) Coarse lamination of strongly oriented tangential kaolinite/illite/berthierine 'stacks' (Fig. 9e–h).
(3) Segregation of berthierine (and siderite) into an outer cortex, with randomly oriented kaolinite forming the core (Fig. 6c,d,e).

*Interpretation*

These types of kaolinite-rich ooid are clearly soft and fragile structures which could not withstand abrasion during rolled movement. They are thus almost certain to have formed by diagenetic or weathering transformation of the other ooid types or by *in situ* precipitation of a mixture of berthierine and kaolinite.

Studies of sandstone porosity reduction show that kaolinite precipitation normally occurs as crystallographically oriented sheets nucleated on detrital muscovites, or as vermicules of stacked platelets. While thick, radial, pore-lining kaolinite cements do form in ooids they are relatively uncommon. The tangential orientation within most B4 ooids suggests the majority of kaolinite inherits microfabric from a stronger primary berthierine ooid, and either uses berthierine plates as a template or is formed by transformation on leaching of iron by pore fluids. The coarser (30 micron) interlayered clay mineral stacks (Fig. 9h) probably represent a low nucleation rate for berthierine and rapid growth to produce coarse globules, followed by 'syntaxial' intergrowth of illite and kaolinite lamellae along the (001) cleavage plane. The microfabrics thus strongly suggest that kaolinite is a secondary mineral and is not a contributor to primary ooid accretion.

Kaolinite may be generated in a number of stages from early diagenesis to eventual subaerial weathering, specific microfabric and mineral associations can be used to determine when the process occured. Early diagenetic B4 ooids often possess a thin external siderite rim (Fig. 6c–e), or may be compacted onto zoned ferroan dolomite or siderite rhombs (Figs 6a, 6b; Fig. 9e–h). The carbonate phase seems synchronous in growth with kaolinite formation, clearly predates much compaction (Fig. 6c) and probably acts as a sink for iron released during degradation of berthierine. Carbon and oxygen isotope studies are needed to pinpoint the diagenetic source of the fixed carbonate ions. B4 ooids produced during subaerial weathering often show a random orientation of kaolinite vermicules throughout the ooid, and generation of iron oxyhydroxides rather than carbonate.

The indicators of kaolinite generation are often masked by synchronous and later growth of other minerals and by extreme flattening and it may be difficult to determine the precursor ooid type. Good examples of B4 ooids occur within the Cleveland Ironstone Formation and the Raasay Ironstone.

*Subclass B5.* Compact berthierine and hematite ooids associated with berthierine replacement of carbonate bioclasts.

Dark green ooids of this subclass occur in great abundance within the lenticular Ardnish Ironstone, L. Jurassic of Skye, Inner Hebrides, Scotland. Peloids and well formed concentrenic ooids are found in a calcite- and hematite-cemented quartz arenite which also contains abundant carbonate bioclasts partially replaced by berthierine. Ooid coarse structure is a subspherical to ovoid outline composed of up to 100 microns thickness of berthierine cortex surrounding a core which is commonly of bioclastic origin but *always* of berthierine composition. In plane-polarized light conspicuous concentric lamination is revealed within the cortex, resembling lamination seen in subclasses B1, B2 and B3. In BEI however, there is a little concentric structure visible, although ooid nuclei and bioclast matrix components may appear bright due to abundant randomly-oriented hematite flakes within the berthierine (Figs 7b,d).

In SEI the microstructure of peloids and ooids is clearly seen to be identical, formed from pore-occluding, randomly oriented sheaved berthierine platelets (Fig.10a, b). Crystallite size

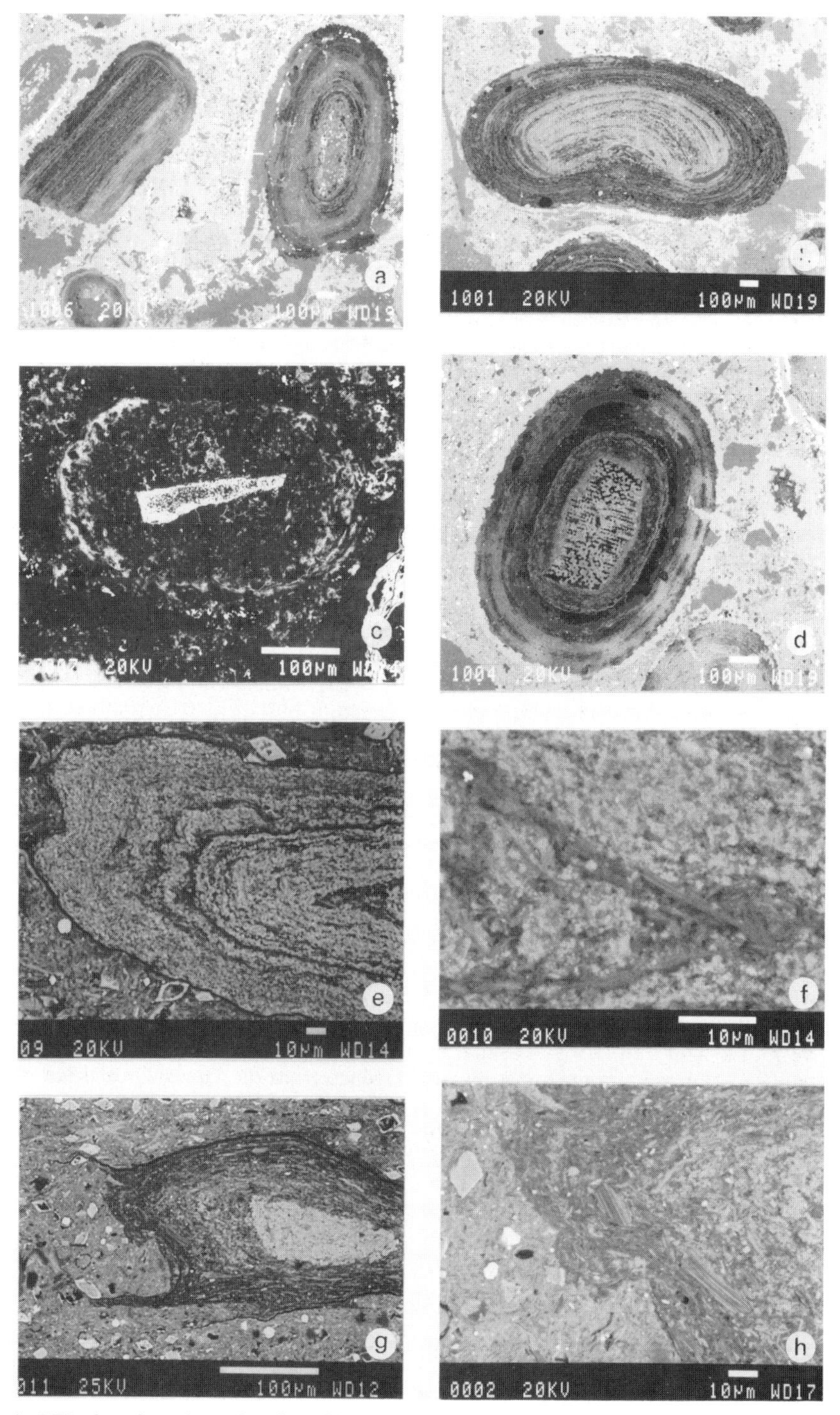


FIG. 9. a – h, BEI of sections through ooids of classes B and C; a – b, calcite replacement, siderite rim and siderite cementation of berthierine-kaolinite ooids, transition from B2-B4-C2-4; c, calcite replacement of phosphatic ooid within ammonite, transition C5-C3; d, berthierine replacement of echinoderm calcite nucleus in B2; e, compacted francolite and phyllosilicate-rich laminae of B4-C5 ooid; e, compacted francolite and phyllosilicate-rich laminae of B4-C5 ooid; f, tangential orientation of berthierine-kaolinite stacks, in B4-C5; g, compacted B4 ooid with C5 core; h, coarse stacks in B4 ooid cortex. a, b, d, from green facies, c, e, f, g, h, from black facies of Raasay Ironstone, Lower Jurassic, Isle of Raasay, Scotland.

does vary between laminae, but there appears to be no tangential fabric. Surprisingly, an excellent 'extinction cross' is obtained on B5 ooids when viewed under crossed polars. This may result from the subspherical laminar bounding surfaces, rather than from tangentially or radially oriented crystals.

The berthierine microstructure of replaced carbonate bioclasts is identical to that of ooid cortices (Fig. 10h). In plane – polarized light primary bioclast textures are still apparent (Fig. 3c–h). A wide range of carbonate compositions are totally replaced; aragonite in gastropods; high-Mg calcite in crinoid and echinoid stereom plates; ?low-Mg calcite in bivalve and ostracod valves. Incomplete replacement is occasionally seen, with bivalve nacreous layers preserved in berthierine while the neighbouring prismatic (or crosslamellar) layer is still in patches calcite. This represents the preservation of a front along which rates of carbonate dissolution and berthierine precipitation were equal. Intermediate stages are best seen in nodosariine forams (Fig. 7f–h) where the test fine structure is both broken by berthierine veins and is partly replaced. Brittle radial fracture of ooid cortices and matrix and infill by pure berthierine is common (Fig. 7e) and may be due to ooid shrinkage on dewatering. Ooids occasionally show concentric lamellar fractures filled by calcite cement, but do not show signs of alteration other than oxidation to iron oxyhydroxides in the thin weathering crust. Siderite is apparently absent from the Ardnish Ironstone.

*Interpretation*

These compact berthierine ooids are probably diagenetic replacements of primary aragonite ooids. Replacement of bioclasts cannot conclusively prove that co-existing ooids were originally of carbonate composition, but it does remove the objections based upon textural and geochemical models which were discussed by Bradshaw *et al.* (1979), Hallam & Bradshaw (1979) and Maynard (1986). There is no evidence that the Ardnish Ironstone was subjected to downward migration of fluids but the underlying sequence does contain palaeokarstic surfaces within oolitic calcarenites and is thus similar to the emergent mixed siliciclastic and carbonate environment suggested by Kimberley (1979). Replacement of echinoderm stereom structure by berthierine also occurs in the Raasay Ironstone (Fig. 9d; Fig. 3h) demonstrating that the process may be relatively common and not necessarily restricted to a specific palaeoenvironment.

It is, however, unwise to postulate diagenetic replacement of carbonate grains unless an identical diagnositc microstructure positively identifies berthierine replacement of co-existing bioclasts. The supposed nubecularian foraminiferan ooids described by Champetier *et al.* (1987) seem lacking in any characterisitic biogenic fabric and their 'chambering' is best explained by compactional detachment of separate concentric laminae of ooids and preferential preservation at the ooid extremities.

*Subclass B6.* Chlorite and silica ooids.

These ooids show a coarse (100 microns) and apparently random intergrowth of iron-rich chlorite, silica, and Fe oxides, each mineral often extending from within the ooid through the outer wall and into the pore space beyond. Ooid margins are thus frequently broken and the oval outline may become indistinct. Internal concentric lamination (rarely present) is interrupted by coarse patches of silica.

*Interpretation*

During burial diagenesis and low grade metamorphism of ironstones the common mineral assemblage of berthierine, kaolinite and siderite is no longer stable and is transformed to a relatively coarse grained mixture of Fe-chlorite, silica and iron oxides. Earlier microfabrics are retained if sufficient chlorite is present and the degree of deformation is low, but patchy silica growth often displaces the phyllosilicate framework. Angular patches of chlorite with random crystallite orientation are common within the ooid rim as pseudomorphs after coarse replacive or displacive siderite. This relationship is also seen in metamorphosed ironstones from the Upper Ordovician of Portugal (T. Young, personal communication) where rhombic patches within the ironstone matrix are now composed of fine chlorite, replacing diagenetic ankerite or siderite crystals similar to those found in the Lower Jurassic Raasay Ironstone (Fig. 9g).

Thus despite considerable modification of mineralogy the texture of metamorphosed ooids may still preserve evidence of diagenetic processes. Good examples of chlorite, hematite and silica dominated ooids are found in the Welsh Ordovician ironstones described by Trythall (this volume) and Trythall *et al.* (1987).

**Class C. Siderite, calcite, apatite, pyrite dominated ooids**

This class of ooids contains a remarkable variety of mineral and textural combinations, all apparently produced by replacement or dis-

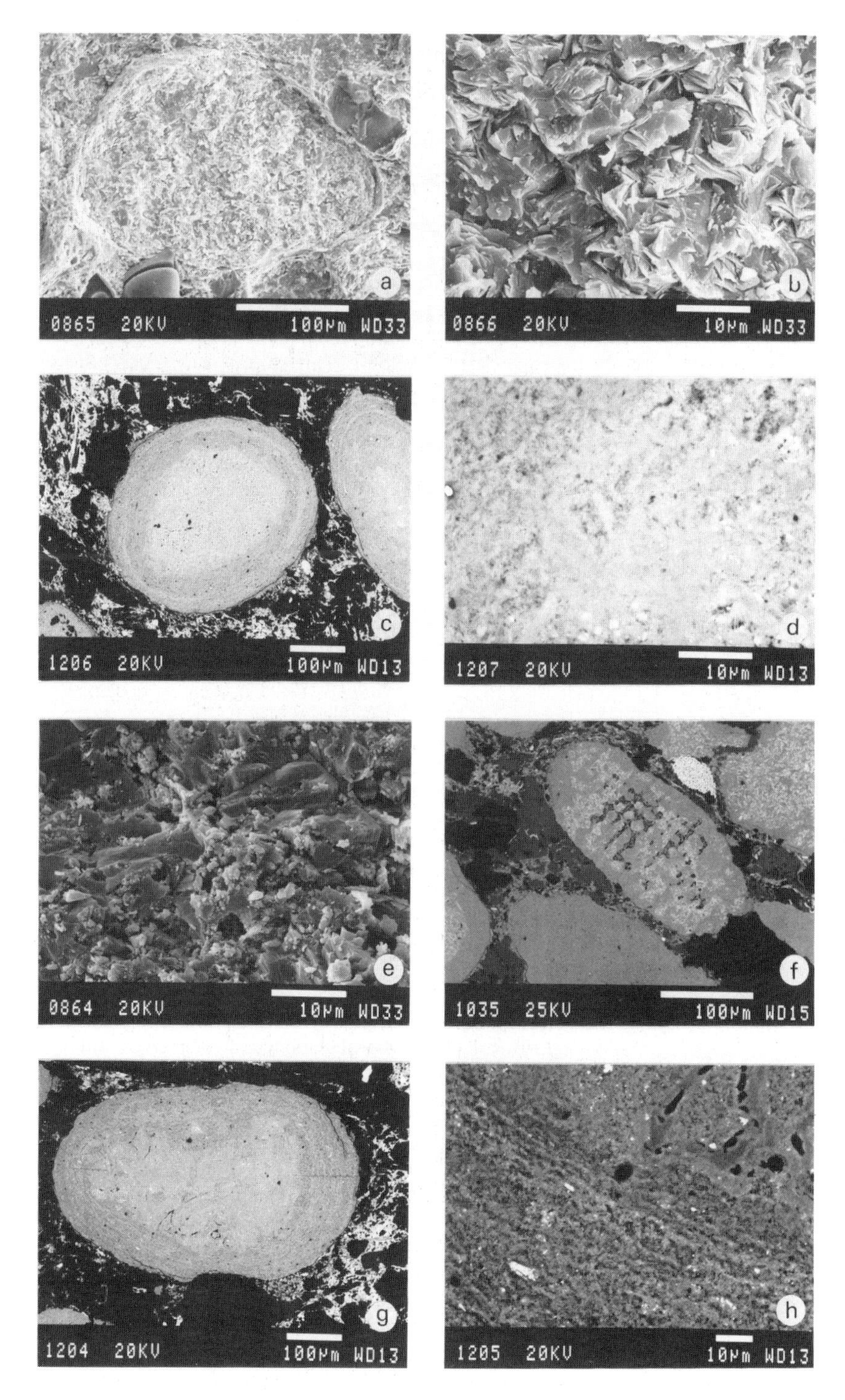


FIG. 10. a, b, e, SEI; c,d,f,g,h, BEI of peloids and ooids of subclass B5; a, with hematite flakes in core and fine berthierine cortex; e–f, berthierine replacement of echinoderm calcite stereom fragment; g,h, ooid with nucleus of echinoderm stereom, both cortex and core of fine berthierine. Ardnish Ironstone, Upper Broadford Beds, Lower Jurassic, Isle of Skye, Scotland.

placement of earlier phyllosilicate ooid constituents. Six major combinations are sufficiently common to be worthy of description.

*Subclass C1.* Coarse siderite displacive or cement growth in ooids.

Coarse euhedral calcian siderite crystals frequently occur within berthierine and kaolinite ooids. They are normally nucleated on a siderite replaced rim and grow inwards to displace the phyllosilicate microfabric. A pore-lining cement radial outward from the ooid surface is also common. Crystals reach 200 microns in length and may eventually become the predominant iron-bearing mineral. Later removal of aluminosilicates often leaves a highly porous and fragile siderite eggshell rock which can be subsequently reinforced by either siderite or calcite cement. The process usually post-dates fine-grained siderite growth and is most common in porous, grain supported ironstones such as that found within burrows in the Avicula Seam, Cleveland Ironstone Formation, L. Jurassic of N. Yorkshire (Fig. 6e, h) and in the green facies of the Raasay ironstone, L. Jurassic of Raasay, Inner Hebrides, Scotland (Fig. 9a,b,d).

*Subclass C2.* Siderite replacement of ooid rim and cortex.

Ooids often possess a thin (10 to 50 microns), compact and finely crystalline siderite rim which has a very smooth external surface but may internally grade into either mixed berthierine and siderite or a dominantly kaolinitic core (Fig. c, e). The entire ooid cortex may be replaced (Fig. 6g) giving a discrete rim and often well defined concentric lamination. The preservation of fine structure despite loss of crystallographic fabric is due to control of siderite growth by the phyllosilicate grain size and porosity of individual laminae. A 'spongy' texture without strong fabric is produced. Siderite usually occurs as 2 to 5 micron rhombs, identical to those in the external matrix and is probably of very early diagenetic origin, filling available pore space. Cemented but brittle ooid rims often fracture to produce sigmoidal spastoliths with plastic deformation of the phyllosilicate and poorly-cemented siderite core (Fig. 6c). Much siderite growth must therefore predate even early compaction. Each group of class B ooids (except the low porosity B5) can probably be modified to C2, with berthierine destruction linked to siderite and kaolinite growth.

This type of ooid is common in matrix-rich ironstones which contain little organic matter and are extensively bioturbated. Examples occur in the Cleveland Ironstone Formation, L. Jurassic of N. Yorkshire.

*Subclass C3.* Calcite replacement of ooid cortex.

Ooid cortices composed of calcite are common in sideritic ironstones. The entire ooid may be a single clean crystal showing uniform extinction, with no concentric lamination, although retaining the oval external outline (see sublcass C4 below). This cannot easily be confused with a primary carbonate ooid as it lacks the distinctive fine structure. Partial replacement often occurs, with loss of fabric in some laminae but retention of an earlier 'spongy' siderite texture in others (Fig. 6g). Rhombic calcite may also grow in phosphate dominated ooids, particularly when they are isolated from the surrounding matrix. Both francolite and berthierine can be replaced (Fig. 9c). Good examples of these ooids occur in the Cleveland Ironstone Formation and Raasay Ironstone.

*Subclass C4.* Coarse calcite cemented ooids.

After phyllosilicate components are removed to generate a siderite 'eggshell' the pore space may be completely filled by growth of coarse siderite (subclass C1) and calcite. The primary ooid textures are usually lost, and the resulting low-porosity ironstone is incapable of further modification unless metamorphosed or subjected to low-pH fluids. Ooid-rich grain-supported beds within the Cleveland Ironstone Formation, the Raasay Ironstone and the Rosedale Ironstone (M. Jurassic of N. Yorkshire) contain abundant C4 ooids. Little textural evidence of their origin and evolution is preserved.

*Subclass C5.* Phosphate dominated ooids.

Francolite is a common mineral in ironstones, occuring both in the sideritic matrix and as well defined cores and laminae of kaolinite and berthierine ooids (Fig. 9e,f,g: Fig. 3i,j). Good examples are found in the Lower Jurassic Raasay Ironstone of the Inner Hebrides, Scotland. A complete spectrum from wholly phyllosilicate to francolite ooids is seen.

Tiny crystals may form up to 80% by volume of individual 20 to 50 micron thick lamellae, which usually show compactional flattening and folding (Fig. 9e). Phosphate ooid cores are more resistant to compaction and are often mantled by a tapering outer cortex of flattened kaolinite and berthierine (Fig. 9g). Vertebrate skeletal phosphate occasionally forms ooid nuclei, but the source and growth mechanism for most of the francolite are uncertain.

*Interpretation*
Association with abundant zoned carbonate rhombs (Fig. 9g) indicates a strong microbial influence on early diagenesis, probably involving prolonged sulphate reduction (Pye 1985) and this may have promoted phosphorus mobility. The origin of similar phosphatic ooids is discussed by Horton *et al.* (1980) who describe goethite-francolite grains from a highly oxidized time equivalent horizon, the condensed Toarcian Cephalopod Bed limestone of Central England.

In the Raasay Ironstone phosphatic ooids are restricted to the dark, organic-rich mud facies, occurring in both the basal unit and as thin intercalations in the overlying green siderite facies. Penecontemporaneous erosion of the sediment may have repeatedly reworked stronger grains, but the more fragile ooids are likely to have achieved their present mineralogy *in situ*. Laminae may be diagenetic increments upon a micronodule rather than either a direct precipitate from seawater or a replacement of primary carbonate or berthierine components. Further research on the origin and diagenesis of the Raasay Ironstone is clearly required.

*Subclass C6.* Pyrite dominated ooids.

Pyrite is usually a minor component of oolitic ironstones but may occasionally become important if a supply of organic matter and sulphate ions became available during the diagenesis of siderite dominated ooids. Anastomosing veins of sulphide penetrate the nodular surface of the Avicula Seam (Cleveland Ironstone Formation) and often terminate in partially pyritized ooids. Similar ooids occur scattered throughout the matrix with no apparent line of fluid access. Coarse euhedral crystals may displace phyllosilicate fabric, or fine-patches apparently replace 'spongy' siderite texture (Fig. 6f). Growth of sulphides therefore usually post-dates berthierine, kaolinite and siderite development and overprints their textures, variably preserving early microfabrics.

## Conclusions

Textures and microfabrics can be used to recognize the sedimentary and diagenetic processes occurring during the evolution of iron-rich ooids. Sequences of mineralogical modification can be easily interpreted from simple relationships such as fabric displacement and grain pseudomorphism. The origin and subsequent mineral transformation of the phyllosilicate microfabric is much more difficult to investigate, particularly due to destruction of these primary textures during pervasive siderite and calcite replacement and compactional deformation.

The most common origin of iron-rich ooids is probably by growth of pure berthierine from an enveloping fluid, producing a random to sub-radial crystallite fabric which is subsequently partially reoriented by rolling compaction to fine individual cortical laminae with sub-tangential berthierine platelets (subclass B2). These pure berthierine ooids can withstand rolling and abrasion but will undergo partial compactional flattening and may be brittly fractured. The high porosity and permeability of the framework makes then vulnerable to reactive fluids, and alteration of coarser laminae to give zones of iron oxyhydroxides and siderite is common.

Concentric lamination of the ooid may in other ooids be a relict primary texture, either reflecting mechanical accretion of an iron oxyhydroxide precursor of the cortex or the replacement of a carbonate ooid by berthierine. In neither of these cases is a strong sub-tangential microfabric generated in the phyllosilicate components of the cortex.

Further integrated studies of iron-rich ooids using optical microscopy, SEI, BEI, TEM and EDS are need to verify the relationships between sedimentary processes, microfabrics and mineralogy. This paper provides a preliminary discussion of the approaches necessary to understand the origin and evolution of such ooids.

ACKNOWLEDGEMENTS: Discussions with T. Young, C. Hughes, R. Knox, C. Curtis, D. Bhattacharyya and H. Shaw have acted as a considerable stimulus to this work and I thank them while solely accepting responsibility for any erroneous conclusions. A. Mackenzie, M. Riley, S. Hughes and M. Hoggins are thanked for their assistance. This research was supported by the Electron Microscopy Services account, Oxford Polytechnic.

## References


ADELEYE, D.R. 1980. Origin of oolitic iron formations — discussion. *Journal of Sedimentary Petrology*, **50**, 1001–1003.

AMBROSI, J.P. & NAHON D. 1986. Petrological and Geochemical differentiation of lateritic iron crust profiles. *Chemical Geology*, **57**, 371–393.

BATHURST, R.G.C. 1976. Carbonate sediments and their diagenesis. *Developments in Sedimentology*, **12**, Elsevier, Amsterdam.

BHATTACHARYA, D.P. 1983. Origin of berthierine in ironstones. *Clays and Clay Minerals*, **31**, 173–182.

—— & KAKIMOTO, P.K. 1982. Origin of ferriferous ooids: an SEM study of ironstone ooids and bauxite pisoids. *Journal of Sedimentary Petrology*, **52**, 849–857.

BRADSHAW, M.J., JAMES, S.J. & TURNER, P. 1979. Origin of oolitic ironstones — discussion. *Journal of Sedimentary Petrology*, **50**, 295–304.

BUBENICEK, L. 1983. Diagenesis of Iron-Rich Rocks. *In:* LARSEN, G. & CHILLINGAR, G.V. (eds) *Diagenesis in Sediments and Sedimentary Rocks, 2. Developments in Sedimentology*, **25**, Elsevier, Amsterdam.

CHAMPETIER, Y., HAMADOU, E. & HAMADOU, M. 1987. Examples of biogenic support of mineralisation in two oolitic iron ores — Lorraine (France) and Gara Djebilet (Algeria). *Sedimentary Geology*, **51**, 249–455.

CHAUVEL, J.J. & GUERRAK, S. 1989. Oolitization processes in Palaeozoic ironstones of France, Algeria & Libya. *In:* YOUNG, T.P. & TAYLOR, W.E.G. (eds) *Phanerozoic Ironstones*, Geological Society, London, Special Publication **46**, 165–174

CURTIS, C.D. & SPEARS, D.A. 1968. The formation of sedimentary iron minerals. *Economic Geology*, **63**, 257–270.

DAHANAYAKE, K. & KRUMBEIN, W.E. 1986. Microbial structures in oolitic iron formations. *Mineralium Deposita*, **21**, 85–94.

DIMROTH, E. 1979. Facies models 16. Diagenetic facies of iron-formation. *Geoscience Canada*, **4**, 83–88.

DINGLEY, D.J. 1981. A comparison of diffraction techniques for the SEM. *In:* O'HARE, A.M.F. (ed). *Scanning Electron Microscopy, 1981/IV*, 273–286, Illinois.

HALLAM, A. 1960. Stratigraphy of the Broadford Beds of Skye, Raasay and Applecross. *Proceedings of the Yorkshire Geological Society*, **32**, 165–184.

—— & BRADSHAW, M.J. 1979. Bituminous shales and oolitic ironstones as indicators of transgressions and regressions. *Journal of the Geological Society*, London, **136**, 157–64.

HARDER, H. 1978. Synthesis of iron layer silicate minerals under natural conditions. *Clays and Clay Minerals*, **26**, 65–72.

—— 1989. Mineral genesis in ironstones: a model based upon laboratory experiments and petrographic observations. *In:* YOUNG, T.P. & TAYLOR, W.E.G. (eds) *Phanerozoic Ironstones*, Geological Society, London, Special Publication **46**, 9–18.

HORTON, A., IVIMEY-COOK, H.C., HARRISON, R.K. & YOUNG, B.R. 1980. Phosphatic ooids in the Upper Lias (Lower Jurassic) of central England. *Journal of the Geological Society, London*, **137**, 731–740.

HUGHES, C.R. 1989. The application of analytical transmission electron microscopy to the study of oolitic ironstones: a preliminary study. *In:* YOUNG, T.P. & TAYLOR, W.E.G. (eds) *Phanerozoic Ironstones*, Geological Society, London, Special Publication **46**, 121–133.

JAMES, H.E. & VAN HOUTEN, F.B. 1979. Miocene goethitic and chamositic oolites, northeastern Colombia. *Sedimentology*, **26**, 125–133.

KIMBERLEY, M.M. 1979. Geochemical distinctions among environmental types of iron formations. *Chemical Geology*, **25**, 185–212.

—— 1980a. The Paz de Rio Oolitic inland-Sea Iron Formation. *Economic Geology*, **75**, 97–106.

—— 1980b. Origin of Oolitic iron formations — reply. *Journal of Sedimentary Petrology*, 50, 1003–1004.

KRINSLEY, D.H., PYE, K. & KEARSLEY, A.T. 1983. Application of backscattered electron microscopy in shale petrology. *Geological Magazine*, **120**, 109–208.

MAYNARD, J.B. 1986. Geochemistry of Oolitic Iron Ores, an Electron Microprobe Study. *Economic Geology*, **81**, 1473–1483.

NAHON, D., CAROZZI, A.V. & PARRON, C. 1980. Lateritic weathering as a mechanism for the generation of ferriginous ooids. *Journal of Sedimentary Petrology*, **50**, 1287–1298.

PARRON, C. & NAHON, D. 1980. Red bed genesis by lateritic weathering of glauconitic sediments. *Journal of the Geological Society, London*, **137**, 689–693.

PYE, K. 1985. Electron microscope analysis of zoned dolimite rhombs in the Jet Rock Formation (Lower Toarcian) of the Whitby area, U.K. *Geological Magazine*, **122**, 279–286.

SIEHL, A, & THEIN, J. 1978. Geochemische Trends in der Minette (Jura, Luxembourg/Lothringen). *Geologische Rundschau*, **67**, 1052–1077.

—— 1989. Minette-type ironstones. *In:* YOUNG, T.P. & TAYLOR, W.E.G. (eds) *Phanerozoic Ironstones*, Geological Society, London, Special Publication **46**, 175–195.

TALBOT, M.R. 1974. Ironstones in the Upper Oxfordian of Southern England. *Sedimentology*, 21, 433–450.

TARDY, Y. & NAHON, D. 1985. Geochemistry of laterites, stability of Al-geothite, Al-hematite and $Fe^{3+}$ kaolinite in bauxites and ferricretes: an approach to the mechanism of concretion formation. *American Journal of Science*, **285**, 865–903.

TRENDALL, A.F. 1983. Introduction. *In:* TRENDALL, A.F. & MORRIS, R.C. (eds). *Iron Formation Facts and Problems*. Elsevier, Amsterdam, 1–12.

TRYTHALL, R.J.B. 1989. The mid-Ordovician oolitic ironstones of North Wales: a field guide *In:* YOUNG T.P, & TAYLOR, W.E.G. (eds) *Phanerzoic Ironstones,* Geological Society, London, Special Publication **46**, 213–220.

——ECCLES, C., MOLYNEUX, S.G. & TAYLOR, W.E.G. 1987. Age and controls of ironstone deposition (Ordovician) North Wales. *Geological Journal,* **22**, 31–43.

WHITE, S.H., SHAW, H.F. & HUGGETT, J.M. 1984. The use of backscattered electron imaging for the petrographic study of sandstones and shales. *Journal of Sedimentary Petrology,* **54**, 487–494.

A.T. KEARSLEY, Department of Geology, Oxford Polytechnic, Gipsy Lane, Oxford. OX3 0BP, UK.

# Oolitization processes in Palaeozoic ironstones of France, Algeria and Libya

J-J. Chauvel and S. Guerrak

SUMMARY: On the basis of the study of Palaeozoic ironstones from France, Algeria and Libya, some oolitization processes are documented, particularly intrasedimentary accretion and 'snow-ball' type accretion. In a single ooid, successive layers of the cortex can have been formed by different processes. Chemical analysis of an ooid cortex can help to reconstruct its history. Chamositic layers are always weathered in their outer part, with the leaching of aluminium and magnesium. Haematitic layers result from iron, and often titanium concentration in the weathered outer parts of primary chamositic layers. Crystallization of siderite appears to be late diagenetic, and often took place at the same time as, or after, the development of quartz overgrowths.

The aim of this paper is not to examine all the possible oolitization processes proposed in the literature, because numerous critical reviews have already been published (e.g. Kimberley 1981, 1983; Van Houten & Bhattacharya 1982; Van Houten & Purucker 1984). We will only try to underline some aspects of the problem on the basis of some features exhibited by Palaeozoic ironstones of France (Brittany, Normandy), Portugal, Algeria and Libya.

The genesis of oolitic ironstones presents several complex problems. Firstly, what is the origin of the concentric oolitic structure? Three models which have been proposed are the pseudomorphing of primary calcareous ooids, the precipitation of iron-rich minerals on suspended nuclei and the mechanical accretion of an iron-rich mud around scattered nuclei. In these three models the oolitic structure developed respectively before, at the same time as, and after the concentration of iron-rich minerals.

The second problem concerns the origin of the iron-rich minerals of the ooids and of the cement/matrix; are they primary phases? A third problem is the accumulation of ooids, and whether they accumulate in their environment of formation, or in one rather different from that in which they originated. The fourth problem is the nature of the iron supply and its origin.

Chemical and mineralogical data must form the main starting point for any genetic model for ironstones. It therefore seems better to attempt to address the problems separately than to seek a global model.

Since Sorby (1856) many genetic models have been proposed, generally with each claiming to be the single possible solution. In this paper we will try to re-examine some Palaeozoic oolitic ironstones, drawing examples from various ages and locations.

## Geological data

### Ordovician ironstones of Brittany and Normandy

Oolitic ironstones of Arenig age occur at four main levels within the lower member of the Armorican Quartzite Formation in Brittany. These ironstones have been previously studied by Cayeux (1909), Caillère & Kraut (1965), Chauvel (1968, 1974) and Joseph (1982). They are interbedded in near-shore (inter- or sub-tidal) sandstones (Joseph 1982).

Important data on the Llanvirn ironstones of Normandy were presented by Cayeux (1909), Caillère & Kraut (1965), Courty (1959, 1964, 1979), Joseph (1982) and Joseph & Beaudoin (1983). Joseph (1982) proposed a near-shore depositional environment of giant sand waves.

Much recent data for the ironstones of Brittany and Normandy appears in the unpublished thesis by Joseph (1982), together with the short paper by Joseph & Beaudoin (1983). These works provide much of the data used here but, in addition, we have also used information from the thin oolitic ironstones of middle Ordovician age (Chauvel, unpublished data; Deunff & Le Corre 1970) and of late Ordovician age (Chauvel & Robardet 1970).

### Palaeozoic oolitic ironstones of Algeria

There are many oolitic ironstones present in the Palaeozoic sedimentary cover of the Algerian Sahara (Guerrak 1987). In the Tindouf Basin there are ironstones of Silurian, Gedinnian, Siegenian and Famennian ages. In the Ougarta Ranges there are Ordovician oolitic ironstones of Tremadoc, Arenig, Llanvirn and Ashgill ages. In Ahnet and Mouydir there are ironstones of

*From* YOUNG, T. P. & TAYLOR, W. E. G. (eds), 1989, *Phanerozoic Ironstones*
Geological Society Special Publication No. 46, pp. 165-173

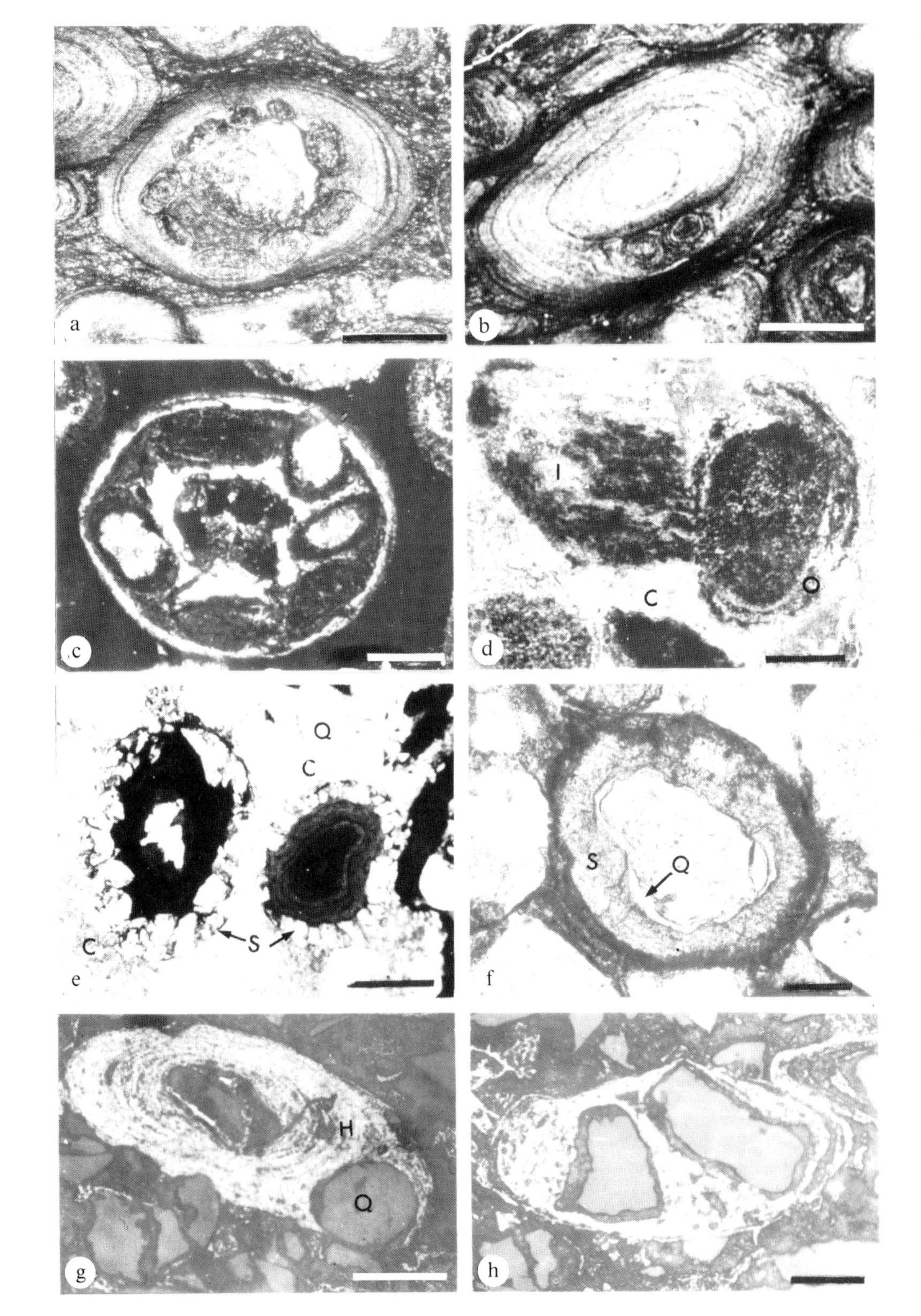
a
b
c
I
C
O
d
Q
C
S
C
e
S
Q
f
H
Q
g
h

Gedinnian, Eifelian, Frasnian and Famennian ages, and in the Tassilis there are ironstones of Gedinnian, Eifelian, Emsian and Famennian ages.

The oolitic ironstones of the Tindouf Basin accumulated on the northern border of the Reguibat Shield, mainly in barrier-island and deltaic environments. In Tassilis the ironstone layers are generally thin, and may have been deposited on a very extended shallow platform. The ironstones of the Ougarta Ranges are probably related to successive transgressions over large flat areas.

### Palaeozoic ironstones of Libya

The Algerian ironstone-rich belt extends eastwards through Libya, with oolitic ironstones occurring at various levels from the lower Ordovician to the upper Tournaisian (Chauvel & Massa 1981).

Some additional data from thin Caradoc oolitic ironstone beds in Portugal have also been used here. For all the samples used here, the green Fe-rich silicates have provided X-ray diffraction data. These silicates are always a 2:1:1 chlorite and in accord with the guidelines of the Nomenclature committees (e.g. Bailey *et al.* 1971) we will use the name chamosite.

## Evidence for oolitization processes

Intrasedimentary accretion and 'snow-ball' accretion are the well evidenced modes of ooid formation in the Palaeozoic ironstones of France, Algeria and Libya.

### Intrasedimentary accretion

This genetic model for iron-rich ooids has been previously suggested for the Palaeozoic ironstones of the Armorican Massif by Caillère & Kraut (1954, 1965) and Chauvel (1968, 1974), with more detailed evidence provided by Joseph (1982) and Joseph & Beaudoin (1983). The same model can also be applied to some Libyan (Chauvel & Massa 1981) and Algerian (Guerrak 1987) examples.

Oolitic ironstones are sometimes associated with silt- and sand-grade detritus, with the clastic grains scattered in a fine-grained chamositic matrix. In these sediments the matrix is frequently arranged concentrically around the grains (Figs. 1d–h). The boundary between the cortical zone and the randomly oriented matrix is generally very gradual.

Ooids with a nucleus constituted by several elements are frequent in some Algerian and Libyan ironstones. In these ooids the cortex coats two or more quartz grains of different optical orientation, or grains of different mineralogical nature (Fig. 1h). Such ooids cannot have been generated in an agitated environment, and strongly suggest intrasedimentary accretion.

It has been demonstrated experimentally (Bucher 1918, Gay 1945) that spherical granules with a concentric structure can be generated from an iron-rich gelatinous precursor. These experiments also produced deformed granules and complex granules (two granules in a single cortex), similar to the spastoliths and complex ooids so frequent in oolitic ironstones. Gay (1945) claimed that iron-rich ooids, devoid of any nucleus, could have been formed by the same accretion process around small pieces of organic matter.

### 'Snow-ball' accretion

Algerian and Armorican ironstones locally exhibit ooids with a structure typical of armoured grains; one layer of the cortex is constituted partially or totally by juxtaposed quartz grains or small ooids (Figs 1a, 1b, 1c & 1g). The oolitic structure of these ooids results from at least two modes of accretion. Firstly the

FIG. 1. (a) Chamositic ooid. One layer is constituted by little ooids pasted by snow-ball accretion process. Sample Ba2, Montreuil-le-Chétif (Sarthe, France); Llandeilo, scale bar : 500 μm transmitted light. (c) Iron oxide ooid. One layer is constituted by little ooids. Sample GJ3, Gour Jiffa, Tindouf Basin, (Algeria); Lochkovian, scale bar: 150 μm transmitted light. (d) Sideritic ooid (O) and intraclast (I) embedded in a chamositic matrix (C). Sample M2, Maleroche Quarry (Ille-et-Vilaine, France), Armorician Quartzite Formation; Arenig; scale bar : 50 μm transmitted light. (e) Ooids partly replaced by siderite (S), embedded with quartz grains (Q) in a chloritic matrix (C). Sample TL 876, Exploration shaft, Libya; Emsian, scale bar : 200 μm transmitted light. (f) The quartz nucleus of this sideritic ooid exhibits a well developed diagenetic overgrowth (Q), Thick sideritic cortex (S). Sample LB 603, Drilling B1-23, (1730m), Libya; middle Ordovician, scale bar: 100μm transmitted light. (g) Quartz grain (Q) embedded in the hematitic cortex (H) of an ooid. Sample TA 114, Fadnoun, Tassilis N'Ajjer, Algeria; Lochkovian, scale bar : 100 um reflected light. (h) The nucleus of this hematitic ooid is constituted by two quartz grains. Sample TA 114, Fadnoun, Tassilis N'Ajjer (Algeria); Lochkovian, scale bar: 50 μm reflected light.

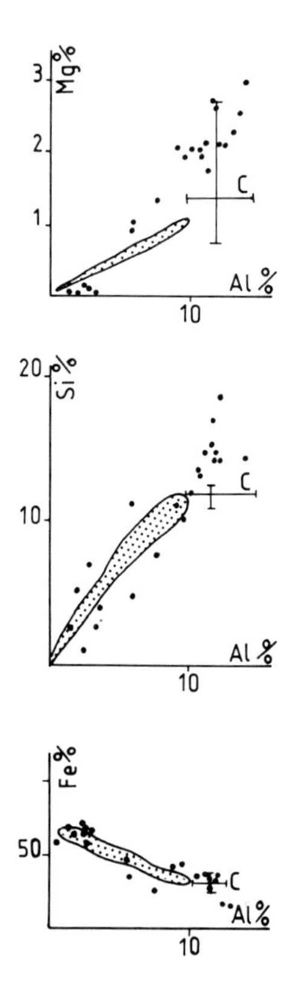


FIG. 2. Mg, Si, Al, Fe (wt. %) correlations in alternating hematitic and chamositic layers of ooids. M30, M31 and M32 drillings — Zemila, southern part of the Tindouf Basin (Algeria) — M30, M31: Llandovery — M32: Pragian. Dotted area: Llanvirn ooids (Normandy) after Joseph & Beaudoin (1983). C: Chamosite (Chrustenic, Windgällen, Schniedefeld and Bas-Vallon — Chauvel (1968). Bars indicate total range.

inner part of the cortex was generated by intrasedimentary accretion (or possibly by another method), and then the quartz grains or small ooids were pasted onto the soft ooid by its rolling on the sediment surface. The outer part of the cortex then developed over the armoured grain during further intrasedimentary accretion.

These examples allow several conclusions. Firstly, that intrasedimentary accretion processes can give rise to concentrically layered cortices around elements scattered in a fine-grained groundmass. They also demonstrate that soft elements, rolling on the sediment surface, can be coated by accretion of particles. In addition they show that the cortex can be generated by different successive processes. There has been a recent suggestion (Champetier *et al.* 1987) that ooids in the Siegenian ironstones of Gara Djebilet (Tindouf Basin, Algeria) are the mineralized tests of the foraminifera *Nubecularia*. The mineralogical transformation of bioclasts is a well known phenomenon, and in some ferruginous limestones bioclasts can be formed of iron-rich minerals. It is impossible to agree with this proposed model for the Gara Djebilet ironstone, however, because the oldest representative of the Nubecularidae (*Palaeonubecularia*) only appeared during the Middle Carboniferous (Moore 1964; Vachard, personal communication).

There is general agreement on the fact that ooids were not generally accumulated in their place of origin, in spite of the variety of the models proposed for the genesis of the ooids. Alternating periods of transportation and deposition resulted in a complex history, documented by the abundance of broken ooids and oolitic intraclasts, often themselves coated in a layered cortex. This observation is emphasised by persistent mineralogical differences between the ooids and their matrix.

The chemical and mineralogical compositional variation of the concentric cortical layers record the successive environments in the history of an ooid. Detailed analysis of this variation is made possible by the electron microprobe. In the following discussion we will use data provided by Joseph (1982) from Llanvirn ooids (Normandy) and recent unpublished analyses of Devonian ooids from the Tindouf Basin.

The main results are summarized in Figs. 2, 3, 4, 5 & 7, and they allow the following conclusions to be drawn:

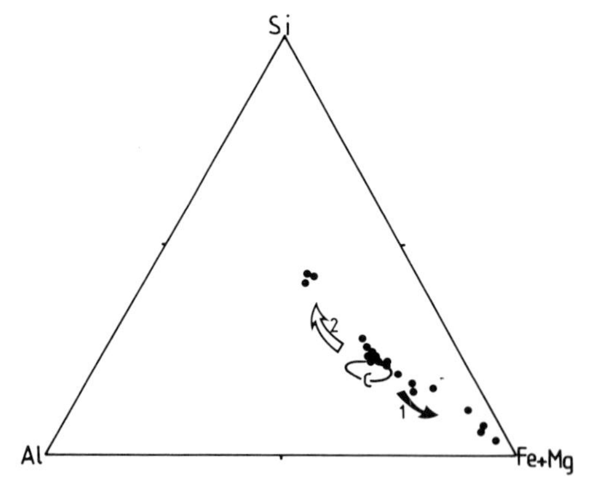


FIG. 3. Composition of alternating hematitic and chamositic layers of ooids. Same samples as for Fig. 1. C: chamosite (analyses as for Fig. 1) 1: Si-Al leaching, 2: Fe leaching.

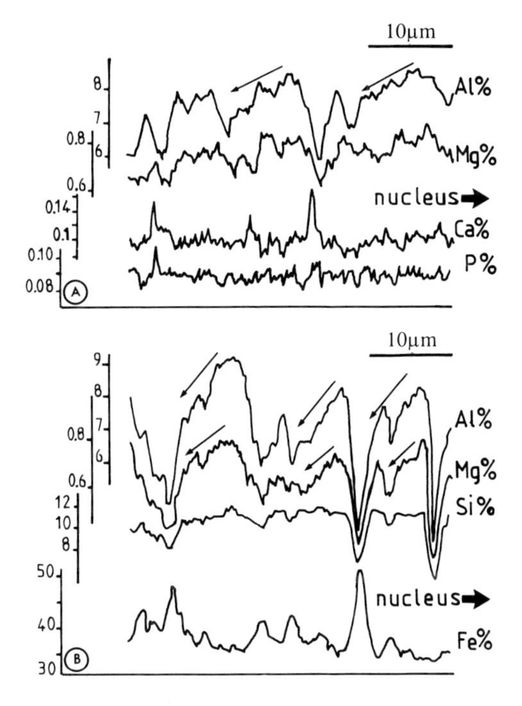


FIG. 4. Chemical records in successive layers of ooids. After Joseph (1982) (A : Fig. 84 p.142 — B : Fig. 80 p.140). Halouze mine (Normandy). Llanvirnian. A: successive chamositic layers. B: alternating haematitic and chamositic layers.

(i) In the hematitic layers Mg, Al and Si are present, with constant Mg/Al and Si/Al ratios in good agreement with those of chamosite (Fig. 2). Chamosite is present, therefore, in the hematitic layers.

This conclusion is in agreement with experiments performed on Armorican (Joseph 1982) and Libyan (Chauvel & Massa 1983) ironstones, where the chamositic framework of the hematite layers was exposed by leaching the iron oxides with sodium dithionite solution.

The iron oxides in cortical layers could have been formed by two processes: either by precipitation of iron oxides within a chamositic framework, or by the oxidation of chamosite. The observed correlations between Fe, Mg, Al and Si can only be explained by the second hypothesis (Figs 2 & 3). The oxidation of iron was accompanied by leaching of other elements, with Mg being more or less stable.

(ii) Each layer of the cortex of an ooid exhibits centrifugal variation of its chemical composition. These variations have already been described by Joseph (1982) and are also shown by analyses of Algerian ooids (Figs 4 & 5). In chamositic layers there is a centrifugal decrease of Al and Mg, and sometimes of Si. The boundary between two successive layers is underlined by the abrupt difference between the low Al and Mg contents of the outer part of one layer, and their high values in the inner part of the next (Fig. 4a).

These centrifugal variations are also clear in hematitic layers, where they are emphasized by the more abrupt decrease in Al and Mg content in the outer part of each layer, accompanied by an equivalently abrupt increase in Fe. The boundary between adjacent layers can be further marked by a thin layer of calcium phosphate, and sometimes by an increase in Ti content (Figs 4, 5 & 6c).

The thickness of cortical layers, as determined by these chemical variations, is about 7 μm for the ooids studied by Joseph and 4 μm for ooids from Algeria.

These centrifugal variations of Al, Mg and Si contents of successive layers in an ooid cortex could be the result either of a mineralogical evolution of the formation of each layer, or of the leaching of the outer parts of layers between successive accretion periods. This second solution seems to be favoured by the frequent accumulation of more stable elements (e.g. Fe, Ti) in the outer parts of layers (Figs 4 & 5). The mineralogical composition and the regular chemical variations of cortical layers are apparently not in good agreement with the hypothesis that oxide and silicate phases could both crystallize from a gel precursor.

The ooids studied here appear to be the result

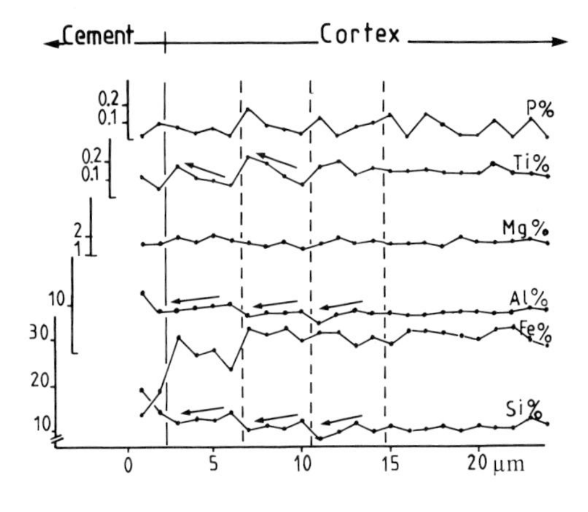


FIG. 5. Chemical records in successive layers of ooids. M30 drilling (−41,10 m). Zemila, southern part of the Tindouf Basin (Algeria) — Llandovery.

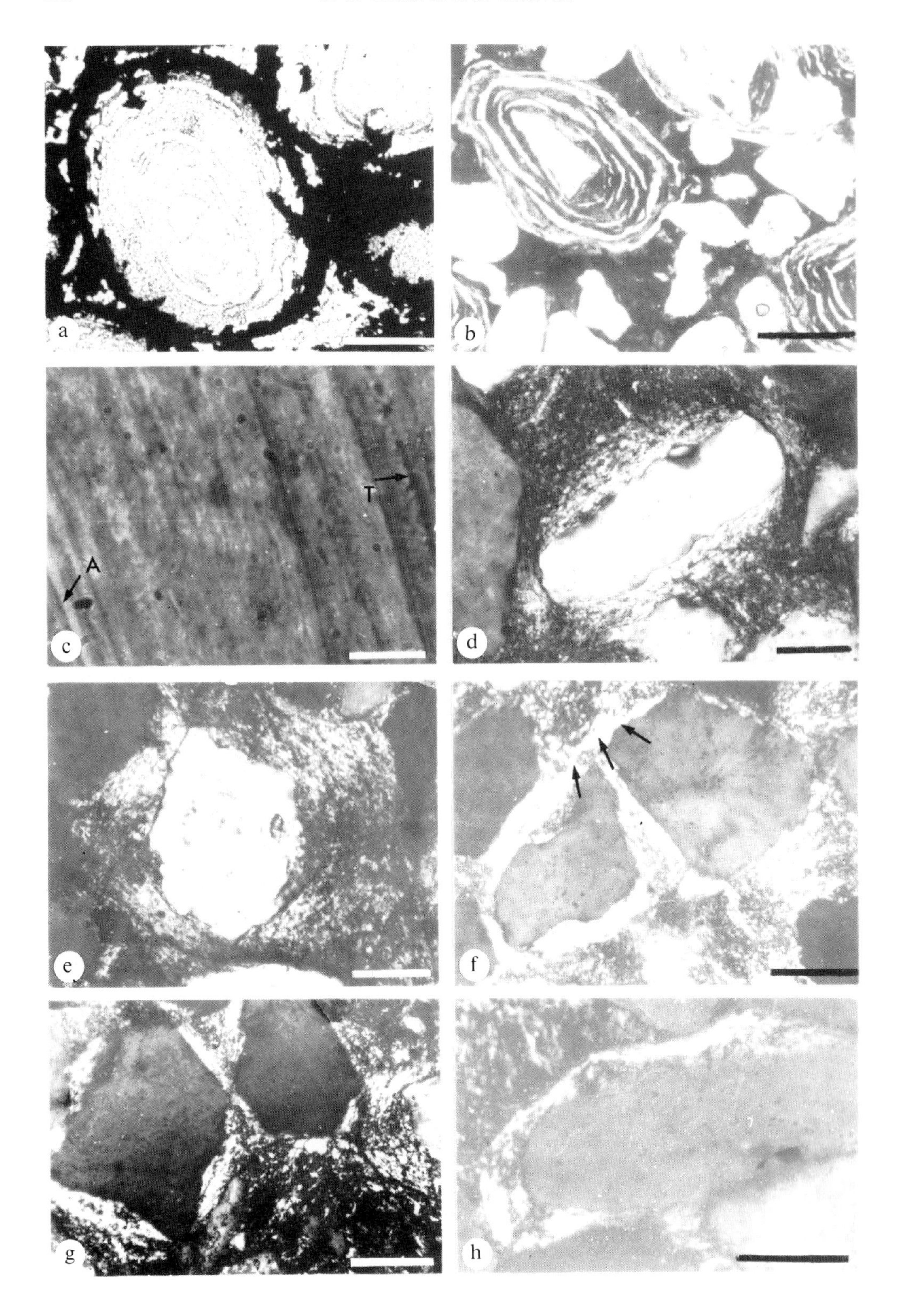
a
b
c
T
A
d
e
f
g
h

of the alternating accretion of chamosite in greater than 10 $\mu$m layers, and the transformation of the outer part of the ooid during the breaks in accretion.

Under moderately oxidizing conditions Al and Mg were variably leached, while under highly oxidizing conditions the intense leaching of Al, Mg and Si was accompanied by the enrichment of stable elements, such as Fe and Ti.

These reactions might also involve the formation of a thin phosphatized film, and it is also possible that the abnormally thick development of such a film could have given rise to the thick phosphatic layers often present in ferriginous ooids. The more or less complete quartz layers present in some ooids (Fig. 6a) could have been formed under reducing conditions by the leaching of Fe, Mg and Al, resulting in an accumulation of silica.

The concentric cortical structure of sideritic ooids is generally poorly preserved, and it is never easy to know if siderite crystallization took place before or after the accumulation of the ooids.

Some ooids exhibit alternating layers of chamosite and siderite, with the siderite layers devoid of Si and Al (Fig. 7). In some ironstones siderite only occurs within the ooids, with the matrix (or cement) mainly composed of chamosite (Fig. 1d). In ironstones with chamositic ooids scattered in a chamositic matrix, the ooids may be surrounded by a thick siderite layer, with the siderite partially or totally replacing the ooids (Fig. 1e). This replacement is irregularly developed, and is always centripetal.

It is not easy to construct a model to account for these particular features, but they suggest that siderite may be the result of transformation of chamosite, comparable with the development of hematite; in oxidizing conditions Al, Si and Mg were leached, leaving hematite, but in reducing, carbonate-rich environments, siderite may form.

In some Libyan ironstones the quartz nuclei of siderite ooids frequently exhibit diagenetic overgrowths (Fig. 6f). Such overgrowths have not been observed in chamositic ooids, and may be the result of the release of $SiO_2$ during the transformation of chamosite to siderite.

In these Palaeozoic ironstones the possibility of late mineralogical and textural transformations must be examined, especially the possible overprinted effects of the berthierine — chamosite transformation.

In the ironstone samples used here we did not find evidence for late transformations. There is general agreement on the fact that chamosite (2:1:1 chlorite) is produced by a late transformation of berthierine (1:1 layer silicate). Evidences for this transformation are found in the fact that chamosite is generally restricted to Precambrian and Palaeozoic ironstones. After Schoen (1964) and Iijima & Matsumoto (1982) the temperature of the berthierine — chamosite transformation could be 130–160°C (Maynard 1986). This conclusion must be used very cautiously for several reasons: (i) Berthierine has been evidenced by X-ray diffraction in various metamorphic Precambrian rocks (Gole 1980; Floran & Papike 1975). (ii) It is not easy and often impossible to distinguish a pure chamositic phase and some mixtures of 2:2:1 chlorite and 1:1 layer silicate on the basis of X-ray diffraction data. (iii) Berthierine and chamosite have the same chemical composition. (iv) Distinction between chamosite and thuringite is based on the $Fe^{3+}/Fe^{2+}$ ratio, but this ratio is not available for the current study in which chemical analyses are obtained by electron microprobe.

## Conclusions

On the basis of the study of these examples, it appears impossible to propose only a single model for the genesis of ooids in oolitic

FIG. 6. (a): Half-moon quartz layers in the chamositic cortex of an ooid. Sample M6, Montflours (Mayenne, France), middle Ordovician; scale bar : 200 μm transmitted light. (b): Leaching of iron-oxides by a sodium dithionite solution reveals the concentric structure of ooids and the chloritic framework (grey) appears. Sample LB 615A, Drilling B1-23 (1679,50m), Libya, middle Ordovician; scale bar : 150 μm transmitted light. (c): Thin layering in the chamositic cortex of an ooid. A thin film of calcium phosphate (A) appears in white. Thin accumulation of Ti-rich granules (T). Sample J. L. Henry, Buçaco, Portugal, Caradoc; scale bar : 10 μm transmitted light. (d): Orientation of a chamositic matrix around a quartz grain. Sample F 39, Brieux (Orne, France), Urville Formation, Llanvirn; scale bar : 50 μm, polarized transmitted light. Orientation of a chamositic matrix around a quartz grain. Sample F 39, Brieux (Orne, France), Urville Formation, Llanvirn; scale bar : 50 μm polarized transmitted light. (f): The orientation of a chamositic matrix developed a common sheath on two quartz grains. Sample F. 39, Brieux (Orne, France), Urville Formation, Llanvirn; scale bar : 50 μm — polarized transmitted light. (g): Orientation of a chamositic matrix around a quartz grain. Sample F 39, Brieux (Orne, France), Urville Formation, Llanvirn; scale bar : 50 μm polarized transmitted light. (h): Orientation of a chamositic matrix around a quartz grain. Sample Md 2, Maleroche Quarry (Ille-et-Vilaine, France), Armorican Quartzite Formation, Arenig; scale bar : 50 μm transmitted light.

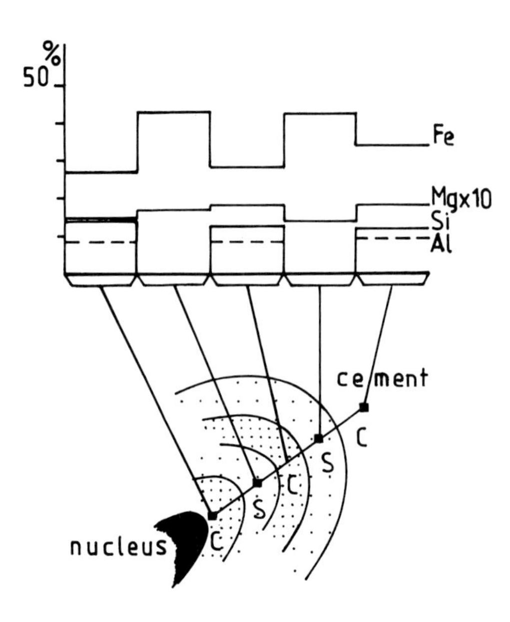


FIG. 7. Chemical records in alternating chamositic and sideritic layers. Exploration shaft. Gara Djebilet, Southern part of the Tindouf Basin (Algeria). Pragian.

ironstones; several successive processes may have been involved in the formation of a single ooid. Here we have documented intrasedimentary accretion and 'snow-ball' mechanical accretion as examples of such processes.

Ooids were not accumulated in their place of formation and they may have been reworked into various sediments prior to their eventual accumulation. During these periods of alternate suspension and deposition the outer parts of ooids became weathered, with the leaching of Al, Si and Mg and the enrichment of iron, either as hematite or siderite. Hematite may have been developed during suspension of reworked ooids in agitated waters and siderite during periods of deposition.

## References


BAILEY, S.W., BRINDLEY, G.W., JOHNS, W.D., MARTIN R.T. & ROSS, M. 1971. Summary of national and international recommendations on clay mineral nomenclature. *Clays and clay minerals*. 19, 129–132.

BUTCHER, W. 1918. Oolites and spherulites. *Journal of Geology*, **26**, 593–609.

CAILLERE, S. & KRAUT, F. 1954. Les gisements de fer du bassin lorrain. *Mémoires du Museum d' Histoire Naturelle de Paris, (C)*, 4.

——&——1965. Les minerais de fer d'âge primaire de Normandie et de l'Anjou. *Mémoire du Museum d'Histoire Naturelle de Paris. Nouvelle série*, **12**, 2.

CAYEUX, L. 1909. *Les minerais de fer oolithiques de France. Tome I: Minerais de fer primaires.* Imprimé Nationale, Paris.

CHAMPETIER, Y., HAMDADOU, E., & HAMDADOU, H., 1987. Examples of biogenic support of mineralization in two oolitic iron ores — Lorraine (France) and Gara Djebilet (Algeria). *Sedimentary Geology*, **51**: 249–255.

CHAUVEL, J. J., 1968. *Contribution à l'étude des minerais de fer de l'Ordovicien inférieur de Bretagne*. Thèse. Universite de Rennes.

——1974. Les minerais de fer de l'Ordovicien inférieur du bassin de Bretagne-Anjou. *Sedimentology, 21*: 127–147.

——& MASSA, D. 1981. Paléozoique de Libye occidentale. Constantes géologiques et pétrographiques. Signification des niveaux ferrugineux oolithiques. *Compagnie Francaise des Petroles, Notes et Mémoires, Paris*, **16**, 25–66.

——,DEUNFF, J. & LE CORRE, C., 1970. Découverte d'une association minerai de fer-microplancton dans l'Ordovicien du flanc nord du Bassin de Laval (Mayenne). *Comptes Rendus de l'Academie des Sciences. Paris, (D)*, **270**, 1219–1222.

——& ROBARDET, M. 1970. Le Minerai de fer de Saint-Sauver-le-Vicomte (Manche). Position stratigraphique. Etude pétrographique. Signification paléogéographique. *Bulletin de la Société Géologique et Minéralogique de Bretagne, (C)*, **2**, 61–71.

COURTY, G. 1959. Contribution à l'étude du minerai de fer de May-sur-Orne (Calvados). *Bulletin de la Société Géologique de France*, **7**, 500–510.

——1964. Esquisse sur la pétrographie et remarques sur la diagenèse de la couche principale du minerai de fer ordovicien normand. *Bulletin de Service de la Carte Géologique de France*, **60**, 95–113.

——1979. Caractéres paléogéographiques du milieu de formation de la couche principale du minerai de fer ordovicien normand. *Annales de la Société Géologique du Nord*, **99**, 481–486.

FLORAN, R. J. & PAPIKE, J.J. 1975. Petrology of the low grade rocks of the Gunflint Iron Formation, Ontario — Minnesota. *Bulletin of the Geological Society of America*. **86**, 1169–1190.

GAY, R., 1945. Etude des phénomènes de diffusion. Applications minéralogiques. 4ème Partie : forme extérieure des concrétions dans les sediments. *Bulletin de la Société Francaise de Minéralogie,* **68,** 60–152.

GOLE, M.J. 1980. Mineralogy and petrology of very-low metamorphic grade Archaean banded iron-formations, Weld Range, western Australia. *American Mineralogist,* **65,** 8–25.

GEURRAK, S. 1987. Paleozoic oolitic ironstones of the Algerian Sahara: a review. *Journal of African Earth Sciences,* **6,** 1–8.

IIJIMA, A. & MATSUMOTO, R. 1982. Berthierine and chamosite in coal measures of Japan. *Clays and clay minerals,* **30,** 264–274.

JOSEPH, Ph. 1982. *Le Mineral de fer oolithique ordovicien du Massif Armoricain : sédimentologie et paléogéographie.* Thèse de Docteur Ingenieur, Paris.

——& BEAUDOIN, B. 1983. Microséquences intra-oolithiques dans le minerai de fer ordovicien normand (Llanvirn). Nouvelle hypothèse de genèse des oolithes ferrugineuses. *Comptes Rendus de l'Académie des Sciences, Paris,* **296,** 1533–1537.

KIMBERLEY, M.M., 1981. Oolitic iron formations. *In:* WOLF, K.H. (Ed.). *Handbook of strata-bound and stratiform ore deposits.* **9,** 25–76.

——1983. Ferriferous ooids. In: PERYT, P. (Ed): *Coated grains.* Springer-Verlag, Heidelberg, 100–108.

MAYNARD, J.B. 1986. Geochemistry of oolitic iron ores and electron microprobe study. *Economic Geology.* **81,** 1473–1483.

MOORE, R.C. 1964. *Treatise on Invertebrate Paleontology. Part C, Protista 2,* The Geological Society of America and the University of Kansas Press, 1–510.

SCHOEN, R. 1964. Clay minerals of the Silurian Clinton ironstones, New York State. *Journal of Sedimentary Petrology* **34,** 855–863.

SORBY, H.C. 1856. On the origin of the Cleveland Hill ironstone. *Proceedings of the Geological & Polytechnic Society of West Riding Yorkshire,* **3,** 457–461.

VAN HOUTEN, F.B. & BHATTACHARYA, D.P. 1982. Phanerozoic oolitic ironstones. Geologic Record and Facies. Model. *Annales Review of Earth and Planetary Science,* **10:** 441–457.

——& PURUCKER, M.E., 1984. Glauconite peloids and chamosite ooids. Favorable factors, contraints and problems. *Earth-Science Reviews,* **20,** 211–243.

J.J. CHAUVEL, Centre Armorician d'Etude Structurale des Socles, Université de Rennes I, Laboratoire de Pétrologie Sédimentaire, 35042 Rennes Cédex, France.

S. GUERRAK, Office Nationale de la Géologie, 18A Avenue Mustapha El Ouali, Alger, Algeria.

# Minette-type ironstones


## A. Siehl & J. Thein



SUMMARY: Oolitic ironstones occur in various sedimentary environments: shallow marine to deltaic, lacustrine, fluviatile and pedogenic. Distinction between formational and depositional environment is not always possible. Most of the marine and fluvial minette-type ironstones consist of reworked ferruginous coated grains deposited in agitated water, but there exist also indicative structural features of *in situ* formation in the supporting medium of lateritic and hydromorphic environments. In the zone of oscillating groundwater repeated leaching and subsequent concretionary precipitation of hydrated ferric oxides take place, according to the prevailing Eh/pH-conditions and microbial activity. The moderate Al substitution of goethite from hydromorphic environments corresponds to the observed range in oolitic ironstones. The authors therefore assume erosion, reworking and subsequent fluviomarine redeposition of soil derived ooids to be the major processes of generating minette-type ironstones. Postdepositional diagenetic changes may convert the aluminous, silica-rich ferric oxides into berthierine in reducing environments if the chemical bulk composition of the primary goethite is similar. Since any aquatic milieu with appropriate fluctuations of Eh and pH can produce ferruginous coated grains, marine iron ooids associated with hardgrounds and areas of low sediment input can also occur. But there, release of ferrous iron, transport in saline interstitial waters and fixation of ferric hydroxides — usually with very low Al-substitution — take place in a much smaller scale, unable of generating the huge iron accumulations of minette-type ore deposits.


Minette-type ironstones are detrital sediments containing typically ooids, pisoids and clasts of silica-rich, aluminous goethite, of hematite, of Al-rich berthierine/chamosite or of a combination of these. They are known since the late Precambrian and occur in various sedimentary environments: the majority of Phanerozoic oolitic ironstones are shallow marine (Wabana, Brittany, Bohemia, Minette, Northampton, Gifhorn, Salzgitter), or brackish-estuarine (Peace River, Kerch), but also lacustrine (Continental Terminal of West Africa, Siderolithique, Chad), fluviatile (North Aral, Turgai) and pedogenic reworked and residual oolitic-pisolitic ironstones (Akkermanovka, Alapaevo, Southern Timan, Serov, Malka: USSR, Lokris: Greece) are described (for references see: Kimberley 1978; Smirnov 1977; Zitzmann 1977).

Most of the marine and fluviatile minette-type ironstones consist of reworked coated grains deposited in agitated water, but there are also indicative structural features of *in situ* formation in a supporting medium in lateritic vadose and hydromorphic environments (Siehl & Thein 1978; Nahon *et al.* 1980). Postdepositional diagenetic changes are often severe in these polyphase detrital associations, which are not in thermodynamic equilibrium at the time of their deposition and, therefore, particularly susceptible to postdepositional transformations. These may obscure the primary structure and mineral content, so that the distinction between formational and depositional environments is not always possible, nor whether goethite or berthierine/chamosite was the primarily formed mineral.

In addition to the enrichment of the major elements Fe, Al and Si, rather high contents of the trace elements P, V, Cr, Zr, Th, and often also of Co, Ni, As, Zn are characteristic of minette-type ironstones. These elements are also typically enriched in hydromorphic soils and residue of lateritic weathering, where their concentration varies with a number of factors including climate, relief, Eh and pH, groundwater movement and the composition of the source rock (Norton 1973; Schellmann 1986). This fact is quoted by Harder (1964) as proof that minette-type ironstones originated from lateritic weathering products, though it is obvious that the set of siderophilic elements will be adsorbed by colloidal iron-hydroxides wherever they may precipitate, even in marine environments. Noteworthy, moreover, is an average Al substitution of 7 to 10 mol% in Minette goethite (Correns & von Engelhardt 1941; Flehmig 1967) and the existence of high Al contents in berthierine oolites (Maynard 1986). Therefore, an environment of high Al concentration and mobility is required during the formation of the ferruginous particles, as it is present, for example, at the sites of ferralitic weathering (Huang & Keller 1972). It seems


*From* YOUNG, T. P. & TAYLOR, W. E. G. (eds), 1989, *Phanerozoic Ironstones*
Geological Society Special Publication No. 46, pp. 175-193

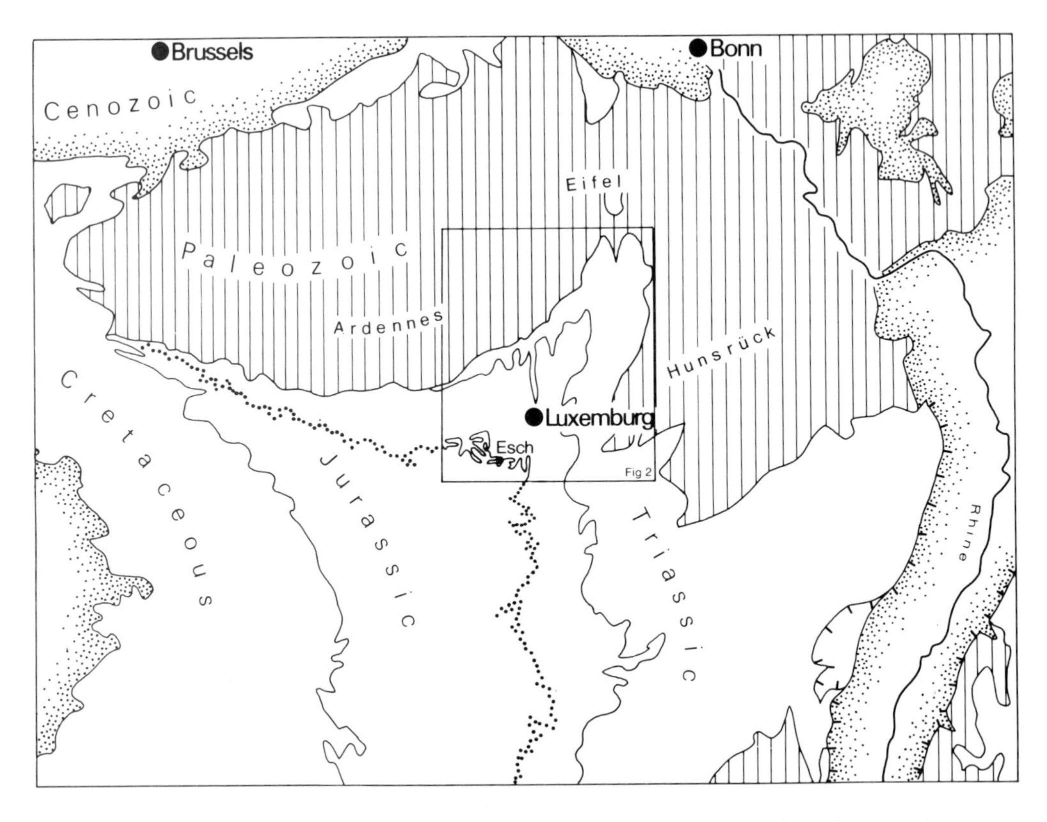


FIG. 1. Geological situation of the Minette ironstone deposit in the northeastern Paris Basin (Luxemburg Gulf). The dotted line indicates the outcropping Aalenian, bearing the upper part of the Minette ores in Luxemburg and Lorraine.

rather surprising that until now there has been no convincing report of ooids *in statu nascendi,* considering the fact that ferruginous ooids originated throughout the Phanerozoic under apparently actualistic conditions, although more commonly in certain periods (Van Houten & Bhattacharyya 1982). The search for such occurrences is usually focused on marine environments, but they might not be the most favourable places to observe ooids growing. We will therefore turn our attention to the terrestrial environment, where, in the zone of oscillating groundwater, repeated leaching and subsequent concretionary precipitation of hydrated ferric and aluminium oxides take place on a very large scale, according to the prevailing Eh/pH conditions and microbial activity.

## Minette deposits of Luxemburg and Lorraine

The Lower to Middle Jurassic sediments of the Minette type locality in the eastern Paris Basin include oolitic ironstones at several horizons in the upper part of up to 15 coarsening upward cycles (Figs. 1 & 2). The sequences are detrital throughout. They show a marked regional and vertical facies variation; in the lower and north-western part berthierine silicate ores predominate, while in the upper and eastern parts calcareous goethite ores are more important. The ironstone facies consists of cross bedded iron oolites and bioarenitic limestones with *Scolithos,* showing typical features of a high-energy environment. It is geochemically characterized by a number of intercorrelated siderophilic elements and by Ca and Sr (Figs 3 & 4). The intercalated, strongly bioturbated fine-grained mud- and siltstone facies, deposited in a lower-energy environment, is marked by a set of lithophilic elements, brought in by the fine grained siliciclastic detritus. The alternation of the two facies builds up a repeated sequence of shallowing upward cycles. Coarse-grained lithoclasts and large biomorpha indicate transgressive events at the base of each cycle. The bioturbated siltstones with *Rhizocorallium* and *Planolites* grade, with increasing iron-rich particle content, into the ironstone facies with a cross-bedded biorudstone on top indicating the shallowest part of the cyclothem (Fig. 5).

The Minette sequence was deposited in a near-shore shallow marine environment (Lucius 1945; Bubenicek 1961, 1971; Thein 1975; Siehl & Thein 1978; Teyssen 1984), reaching its maximum thickness of 60 m in the subsiding Luxemburg Gulf of the Jurassic sea, bordered to the north and east by deeply weathered lowlands of the variscan folded basement and its Mesozoic cover. The regional context of the iron ore deposit suggests a palaeogeographic setting comparable with that of many other Mesozoic and Cenozoic marine ironstones; a shallow tidal-influenced inland sea at low palaeolatitude (35°N), surrounded by peneplained Palaeozoic terrains and situated in a cratonic position. There is no direct evidence of the exact position and nature of the coastlines, nor of the type of weathering on land, but from the fine-grained low-quartz detritus supplied, the existence of a low-relief terrain covered by a thick cap of iron hydroxide- and kaolinite-rich subtropical weathering products can be concluded. Repeated reworking of these hydromorphic soils and lacustrine mudflats probably produced terrestrial preconcentrations of ferruginous microconcretions and duricrust fragments. Washed into the sea and winnowed in agitated water, the soil-derived clasts were finally accumulated as marine ironstone. At the end of the deposition of the ironstone-bearing succession, erosion apparently cut deeper, because quartz sand occurs in the detrital record and dilutes the ferruginous components in the uppermost goethitic ironstone beds. According to Bubenicek (1971, see also IRSID-Atlas 1967) the Minette sequence was deposited in the vast delta system of a river draining the region of the present Eifel mountains. A brackish mixing zone of fresh and marine waters was supposed to be the site of goethite ooid formation, their enrichment to minable concentrations took place

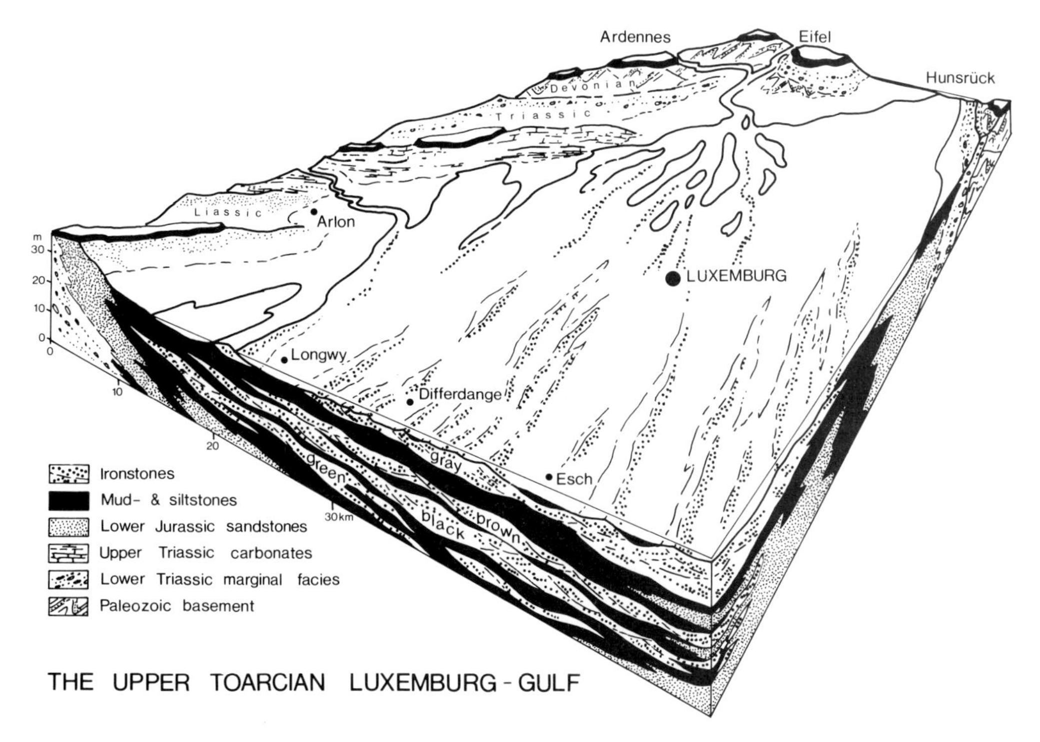


FIG. 2. Sketch diagram displaying the tentative palaeogeographic situation of the Luxemburg Gulf during the deposition period of the 'couche grise', Toarcian. The Hercynian basement of the Ardennes, Eifel and Hunsrück and its Lower Mesozoic cover are deeply peneplained and capped by a thick laterite duricrust. Rivers draining the area transport the reworked hydromorphic soils containing ferriclasts, pisoids and ooids into the shallow sea, where they settle together with bioclasts and siliciclasts in NE/SW elongated sand bars. They interfinger laterally with muddy siltstones of quiet water areas.

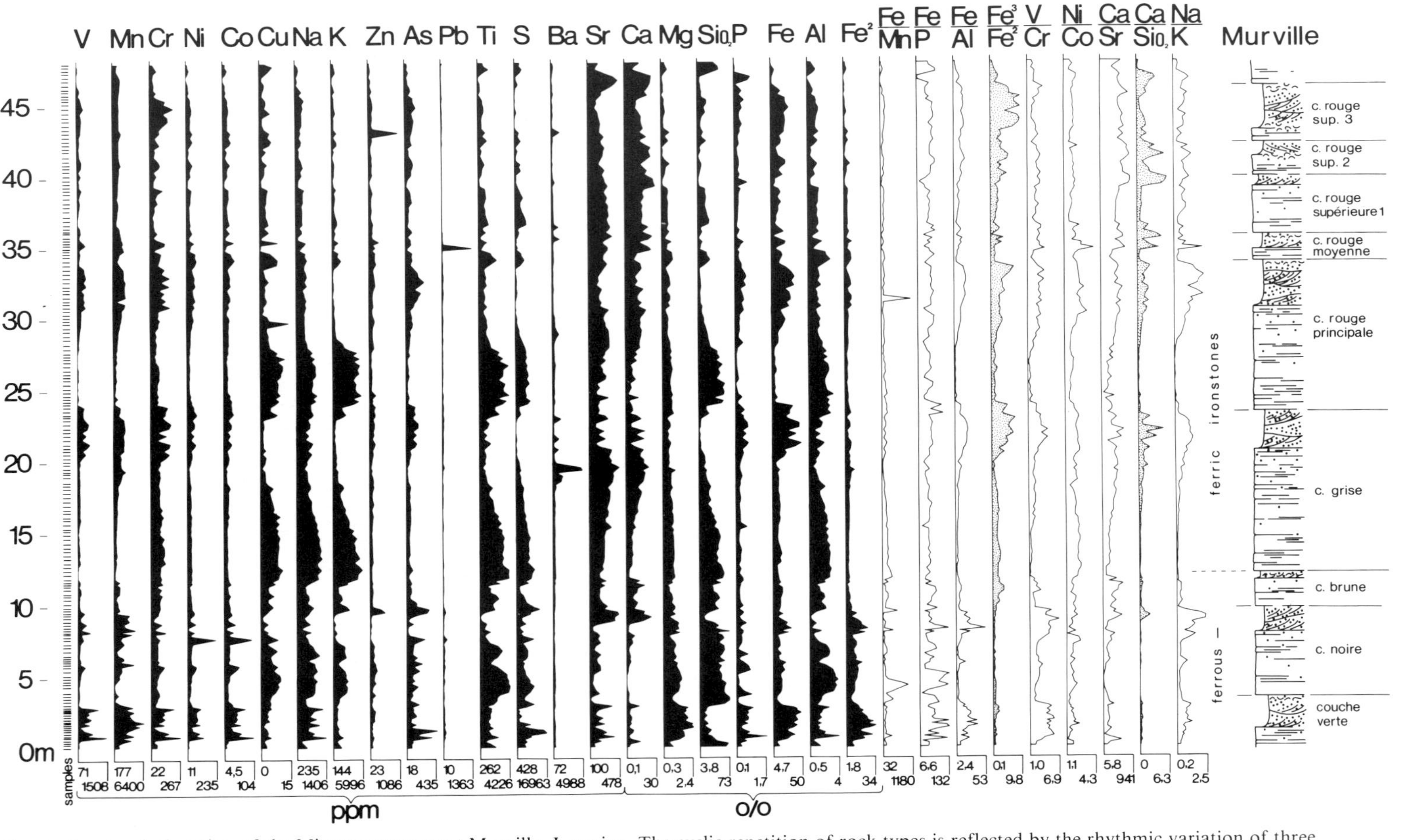


FIG. 3. Geochemical section of the Minette sequence at Murville, Lorraine. The cyclic repetition of rock types is reflected by the rhythmic variation of three groups of intercorrelated elements, representing the three main lithologic facies: Muddy siltstone, ironstone, limestone (see also Fig. 4). Note the superimposed long term trend with high Si and $Fe^{2+}$ values in the lower part (berthierine bearing siliceous ores) and high Ca and $Fe^{3+}$ contents at the top of the section (goethitic calcareous ores).

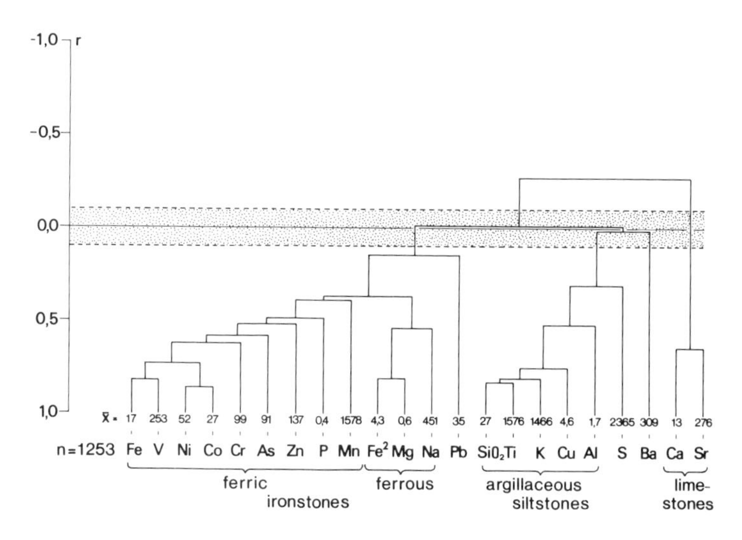


FIG. 4. Cluster analysis dendrogram of the geochemical composition of Minette ironstones. The three major element clusters correspond to the main lithologic facies: *ironstones* with the siderophilic elements Fe, V, Ni, Co, As, Zn, P, Mn, and a distinct special group for the siliceous berthierine facies ($Fe^{2+}$, Mg, Na). *Argillaceous siltstones,* characterized by the lithophilic elements Si, Ti, K, Cu, Al carried by the siliciclastic terrigenous input, and *limestones* with the biogenic Ca, Sr group. ($n$ = number of analysed samples, $x$ = mean values in per cent and ppm; see Fig. 3. Correlations in the stippled area are not significant.)

during periodic regressions. Nowhere in the Minette, however, can a distinct terrestrial facies be recognized, nor is there any sign of fresh water influence. The fauna is fully marine throughout and no deltaic sediments can be observed. Thein (1975) interpreted the Minette series as intertidal sediments, Siehl & Thein (1978) and Teyssen (1984) as subtidal. The relief of the sedimentation area was accentuated by wave- and current-dominated offshore shoals, the deposition site of the ironstones. The iron-oolite/bioarenite sand bars were elongated in the NE/SW direction, controlled by tidal currents and probably also by a submarine low relief caused by synsedimentary faults running in the same direction. The bars die out towards the SW with increasing water depth, where they grade into highly bioturbated mud- and siltstones. Occasionally intercalated lumachelle layers may be observed, which can be traced as good marker horizons over large distances and most probably represent storm layers.

In all the sediments the ferruginous ooids occur in a clastic assemblage, accompanied by other detrital grains like ferriclasts and quartz grains, so that they appear as definitely allochthonous. The same applies to the goethite ooids and ferriclasts reported from a site comparable to that of the Bubenicek model: the Mahakam Delta in Kalimantan (Allen *et al.* 1979). They have been found there in water depths of 1–4m, imbedded in unconsolidated delta-front clays as solid clastic particles together with quartz grains of approximately the same grain size (0.2mm). All were most probably brought in as clastic grains by the river.

## The terrestrial origin of iron

The discussion about the origin of iron ooids found in marine sequences is closely connected with the question of the source and the mode of transportation of iron (Siehl & Thein 1978; Gygi 1981; Maynard 1983; see also Fig. 6). The extremely low solubility of iron, aluminium and some of the associated trace elements (Ti, V, Cr, Zr, Th) exclude an analogy with the formation of carbonate oolites. There is no doubt that the ironstone element combination must have been carried into the sea by terrigenous clastic particles of various grain sizes, from clay fraction (Carrol 1958) to coarse ferruginous gravels (Kölbel 1944), where it may be mobilized during early diagenesis under reducing conditions caused by the decay of organic matter, and subsequently precipitated at the redox interface within the sediment or at its surface. It is unlikely, however, that the iron concentration and transport rate in the saline interstitial waters of compacting sediments is high enough to produce the iron accumulations known from the large ore deposits.

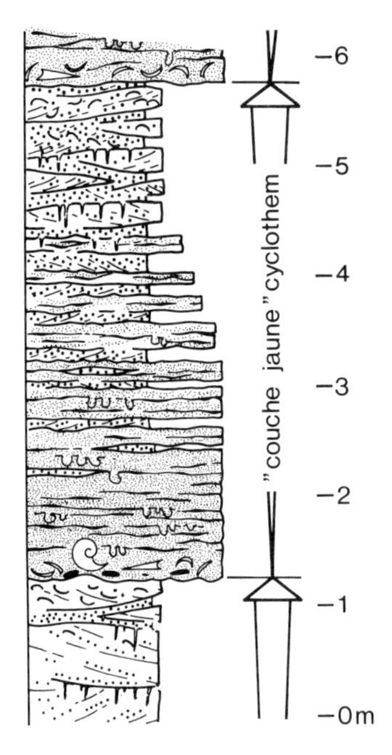


FIG. 5. Asymmetric coarsening upward sequence of a typical shoaling Minette cyclothem. The basal bed contains abundant biomorpha of marine fossils (ammonites, belemnites, pelecypoda) and conglomerate-size intraclasts. The overlying linsen- and flaser-bedded muddy siltstone facies, poor in ooids, is strongly bioturbated by sediment feeders. A ripple-bedded transition zone grades into the large scale cross-bedded ironstone facies, rich in goethitic particles and disturbed by vertical burrows of suspension feeders. The cycle generally terminates with a cross-bedded biorudite.

Quite a different situation exists in the deeply weathered rocks of subtropical regions with distinct alternations of humid and dry seasons. Oscillating groundwater tables and redox interfaces with extensive lateral groundwater flow cause the intensive dissolution of iron, which may be transported over large distances. Huang & Keller (1972) point to the high solubility of alumimium in the zone of intense leaching, under the influence of organic acids. The distance of transport is, however, much shorter than that of iron under the same conditions.

Repeated leaching and the subsequent precipitation of large amounts of iron and aluminium hydroxides in the hydromorphic zone, cause the formation of ferruginous nodules, pisoids and ooids. These processes are influenced by the activity of bacteria, some of them causing reduction and mobilization of iron, the others using in turn ferrous iron as a source of energy and precipitating ferric hydroxides (Aristiovskaja & Zavarzin 1971).

In the vadose soil-forming environment of tropical and subtropical areas *in situ* formation of ferruginous coated grains is a common feature and the fabric of the pisolitic ironstone gravels developing here is very similar to the well known carbonate vadolites of the same environment (Dunham 1969; Peryt 1983; see also Fig. 7a,b). Nahon *et al.* (1980) described in detail the gradual replacement of marine Mesozoic–Cenozoic sediments by aluminous goethite and hematite during the formation of the Continental Terminal of Western Africa (Fig. 8a, b). Intraclasts and broken pisolitic nuclei are coated during wet periods by colloidal laminae with varying iron and aluminium content. During dry periods the grains shrink, to form radial and circumgranular desiccation cracks. Alternating wet and dry cycles produce polyphase fracturing and coating (Adeleye 1975). Gravitational instability of the structure causes settling and rotation of fragments, as well as polygonal fitting due to compaction while the colloidal coatings are soft. Ooids often share common coatings. Evidence of vadose influence on iron mobilization, transport and precipitation is widespread in the ferricretes on top of the tropical weathering profiles of iron-rich parent rocks. The Palaeogene 'Siderolithique' of western Europe provides an excellent example (Thiry & Thurland 1985; see also Fig. 8c). They also may be recognized, according to the literature, in most of the continental iron ore deposits of the USSR quoted above, as well as in the oolitic-pisolitic deposits of Iraq (Skocek *et al.* 1971) and Syria (El Sharkawi *et al.* 1976), and in the pisolitic manganese deposits of Chiatura, USSR (Bolton & Frakes 1985) and Groote Eyland, Australia (Frakes & Bolton 1984).

Other examples for pisolitic-oolitic palaeosoils are: Palaeogene clayey bean ores on the karstic surface of Mesozoic limestones of Southern Germany (Muschelkalk, Malm; Eichler 1961), and the Triassic and Lower Jurassic palaeosoils in Israel and Jordan (Bandel 1981; Goldbery 1982; Goldbery & Beyth 1984; Fig. 7c,d). These lateritic and bauxitic soils occasionally are reworked and rest as 'laterite derivative facies' on a karstified unconformity surface, or they may be found as concentrations of pisoids in channel fills (Valeton *et al.* 1983). Very often in one pisolitic-oolitic deposit, a transition may be observed between the weathering products *in situ,* immediately on a weathered surface, and various stages of reworking and transportation into local depressions and basins. Examples are described by Valeton (1972) from the Lower

Cretaceous bauxites in Southern France (Fig. 7e,f), by Sokolova (1964) from the pisolitic Lower Jurassic ironstones of Malka, USSR, by Maksimovich (1975) from the Lower Cretaceous oolitic nickel–iron ores of the Lokris area in Greece, by Kleinsorge *et al.* (1960) from oolitic laterites and their derivates west of Darfur, Sudan. The Oligocene fluviatile to lacustrine ferruginous oolites and pisolites of the Northern Aral region (Formosova 1959, Davidson 1961) and the Turgai depression, Kasakhstan (Yanitzki 1960, Teterev 1975) are also redeposited continental weathering products. The ore bodies are ribbon-like channel fillings with a length of up to 100 km and a width of 2–8 km, passing from fluvial through deltaic into lacustrine basin deposits. They are intercalated with fluvial sands and clays containing wood fragments and coal lenses. Upstream the clastics are coarser and include quartz and ferriclast gravels, all bearing dispersed ooids. Usually a diagenetic sequence can be found with an upper section of brown goethite ores and a lower one below groundwater level with berthierine and siderite. These deposits with their relatively high concentrations of Al, Si, Ti, V, Cr and P are clearly to be classified as minette-type ironstones. The Pliocene oolitic-pisolitic iron ore deposit of Kerch, deposited in brackish environment (Putzer 1943, Smirnov 1977), also has similar characteristics.

Terrestrial oolitic and pisiform ironstones associated with laterites occur also in West Africa, a continential area with intense ferallitic weathering during the Tertiary. The fluviatile–lacustrine facies association of the 'Contental Terminal' of the subsaharan Chad and Iullemmeden Basins contain layers of goethite, hematite, and, in lower parts of the sections, also iron silicate oolites (Lang *et al.* 1986). The ooids had been reworked and fragmented when solid, and sorted during transportation prior to sedimentation (Fig.8b), giving the usual appearance of an oolitic ironstone. All transitions to ooids and pisoids formed *in situ* can be recognized, together with polygonal fitting of collapsed, plastically deformed coatings (Fig. 8a). The sequence rests with a distinct boundary on marine sediments of Palaeocene to Lower Eocene age, deposited in a restricted shallow inland sea and bordered by a fringe of continental series (Lang *et al.* 1986). The marine sediments also contain oolites, which

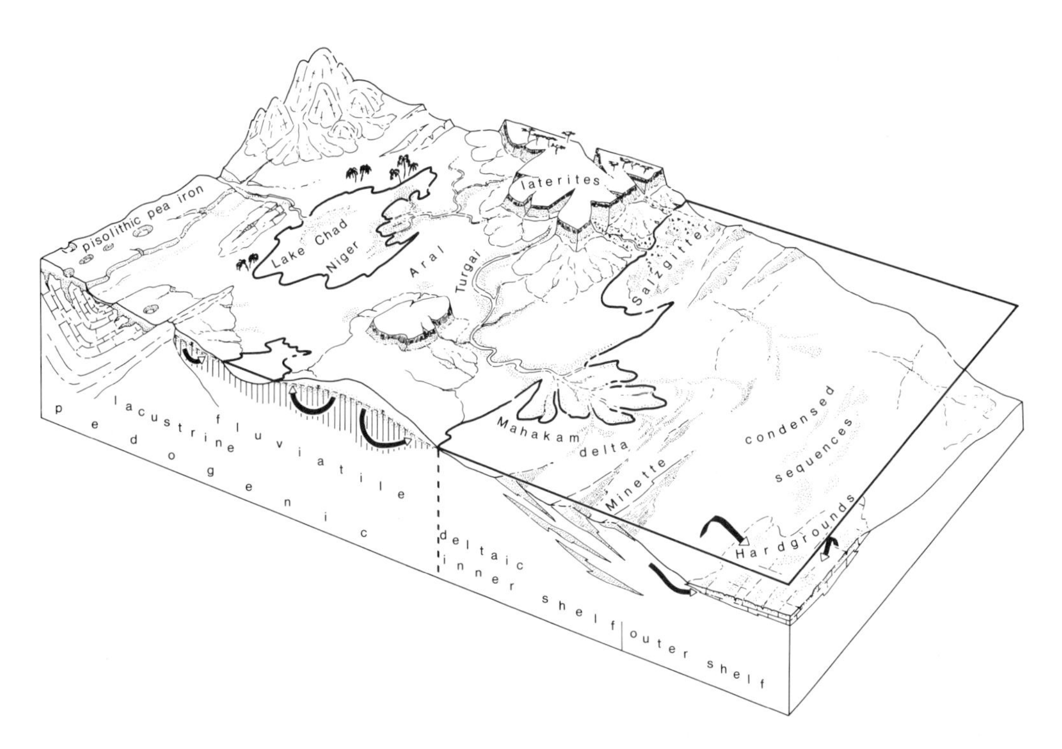


FIG. 6. Environments of iron ooid and pisoid formation and loci of sedimentation and accumulation. Type cases are indicated for pedogenic, fluviatile, lacustrine and marine ironstones.

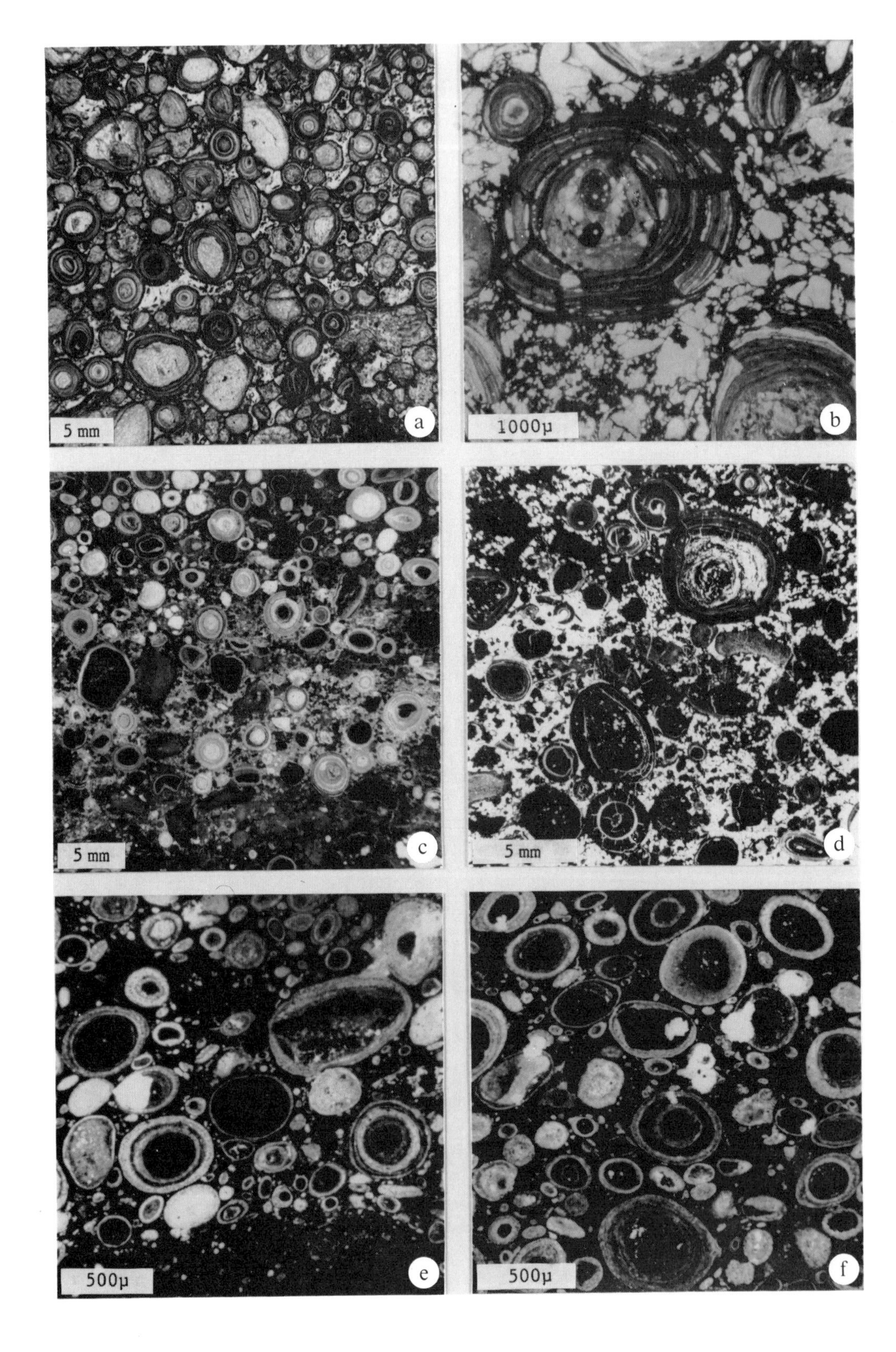
5 mm
a
1000μ
b
5 mm
c
5 mm
d
500μ
e
500μ
f

may well have been derived from the eroded surrounding fluvio-lacustrine mudflats and lowlands and the lateritic cover of the topographic highs. Grain structures and mineral associations of the oolites of the 'Continental Terminal' can hardly be distinguished from those accumulated in the marine environment (Jones 1958, 1965, Adeleye 1973). Repeated solution and reprecipitation of iron and aluminium have overprinted the texture of all these rocks in the zone of fluctuating groundwater as well as in the vadose environment. Jones (1965) stated ' . . . field evidence often cannot distinguish between a sedimentary ooilte and an oolitic laterite'. Similar observations led Du Preez (1954) to the assumption that the pisoids and ooids were formed in lowland position during the process of lateritization by rhythmic accretion of concentric shells of colloidal hydrated ferric oxides.

In the discussion about the origin of ferruginous ooids in this region, the occurrence of well sorted oolites lying on muddy lake deposits around the present Chari river delta of Lake Chad (Lemoalle & Dupont 1973 ; Mathieu 1978) must also be considered. The age of the grains as well as their mode of formation and concentration remain uncertain. Whereas Lemoalle & Dupont favour an autochthonous lacustrine precipitation of goethite at the sediment-water interface, Pedro *et al.* (1978) observed that goethite is not stable in the present lacustrine environment. A diagenetic evolution of hard dark goethitic grains to fissile, irregular surfaced nontronite pellets takes place. This transformation is further advanced with increasing distance from the delta. So it can be concluded that the ooids were brought in by the Chari river from another environment. They probably formed as soil concretions in the sediments of the coastal and deltaic mudflats, on which hydromorphic pedogenesis took place during periods of lowered groundwater level. During subsequent erosion and reworking of the soils, the ferruginous grains were washed out and transported by the rivers into the lake.

In near-shore marine to brackish environments of the shallow Late Cretaceous inland sea of Upper Egypt, precipitation and concentration of iron oolites (Bhattacharyya 1980) took place in a palaeogeographic position similar to that of the Tertiary in subsaharan West Africa. Here peri-marine low level latosol development, which underwent reworking and redeposition is well established by Germann *et al.* (1987). These authors confirm the convergence of ooids and pisoids both in marine and continental sediments, but they believe in the theoretical 'snowball' model of mechanical accretion of coagulated Fe–Al colloids in low-energy sedimentary environments for the formation of iron ooids, which, however, has no actualistic example. They rule out the suggestion that the ooids and ferriclasts deposited nearshore could be detrital products of pedogenesis, but they do not put forward a convincing argument for this rejection.

The Lower Cretaceous ferriclastic-oolitic ironstones of Salzgitter north of the palaeoslope of the Harz mountains (Bottke *et al.* 1969; Fig. 9) were accumulated in an entirely marine environment. Reworked weathered siderite concretions washed out from Lower to Middle Jurassic clays were piled up in deep local troughs caused by synsedimentary salt tectonics (Kölbel 1944; Kolbe 1962). The time of their sedimentation corresponds to a stratigraphic gap between the Oxfordian and Aptian in the Goslar region marked by the occurrence of partially pisolitic lateritic soils on a karstic surface on top of the Korallenoolith (Valeton 1957). The soils are covered by Hilssandstein, the caprock of the marine iron ores. In regions of high relief, the submarine detrital fans consist of poorly sorted large fragments of goethite crusts in a clay matrix containing fine grained ooids; farther to the north, towards the basin, they grade, due to sorting, into oolitic ironstones with fine-grained ferriclasts.

Al substitution in the ferriclasts is low (1–3 mol%) compared to that of the ooids (7–11 mol%, Fig. 10). The former are characterized by

FIG. 7. (a) Tertiary pea iron from a karst sink (La Sauvage, Luxemburg). Large pisoids and coated fragments of broken and partically reworked ferruginous crusts with alternate layers of goethite and kaolinite, formed *in situ* in a vadose environment (polished section). (b) Close-up of same sample. Goethite-kaolinite pisoid with clay and quartz as nucleus. Compaction during desiccation phases caused fractionation, the fracture surfaces are coated by secondary goethite. The fragments may serve as nuclei for younger pisoids. (c) Lower Jurassic 'laterite derivative facies' (Wadi Zarqa, Jordon). Goethite-hematite-kaolinite pisoids, ooids and coated grains of a reworked paleosol (polished section). (d) Close up of same sample. Note badly sorted coated particles. Radial cracks caused by desiccation. (e) Allochthonous lateritic bauxite from an Early Cretaceous subaqueous mudflow (Brignoles, Southern France). Poorly sorted pisoids and ooids with rhythmic concentric precipitates of boehmite and goethite in an argillaceous matrix. (f) Iron-rich bauxite (Cretaceous, Bosanska Krupa, Bosnia). Similar structure to sample in (e).

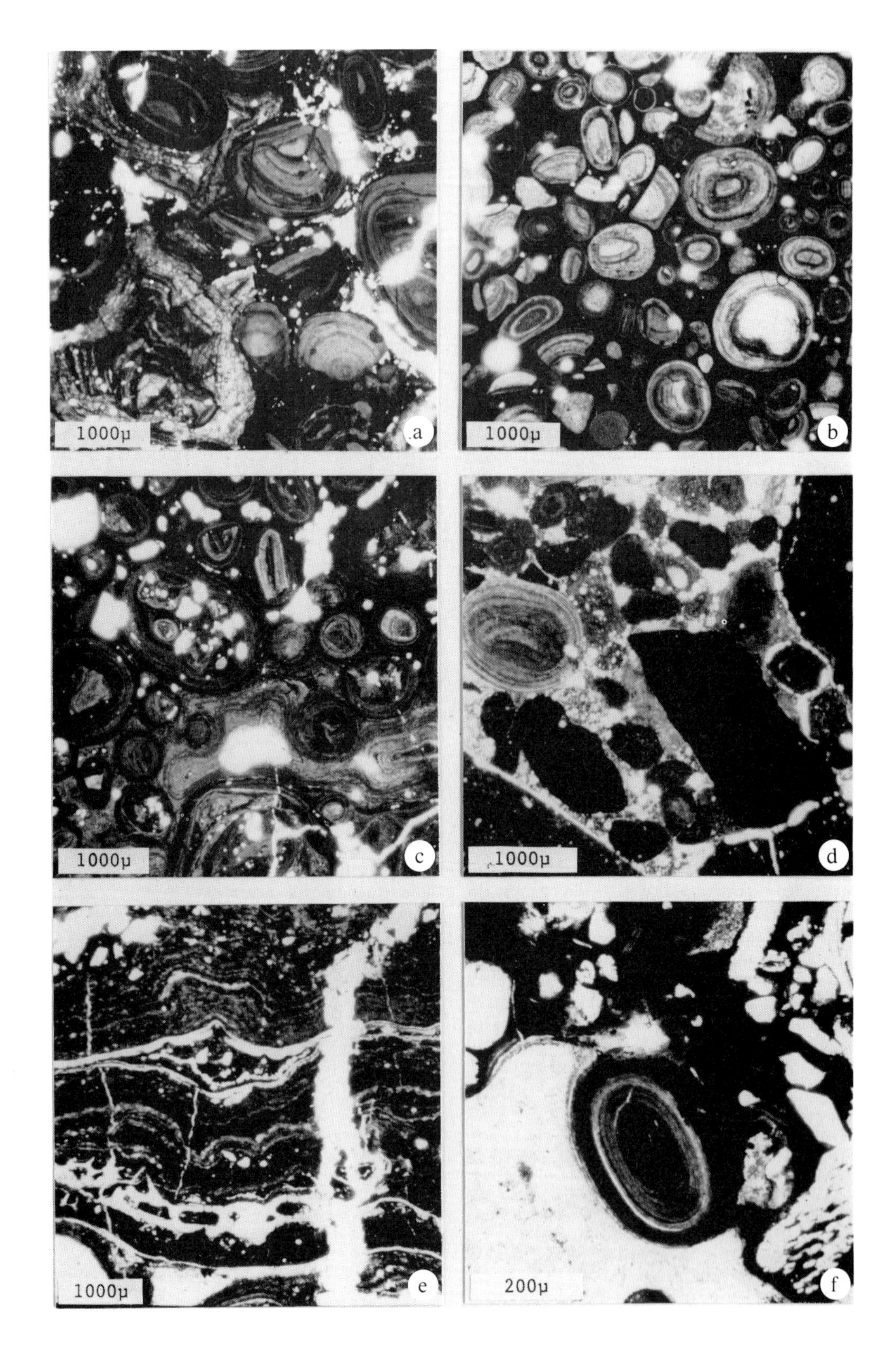
1000µ
a
1000µ
b
1000µ
c
1000µ
d
1000µ
e
200µ
f

the typical elements of a marine calcareous environment like Mn, P, Ca and Sr (Fig. 11). On the contrary, the ooids have higher concentrations of typical elements of the weathering cycle Al, Ti, V, Cr. The exposure time to lateritic weathering of the solid clay-siderite concretions was too short to alter drastically the geochemical composition. The ooids were newly formed in the hydromorphic soils and thus reflect the increased concentration of dissolved aluminium in that environment.

The Upper Cretaceous ferriclast deposits of Lower Saxony (Bülten, Lengede-Broistedt) are, similar to the Lower Cretaceous Salzgitter deposits, consisting of reworked Lower Cretaceous iron-rich clay-siderite sediments (Ferling 1953; Fehlau 1973). Here ooids are absent and glauconite grains occur in the matrix. These ironstones originated from very fast erosion along marine cliffs with immediate redeposition in an entirely marine environment, without an intervening period of lateritic weathering.

## Discussion

We propose to fit together the ironstone occurrences mentioned from various environments, to a single genetic model of supergene iron enrichment by repeated solution and precipitation as silica- and aluminium-enriched iron-hydroxides, forming a diversity of encrustations and globular concretions, and their subsequent mechanical enrichment. The preserved stages of reworking, transport, sorting and redeposition in the geological record represent the different types of oolitic-ferriclastic ironstone deposits. After deposition, diagenetic evolution may convert the hydroxides into berthierine provided that the primary chemical bulk composition of the silica-rich, aluminous goethite ooids corresponds to that of the resulting berthierine and that a reducing environment is created by the degradation of organic matter (Brindley 1951; Harder 1957; Schellmann 1966; Velde 1985). Slight variations of the Fe–Al–Si proportions and of the organic carbon content control the microenvironment within the ooids and may result in alternating layers of unchanged goethite and neo-formed berthierine within one single ooid. In the deepest and southernmost regions of the Minette area in Lorraine, goethite is completely recrystallized to berthierine, though the sedimentological features of a well aerated, normal marine environment are preserved, like cross-bedding and intense bioturbation. So there is clear evidence for the secondary, diagenetic nature of the iron silicate facies (Bubenicek 1961, 1971).

Bhattacharyya (1980) and Bhattacharyya & Kakimoto (1982) proposed the formation of iron ooids by accretionary accumulation of goethite and kaolinite flakelets by the ooid rolling over a muddy sediment surface. The adhesion is reinforced by the opposite electrical charge of clay- and hydroxide particles. In this way the ooid-constructing subgrains get a tangential orientation, as can be observed under the SEM (Fig 12b,c). Fe-substitution of kaolinite under reducing conditions causes diagenetic conversion to berthierine, while the delicate primary concentric ooid structure will be preserved. This 'semidynamic' ooid formation, intermediate between the Bahama-type carbonate ooid formation and the *in situ* concretion of ooids in a supporting medium, suffers from some deficiencies; there is no actualistic example found for this way of ooid formation. The ooids should have agglutinated terrigenous or biogenic clastic particles between the coatings, which, however, are not found. A tangential orientation of the submineral flakelets, one of the major arguments of this way of ooid formation can, however, also be acquired during *in situ* growth in a supporting soil medium by pushing the soil substance outwards, giving an orientation parallel to the surface of the grain to the flat

FIG. 8. (a) Lacustrine ironstone of the Continental Terminal (say, Niger valley). Goethitic pisoids and ooids grown in several generations with periodical cracking and reworking. The outer coatings of goethite and kaolinite display plastic deformation. All features are indicative of *in situ* formation in a vadose environment. (b) Same locality. Allochthonous, reworked and sorted ooids and pisoids. (c) Tertiary ironstone ('Siderolithique', Ehlange, Luxemburg). Poorly sorted, *in situ* formed goethite ooids, pisoids and coated grains of a meteorically influenced ferricrete, partially sharing common coatings. Note the irregular laminated goethite crust, coating a large void in the lower part of photograph. (d) Lower Cretaceous oolitic-ferriclastic ironstone (Salzgitter-Haverlahwiese, Lower Saxony). Proximal facies of a breccia of laterite derived goethite duricrust fragments with rare pisoids and ooids. The distal basinal facies is merely oolitic. (e) Bajocian hardground (Differdange, Luxemburg) with goethite crusts, stromatolitic laminae, partly overgrowing bryozoans; originated in a shallow marine, well aerated environment. (f) Same locality. Marine goethitic ooid in hardground crust on limestone. Note the corroded and goethite impregnated bioarenite particles and the primary ellipsoidal shape of the ooid.

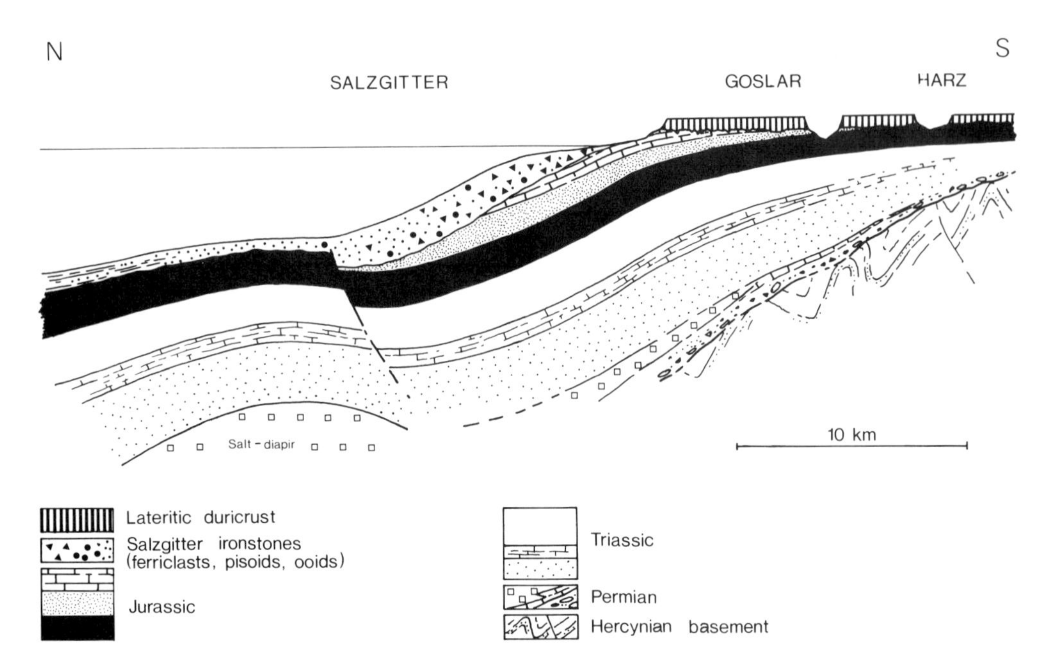


FIG. 9. Tentative palaeogeological section for Lower Cretaceous times of Salzgitter-type ironstone formation. Ferriclasts, derived from weathered clay-siderite concretions of Lower and Middle Jurassic strata as well as pisoids and ooids derived from lateritic soils are washed into the nearby sea, where they are sorted according to water agitation and grain size into near-shore *ferriclastic,* intermediate mixed *ferriclastic-oolitic,* and distal *oolitic ironstones.*

mineral particles. In contrast to the observations of Bhattacharyya & Kakimoto (1982), our SEM analyses have shown that a concentric orientation can be found in many ooids and pisoids, independently of their mode of formation (Figs. 10b,c, 13a–f). The primary structure is often, however, altered by diagenetic overprint.

Von Freyberg (1962) and Schellmann (1969) propose the growth of iron ooids in the supporting medium of marine muds, where a certain mobilization of iron takes place in reducing environments. Gygi (1981) modified this idea when he transfered the locus of ooid growth either to the sediment surface or below a thin sediment cover. He tried to explain iron ooids with alternating ferric–ferrous coatings in the Jurassic of northern Switzerland by this dual mode of formation. In the oxidizing bottom water environment, goethite is precipitated around a nucleus, while in the mildly reducing sub-bottom conditions, berthierine will form the next layer. The more or less complete absence of detrital sedimentation as on hardgrounds or in non-sequences, is a prerequisite of this model. In this context, however, Gygi (1981) and Gehring (1986) drew attention to the close relationship of hardgrounds and condensed horizons and the formation of iron ooids. The growth of hydroxide crusts in oxygen-rich shallow water areas is wide spread, not only in the Swiss Jura, but also in many other examples of reduced detrital sedimentation. The Bajocian of the type locality at Bayeux (NW France) and other hardgrounds in Bajocian carbonates in the Paris Basin provide good examples. Fig. 8e, f display stromatolite-like goethite crusts, probably precipitated under microbial participation, the impregnation of biogenic carbonates, and also truely marine goethite ooids. These bio- and hydrochemically precipitated hydroxide coatings often encrust sessile benthonic organisms like bryozoans and foraminifera. Similar features may also occasionally be observed in the Minette ironstones, where diagenetic mobilization of iron in the sediment and precipitation at the redox interface at, or just below the sediment-water boundary, may well occur in protected areas of the basin. The low content of dissolved iron in sea water and saline interstitial waters, even if reducing, yield, however, only small amounts of ooids, which hardly can build up the large minette-type ironstone deposits. These crusts and related ooids show very low Al substitution in the goethite (max. 6 mol%, Fig. 10) and a geochemical composition which reflects a marine mobilization (high values of Mn, P, Na, V), rather than resembling that of the minette-type ooids (Fig. 11).

Dahanayake *et al.* (1985) quote the Minette ironstones as an example of *in situ* marine biogenic formation of ferruginous coated grains and stromatolites within microbial mats in intertidal to shallow subtidal environments. Some of their stated indications are somewhat misleading and in contradiction to the observed facts. The pictured sedimentary fabric of randomly oriented ooids and other components like bioclasts and ferriclasts was definitely produced in a high energetic environment. The small-scale oblique stratification is real cross bedding and not a 'false cross lamination' produced by fungal mats. All the ooids are reworked clastic particles, both in the ironstone facies and in the bioturbated muddy siltstone facies, where they occur together with clastic quartz grains and bioclasts. The illustrated globular voids at the centre of ooids are merely their broken-out inner shells and give no evidence of the supposed formation of coatings around gas bubbles or drops of physiological liquids. On the other hand, bored reworked coated grains are rather common in the Minette, as described by Bender (1951) from Liassic iron oolites. The filamentous skeleton revealed by etching (Dahanayake *et al.* 1985) was caused most probably by this phenomenon. Certainly microbial iron precipitation onto pre-existing ferruginous particles might have also taken place occasionally in quiet marine environments of the Minette depositional area during periods of reduced sedimentation (Gehring 1986, Dahanayake & Krumbein 1986), even if there are no indications of *in situ* growth of biogenic ferruginous mats.

The idea of metasomatic transformation of calcium carbonate ooids by iron-rich groundwaters into goethite and iron-silicate proposed by Kimberley (1978, 1979, 1980a 1980b; see also discussions by Binda & Moltzer 1979, Bradshaw *et al.* 1980, Adeleye 1980) is only of marginal interest. The lack of relicts of calcitic ooids in ironstones is only one of the weak points of this hypothesis. Geochemistry provides one of the major counter-arguments: no Al substitution can be observed in groundwater goethite as our own analyses (Fig. 10) and data of Fitzpatrick & Schwertmann (1982) have revealed. Compared with Minette ooids they are depleted in elements such as Al, Si, Ti, Cr, Co, Zr, Th, with very low

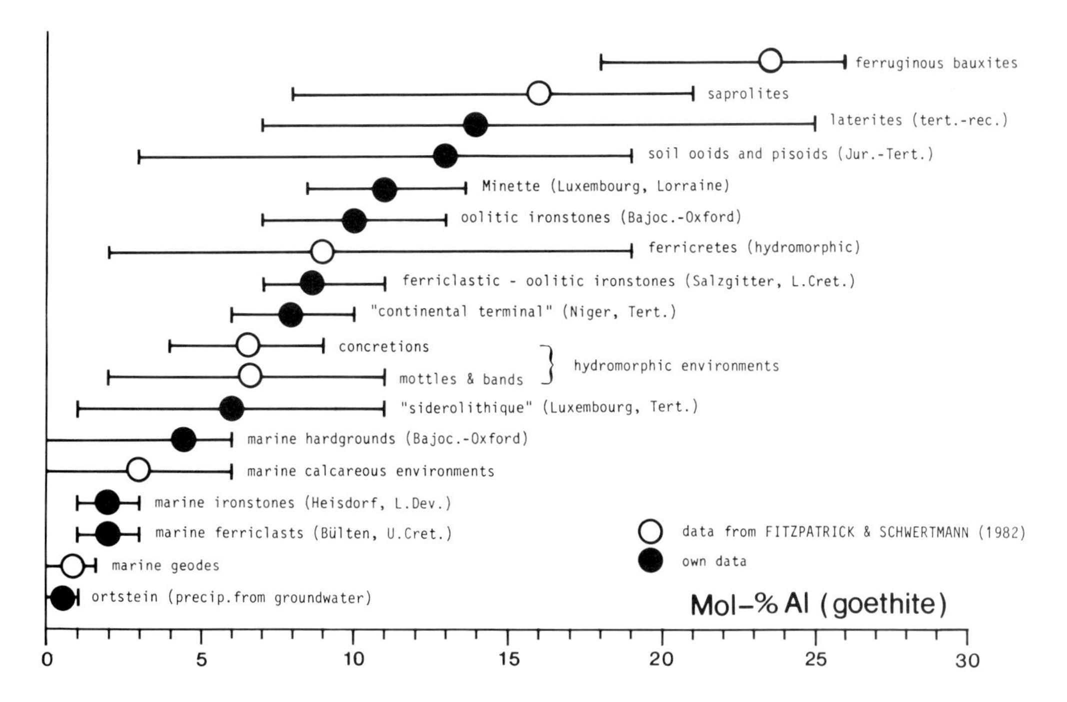


FIG. 10. Aluminium substitution in goethites of different environments of formation. Mol% values were determined by means of X-ray diffraction analysis of $d(111)$ spacing. For supplementary information, data of Fitzpatrick & Schwertmann (1982) are added to our own results.

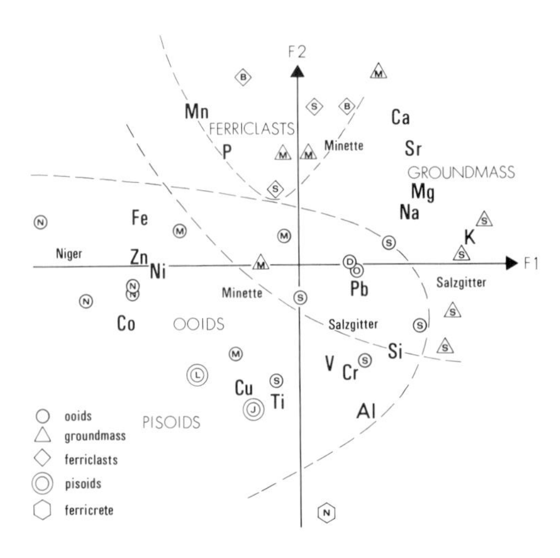


FIG. 11. Evaluation of geochemical data of ooids, pisoids, ferriclasts and groundmass in different ironstones by correspondance factor analysis. In the projection plane of factors 1 and 2 a discrimination between pisoids and ooids on one side, and groundmass so as ferriclasts on the other can be observed. Whereas pisoids and ooids are characterized by the typical laterite elements Fe, Ni, Co, Cu, Ti, V, Cr, Al, ferriclasts are dominated by Mn and P. The indicative elements for the groundmass are Ca, Sr, Na, Mg and K.

Letters in Symbols: B: Upper Cretaceous ferriclast iron ores (Bulten, Lower Saxony); D: Dogger Macrocephalus-oolite (Porta, Westfalia); J: Lower Jurassic soil pisoids (Wadi Zarqa, Jordan); L: Tertiary soil pisoids (La Sauvage, Luxemburg); M: Minette calcareous goethite ores (Rumelange, Luxemburg); N: Continental Terminal (Say, Niger); O: Oxford Schellenbrücke oolite (NW–Switzerland); S: Lower Cretaceous Salzgitter ores (Haverlahwiese, Lower Saxony).

water solubility. Locally, impregnation of carbonate ooids and grains may occur, as shown in hardgrounds and near the boundary between adjacent units of ironstones and oolitic carbonates.

We think the answer to Gygi's (1981) pointed question in the title of his article is: oolitic iron formation takes place in marine *and* non-marine environments. For the most part it is not marine, as we have tried to show in our synoptic sketch of different sites of ooid formation (Fig. 6). Are there diagnostic means to determine the origin of ooids? Ultrastructure seems not to be specific to the environment. Geochemistry and especially Al substitution in goethite could be helpful in recognizing soil-derived ooids, but more data have to be collected. The same applies to palaeotemperature determination by means of oxygen isotope analysis (Yapp 1987), which might provide a method to distinguish between low-temperature marine goethite and terrestrial precipitates in tropical and subtropical soils.

ACKNOWLEDGEMENTS: The investigations have in part been funded by the Deutsche Forschungsgemeinschaft. The authors would like to thank all the colleagues having supplied them with ironstones from different parts of the world, especially J.P. Descaves (Mezieres-les-Metz), R. Gygi (Basel), K. Bandel (Hamburg), U. Hennicke and A. Schulz (Cologne). They are very grateful to anonymous reviewers for improving their manuscript.

FIG. 12. (a) Goethitic ironstone (Couche grise, Esch, Luxemburg). Coarse grained ooids and pisoids are badly sorted, partly broken and recoated. Note irregularly coated grain in upper part of photograph. (b) Goethitic ironstone (Couche grise, Rumelange, Luxemburg). Well sorted ooids with nuclei of fragmented coated grains. Berthierine cement indicates mobilization of ferrous iron during diagenesis. (c) Same locality as 1. Ultrastructure of a goethitic ooid with nucleus of unstructured ferriclast (SEM photograph). (d) Close up of photo 3. Concentric sheaths show tangential orientation of the platy geothite subgrains, measuring about 5 μm in size and 0.2 – 0.5 μm in thickness. (e) Berthierine ironstone (Couche noire, Differdange, Luxemburg). Berthierine ooids, derived from primary goethitic ooids display a delicate concentric structure (crossed nicols). (f) Berthierine ironstone (Couche verte, Hayange, Lorraine). Berthierine ooids deformed by diagenetic alteration and compaction to irregular formed 'spastoliths'. Fibrous berthierine rim cement indicates a secondary late mobilization of ferrous iron. Note large bioclast in upper left part (crossed nicols).

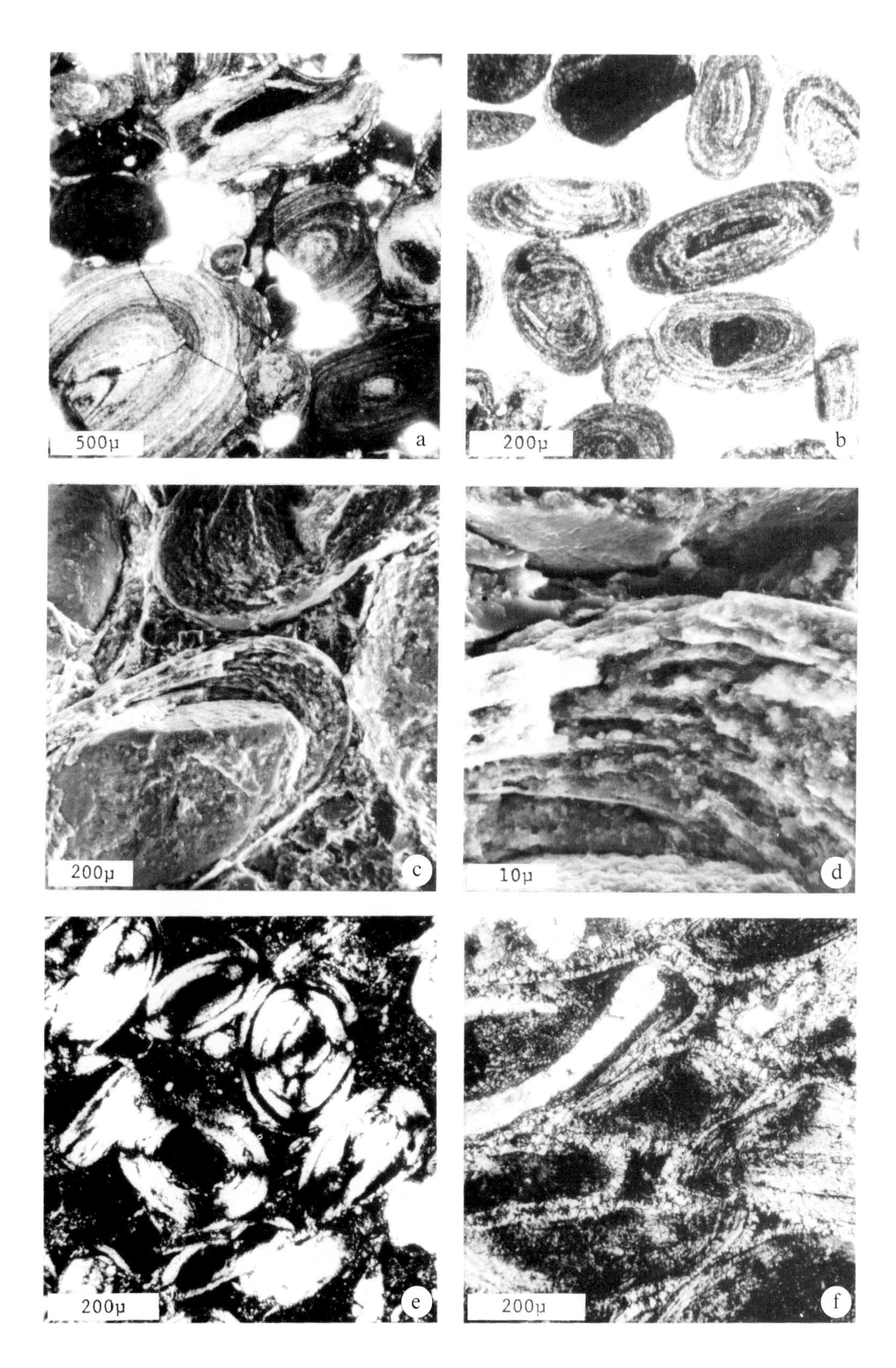
500μ
a
200μ
b
200μ
c
10μ
d
200μ
e
200μ
f

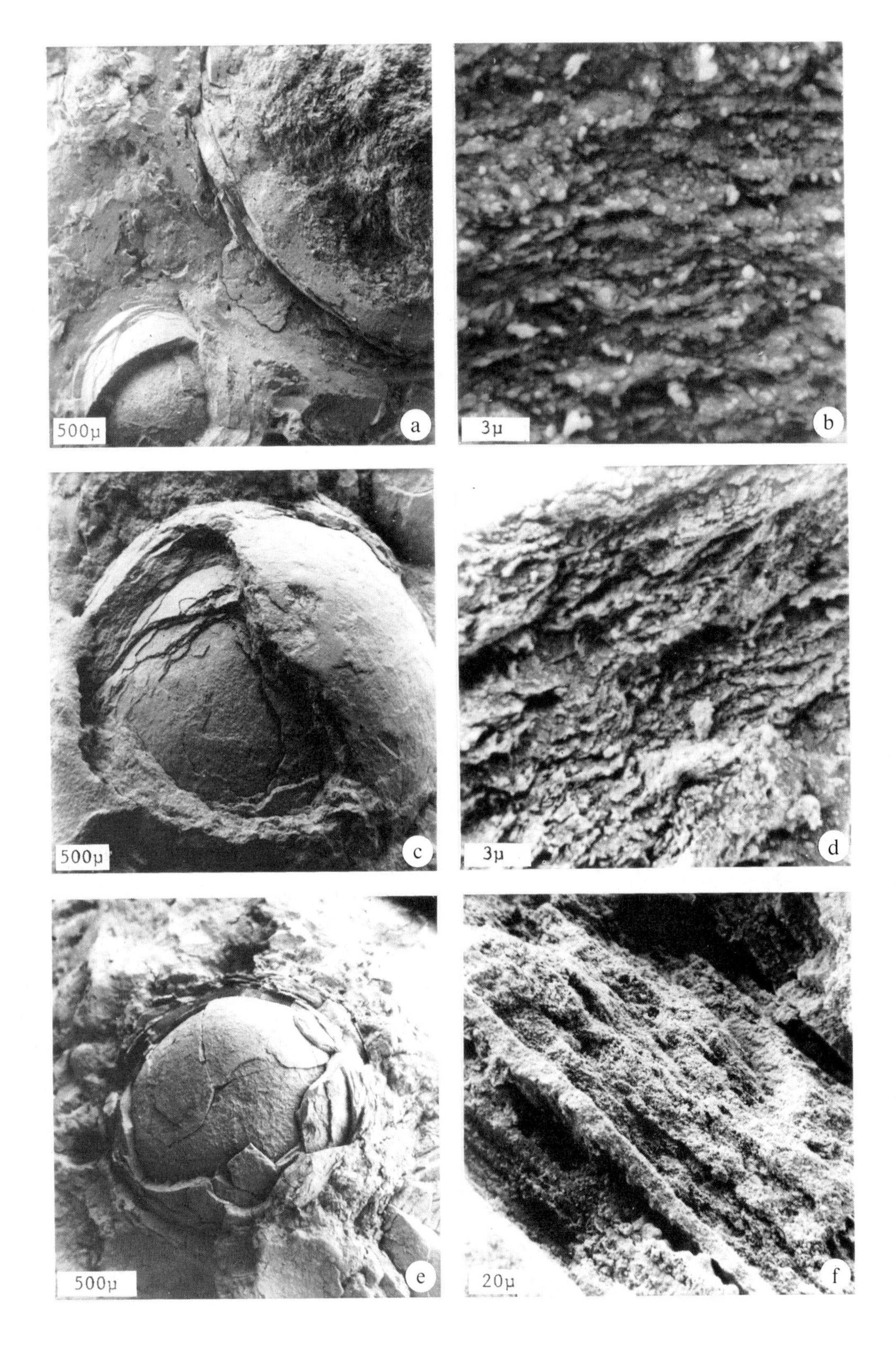
500μ
a
3μ
b
500μ
c
3μ
d
500μ
e
20μ
f

## References


ADELEYE, D.R. 1973. Origin of ironstones, an example from the middle Niger Valley, Nigeria. *Journal of Sedimentary Petrology,* **43,** 709–727.

——1975. Derivation of fragmentary oolites and pisolites from desiccation cracks. *Journal of Sedimentary Petrology,* **45,** 794–798.

——1980. Origin of oolitic iron formations-discussion. *Journal of Sedimentary Petrology,* **50,** 1001–1003.

ALLEN, G.P., LAURIER, D. & THOUVENIN, J. 1979. Etude sedimentologique du delta de la Mahakam. *Notes Memoirs, Companie de la Francais Petroles* **15.**

ARISTOVSKAYA, T.V. & ZAVARZIN, G.A. 1971. Biochemistry of Iron in Soil; *In:* MCLAREN, A.D. & SKUJINS, J. (eds) *Soil Biochemistry, Dekker, New York,* 385–408.

BANDEL, K. 1981. New stratigraphical and structural evidence for lateral dislocation in the Jordan rift valley connected with a description of the Jurassic rock column in Jordan. *Neues Jahrbuch fur Geologie Palaeontologie.* **161,** 271–308.

BENDER, F. 1951. Fossile Pilze aus einem Eisenoolithhorizont des Lias in Wurttember. *Palaeontographica,* **91B** 152–158.

BHATTACHARYYA, D.P. 1980. *Sedimentology of the Late Cretaceous Nubia Formation at Aswan, Southeast Egypt, and the Origin of the Associated Ironstones.* Ph.D. thesis, Princeton University.

——& KAKIMOTO P.K., 1982. Origin of ferriferous ooids: An SEM study of ironstone ooids and bauxite pisoids: *Journal of Sedimentary Petrology,* **52,** 849–857.

BINDA, P.L. & MOLTZER, J.G., 1979. Origin of oolitic iron formations: discussion. *Journal of Sedimentary Petrology,* **49,** 1351–1353.

BOLTON, B.R. & FRAKES, L. 1985. Geology and genesis of manganese oolite, Chiature, Georgia, USSR. *Bulletin of the Geological Society of America,* **96,** 1398–1406.

BOTTKE, H. *et al.*, 1969. sammelwerk deutsche Eisenerzlagerstatten. II.Eisenerze im Deckgebirge (Postvaristikum). 1. Die marinsedimentaren Eisenerze des Jura in Nordwestdeutschland. *Beiheft geologishes Jahrbuch,* **79,** 391p.

BRADSHAW, M.J., JAMES, S.J., & TURNER, P. 1980. Origin of oolitic ironstones — a discussion *Journal of Sedimentary Petrology,* **50,** 295–299.

BRINDLEY, G.W., 1951. The crystal structure of some chamosite minerals. *Mineralogical Magazine,* **29:** 502–525.

BUBENICEK, L., 1961. Recherches sur la constitution et la repartition du minerai de fer dans l'Aalenien de Lorraine. *Science de la Terre,* **8,** 5–204.

——1971. Geologie du gisement de fer de Lorraine. *Bulletin de Centride Recherches de Pau, Societe Nationale des Petroles d'Aquitaine,* **5,** 223–320.

CARROLL, D., 1958. Role of clay minerals in the transportation of iron. *Geochimica et Cosmichimica Acta,* **14,** 1–27.

CORRENS, C.W. & V. ENGELHARDT, W. 1941. Rontgenographische Untersuchungen uber den Mineralbestand sedimentarer Eisenerze. — *Nachrichten fur Akadamie Wissenschaften, Gottingen, Mathemtik-Physik,* **2,** 131–137.

DAHANAYAKE, K., GERDES, G. & KRUMBEIN, W.E. 1985. Stromatolites, oncolites and oolites biogenically formed in situ. *Naturwissenschaften,* **72,** 513–518.

——& KRUMBEIN, W.E. 1986. Microbial structures in oolitic iron formations. *Mineralium Deposita,* **21,** 85–94.

DAVIDSON, C.F. 1961. Oolitic ironstones of fresh-water origin — a review of a Russian monograph. *The Mining Magazine,* **104,** 158–159.

DUNHAM, R.J. 1969. Vadose pisolite in the Capitan Reef (Permian), New Mexico and Texas. *Special Publication 14 of the Society of Economic Palaeontologists and Mineralogists.* **14:** 182–191.

DU PREEZ, J.W. 1954. Notes on the occurrence of oolites and pisolites in Nigerian laterites. *19th International Geological Congress, Algier,* **21,** 163–169.

EICHLER, J. 1961. Mineralogische und geologische Untersuchungen von Bohnerzen in Baden-Wurttemberg, besonders der Vorkommen bei Liptingen, Kreis Stockach. *Neues Jahrbach fur Mineralgie,* **97,** 51–111.

EL SHARKAWI, M.A., MAHFOUZ, S and EI DALLAL, M.M.N. 1976. The pisolitic ironstone of Gdeidet Yabous and Naba Barada localities, Zebdani District, Syria. *Chemie der Erde,* **35,** 241–250.

FEHLAU, K. P. 1973. Sedimentpetrologie der Trummereisenerz-Lagerstatte von Bulten-Altenstedt (Oberkreide, NW-Deutschland). *Mitteilungen fur Geologie und Palaeontologie, Universitat Hamburg,* **42,** 81–160.

FERLING, P. 1958. Mineralogische, petrographische, fazielle und chemische Untersuchung der Brauneisen-Trummerezlagerstatte von Lengede-Broistedt. *Geologisches Jahrbach,* **75,** 555–590.

FITZPATRICK, R.W. & SCHWERTMANN, U. 1982. Al-substituted goethite — an Indicator of pedogenic and other weathering environments in South Africa. *Geoderma,* **27,** 335–347.

FLEHMIG, W. 1967. *Zur Erklärung des*


FIG. 13. (a) Tertiary pea iron (Lasauvage, Luxemburg). Ultrastructure of a pedogenic pisoid, with tangentially arranged platy flakelets (SEM photograph). (b) Close-up of same sample. Size of platelets 1 μm, thickness 0.1 μm (SEM photograph). (c) Lower Jurassic 'laterite derivative facies' (Wadi Zarqa, Jordan). Ultrastructure of a pedogenic pisoid showing concentric sheaths (SEM photograph). (d) Close-up of same sample. Tangential orientation of mineral platelets, 1 to 2 μm in size 0.2 μm thick (SEM photograph). (e) Lacustrine ironstone of the Continental Terminal (say, Niger valley). Ultrastructure of pisoid with tangential fabric of platy grains (SEM photograph). (f) Close-up of same sample (SEM photograph).

Kieselsäuregehalt in Nadeleisenerzooiden. PhD-Thesis, Munster University.

FORMOSOVA, L.N. 1959. Zheleznye rudy severnogo priarala. (Iron ores of the Northern Near-Aral region). *Trudy Geologicii Institut Moskva,* **20**.

FRAKES, L.A. & BOLTON, B.R. 1984. Origin of manganese giants: sea-level change and anoxic-oxic history. *Geology,* **12**, 83–86.

FREYBERG, B.V. 1962. Eisenerzlagerstatten im Dogger Frankens. *Geologisches Jahrbuch,* **79**, 207–254.

GEHRING, A.U., 1986. Mikroorganismen in kindensierten Schichten der Dogger/Malm-Wende im Jura der Nordostschweiz. *Eclogae Geologicae Helvetiae,* **79**, 13–18.

GERMANN, K., MUCKE, A., DOERING, T. & FISCHER, K. 1987. Late cretaceous laterite-derived sedimentary deposits (oolitic ironstones, kaolins, bauxites) in Upper Egypt. *Berliner geowissenschafen,* **75**, 727–758.

GOLDBERY, R. 1982. Paleosols of the lower Jurassic Mishhor and Ardon formations ('Laterite derivative facies'), Makhtesh Ramon, Israel. *Sedimentology,* **29**, 669–690.

——& BEYTH, M., 1984. Lateritization and ground water alteration phenomena in the Triassic Budra formation, south-western Sinai. *Sedimentology,* **31**, 575–594.

GYGI, R.A., 1981. Oolitic iron formation: marine or not marine? *Eclogae Geologicae Helvetiae,* **74**, 233–254.

HARDER, H., 1957. Zum Chemismus der Bildung einiger sedimentarer Eisenerze. *Zeitschift fur deutscher geologische Gesellschaft,* **109**, 69–72.

——1964. Geochemische Unterscheidung genetisch verschiedener Marin-sedimentarer Eisenerz-lagerstatten. *Bereich fur geologische Gesellschaft. DDR,* **9**, 475–478.

HUANG, W. & KELLER, W.D. 1972. geochemical mechanics for the dissolution, transport, and deposition of aluminium in the zone of weathering. *Clays & Clay Minerals,* 69–74.

Institut de Recherches de la Siderurgie (IRSID) 1967. *Atlas Geologique du gisement de fer de Lorraine.*

JONES, H.A. 1958. The oolitic ironstones of the Agbaja Plateau, Kabba Province. *Records of the Geological Survey of Nigeria, 1955,* 20–43.

——1965. Ferruginous oolites and pisolites. *Journal of Sedimentary Petrology,* **35**, 838–845.

KIMBERLEY, M.M. 1978. Paleoenvironmental classification of iron formations. *Economic Geology,* **73**, 215–229.

——1979. Origin of oolitic iron formations. *Journal of Sedimentary Petrology,* **49**, 111–132.

——1980a. Origin of oolitic iron formations—reply. *Journal of Sedimentary Petrology,* **50**, 299–302.

——1980b. Origin of oolitic iron formations—reply. *Journal of Sedimentary Petrology,* **50**, 1003–1004.

KLEINSORGE, H., KREYSING, K., ECKHARDT, F.J., FESSER, H. & GUNDLACH, H. 1960. Uberein Vorkommen von oolithischen Eisenerzen in der Nubichen Serie der Provinz Kordofan, Republik Sudan. *Zeitschift fur deutscher geologische Gesellshaft,* **112**, 267–277.

KOLBE, H. 1962. Die Eisenerzkolke im Neokom-Eisenerzgebiet Salzgitter. *Mitteilungen fur geologie Staats Institut, Hamburg,* **31**, 276–308.

KOLBE, H., 1944. Die tektonische und Paleogeographische Geschichte des Salzgitter Gebietes. *Abhadlungen R.A. Bodenfer N.F.,* **207**

LANG, J., KOGBE, C., ALIDOU, S., ALZOUMA, K., DUBOIS, D., HOUESSOU, A. & TRICHET, J., 1986. Le Siderolithique du Tertiaire ouestaricain et le concept due Continental terminal. *Bulletin Société geologique de la France,* **8**, 605–622.

LEMOALLE, J. & DUPONT, B., 1973. Iron-bearing oolites and the present conditions of iron sedimentation in Lake Chad (Africa). *In*: AMSTUTZ, G.C. & BERNARD, A.J. (eds) *Ores in sediments.* Springer Berlin 167–178.

LUCIUS, M., 1945. Die Luxemburger Minette-formation und die jungeren Eisenerzbildungen unseres Landes. Beitrage zur Geologie von Luxemburg. *Publique Service geologique de Luxembourg,* **4**, 350p.

MAKSIMOVICH, Z. 1975. Nickel clay minerals in some laterites, bauxites and oolitic iron ores. *6th Conference of Clay Minerology and Petrology, Praha,* 119–134.

MATHIEU, P. 1978. Decouverte d'oolithes ferruginouses en strati-graphie sous le delta actuel due Chari (Tchad). *Cahiers d'ORSTOM, series Geologique,* **10**, 203–208.

MAYNARD, J.B. 1983. *Geochemistry of sedimentary ore deposits.* Springer, New York.

——1986. Geochemistry of oolitic iron ores, an electron microprobe study. *Economic Geology,* **81**, 1473–1483.

NAHON, D., CAROZZI, A.V. & PARRON, C. 1980. Lateritic weathering as a mechanism for the generation of ferruginous ooids. *Journal of Sedimentary Petrology,* **50**, 1287–1298.

NORTON, S.A. 1973. Laterite and bauxite formation. *Economic Geology,* **68**, 353–361.

PEDRO, G., CARMOUZE, J.P. & VELDE, B. 1978. Peloidal nontronite formation in recent sediments of Lake Chad. *Chemical Geology,* **23**, 139–149.

PERYT, T.M. 1983. Vadoids. *In*: PERYT, T.M. (ed.) *Coated grains.* Springer, Berlin, 437–449.

PUTZER, H. 1943. Die oolithischen Brauneisenerz-Lagerstatten der Kertsch-Halbinsel. *Zeitschift fur Angewandte Mineralogie,* **4**, 363–378.

SCHELLMANN, W. 1966. Sekundare Bilding von Chamosit aus Goethit. *Zeitschrift fur Erzbergbau und Metallhuttenwissenschaften,* **19**, 302–305.

——1969. Die Bildungsbedingungen sedimentarer Chamositund Hamatit-Eisenerze, am Beispiel der Lagerstatte Echte. *Neues Jahrbuch fur Mineralogie,* **111**, 1–31.

SCHELLMANN, W. 1986. On the geochemistry of laterites. *Chemie der Erde,* **45**, 39–52.

SIEHL, A. & THEIN, J. 1978. Geochemische Trends in der Minette (Jura, Luxemburg/Lothringen). *Geologische Runschau,* **67**, 1052–1077.

SKOCEK, V., AL-QUARAGHULI, N. & SAADALLAH, A.A. 1971. Composition and sedimentary structures of iron ores from the Wadi Husainya area, Irq. *Economic Geology,* **66**, 995–1004.

SMIRNOV, V.I. 1977. *Ore deposits of the USSR* Pitman, London 6–113.

SOKOLOVA, E.E. 1964. *Physicochemical investigations of sedimentary iron and manganese ores and associated rocks. Israel Programme of Scientific Translations, Jerusalem.*

TETEREV, G.M. 1975 (ed). *Geologiya SSSR, 34, Turgayskiy progib, poleznye iskopaemye.* Akademiya Nauk, Moscow.

TEYSSEN, T.A.L. 1984. Sedimentology of the Minette oolitic ironstones of Luxembourg and Lorraine: a Jurrasic subtidal sandware complex. *Sedimentology,* **31**, 195–211.

THEIN, J. 1975. Sedimentologisch-stratigraphische Untersuchungen in der Minette des Differdnger Beckens (Luxemburg). *Publique Service de geologre, Luxemburg,* **24**, 60p.

THIRY, M. & THURLAND, M. 1985. Paleotoposequences de sols ferrugineux et de cuirassements siliceux dans le Siderolithique du nord du Massif central (Bassin de Montlucon-Domerat). *Geologie de la France,* **2**, 175–192.

VALETON, I. 1957. Lateritische Verwitterungsboden zur Zeit der jungkimmerischen Gebirgsbildung im nordlichen Harzvorland. *Geologische Jahrbuch,* **73**, 149–164.

——1972. *Bauxites — Developments in Soil Science* **1**, Elsevier, Amsterdam.

——STUTZE, B. & GOLDBERY R. 1983. Geochemical and mineralogical investigations of the lower Jurassic flint-clay bearing Mishho and Ardon formations, Makhtesh Ramon, Israel. *Sedimentary Geology,* **35**, 105–152.

VAN HOUTEN F.B. & BHATTACHARYYA, D.P. 1982. Phanerozoic oolitic ironstones—geologic record and facies: *Annual Review of Earth and Planetary Science,* **10**, 441–458.

VELDE, B., 1985. *Clay minerals: A physico-chemical explanation of their occurrence. Developments in sedimentology* **40**. Elsevier, Amsterdam.

YANITZKI, A.L., 1960. Oligocenovye oolitovye zheleznye rudy severnogo Turgua i itch genezis (Oligocene oolitic iron ores of northern Turgai and their genesis). *Trudy Institat Geologica Mestorozhdenii, Mineralogii i Geochmii,* **37**, 1–219.

YAPP, C.J., 1987. Oxygen and Hydrogen isotope variations among goethites ($\alpha$-FeOOH) and the determination of paleotemperatures. *Geochimiaet Cosmochimica Acta,* **51**, 355–364.

ZITZMANN, A. (ed.), 1977. *The iron ore deposits of Europe and adjacent areas.* BGR, Hannover.

SIEHL, A., Geologisches Institut der Universität Bonn, Nussallee 8, D 5300 Bonn 1, West Germany.

THEIN, J., Westfälische Berggewerkschaftskasse, Herner Strasse 45, D 4630 Bochum, West Germany.

# Case Studies

# Time and space distribution of Palaeozoic oolitic ironstones in the Tindouf Basin, Algerian Sahara

S. Guerrak

SUMMARY: The Tindouf Basin is located, mainly in Algeria, in the SW part of the Saharan Platform. It comprises Cambro-Ordovician to Carboniferous marine formations overlain by the continental Cretaceous and Pliocene Hamada cover. On the southern limb of the Tindouf Basin several oolitic ironstone deposits occur within Silurian and Devonian sediments. They occupy an area of 500 x 40 km and have an east–west trend. These oolitic ironstone deposits are part of the major North African oolitic ironstone belt, extending more than 3000 km, from Zemmour to Libya. These occurrences are of the LOID type (Local Ironstone Deposition), and contain more than 10000 million tonnes of ironstone reserves (1500 million tonnes in the Silurian and 9200 million tonnes in the Devonian).

The various ironstones were formed in shallow marine environments such as barriers and deltas. The major features of these deposits are: (i) a palaeogeographical control of the sediment which was mainly formed on the flanks of uplifts; (ii) the occurrence of the ironstones at the top of coarsening-upward sequences, mostly located towards the end of major regressive cycles, particularly in the Pragian and Famennian; (iii) ooids developed in quiet environments such as lagoons or embayments, within an iron-rich mud; (iv) a southern, relatively close, iron source area and; (v) a palaeolatitudinal distribution giving successively cold and temperate climates.

The Palaeozoic history of the basin involved pericratonic sedimentation on the borders of a large epicontinental sea.

## Structural framework, stratigraphic record and palaeogeography

In the south-western part of the Algerian Saharan Platform, several oolitic ironstone deposits occur within the Silurian–Devonian formations along the southern limb of the Tindouf Basin. These oolitic ironstones constitute the major branch of the North African Palaeozoic oolitic ironstone belt, which can be followed from Rio de Oro (Sougy 1964), through Morocco (Destombes *et al.* 1985) and Algeria (Guerrak 1987) to Libya (Goudarzi 1970; Chauvel & Massa 1981; Van Houten & Karazek 1981). This belt contains Ordovician to Devonian deposits, developed during cold to temperate climates.

The Tindouf Basin is a post-Silurian WSW–ENE trending asymmetrical synclinorium, covering 120 000 km$^2$ (Fig. 1). It is bounded to the north by the Anti-Atlas, to the south by the Reguibat Shield, to the west by the El Aioun depression, and to the east by the Erg Chech depression.

Numerous oolitic ironstone lenses, interbedded with argillaceous and arenaceous deposits, are found in the two sub-basins, Djebilet and Iguidi, of the nearly sub-horizontal southern flank (Fig. 2). This limb contrasts with the folded northern flank, devoid of ironstones.

During Siluro-Devonian times, the Tindouf Basin was marked by epicontinental sedimentation. The thickness of the Siluro-Devonian sediments is about 1000m in the south and about 14 000m in the north. This can be observed on the isopach maps (Fig.3) drawn partly from unpublished petroleum exploration data (Graverot & Planchon 1964, private communication).

Furthermore, the isopach maps suggest an approximate location of the early continental areas, and therefore the probable palaeoslopes. The emergent areas appear to have been very close to the present southern limit of the Siluro-Devonian deposits, except during upper Devonian times, when they were probably located farther to the south. There are no available data on thickness variations in the Taoudenni Basin, as the data are considered confidential. The clastic sediments of the Tindouf Basin were derived from a southern source on the southern flank, and from a northern source on the northern flank.

This north-western passive margin of Gondwanaland was affected by several weak tectonic movements, which have previously been related to the Caledonian and Hercynian orogenies (Gevin 1960; Liouville & Graverot 1963, private communication; Hollard 1967). According to Black *et al.* (1979), Poole *et al.* (1983), Lesquer *et al.* (1984) and Wendt (1985), however, four tectonic periods can be recognized:

*From* YOUNG, T. P. & TAYLOR, W. E. G. (eds), 1989, *Phanerozoic Ironstones*
Geological Society Special Publication No. 46, pp. 197-212

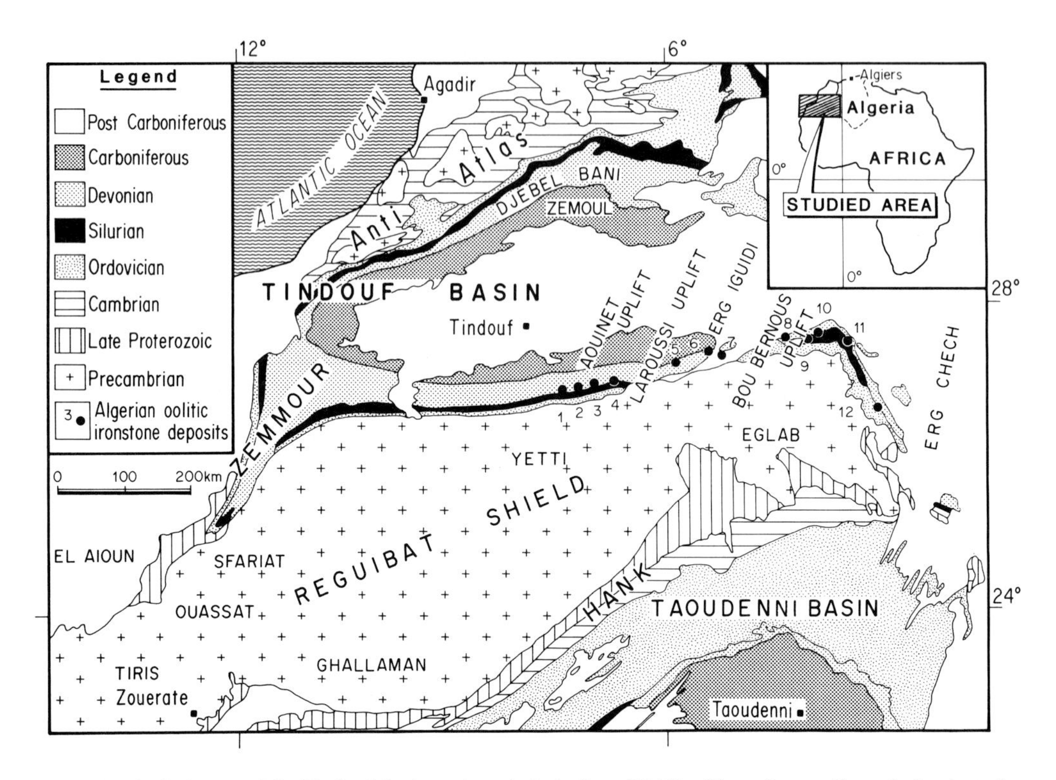


FIG. 1. Geological map of the Tindouf Basin region. 1, 2, 3, Gara Djebilet West, Centre, East; 4, Aouinet; 5, Oguilet Laroussi; 6, Mecheri Abdelaziz; 7, Zemila; 8, Nba; 9, Tguililet el Hamra; 10, Fedj Mtaigat; 11, Fedj Mlehas; 12, Gour Jiffa.

(i) The Pan-African Event, which was the result of the collision between the old West-African Craton (2000 Ma) and the trans-saharan Pan-African Belt about 600 million years ago.

(ii) The Caradoc Event which was synchronous with the African Taconic Event, clearly expressed in the Rockelides, and to a lesser extent in the Mauritanides (Lecorché 1983). The widespread intra-Caradoc unconformity can be related to this phase.

(iii) The Late Silurian – Early Devonian Event, which can be related to a weak Pre-Acadian Event (formerly Caledonian), essentially involving compressive intraplate movements. During this event a 'disintegration' phenomenon can be invoked, which is in agreement with the

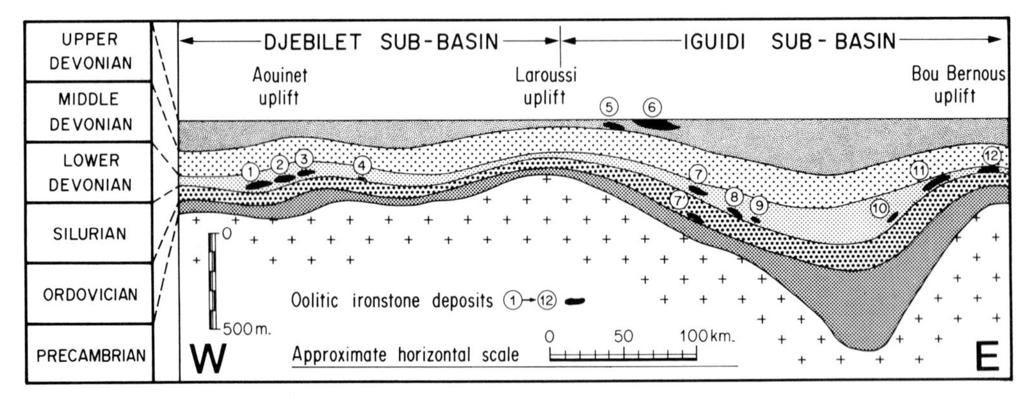


FIG. 2. Stratigraphic section of the southern limb of the Tindouf Basin showing the oolitic ironstone deposits: Numbers as for Fig. 1, also 7, Zemila.

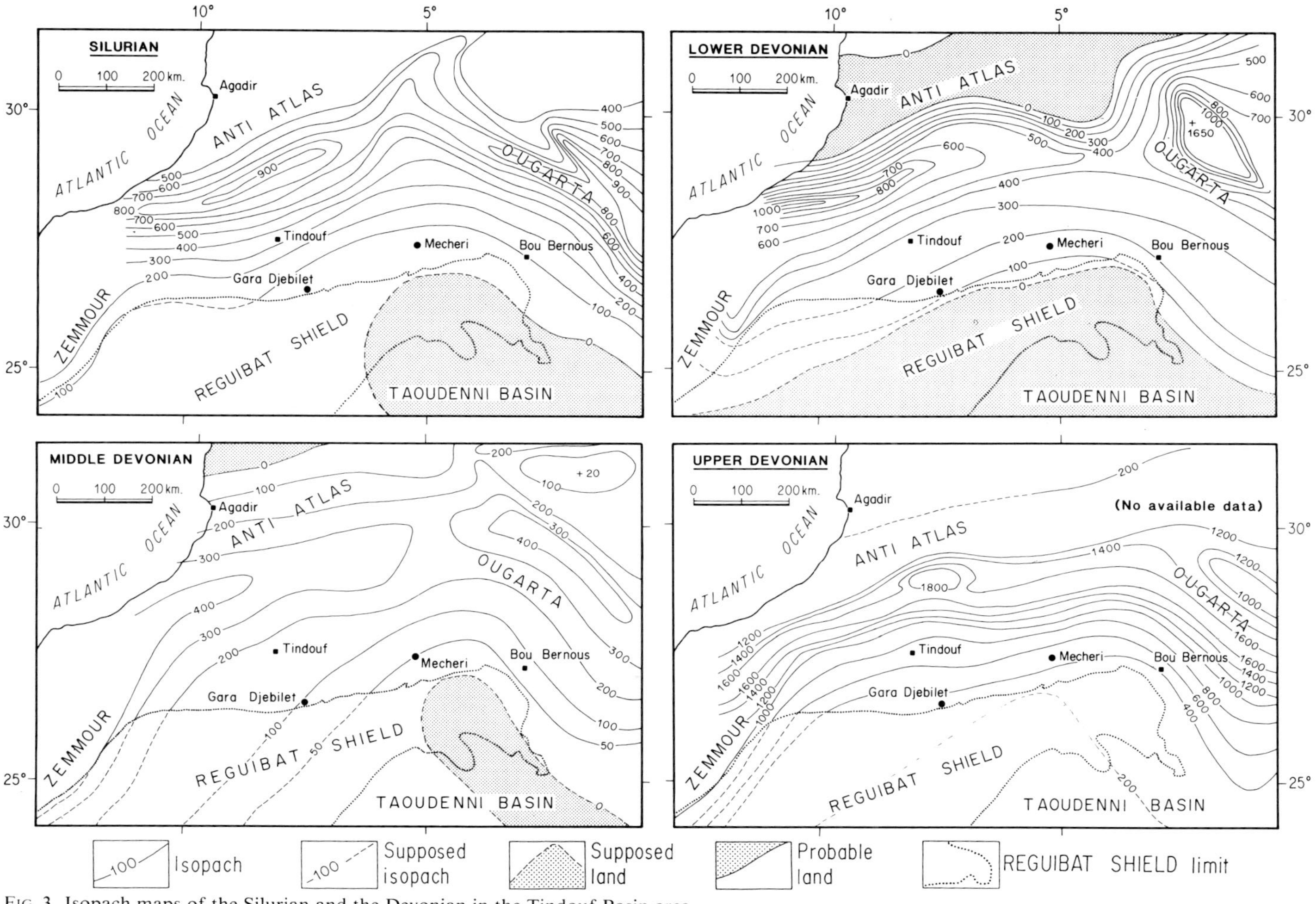


FIG. 3. Isopach maps of the Silurian and the Devonian in the Tindouf Basin area.

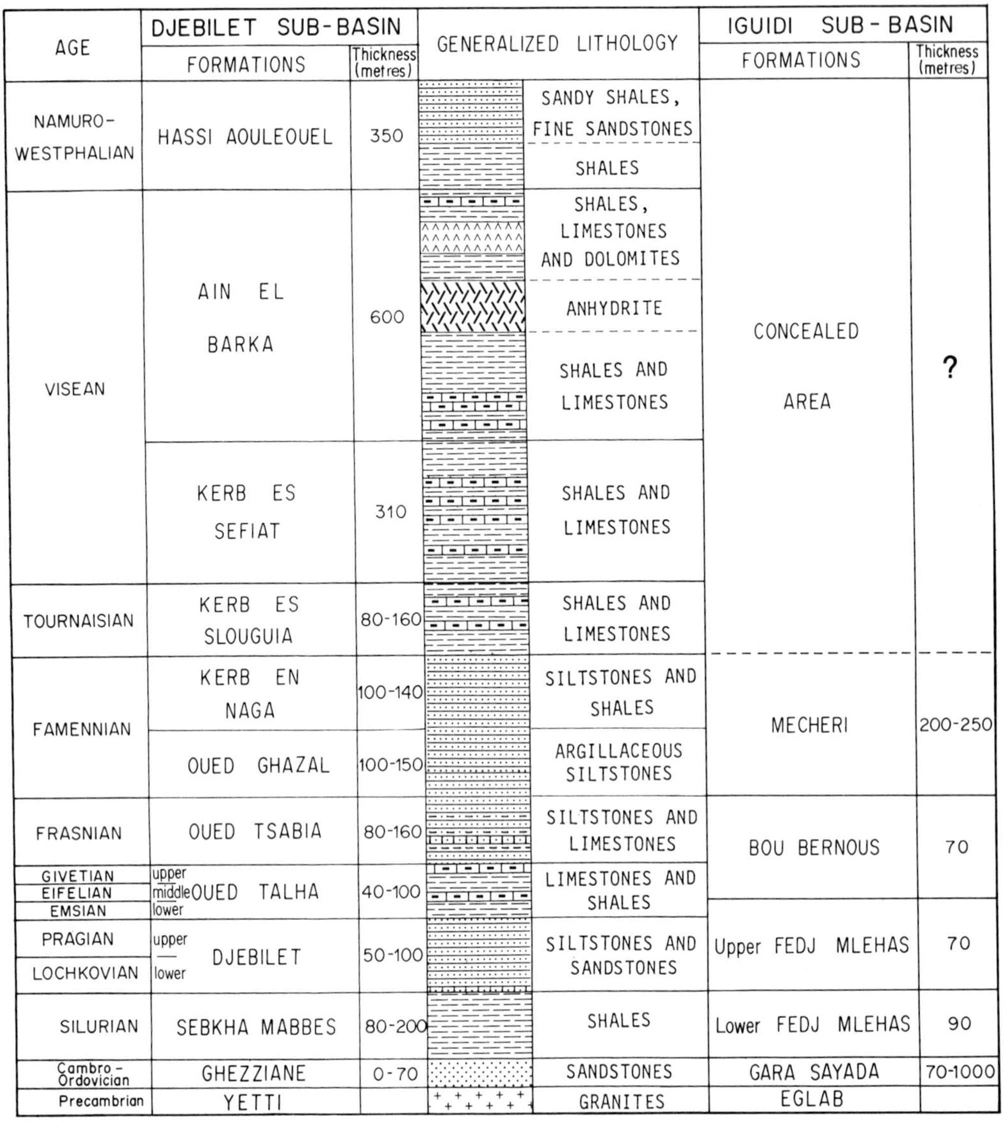

| AGE | DJEBILET SUB-BASIN FORMATIONS | Thickness (metres) | GENERALIZED LITHOLOGY | IGUIDI SUB-BASIN FORMATIONS | Thickness (metres) |
|---|---|---|---|---|---|
| NAMURO-WESTPHALIAN | HASSI AOULEOUEL | 350 | SANDY SHALES, FINE SANDSTONES<br>SHALES | CONCEALED AREA | ? |
| VISEAN | AIN EL BARKA | 600 | SHALES, LIMESTONES AND DOLOMITES<br>ANHYDRITE<br>SHALES AND LIMESTONES | | |
| | KERB ES SEFIAT | 310 | SHALES AND LIMESTONES | | |
| TOURNAISIAN | KERB ES SLOUGUIA | 80-160 | SHALES AND LIMESTONES | | |
| FAMENNIAN | KERB EN NAGA | 100-140 | SILTSTONES AND SHALES | MECHERI | 200-250 |
| | OUED GHAZAL | 100-150 | ARGILLACEOUS SILTSTONES | | |
| FRASNIAN | OUED TSABIA | 80-160 | SILTSTONES AND LIMESTONES | BOU BERNOUS | 70 |
| GIVETIAN<br>EIFELIAN<br>EMSIAN | upper<br>middle OUED TALHA<br>lower | 40-100 | LIMESTONES AND SHALES | | |
| PRAGIAN<br>LOCHKOVIAN | upper<br>—<br>lower DJEBILET | 50-100 | SILTSTONES AND SANDSTONES | Upper FEDJ MLEHAS | 70 |
| SILURIAN | SEBKHA MABBES | 80-200 | SHALES | Lower FEDJ MLEHAS | 90 |
| Cambro-Ordovician | GHEZZIANE | 0-70 | SANDSTONES | GARA SAYADA | 70-1000 |
| Precambrian | YETTI | | GRANITES | EGLAB | |

FIG. 4. Stratigraphical column of the Tindouf Basin region.

observations of Wendt (1985) in the Hercynian terranes of Morocco. The induced distensive movements were responsible for the transgressions, while the compressive movements, due to rearrangement of the previously disintegrated blocks, resulted in epeirogenic activity and regressions.

(iv) The final configuration of the area was the result of epeirogenic movements induced by the Variscan Event (Hercynian), Late Devonian–Carboniferous in age, and well developed from Senegal to Morocco.

The main effects of the various different tectonic events in the Tindouf Basin can be summarized as follows: (i) Pan-African: reactivation and creation of N–S basement faults and basement folding, parallel with the Pan-African suture which limits the West African Craton and the Pan-African mobile belt; (ii) Caradoc: development of N–S ridges, parallel with the Pan-African suture zone, revealed by discordance and pinching out

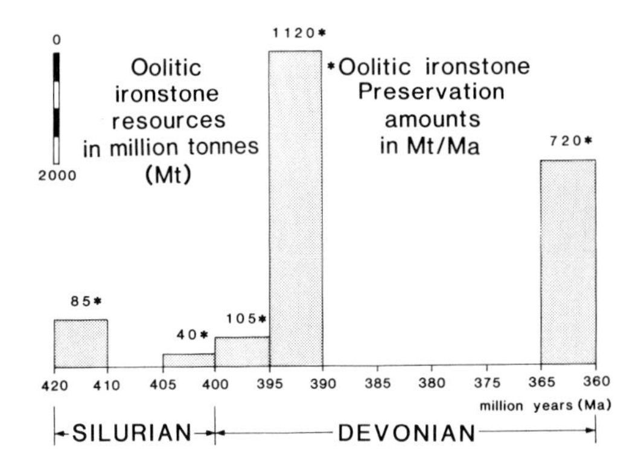


FIG. 5. Oolitic ironstones resources and amounts preserved through Silurian and Devonian Periods, in the Tindouf Basin.

of post Caradoc formations; (iii) Pre-Acadian (Caledonian): from west to east, three uplifts (Aouinet, Laroussi and Bou Bernous) appeared in the Tindouf Basin separated by the Djebilet and the Iguidi sub-basins. The Djebilet sub-basin was very shallow, while the Iguidi sub-basin constituted a subsiding trough, previously developed during Ordovician times, with local deposition of about 1000m of sandstones; (iv) Variscan (Hercynian): weak epeirogenic activity and distensive movements developing regional transgressions.

Stratigraphic correlation and terminology has been proposed for the whole area (Fig. 4), based on Gevin (1960), Graverot & Planchon (1964, private communication) and on my recent investigations, particularly of the Silurian and Devonian sediments.

While there are notable facies and thickness variations in the Middle Devonian formations, the Silurian, Lower and Upper Devonian sediments are thinned-out around the uplift zones. This can be particularly observed in the following instances: (i) the Silurian deposits disappearing to the east on the Bou Bernous uplift, where lower Devonian directly overlies Ordovician sandstones; (ii) the Silurian reduced to about 25 m on the Laroussi uplift; (iii) the Lochkovian which disappears on the Laroussi uplift, and (iv) the Pragian, very thin (2 m) in the Fedj Mlelas area (Bou Bernous uplift) and in the Zemila zone (about 15m in boreholes).

## The oolitic ironstones

In this section an attempt is made to quantify the oolitic ironstone sedimentation and to describe in brief its field relationships, to enable a better understanding of the palaeoenvironment and stratigraphy of the oolitic ironstones.

### Quantity of oolitic ironstone

Considering a mean density value of 3.75 for the oolitic ironstone, and an average width (for the non-drilled deposits) of 1 km, the possible tonnage of these deposits in the study area is estimated to be about 10500 million tonnes: 9500 million tonnes within Devonian sediments and 1000 million tonnes within Silurian sediments. The amounts of oolitic ironstones preserved, expressed in million tonnes per million years, are shown in Fig. 5 (absolute dating after Odin & Gale 1982 and Odin 1985). Pragian and Famennian appear as the more favourable periods for oolitic ironstone sedimentation, with 1120 and 720 million tonnes per million years respectively.

### Field relations

During the Silurian and Devonian Periods, the sedimentation in the Tindouf Basin was characterized by the deposition of oolitic ironstones, essentially interbedded with Llandovery, Ludlow, Lochkovian, Pragian and Famennian sediments (Figs. 1, 2 & 6).

#### *Gara Djebilet*

This deposit occurs as three large separate lenses: Gara West (Fig.7a), Gara Centre and Gara East, extending along strike for about 60 km.

This deposit is interbedded with the argillaceous to sandy sediments of the Upper Djebilet Formation of Pragian age (Guerrak, 1988a). Fig. 8 shows a stratigraphical and sedimentological log where the main elements of this coarsening-upward sedimentary sequence are documented.

#### *Mecheri Abdelaziz*

This large deposit consists of three zones (West, Centre, East) and is interbedded with Upper Famennian argillaceous rocks (including siltstones) (Fig. 7b). Sixteen oolitic ironstone lenses of different sizes are recognized in coarsening-upward units, towards the top of a major regressive sequence (Guerrak & Chauvel 1985) (Fig. 9). The sedimentary cyclicity has been reconstructed by Markov chain analysis (Krumbein & Dacey 1969; Selley 1970; Schwarzacher 1975; Harms *et al.* 1982) using numerous boreholes and has resulted in the

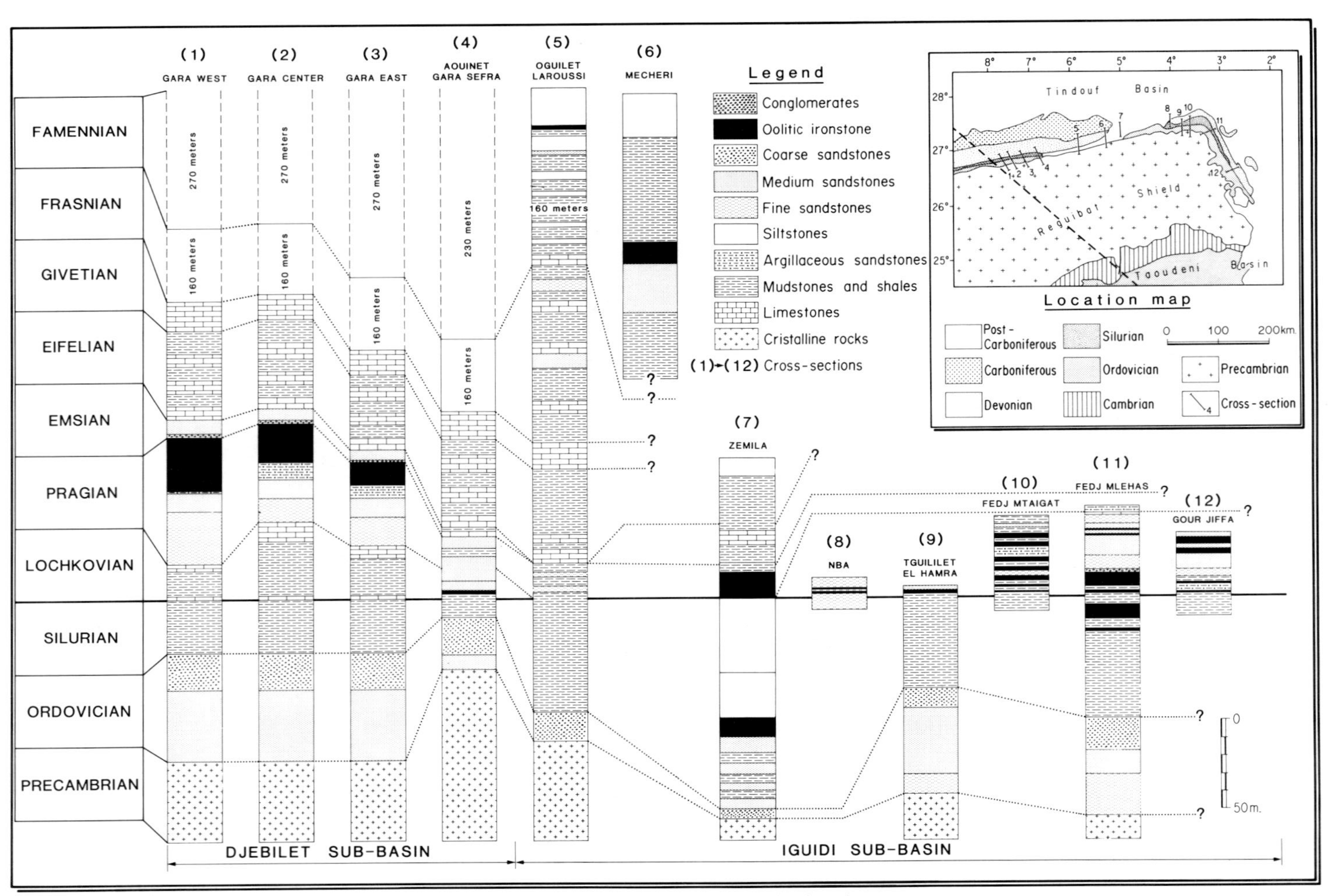


FIG. 6. Stratigraphical correlation of the ironstone deposits in the southern flank of the Tindouf Basin.

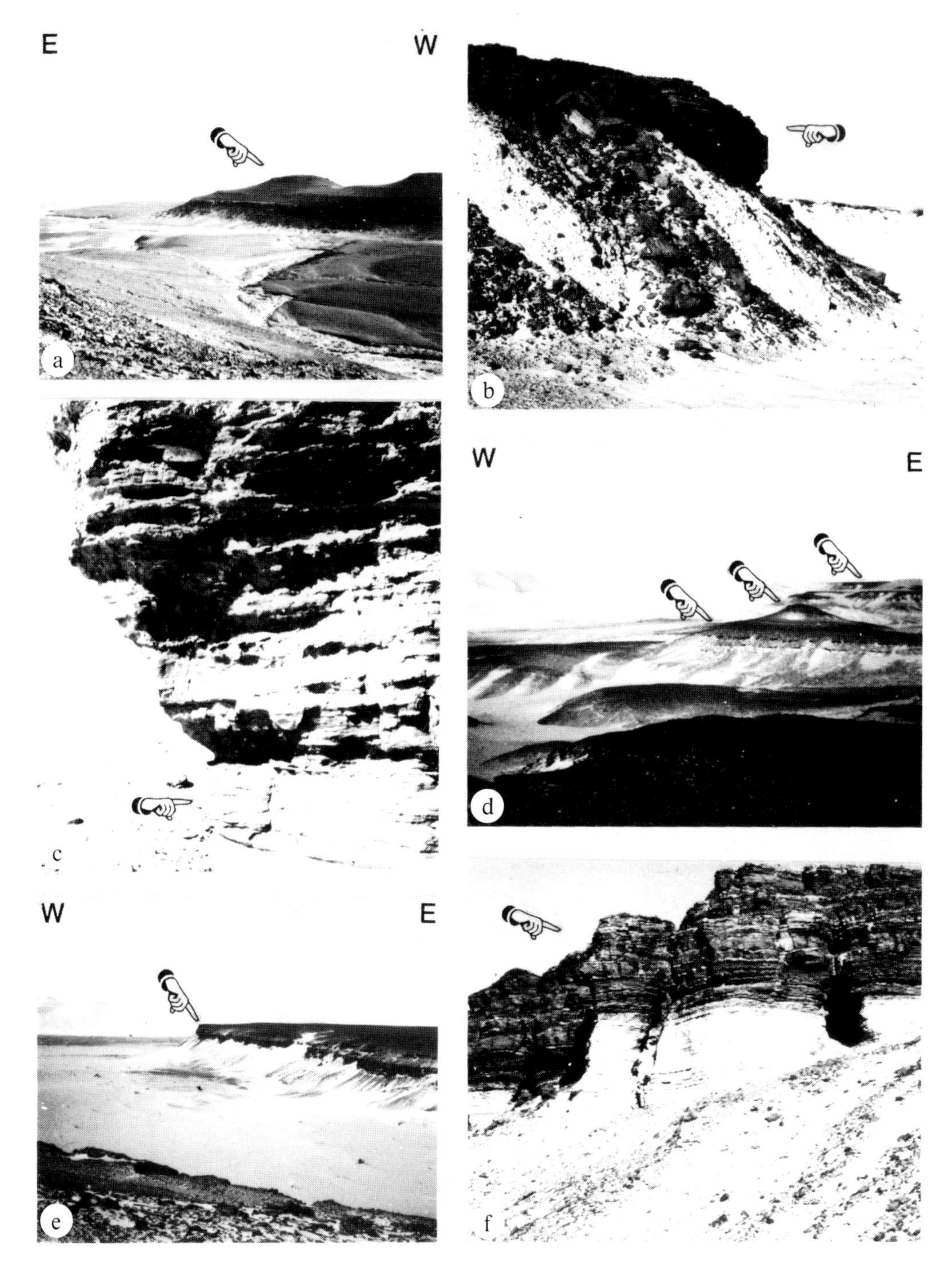


FIG. 7. (a) Northern side of Gara Djebilet West. The hill is capped by an oolitic ironstone lens. (b) Mecheri Abdelaziz West (Ironstone Lens 8): at the base of the cliff trough cross beddings occur. (c) Fedj Mlehas. Silurian oolitic ironstone lenses caping green shales. (d) Fedj Mtaigat: Silurian oolitic ironstone (Lens 3). (e) Fedj Mlehas: lenses 2 (Silurian) 3 & 4 (Lochkovian) are located at the top of hills surrounded by sand dunes. (f) Fedj Mlehas. Lens 4 (Lochkovian) — laminated oolitic ironstone overlying argillaceous siltstones and shales.

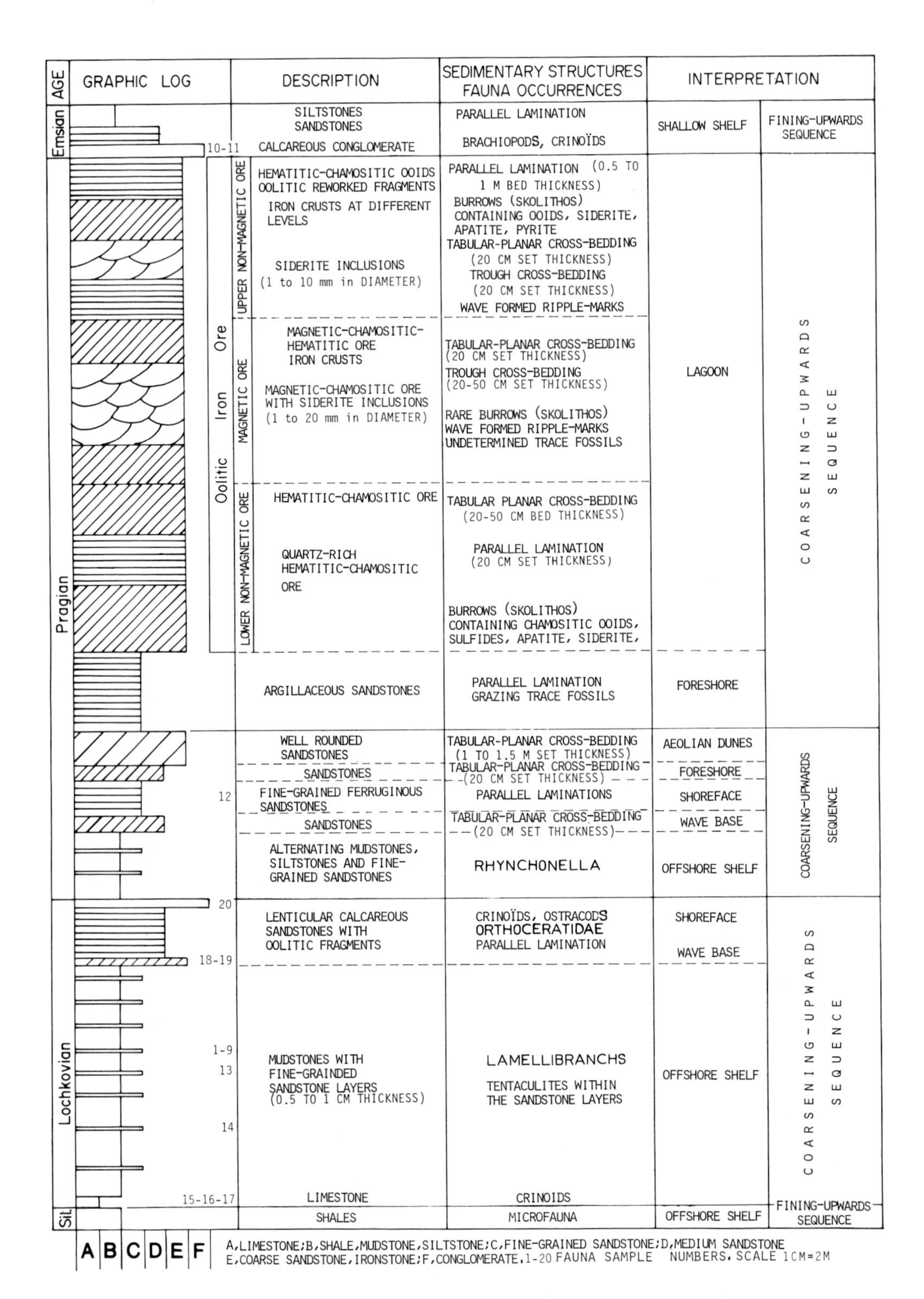


FIG. 8. Detailed lithostratigraphic column of the Gara Djebilet deposit.

identification of three main sedimentary sequences (S1, S2, S3) (Fig. 10).

*Fedj Mlehas*

Six flat-lying oolitic ironstone lenses of various sizes are exposed in the eastern part of the Iguidi sub-basin, and are surrounded by the important sand dunes of the Erg Chech (Fig. 7c, e, f). They occur within Late Silurian (Lenses 1 and 2) and Lochkovian (Lenses 3,4,5 and 6) argillaceous, sandy and conglomeratic sediments (Fig. 11). The lenses are of various shapes, expressing the irregularities of the Late Silurian and Lochkovian palaeotopographies. Numerous oolitic conglomerates are developed in the depositional area, with ooliths in the matrix and in the intraclasts. The oolitic ironstone lenses are often capped by ferruginous crusts of various extent, which correspond to the top of coarsening-upward units.

*Gour Jiffa*

Five oolitic ironstone lenses of different sizes (Fig.12), occur 50 km south of Fedj Mlehas, over an area of about 50 sq km (10 x 5). They are interbedded mostly with sandstones and mudstones, and are located at the top of coarsening-upward sequences. The recorded sedimentary structures are those of very shallow deposits, with occurrence of bioturbation (Skolithos), ripple-marks and large-scale cross bedding. Traces of emersion can be found in the form of remnants of ferruginous crusts, overlying shales or ferruginous sandstones.

*Fedj Mtaigat*

Over an area of 10 × 7 km, eight oolitic ironstone lenses have been recorded in the Lochkovian mudstones, argillaceous sandstones and fine- grained sandstones (Fig. 7d). Ripple-marks, mud-cracks and trace fossils are widespread throughout a succession of thin coarsening-upward units (few metres thick) capped by oolitic ironstones. These ironstones are sometimes conglomeratic, and some lenses contain trough cross-bedding.

*Zemila*

Two oolitic ironstone beds are recognized within the Ludlow–Pridol argillaceous sediments and Lochkovian sandy sediments from boreholes drilled over an area of 240 km2. The oolitic ironstones are located at the top of coarsening-upward sequences. Plant roots occur within the Lochkovian ferruginous coarse-grained sandstones.

*Other oolitic ironstone occurrences*

The Oguilet Laroussi oolitic ironstone occurrence represents a part of a large concealed body and is situated at the top of a minor coarsening upward sequence, which is included within the largest major regressive cycle of the Famennian.

Aouinet, Tguililet el Hamra and Nba, are three small oolitic ironstone deposits developed in the lower part of the Lochkovian. They are interbedded with mudstones and sandstones and occur at the top of small coarsening – upward sequences.

**Palaeoenvironments**

Taking into consideration the associational characteristics, as well as the main sedimentological features, the palaeoenvironments are reconstructed as indicated below:

(i) Gara Djebilet: the deposits were formed on barrier islands bordered by an inner lagoon or a shallow embayment, and developed on the borders of an epicontinental sea. The first coarsening-upward sequence, Lochkovian in age, (Fig. 8), can be interpreted as a seaward prograding sequence, with an offshore shelf, overlain by an incomplete shoreface. The second sequence, Pragian in age, is coarsening-upwards and appears as a complete barrier-island ending with aeolian dunes. These dunes are very significant, containing large tabular-planar cross-bedding and fossil vegetation. Over the aeolian dunes, the last coarsening – upward sequence starts with an inner foreshore zone and ends with the oolitic ironstone, interpreted as lagoonal deposit. Discontinuities between each major ironstone lens correspond to inlets which separate laterally each branch of the island-barrier.

(ii) Mecheri Abdelaziz: the lenticular bodies are located between two distributary bars, the lower one overlying a delta front channel. Therefore the oolite deposits appear contained within deltaic prograding bodies, each one overlapping the other, with various angles, but always deeping to the North.

(iii) Fedj Mlehas and Gour Jiffa deposits present paleoenvironments similar to Gara-Djebilet. They consist of barrier-island systems developed parallel to the coast. The oolite bodies appear as flat lenses extended for several kilometres, located at the top of coarsening-upward sequences.

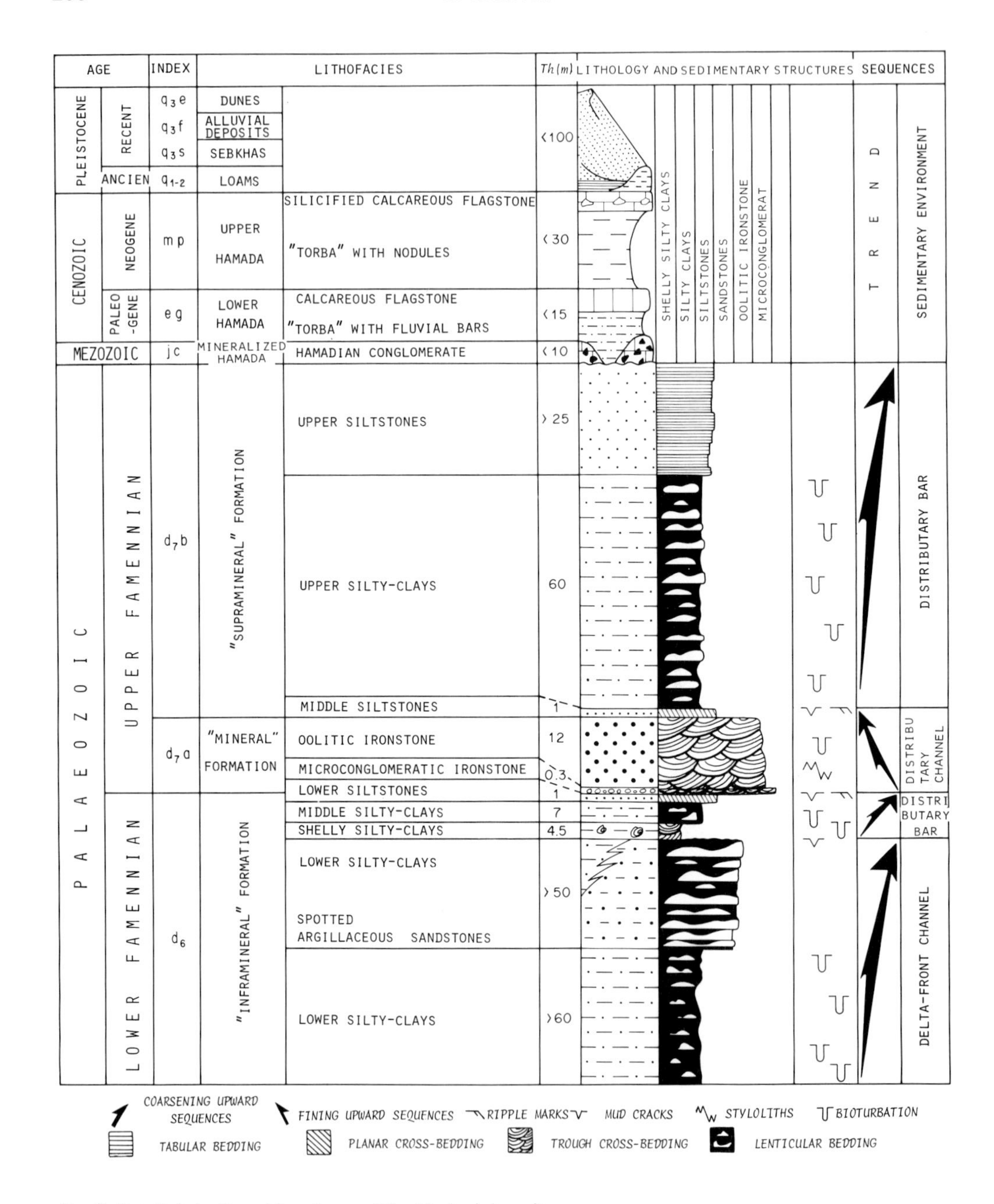


FIG. 9. Detailed stratigraphic column of the Mecheri deposit.

(iv) All the other oolitic ironstone deposits are shallow shelf deposits of beach or barrier-island separating a lagoon or shallow enbayment from the open sea. Numerous traces of emersion are underlined by the occurrence of ferruginous crusts in different levels of the ironstone, especially at its top.

**Petrology and Mineralogy**

Nomenclature used follows Guerrak (1987), and recognizes two types of deposits: LOID type — local ironstone deposit (all the OIS of the Tindouf Basin seem to be of the LOID type) and EXID type — extensive ironstone deposit.

The textural analysis of the oolitic ironstones

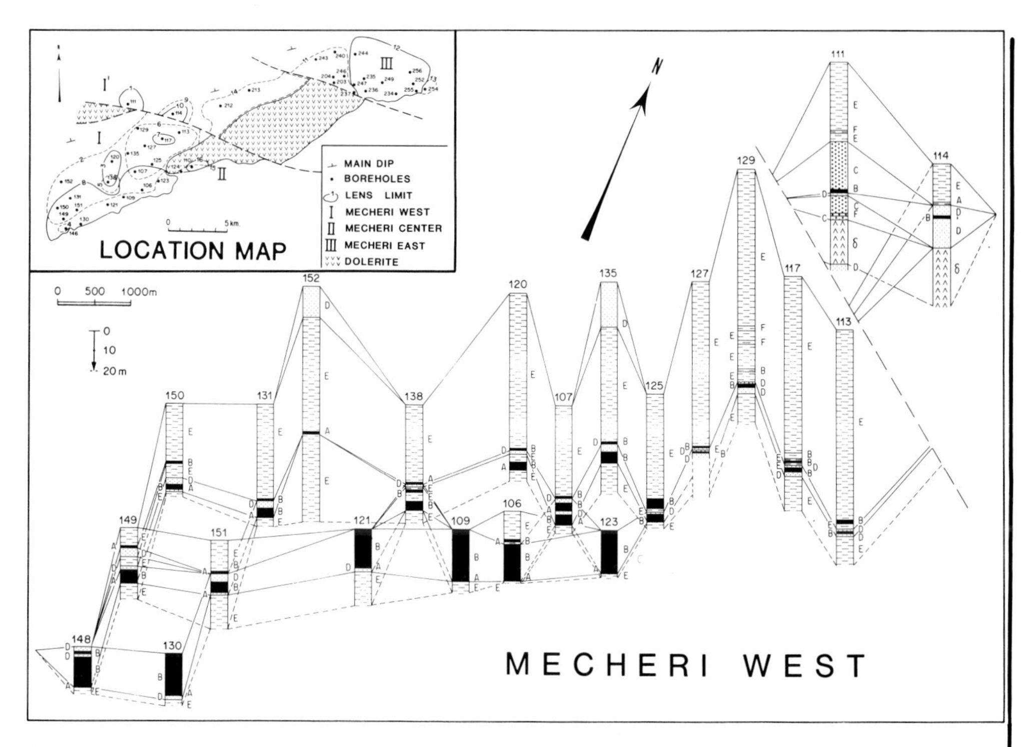


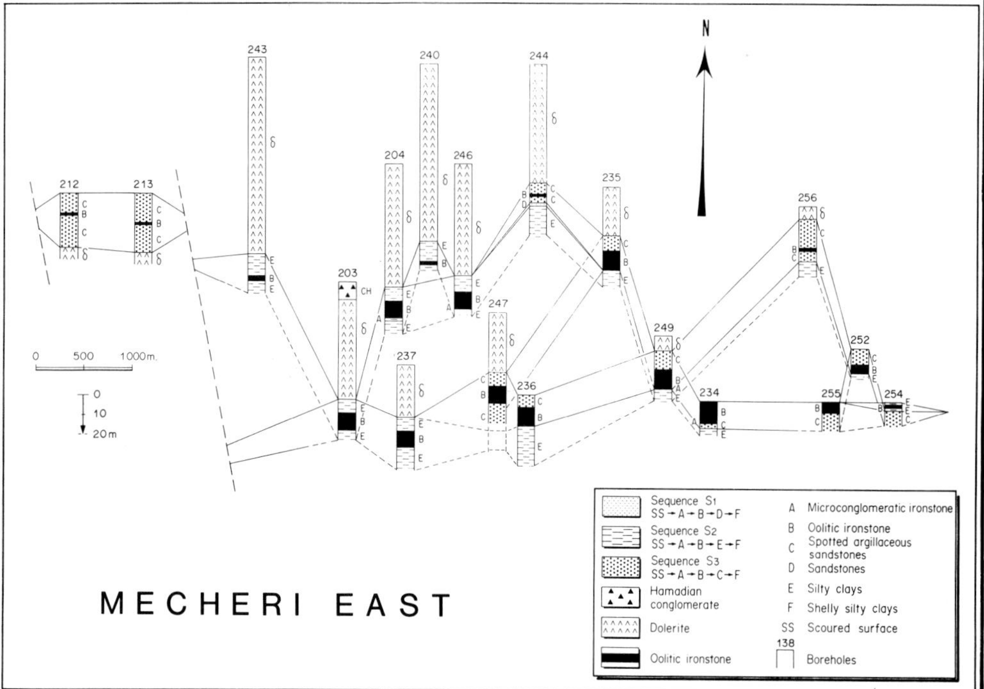


FIG. 10. Detailed correlations of the Mecheri deposit.

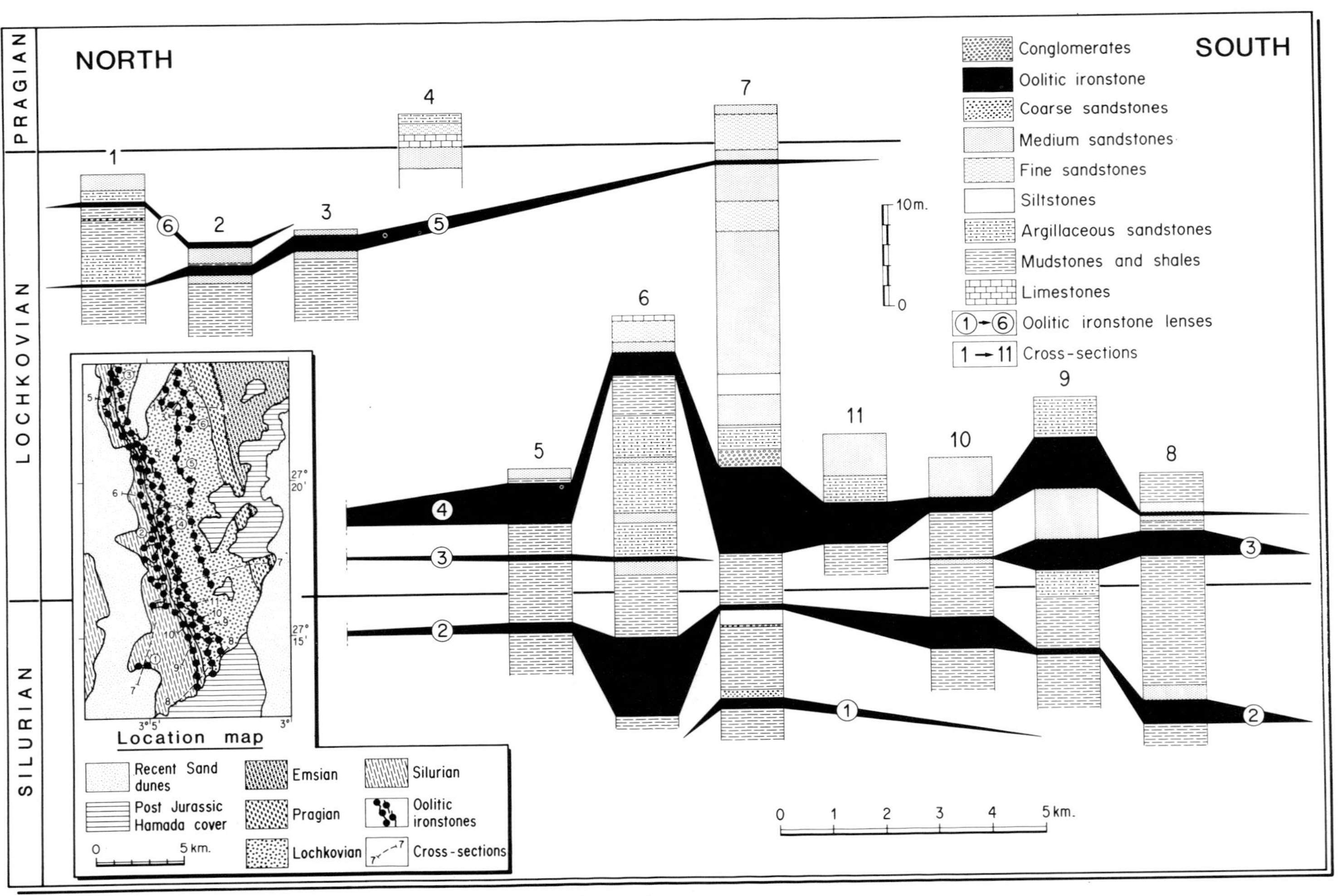


FIG. 11. Correlation of the ironstone lenses in the Fedj Mlehas deposit.

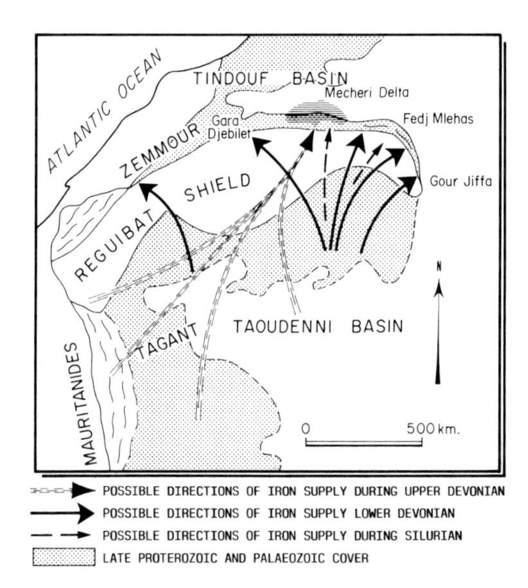


FIG. 12. Possible directions of iron supply during the Silurian and Devonian.

has been performed on several hundreds of borehole and outcrop samples. The oolitic ironstone facies are defined on the basis of the distribution of ooliths within the matrix. The cement is a postdepositional chemical precipitate in the interstices binding ooliths and matrix, or only ooliths. When the matrix is quartz-rich, the facies is named non-detrital.

Five main oolitic ironstone facies have been distinguished (Guerrak & Chauvel 1985): (i) a facies with scattered ooliths within a quartz-rich matrix (detrital OIS facies: FOD): (ii) a facies with scattered ooliths within a quartz-poor (chamositic, calcitic, or iron-oxide-rich) matrix (non-detrital OIS facies: FOND), (iii) a facies consisting of joined ooliths and pore-filling cementing (cemented facies: FOC); (iv) a miconglomeratic facies (FMC); and (v) a conglomeratic facies (FC). These oolitic ironstones can be classified in the Clinton type of Gross (1965) and in the SCOS-IF type (sandy, clayey and oolitic shallow-inland-sea-iron formation) of Kimberley (1978).

# Model of deposition

## Source, transport and accumulation

The mechanism of transport of iron and origin of the different ironstone deposits of the southern flank of the Tindouf Basin are probably similar. Despite the lack of sufficient data, source areas could be located in the western part of the Reguibat Shield or in the positive zone of the Taoudenni Basin (Villemur 1967) (fig. 12).

A major part of the iron could have been carried by rivers,by way of clays (Fe associated in the clay structure) or clastic particles enveloped with oxide films (Carroll 1958; James 1966). The leaching of crystalline rocks, mudstones and ferruginous sandstones could have been the starting point of the iron preconcentration. There are no data on the fluvial network during Silurian and Devonian times in this region, because of the lack of fluvial deposits. The absence of fan deposits and the configuration of isopach maps allow one to indicate a weak slope.

The oolitic accretion mechanisms can be related to intrasedimentary processes as developed by Chauvel & Guerrak (1988). The evidence found in the Saharan deposits of accretion within quiet conditions corroborate the previous observations of Caillère & Kraut (1953); Rohrlich *et al.* (1969); Guerrak & Chauvel (1985); Van Houten & Purucker (1985). The ooids developed within an iron-rich mud, supposed to be either berthierine-rich (Van Houten & Purucker 1984) or kaolinite-rich (Bhattacharyya 1983). This accretion requires peculiar oxidation–reduction potential (Eh) and hydrogen ion activity (pH), Krumbein & Garrels (1952) and Garrel & Christ (1965). Magnetite could occur under strongly reducing conditions ($-0.8 < Eh < -0.3$ and $pH < 8$) as those existing for example in the 'Anoxic–Non Sulfidic–Post Oxid' environments defined by Berner (1981) and Maynard (1982).

## Palaeogeographic distribution

The sedimentation in the Tindouf Basin was controlled by alternating regressive and transgressive episodes, particularly during Silurian–Devonian times. Four episodes are recognized, namely:

(i) Silurian: transgression, (ii) Lochkovian–Pragian: regression, (iii) Emsian–Eifelian–Frasnian: transgression, (iv) Famennian: regression.

The transgressive phases corresponded to quiet sedimentation periods with deposition of shales, fine-grained sediments and limestones, the regressive phases to disturbed periods, with great variation in the type of sediment deposited and occurrence of coarser grained sediments.

Most of the oolitic ironstones were formed

during the regressive phases, except the Silurian ones, which nevertheless are linked to coarsening- upward sequences, corresponding to subordinate regressive trends. This implies that the epeirogenic movements which have resulted in the rising of positive zones aided the regressive conditions.

The timing of the large deposits of Gara Djebilet and Mecheri, in the late stages of major regressive phases, corroborates the similar observations of Hallam & Bradshaw (1979) in the NW European Jurassic.

Similar relations have been observed in the EXID deposits of the Tassilis N'Ajjer and the Illizi Basin (Central Sahara) (Guerrak 1988b).

The pericratonic sedimentation in the Tindouf Basin can be considered as a multipulse sedimentation, each pulse corresponding to a coarser-grained sedimentary supply. The relationship between detrital sedimentation and oolitic ironstone, clearly shows that the latter are inversely proportional to the detrital influx. This has been previously observed in Ordovician oolitic ironstones of the Ougarta ranges, NW Sahara (Guerrak 1988a,b), in the NW European Jurassic ironstones (Hallam 1975), and discussed more generally by Van Houten & Bhattacharyya (1982) for Phanerozoic ironstones. On the other hand, the distribution of the oolitic ironstones is related to the uplifts disturbing the relatively regular palaeotopography of the very shallow sea floor which trapped the ironstones which developed on their flanks. Accordingly, the main oolitic ironstone deposits are located in the following areas: (i) Gara Djebilet on the western flank of the Aouinet uplift; (ii) Mecheri (and Zemila + Laroussi) on the eastern flank of the Laroussi uplift (the western flank being concealed by sand dunes of the Iguidi Erg, could contain other ironstone occurrences) and (iii) Fedj Mlehas on the western flank of the Bou Bernous uplift.

The minor deposits occur both more centrally and more peripherally to the uplifts.

Following the various models depicting Devonian and Silurian continental reconstructions (Morel & Irving 1978; Scotese *et al.* 1979; Bambach *et al.* 1980; Livermore *et al.* 1985; Salmon *et al.* 1987, Hargraves & Van Houten 1985), it appears that N Africa was located between latitudes of 30° and 60° S, which corresponds to the present-day temperate zone. This clearly shows that the southern source for the iron was situated in a colder region and cannot be related to hot climate. However, Heckel & Witzke (1979) and Van Houten & Bhattacharyya (1982), have invoked the existence of warm currents during Siluro-Devonian times to explain Gondwanan reefs and oolitic ironstones; this seems to be untenable taking into consideration the palaeolatitude of Gondwana during the Silurian and Devonian.

ACKNOWLEDGEMENTS: The National Mining Research Company of Algeria (EREM) provided field support and chemical analyses. I gratefully acknowledge the help of A. Slougui, F. Baba Ahmed, A. Belaid and P. Riffault. I have greatly benefited from the advice and criticism of J.J. Chauvel who reviewed an earlier version of this paper. I am grateful to my colleges M. Robardet and B. Mahabaleswar for critical comments and to M.P. Bertrand for her assistance.

## References

BAMBACH, R.K., SCOTESE, C.R. & ZIEGLER, A.M. 1980. Before Pangea. The geographies of the Paleozoic word. *American Scientist,* **68,** 2–38.

BERNER, R.A. 1981. A new geochemical classification of sedimentary environments. *Journal of Sedimentary Petrology.* **51,** 339–365.

BHATTACHARYYA, D.P. 1983. Origin of Berthierine in ironstones. *Clays and Clay Minerals,* **31,** 173–182.

BLACK, R., CABY, R., MOUSSINE-POUCHKINE, A., BAYER, R., BERTRAND, J.M., BOULLIER, A.M., FABRE, J. & LESQUER, A. 1979. Evidence for late Precambrian plate tectonics in West Africa. *Nature.* **278,** 223–227.

CALLIERE, S. & KRAUT, F. 1953. Considérations sur la genèse des minerais de fer oolithiques lorrains. *19th International Geological Congress, Algiers (1952).* **10,** 101–117.

CARROLL, D. 1958. Role of clay minerals in the transportation of iron. *Geochimica et Cosmochimica Acta.* **14,** 1–27.

CHAUVEL, J.J., & GUERRAK, S. 1989. Oolitization processes in Palaeozoic ironstones of France, Algeria and Lybia. *In*: T.P. YOUNG & W.E.G. TAYLOR (eds) *Phanerozoic Ironstones.* Geological Society, London, Special Publication, **46,** 165–174.

——& MASSA D. 1981. Paléozoique de Libye occidentale. Constantes géologiques et pétrographiques. Signification des niveaux ferrugineux oolithiques. *Notes et Mémoires, Compagnie Francaise des Pétroles.* **16,** 25–66.

DESTOMBES, J., HOLLAND, H. & WILLEFERT, S. 1985. Lower Palaeozoic Rocks of Morocco. *In:* HOLLAND, C.H. (ed.) *Lower Palaeozoic of north-western and north-central Africa.* John Wiley & Sons, London, 91–336.

GARRELS, R.M. & CHRIST, C.L. 1965. *Solutions,*

minerals and equilibria. Harper & Row, New-York.

GEVIN, P. 1960. Etudes et reconnaissances géologiques sur l'axe cristallin Yetti-Eglab et ses bordures sédimentaires. *Bulletin du Service de la Carte Géologique de l'Algérie,* **23,** 1–328.

GOUDARZI, G.H. 1970. Geology and Mineral Resources of Libya. A reconnaissance. *U.S. Geological Survey. Professional Paper,* **660.**

GROSS, G.A. 1965. Geology and Evaluation in Canada. Volume I. General Geology and Evaluation of Iron Deposits. *Geological Survey of Canada. Economic Geology, Report 22.*

GUERRAK, S. 1987. Paleozoic oolitic ironstones of the Algerian Sahara: a review. *Journal of African Earth Sciences,* **6,** 1–8.

——1988a. Geology of the Early Devonian oolitic iron ore of the Gara Djebilet field. Saharan Platform, Algeria. *Ore Geology Reviews,* (In Press).

——1988b. Paleozoic multipulse cratonic sedimentation and associated oolitic iron-rich deposits, Tassilis N'Ajjer and Illizi Basin, Saharan Platform, Algeria. *Eclogae Geologicae Helveticae,* **81,** 457–485.

——1988c Ordovician ironstone sedimentation in Ougorta Ranges: northwest Sahara (Algeria). *Journal of African Earth Sciences,* **7,** 657–678.

——& CHAUVEL, J.J. 1985. Les minéralisations ferriféres du Sahara algérien. Le gisement de fer oolithique de Mecheri Abdelaziz (bassin de Tindouf). *Mineralium Deposita,* **20,** 249–259.

HALLAM, A. 1975. *Jurassic environments.* Cambridge University Press, Cambridge.

——& BRADSHAW, M.J. 1979. Bituminous shales and oolitic ironstones as indicators of transgressions and regressions. *Journal of the Geological Society, London,* **136,** 157–164.

HARGRAVES, R.B. & VAN HOUTEN F.B. 1985. Paleogeography of Africa in early-middle Paleozoic: paleomagnetic and stratigraphic constraints and tectonic implications. *13th Colloquium of African Geology, St. Andrews. Occasional Publication,* **3,** 160–161.

HARMS, J.C., SOUTHARD, J.B. & WALKER, R.G. 1982. Stratification sequences: principles and approaches to interpretation. In: *Structures and sequences in clastic rocks: Short course 9. Society Economic Paleontologists & Mineralogists,* 4/1–4/19.

HECKEL, P.H. & WITZKE, B.W. 1979. Devonian world palaeogeography determined from distribution of carbonates and related lithic palaeoclimatic indicators. In: HOUSE, M.R., SCRUTTON, C.H., & BASSETT, M.S. (eds). *The Devonian System. Special Paper in Palaeontology,* **23,** 99–123.

HOLLARD, H. 1967. Le Dévonien du Maroc et du Sahara Nord-Occidental. In: OSWALD, D.H. (ed.). *International Symposium on the Devonian System. Calgary.* **I,** 203–244.

JAMES, H.L. 1966. Chemistry of the iron-rich sedimentary Rocks. *U.S. Geological Survey Professional paper,* **440.**

KIMBERLEY, M.M. 1978. Paleoenvironmental classification of iron formations. *Economic Geology,* **73,** 215–229.

KRUMBEIN, W.C. & DACEY, M.F. 1969. Markov Chains and embedded Markov Chains in Geology. *Mathematical geology,* **1,** 79–96.

——& GARRELS, R.M. 1952. Origin and classification of chemical sediments in terms of pH and oxydation-reduction potential. *Journal of Geology,* **60,** 1–33.

LECORCHÉ, J.P. 1983. Structures of the Mauritanides. *In*: SCHENK, P.E. (ed.). *Regional trends in the Geology of the Appalachian – Caledonian – Hercynian – Mauritanide orogen.* Reidel, U.S.A. 347–353.

LESQUER, A., BELTRAO, J.F. & de ABREU, F.A.M., 1984. Proterozoic links between northeastern Brasil and West Africa : a plate tectonic model based on gravity data. *Tectonophysics,* **110,** 9–26.

LIVERMORE, R.A., SMITH, A.G. & BRIDEN, J.C. 1985. Palaeomagnetic constraints in the distribution of continents in the late Silurian and early Devonian. *Philosophic Transactions of Royal Society of London,* **B. 309,** 29–56.

MAYNARD, J.B. 1982. Extension of Berner's 'New geochemical classification of sedimentary environments' to ancient sediments. *Journal of Sedimentary Petrology.* **52,** 1325–1331.

MOREL, P. & IRVING, E. 1978. Tentative Paleocontinental maps for the early Phanerozoic and Proterozoic. *Journal of Geology,* **86,** 5, 535–562.

ODIN, G.S. 1985. Remarks on the numerical scale of Ordovician to Devonian times. In: SNELLING, N.J. (ed.). *The chronology of the Geological Record.* Geological Society, London, Memoir, **10,** 93–98.

——& GALE, N.H. 1982. Mise à jour de l'échelle des temps calédoniens et hercyniens. *Comptes Rendus de l'Académie des Sciences.* **294,** 453–456.

POOLE, W.H., MCKERROW, W.S., KELLING, G. & SCHENK, P.E. 1983. A stratigraphic sketch of the Caledonide — Appalachian — Hercynian Orogen. *In:* SCHENK, P.E. (ed.). *Regional trends in the geology of the Appalachian — Caledonian — Hercynian — Mauritanide orogen.* Reidel, U.S.A. 75–111.

ROHRLICH, V., PRICE, N.B. & CALVERT, S.E. 1969. Chamosite in the recent sediments of Loch Etive, Scotland. *Journal of Sedimentary Petrology,* **39,** 624–631.

SALMON, E., MONTIGNY, R., EDEL, J.B., PIQUE, A., THUIZAT, R. & WESTPHAL, M. 1987. A 140 Ma K/Ar age for the Msissi Norite (Morocco) : new geochemical and paleomagnetic data. *Earth and Planetary Science Letters.* **81,** 265–272.

SCHWARZACHER, W. 1975. *Sedimentation models and quantitative stratigraphy*: Developments in Sedimentology, **19,** Elsevier, Amsterdam.

SCOTESE, C.R., BAMBACH, R.K., BARTOW, C., Van DER VOO, R. & ZIEGLER, A.M. 1979. Paleozoic base maps. *Journal of Geology,* **87,** 217–277.

SELLEY, R.C. 1970. Studies of sequences in sediments using a simple mathematical device. *Quarterly*

*Journal of the Geological Society of London,* **125,** 557–581.

SOUGY, J. 1964. Les formations paléozoïques du Zemmour Noir (Mauritanie Septentrionale). *Annales de la Faculté des Sciences de Dakar,* **15.** 1–695.

VAN HOUTEN, F.B. & BHATTACHARYYA, D.P. 1982. Phanerozoic oolitic ironstones. Geologic Record and Facies Model. *Annual Review of Earth and Planetary Sciences,* **10,** 441–457.

—& KARAZEK, R.M. 1981. Sedimentologic framework of late Devonian oolitic iron formation, Shatti Valley, West Central Libya. *Journal of Sedimentary Petrology,* **51,** 415–427.

—& PURUCKER, M.F. 1984. Glauconitic peloids and chamositic ooids. Favorable factors, constraints, and problems. *Earth-Science Reviews,* **20,** 211–243.

—&—1985. On the origin of glauconitic and chamositic granules. *Geo-Marine Letters,* **5,** 47–49.

VILLEMUR, J.R. 1967. Reconnaissance géologique et structurale du Nord du Bassin de Taoudeni. *Mémoires du Bureau de Recherches Géologiques et Minières. Paris.* 51.

WENDT, J. 1985. Disintegration of the continental margin of the northwestern Gondwana. Late Devonian of the eastern Anti-Atlas (Morocco). *Geology,* **13,** 815–818.

S. GUERRAK, Centre Armoricain d'Etude Structurale des Socles, Université de Rennes I, Laboratoire de Pétrologie Sédimentarie, 35042 Rennes Cedex, France.

Present address: Geological Survey of Algeria, Office Nationale de la Géologie, 18A Avenue Mustapha, El-Ouali, Alger, Algeria.

# The mid-Ordovician oolitic ironstones of North Wales: a field guide

**R.J.B. Trythall**

This guide to the mid-Ordovician oolitic ironstones of North Wales describes the more readily accessible localities, illustrating the main ironstone types and showing their depositional and post-depositional features. The ironstones are exposed entirely in abandoned mines and adits (Fig. 1). Other ironstones localities are mentioned where pertinent and are located by a grid reference number only.

Within the Lower Palaeozoic North Wales Basin (Kokelaar *et al.* 1984) oolitic ironstones occur as discontinuous lenses within fine-grained shelf siliciclastic sediments. These ironstones formed the study of a PhD thesis (Trythall 1988), and have been previously studied by Trythall *et al.* (1987), Weinberg (1973), Pulfrey (1933) and Strahan *et al.* (1920). There are two phases of ironstone deposition in the North Wales Basin (Trythall *et al.* 1987), a minor Upper Arenig phase (Beckley 1987) and, the subject of this field guide, a mid-Ordovician (upper Llanvirn to basal Caradoc) phase. The majority of the mid-Ordovician ironstones were formed during the *teretiusculus* Biozone regression or subsequent *gracilis* Biozone transgression. They were deposited on shallow water shoals formed by synsedimentary faulting, above a stratigraphic hiatus (Trythall *et al.* 1987).

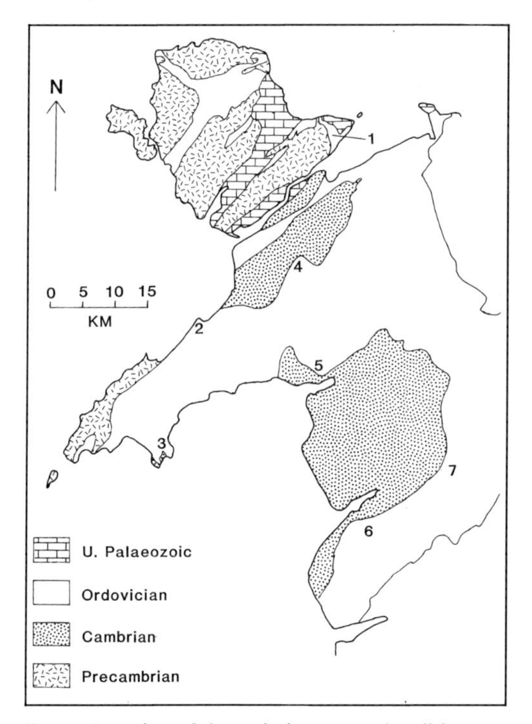


FIG. 1. Location of the main ironstones localities mentioned in the text: 1, Bryn Poeth; 2, Trefor; 3, Pen y Gaer; 4, Betws Garmon; 5, Tremadog; 6, Ffordd Ddu, 7, Cross Foxes.

During the closure of the Iapetus Ocean the Welsh Basin was uplifted and deformed, forming major north-east to south-west trending folds with an axial planar cleavage (Coward & Siddans 1979). Faulting predominantly follows the same trend. Maximum metamorphic grades attained in North Wales were low greenschist facies in Central Snowdonia and prehnite–pumpellyite facies elsewhere (Bevins & Rowbotham 1983; Roberts & Merriman 1985).

Mining of the mid-Ordovician oolitic ironstones occurred throughout most of the 19th century, continuing at some localities to the end of the First World War (Strahan *et al.* 1920). They were mostly worked at a shallow level, as weathering reduced the phosphorus and sulphur contents. However, a few ironstones were mined to a deeper level (Betws Garmon and Cross Foxes) although the hardness of the ore was detrimental to its market value. Approximately 300 000 tons of iron ore were removed from North Wales, of which one third came from Betws Garmon, most being shipped out to the South Wales steel works.

The ironstones are distinguished from the adjacent fine-grained siliciclastic sediments by their darker colour and poorly cleaved or uncleaved nature. The general sequence is of basal chamositic mud-ironstones[1] followed by ooidal and oncoidal pack-ironstone lenses in chamositic mud-ironstone, all interbedded with mudstones, siltstones and shales. This is followed by an upper massive ooidal pack-ironstone, reverting abruptly to clastic mudstones. The upward increasing ooid content and decreasing mud content suggest progressive current winnowing, although cross bedding is not seen. Trythall (1988) has shown that the ironstones were originally worked by tidal currents, but that bioturbation subsequently destroyed primary sedimentary textures and

[1] The ironstone terminology used herein conforms to the proposals in the introduction to this volume. An ironstone is defined as a rock over 15% Fe.

*From* YOUNG, T. P. & TAYLOR, W. E. G. (eds), 1989, *Phanerozoic Ironstones*
Geological Society Special Publication No. 46, pp. 213-220

structures. A strong variation of sorting of the ooids can be seen in the ooidal pack-ironstones across the North Wales Basin. Those nearest to the Irish Sea Landmass (Anglesey and Llŷn) have a poor ooid grain-size sorting and also contain oncoids, detrital quartz and feldspar. The Snowdonia and Cadair Idris ironstones are better sorted with a smaller average ooid grain-size and do not contain oncoids or detritus.

Within the ironstone sequence, early diagenetic phosphate nodules up to 10 cm in length are a conspicuous feature. In the ooidal pack-ironstones they are amoeboid in shape with diffuse margins, but in the underlying muddier facies they are more rounded with sharp well defined margins and at some localities show a pronounced bedding alignment. When fresh the nodules are dark grey and show little internal texture, but on weathered surfaces they are white, displaying randomly scattered ooids, oncoids, inarticulate brachiopod fragments and sponge spicules.

The dominant mineral in the ironstone is 14Å chamosite, with subsequent replacements by siderite, magnetite, pyrite, stilpnomelane and hematite, formed by diagenesis, metasomatism and hydrothermal alteration.

*Geological maps:* 1:50 000 sheet **106** Bangor, 1:50 000 special sheet Anglesey, 1:25 000 special sheet Central Snowdonia.

*Topographical maps:* Ordnance Survey 1:50 000 sheets **114** Anglesey, **123** Llŷn Peninsula, **115** Snowdon, **124** Dolgellau.

**1 Bryn Poeth** [SH6016 7958]

The Ordovician sequence exposed in southeast Anglesey has been described by Greenly (1919) and Bates (1972), consisting of Arenig grits resting unconformably on Precambrian schists, succeeded by Upper Arenig shales (Beckley 1987). Greenly (1919) noted the absence of the *bifidus* Biozone in this area, as the next exposure in the stratigraphic sequence is the Bryn Poeth ironstone with an upper Llanvirn graptolite fauna (Greenly 1919 p. 432).

There are three different ironstone types on Anglesey (Greenly 1919). The first are oolitic ironstones, seen at Bryn Poeth but also at Tynyronen [SH 4327 7932]. The second are 'feebly oolitic ferrified grits' found at the base of the *gracilis* Biozone transgression in Central Anglesey (Fferam Uchaf [SH 3626 8675]). Lastly are the 'feebly oolitic siderite ironstones' on the north coast of Anglesey of *gracilis* Biozone age (Porth Padrig [SH 3756 9434]). These different ironstone types, along with the high detrital quartz and feldspar content, can be related to the proximity of the adjacent Irish Sea Landmass.

*Drive to Anglesey and take the A545 to Beaumaris and then the B5109 to Llangoed. On entering the village take the first marked left turn, The ironstone is on the left, next to the layby, 800m along this lane. Ask for permission at Carwad Farm [SH 585 790] 1.5km away.*

The Bryn Poeth ironstone is uncleaved and undeformed and therefore shows the best preserved sedimentary structures of all the ironstone localities in North Wales. The vertical sequence exposed is shown in Fig. 2. Bedding at this locality strikes 200° and dips 10° W. In the lowest part of the sequence (Fig. 2; section B) ooids can be found in the lower clastic units. The lenses of ooidal pack-ironstones in the chamositic mud-ironstones also contain some angular detrital quartz and feldspar grains, all indicating reworking of ooids before deposition. Graptolites and inarticulate brachiopods are found in this sequence, and burrow mottling of this ironstone indicates that there was some bioturbation of these sediments, but did not disturb the oolitic beds. The ooidal pack-ironstones are poorly sorted containing detrital quartz and feldspar. Current winnowing of the ironstone is indicated by thin siderite cemented ooidal grain-ironstones (now weathered to a strong orange colour) and by the scour and fill structure in the ooidal wacke-ironstones.

The top ooidal pack-ironstone (Fig. 2; section A) is lithologically heterogenous, poorly sorted and contains early diagenetic siderite. Within it are thin (30 cm) beds of fine-grained ooidal pack-ironstone, with a high detrital quartz and feldspar content. The ironstone is succeeded by shales with diagenetic phosphate nodules.

**2 Trefor [SH3715 4747 to SH3672 4739]**

Little is published on the geology of the Trefor region, the main works are by Tremlett (1962) and Roberts (1979). The sediments adjacent to the ironstone on the Trwyn y Tâl peninsula are fine-grained laminated sandstones, siltstones and silty mudstones of possible Arenig age (Roberts 1979), although Trythall *et al.* (1987) assign a Llandeilo age to the ironstone based on acritarchs collected.

*Follow the signs to Trefor off the A499. At the entrance to the village take the right hand turn marked to the beach and park at the end of the road. Walk past the pier to the first ironstone locality [SH 3715 4747] on the Trwn y Tâl peninsula.*

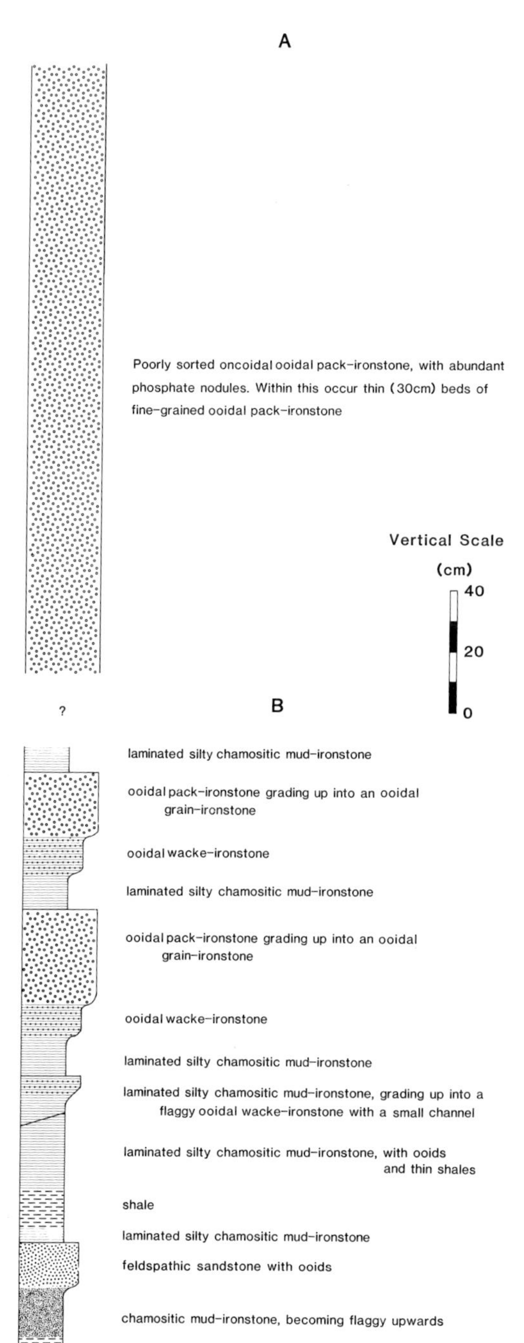


FIG. 2. Vertical section of the exposures at Bryn Poeth. Section A is the upper part of the ironstone sequence exposed on the southwest side and section B is the lower part on the eastern side of the quarry.

Oncoids and chamositic stromatolites are particularly prominent at Trefor, best exposed at the more westerly localities. However, deformation and hydrothermal alteration have affected the ironstone. This is marked by quartz and pyrite veins and chlorite pressure solution seams which tend to remove the sedimentary features of the ironstone. This can be seen in the two easterly exposures of the ironstone [SH 3715 4747, SH 3687 4741].

The most westerly ironstone locality [SH 3675 4739] consists of three parallel trenches, where the bedding strikes 234° with vertical dip. The northernmost trench exposes 2 m of ooidal pack-ironstone with cleaved pale grey mudstones above the ironstone exposed along the south wall. The ooidal pack-ironstone is poorly sorted, containing abundant oncoids and stromatolite crusts. Oncoids are up to 4 cm in size and show a wide variety of shapes. They are composed of chamosite and phosphate lamellae around a chamositic core. A thin (10 cm) oncolite bed is present at this locality, with partial replacement of the chamositic oncoids by hematite.

**3 Pen y Gaer** [SH2984 2829 to SH2992 2821]

In the St Tudwals peninsula, Cambrian sediments are overlain unconformably by the Tudwals Sandstones succeeded by the Llanengan Mudstones (Fig. 3; Nicholas 1915) both of *hirundo* Biozone age (Beckley 1987). These are unconformably overlain in turn, although with little apparent discontinuity, by the Hen dy Capel Ironstone and Pen y Gaer Mudstones (Crimes 1969). Originally this unconformity was interpreted as a thrust zone (Nicholas 1915) similar to that found at Tremadog (Fearnsides 1910) as similar tectonic disturbance can be seen within the ironstones in both areas. Trythall *et al.* (1987) have given an age for this ironstone as early Middle Llandeilo in the *gracilis* Biozone. The St Tudwals region is bounded just to the north by a thrust zone (Crimes 1969; Roberts & Merriman 1985), where Tremadoc and Arenig sediments of different character to those found in the St Tudwals region occur to the north of this thrust.

*Take the A499 to Abersoch and then follow the signs to Llanengan. On entering the village take the right hand turn marked to Llangian. Just before the bridge over the Afon Soch, 1 km along this road, park on the right and walk up to Pen y Gaer.*

The ironstone in the St Tudwals demonstrates

the effects of contemporaneous faulting on its formation. The Pen y Gaer ironstone has been used as the key section for this area as it shows the least effects of tectonic disturbance. The ironstone sequence is exposed in two pits, the lower sequence on the side of the hill [SH 2984 2829], the upper sequence of the ironstone is seen on the top of Pen y Gaer [SH 2992 2821] (Fig. 3). Bedding in the lower sequence dips 40° to the northwest, although the ironstone sequence is overturned, on the limb of a major fold. A minimum thickness of 16 m is estimated for this ironstone.

In the lower sequence, the disturbed beds are of particular interest. They consist of a variety of rock types, including ooidal wacke-ironstones, cleaved pale grey mudstones, lenses of oncoidal wacke-ironstone and silty mudstones. The latter contain reworked and broken ooids and phosphate nodules, ripped-up rafts of chamosite mud, ripped up and deformed laminated silty mudstones and volcanic clasts, all suggesting a debris flow deposit. Throughout this lower sequence the ooids and phosphate nodules in the oolitic ironstones show strong evidence of being reworked and broken, some of the ooidal pack-ironstones take on a flaggy nature.

*Between Llanengan and Sarn Bach is the Hen dy Capel ironstone [SH 3002 2714]. Ask for permission of entry from Mr M. Roberts, Bodorwel, opposite the quarry.*

When compared with the Pen y Gaer ironstone, a marked facies change is apparent. Hen dy Capel exposes metre-thick lenses of reworked ironstone, containing chamositic stromatolite crusts and abundant oncoids, within slumped and deformed mudstones containing ripped-up sedimentary clasts (the 'crushed shales' of Nicholas 1915). These changes occur along strike toward a major north–south trending fault (Fig. 3), which has been shown to have been active in the Cambrian (Nicholas 1915). The ironstone does not occur on the other side of this fault.

## 4 Betws Garmon [SH 5318 5684 to 5453 5802]

Until recently, the Betws Garmon ironstone and adjacent sediments were thought to be of Cambrian age. However, Reedman *et al.* (1983) have shown that the sequence is of Ordovician age (the Nant Ffrancon Formation) and acritarchs collected from the ironstone (Trythall *et al.* 1987) indicate a Lower or Middle Llandeilo age. The base of the Ordovician here is marked by the Arenig Graianog Sandstone. Trythall *et al.* (1987) have suggested that the presence of a

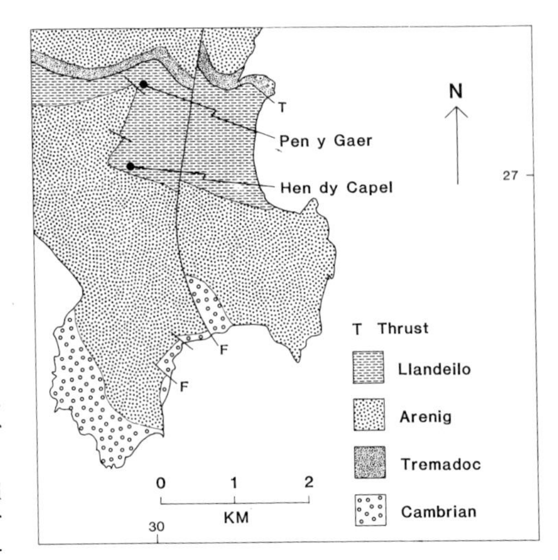


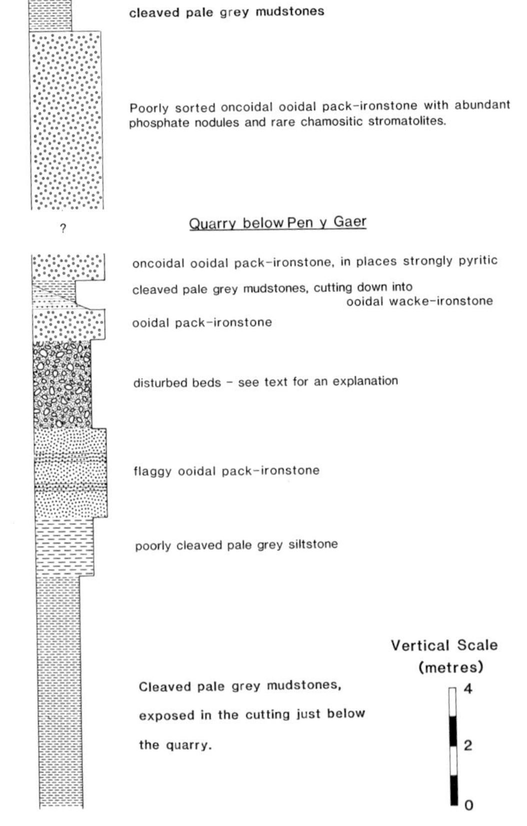


FIG. 3. Geology of the St Tudwals peninsula (after Crimes 1969) and vertical section through the ironstone exposures on Pen y Gaer. The lower section is exposed on the side of Pen y Gaer hill on the northeast face of the quarry, the upper section on the top of Pen y Gaer.

Llandeilo ironstone above Arenig sediments indicate the presence of a stratigraphic hiatus beneath the ironstone.

*On the A4085, 350m southeast of Betws Garmon village, park in the lay-by on the left-hand side next to Ystrad Farm and ask there for permission to study the ironstone mine, the adits of which can clearly be seen running up the side of the hill.*

The principal feature of interest at Betws Garmon is the sedimentary thickening of the upper ooidal pack-ironstone toward the north-east while towards the southwest the ironstone thins out entirely. The lower mud-supported units of the ironstone are lithologically variable, but always grade up into the ooidal pack-ironstone, which is well sorted and contains no oncoids. The top of the ironstone is abrupt and marked either by cleaved mudstones, oncolitic mudstones or intensely weathered shales (because of their high sulphide content).

However the ironstone has subsequently been replaced by magnetite, but only in the pack-ironstone, in the form of magnetite ooids in a chamosite matrix. Magnetite replacement does not occur in the ironstone to the south of the A4085 road. Subsequently the ironstone has also undergone hydrothermal alteration, indicated by quartz, siderite, hematite and pyrite veins, the pyrite veins now strongly weathered.

Proceed to locality 1 (Fig. 4) just beyond Ystrad Farm. The bedding here strikes 256° and dips 44° E. A full ironstone sequence is exposed, being abruptly succeeded by strongly weathered mudstones. Follow along the track and climb the hill up the old ore truck ramp to the second level to the right (locality 2; Fig. 4), where bedding is vertical. Pyrite replacement is much stronger at this locality, both in the ooidal pack-ironstone and the underlying muddy units. Continue up the ramp to the level just above the old winding house (locality 3; Fig. 4) where the bedding dips steeply to the south-east. Here it can be seen that the ooidal pack-ironstone has increased in thickness from 2 m to 4 m. Further up the hill the bedding becomes less steep, to about 45° SE, and the ooidal pack-ironstone continues to thicken to 5–6 m.

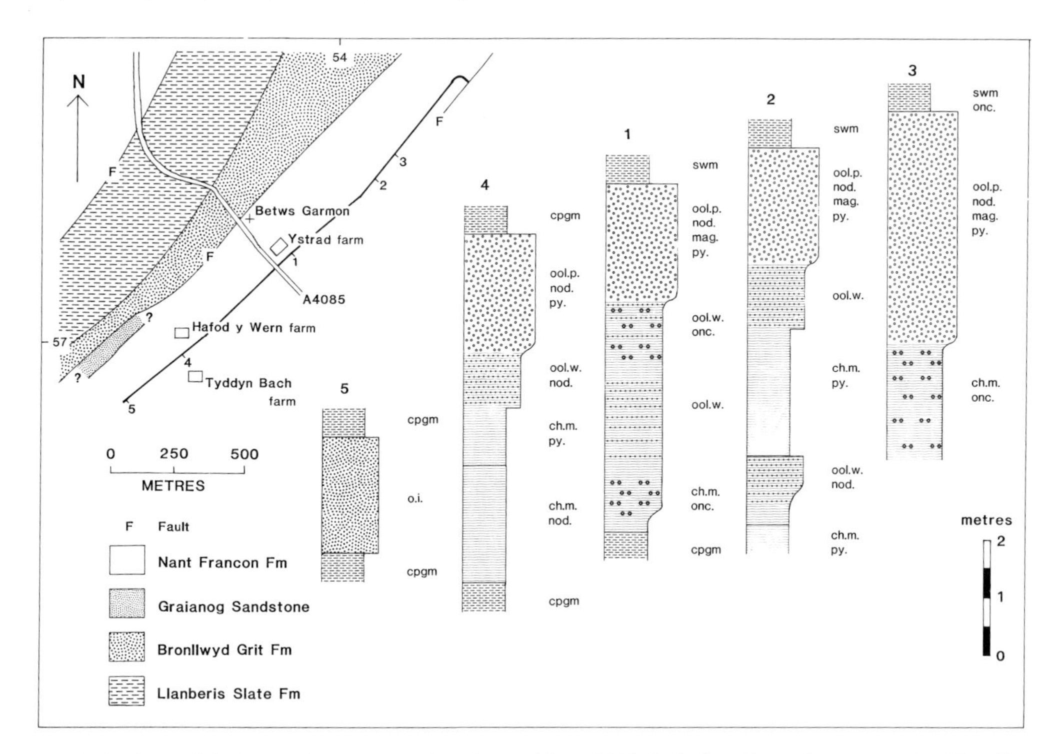


FIG. 4. Geology of the Betws Garmon area, based on BGS 1:10 000 field slips (formation names after Howells *et al.* 1985) and vertical sections exposed along the Betws Garmon ironstone. Captions: cpgm, cleaved pale grey mudstones; ch.m., chamositic mud-ironstones; ool.w., oolitic wacke-ironstones; ool.p, oolitic pack-ironstones; o.i., oolitic ironstone; swm, strongly weathered mudstones; nod., nodules; onc., oncoids; mag., magnetite; py., pyrite.

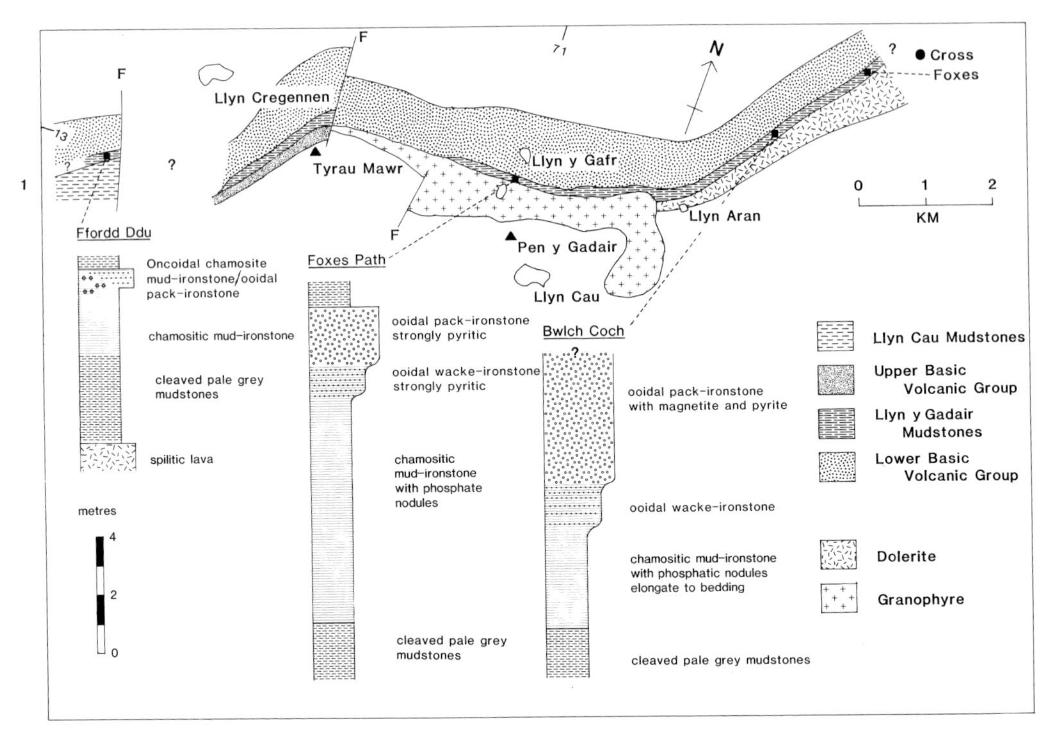


FIG. 5. Geology of the Cadair Idris range (after Cox 1925), showing the outcrop of the Llyn y Gadair mudstones, which contain the ironstone, and vertical sections through the ironstone across the range.

*Take the A4085 back through Betws Garmon village and turn left immediately after the bridge along the track to Hafod y Wern and Tyddyn Bach Farms. Obtain permission for access from Ystrad Farm.*

Locality 4 (Fig. 4) is just south of the track between the two farms. Bedding here is vertical. The ooidal pack-ironstone contains no magnetite and is strongly sideritic due to hydrothermal alteration. Both chamosite matrix and ooids are replaced by siderite associated with pyrite and minor other sulphides, similar to the Aber ironstone [SH 6702 7250]. Locality 5 (Fig. 4), 200 m further up the hillside, is the last ironstone exposure and further up the hillside the ironstone cannot be traced.

**5 Tremadog** [SH 5527 4028]

In the Tremadog region an ironstone occurs intermittently along strike for 3 km as a series of small workings from Tyddyn Deucwm [SH 5442 4099] to Ynys Galch [SH 5680 3929]. It is probable that the ironstone at Rhyd [SH 6285 4093] is at the same horizon, occuring 8 km to the east of the principal iron mine at Tremadog, which is of basal Caradoc age (Trythall *et al.* 1987).

In this region Cambrian sediments outcrop to the south of the ironstone and it was originally thought (Fearnsides 1910) that major thrusting had brought Caradoc sediments, with the ironstone near the base, over Cambrian sediments. However, Smith (1987) has shown that no thrusting has occurred and that the tectonic features noted by Fearnsides (1910) can be explained by slumping of semi-lithified sediment induced by tectonic disturbance.

*Park in the road leading up to the primary school off the A487 at the west end of Tremadog. Carry on along the track past the school and the mine is on the left.*

The general dip of the rocks in this region is 20° – 30° to the northwest. The lowest rocks in the sequence are exposed on the southern side of the Tremadog mine. They comprise cleaved pale grey mudstones that show small-scale slump folding, overlain by 'disturbed' ooidal wacke-ironstones. The latter contain ripped up sedimentary clasts and lenses of reworked phosphate nodules, which are rounded and elongate parallel to bedding. Oncoids and chamositic stromatolites are common in the facies and are also seen 250 m away at Pensyflog [SH 5619 3958].

The upper ooidal pack-ironstone, now virtually mined out, has been broken and rotated into 'pods' during slumping, with mudstones injected in between the pods. This ooidal pack-ironstone has been extensively replaced by magnetite, with some pyrite. Large quartz and dolomite veins, with strongly weathered veins of pyrite, cut the mudstones between the 'pods'.

**6 Ffordd Ddu** [SH 6477 1283] **and 7 Cross Foxes** [SH7597 1639]

The geology of the Cadair Idris area has been outlined by Cox (1925) and Ridgway (1976). The age of the ironstone is uncertain as no acritarchs were found (Trythall *et al.* 1987) but is thought to be of Llandeilo age (Cox 1925). The ironstone outcrops on the northern slopes of Cadair Idris within a sequence of cleaved pale grey mudstones (the Llyn y Gadair Mustones) which overlie a thick sequence of spilitic pillow lavas and ashes (Fig. 5).

The ironstone varies markedly in thickness across the Cadair Idris range (Fig. 5), thickest in the east where no oncoids occur and average ooid size is small (Cross Foxes and Bwlch Coch) and thinnest in the west (Ffordd Ddu) where oncoids and detrital quartz are present. This is complemented by the Llyn y Gadair Mudstones which thin eastwards toward Cross Foxes (Davies 1956). Further west the ironstone is absent until the Llanegryn Fault where on its west side some 50 ft of oolitic and nodular beds occur (Jones 1933). Trythall *et al.* (1987) have suggested that syndepositional movement on the Llanegryn Fault controlled ironstone deposition.

*From the centre of Dolgellau, drive west and take the first marked left turn signposted to Cadair Idris. Follow this road for 10 km along to the end, where a cross roads of farm tracks occurs. Park here and walk up to the left and the Ffordd Ddu ironstone is 450 m along on the right.*

The base of the sequence at Ffordd Ddu is exposed on the hill above the pit, where cleaved pale grey mudstones rest on spilitic lavas (Fig. 5). Bedding strikes 068° and dips 36°S. Above these mudstones occur oncoidal chamositic mud-ironstones with detrital (2–3 mm) quartz grains, lateral equivalents of the ooidal wacke-ironstones exposed within the pit (Fig. 5). Due to a small fault parallel to the strike, the spilitic lavas are repeated just above the ironstone pit, and the majority of the exposure within the pit is of cleaved pale grey mudstones, gradually darkening upward into chamositic mud-ironstones. Thin beds of ooidal wacke-ironstone outcrop at the top of the sequence on the west side.

*To reach the Foxes Path exposure [SH 7104 1378], climb the footpath opposite the Lake Gwernan Hotel [SH 704159] toward the summit of Cadair Idris. The ironstone exposures lie to the left of the path between Llyn y Gafr and Llyn y Gadair.*

Here, thick nodular chamositic mud-ironstones occur below a sulphide-rich ooidal pack-ironstone, now strongly weathered (Fig. 5). This alteration is due to the intrusion of the granophyre above the ironstone.

*To reach the Cross Foxes locality take the A470 from Dolgellau toward Welshpool and Machynlleth. Where the A487 joins the A470 by the Cross Foxes Hotel, follow the road opposite to the hotel. Take the first left hand turn 1 km along the road and follow this lane for 800 m until you reach the ironstone. Bwlch Coch [SH 7495 1557] is 1.5 km toward the southwest from the Cross Foxes ironstone.*

The section for Bwlch Coch, where a more complete section is exposed, is given in Fig. 5. However, the more accessible Cross Foxes exposures adequately illustrate the sedimentary differences when compared to the other Cadair Idris ironstones. At Cross Foxes, the bedding dips at 50–70° toward the southeast. Strongly weathered shales just above spilitic lavas lie below approximately 7 m of fine-grained ooidal pack-ironstone (average ooid size 0.25 mm). Oncoids and detrital quartz do not occur. Northeast of Cross Foxes the ironstone abruptly disappears, probably due to the influence of other contemporaneous faults present in the area (Fitches & Campbell 1987).

In the eastern part of Cadair Idris a dolerite sill has metasomatized adjacent sediments (Davies 1956) including the Cross Foxes and Bwlch Coch ironstones. The oolitic pack-ironstone has been partially replaced by magnetite, stilpnomelane and pyrite, with magnetite replacing the chamositic matrix.

ACKNOWLEDGEMENTS: This work was undertaken while in receipt of a Bedfordshire County Council research studentship. I would like to thank S. Molyneux and M. Smith for their help in this project, and T. Reedman and C. Eccles for comments on an early draft of this guide.

## References

BATES, D.E.B. 1972. The stratigraphy of the Ordovician rocks of Anglesey. *Geological Journal,* **8,** 29–58.

BECKLEY, A.J., 1987. Basin development in North Wales during the Arenig. *Geological Journal Thematic Issue,* **22,** 19–30.

BEVINS, R.E. & ROWBOTHAM, G., 1983. Low grade metamorphism within the Welsh sector of the Paratectonic Caledonides. *Geological Journal,* **18** 141–167.

COWARD, M.P. & SIDDANS, A.W.B., 1979. The tectonic evolution of the Welsh Caledonides. *In:* HARRIS, A.L., HOLLAND C.H. & LEAKE, B.E. (eds) *The Caledonides of the British Isles — reviewed.* Geological Society, London, Special Publication **8,** 187–198.

COX, A.H. 1925. The geology of the Cader Idris range (Merioneth). *Quarterly Journal of the Geological Society of London,* **81,** 539–592.

CRIMES, T.P. 1969. *The stratigraphy, structure and sedimentology of some of the Pre-Cambrian and Cambro-Ordovician rocks bordering the Southern Irish Sea.* PhD thesis, University of Liverpool.

DAVIES, R.G. 1956. The Pen y Gader dolerite and its metasomatic effect on the Llyn y Gader sediments. *Geological Magazine,* **93,** 153–172.

FEARNSIDES, W.G. 1910. The Tremadoc slates and associated beds of south east Carnarvonshire. *Quarterly Journal of the Geological Society of London,* **66,** 142–188.

FITCHES, W.R. & CAMPBELL S.D.G. 1987. Tectonic evolution of the Bala Lineament in the Welsh Basin. *Geological Journal Thematic Issue,* **22,** 131–153.

GREENLY, E. 1919. *The Geology of Anglesey.* Memoirs of the Geological Survey of Great Britain, Vol. 2.

HOWELLS, M.F., REEDMAN, A.J. & LEVERIDGE, B.E. 1985. *Geology of the country around Bangor.* Explanation of the 1:50 000 sheet, British Geological Survey, sheet 106, England and Wales.

JONES, B. 1933. The geology of the Fairborne – Llwyngwril district, Merioneth. *Quarterly Journal of the Geological Society of London,* **89,** 145–171.

KOKELAAR, B.P., HOWELLS, M.F., BEVINS, R.E., ROACH, R.A. & DUNKLEY, P.N. 1984. The Ordovician marginal basin of Wales. *In:* KOKELAAR, B.P. & HOWELLS, M.F. (eds.) *Marginal Basin Geology: volcanic and associated sedimentary and tectonic processes in modern and ancient marginal basins.* Geological Society, London, Special Publication, **16,** 245–269.

NICHOLAS, T.C. 1915. Geology of the St Tudwals peninsula. *Quarterly Journal of the Geological Society of London,* **71,** 83–143.

PULFREY, W. 1933. The iron ore oolites and pisolites of North Wales. *Quarterly Journal of the Geological Society of London,* **89,** 401–430.

REEDMAN, A.J., WEBB, B.C., ADDISON, R., LYNAS, B.D.T., LEVERIDGE, B.E. & HOWELLS, M.F. 1983. The Cambrian–Ordovician boundary between Abert and Betws Garmon, Gwynedd, North Wales. *Report of the Institute of Geological Sciences,* 83/1, 7–10.

RIDGEWAY, J. 1976. Ordovician palaeogeography of the southern and eastern flanks of the Harlech Dome, Merionethshire, North Wales. *Geological Journal,* **11,** 121–136.

ROBERTS, B. 1979. *Geology of Snowdonia and Llyn: an outline and field guide.* Adam Hilger, Bristol.

——& MERRIMAN, R.J. 1985. The distinction between Caledonian burial and regional metamorphism in metapelites from North Wales: an analysis of isocryst patterns. *Journal of the Geological Society, London,* **142,** 615–624.

SMITH, M. 1987. The Tremadoc 'thrust' zone in southern Central Snowdonia. *Geological Journal Thematic Issue,* **22,** 119–129.

STRAHAN, A., GIBSON, W., CANTRIL, T.C., SHERLOCK, R.L. & DEWEY, H. 1920. *Pre Carboniferous and Carboniferous bedded ores of England and Wales.* Memoirs Geological Survey — Special report on the mineral resources of Great Britain, Volume 13.

TREMLETT, W.E. 1962. The geology of the Nefyn-Llanaelhaiarn area of North Wales. *Liverpool & Manchester Geological Journal,* **3,** 157–176.

TRYTHALL, R.J.B. 1988. *The Mid-Ordovician oolitic ironstones of North Wales.* PhD thesis, CNAA, Luton CHE.

——ECCLES, C. MOLYNEUX, S.G. & TAYLOR, W.E.G. 1987. Age and controls of ironstone deposition (Ordovician) North Wales. *Geological Journal Thematic Issue,* **22,** 31–43.

WEINBERG, R.M. 1973. *The petrology and geochemistry of the Cambro-Ordovician ironstones of North Wales.* DPhil thesis, University of Oxford.

R.J.B. TRYTHALL, Faculty of Applied Sciences, Luton College of Higher Education, Park Square, Luton, Beds. LU1 3JU, UK.

# The origin of the Lower Jurassic Cleveland Ironstone Formation of North-East England: evidence from portable gamma-ray spectrometry

K.J. Myers

SUMMARY: The distribution of K, U and Th in the Cleveland Ironstone Formation, measured by gamma-ray spectrometry, is found to be useful in investigating the origin of oolitic ironstones. The oolitic ironstone seams are enriched in Th and depleted in K relative to the interbedded mudstones which show a close correlation between Th and K. The enrichment of Th in the ironstone seams represents a primary depositional rather than a diagenetic process, and the Th-enriched material must have been transported to the site of ironstone deposition from elsewhere. The values of Th (20–45 ppm) and the Th/K ratios (20–100) recorded in the ironstone seams are consistent with the original Th-enriched material being kaolinitic/bauxitic clay from a lateritic weathering source.

A portable gamma-ray spectrometer has been used to measure the potassium, uranium and thorium contents of the Lower Jurassic Cleveland Ironstone Formation in Yorkshire. Portable gamma-ray spectrometry is a rapid method of measuring the concentrations of the three main naturally occurring radioelements, K, U and Th, on field exposures (Løvborg 1971; Myers and Wignall 1987). The work presented here represents part of a wider project to investigate the applications of field gamma-ray spectrometry in sedimentological studies (Myers 1987).

Measurements were taken through the whole of the Pleinsbachian Cleveland Ironstone Formation (Howard 1985) exposed on wave-cut platforms at Staithes, North-East Yorkshire (Figs 1 and 2). The formation consists dominantly of shale, argillaceous siltstone and silty sandstone with interbedded thin seams of siderite and berthierine oolitic ironstone. The prominent sedimentary cyclicity shown by the formation has been described in detail by Rawson *et al.* (1982) and Howard (1985) who defined a type-1 cyclicity consisting of a coarsening of upward siliciclastic sequence capped by a thin oolitic ironstone seam and a type-2 cyclicity which is similar to type 1 but which is capped by a non-oolitic sideritic mudstone. Oolitic ironstone deposition is terminated diachronously across the basin (Howard 1985). At Staithes the top of the *apyrenum* subzone marks the end of ironstone deposition and the overlying shales of the *hawskerense* subzone are lithologically and chemically similar to the lower Toarcian Whitby Mudstone Formation (Catt *et al.* 1971). Some 20 km inland to the west at Great Ayton (Fig. 2), however, ironstone deposition continued to the end of the *hawskerense* subzone (Howard 1985).

There was little or no change in the provenance of the detrital constituents of the Cleveland Ironstone Formation with time. They were derived from a low-lying Pennine land area to the north-west composed mainly of Upper Palaeozoic sediments (Chowns 1966; Catt *et al.* 1971; Howard 1984).

Thorium is a particularly useful trace element to study in ironstones because of its chemical immobility, which means it is unlikely to be greatly affected by diagenetic processes and so its distribution will reflect primary processes of formation. Thorium is also known to be enriched in bauxites and other products of lateritic weathering (Adams & Richardson 1960) and so could act as a tracer element for lateritic weathering in the sourcelands of oolitic ironstones.

The idea of lateritic weathering providing kaolinite coated by iron hydroxides as a source for oolitic ironstones has been long discussed (e.g., Hallam 1966, 1975; Van Houten & Bhattacharyya 1982; Van Houten & Purucker 1984). Carrol (1958) showed that iron oxides can be transported by kaolinite, and Bhattacharyya (1983) and Velde (1985) have shown that berthierine is an early diagenetic alteration of kaolinite. Catt *et al.* (1971) proposed lateritic weathering as the source of the Cleveland Ironstone. The purpose of this paper is to provide independent evidence for this idea.

*From* YOUNG, T. P. & TAYLOR, W. E. G. (eds), 1989, *Phanerozoic Ironstones*
Geological Society Special Publication No. 46, pp. 221-228

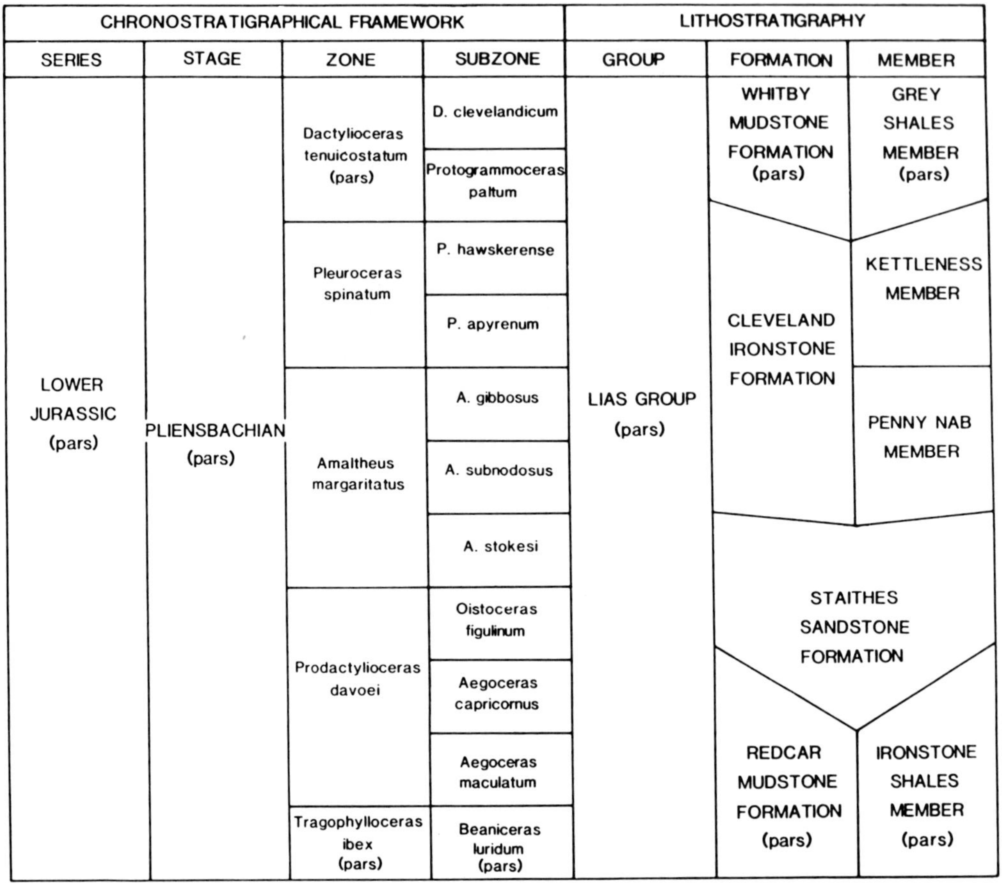

| CHRONOSTRATIGRAPHICAL FRAMEWORK | | | | LITHOSTRATIGRAPHY | | |
|---|---|---|---|---|---|---|
| SERIES | STAGE | ZONE | SUBZONE | GROUP | FORMATION | MEMBER |
| | | Dactylioceras tenuicostatum (pars) | D. clevelandicum | | WHITBY MUDSTONE FORMATION (pars) | GREY SHALES MEMBER (pars) |
| | | | Protogrammoceras paltum | | | |
| | | Pleuroceras spinatum | P. hawskerense | | | KETTLENESS MEMBER |
| | | | P. apyrenum | | CLEVELAND IRONSTONE FORMATION | |
| LOWER JURASSIC (pars) | PLIENSBACHIAN (pars) | Amaltheus margaritatus | A. gibbosus | LIAS GROUP (pars) | | PENNY NAB MEMBER |
| | | | A. subnodosus | | | |
| | | | A. stokesi | | | |
| | | Prodactylioceras davoei | Oistoceras figulinum | | STAITHES SANDSTONE FORMATION | |
| | | | Aegoceras capricornus | | | |
| | | | Aegoceras maculatum | | REDCAR MUDSTONE FORMATION (pars) | IRONSTONE SHALES MEMBER (pars) |
| | | Tragophylloceras ibex (pars) | Beaniceras luridum (pars) | | | |

FIG. 1. Stratigraphic framework of the Cleveland Ironstone Formation (from Howard (1985).

## Field techniques

Per cent K, ppm U, ppm Th and total gamma radioactivity were measured in the field with a standard Geometrics G410A portable gamma-ray spectrometer, (see Myers (1987) and Myers & Wignall (1987) for details of the field procedure). Measurements were taken on the broad wave-cut platforms to the south-east of Staithes harbour (NZ 794 183), which are ideal for portable gamma-ray spectrometry. Measurements were spaced at approximately every 50 cm of sediment thickness, and more often where there was more frequent lithological variation, as in the Pecten ironstone seam *apyrenum* subzone. The measurement counting time varied from 2–6 minutes depending on the radioactivity of the rock. This was found to maintain a precision of better than ± 10% for each radio-element (Myers 1987). Radio-element contents and ratios are plotted against lithology in Figs 3 and 4.

## K, U and Th geochemistry in fine-grained marine sediments

Potassium tends to be carried in the sedimentary cycle as a major component of detrital minerals, such as K-feldspar and micas and as a component of soil clays, particularly illite. Some K is also carried in solution (Heier & Billings 1972).

Thorium exists in a single valence state, $Th^{4+}$ and is considered to be effectively insoluble in natural waters (Langmuir & Herman 1980), due to a combination of slow solution rates, paucity and insolubility of Th-bearing minerals and strong adsorption of Th by colloidal materials. Consequently, Th is transported as a component of detrital resistate minerals such as monazite, rutile and thorianite or adsorbed onto natural colloidal-sized materials, particularly clays and oxyhydroxides (Langmuir & Herman 1980). The amount of dissolved Th in seawater is about 0.00064 ppm (Langmuir & Herman 1980) with a very low residence time of less than 100 years (Cochraine *et al.* 1986). Neither K nor Th substitute in the carbonate lattice, so pure carbonates contain negligible amounts of K and Th.

Unlike Th, U can exist in more than one valence state, the most important being $U^{4+}$ and $U^{6+}$. The ionic radius of $U^{4+}$ is very similar to that of $Th^{4+}$ and the two forms behave similarly, being transported in resistate minerals and adsorbed onto clays. However, $U^{4+}$ is readily oxidized to $U^{6+}$ in the surficial environment, and can form the soluble $UO_2^{2+}$ uranyl ion. The differential mobility of U and Th in oxidizing environments means that the Th/U ratio can be a useful environment indicator (Adams & Weaver 1958).

## Results

Profiles of total radioactivity, %K, ppmU and ppm Th are plotted against lithology in Fig. 3. K and Th are closely correlated ($r$ = 0.84, $n$ = 35) in the muds and silts between the ironstones but show no correlation in the ironstone seams themselves. Th is enriched and K depleted in each of the oolitic ironstones relative to the mudstones above and below. The highest Th values were recorded in the Pecten and Main Seam ironstones of the *apyrenum* subzone. The saw-tooth pattern of K and Th concentrations in the Pecten seam (Fig. 3) contrast with the steady Th/K ratio shown in Fig. 4. The variation in absolute Th and K values can be explained by differential compaction. Early siderite cementation in the hard bands of the Pecten seam preserves an initial porosity, thus diluting the measured values of Th and K whilst preserving the original Th/K ratios (T. P. Young, personal communication).

The Th/K ratios plotted in Fig. 4 show very clearly the differences between the mudstone and oolitic ironstone facies. The mudstones, siltstones and siderite mudstones (hereafter termed facies 1) have Th/K ratios in the range 6–8. The oolitic ironstones (facies 2) have Th/K ratios in the range of 20–90, though most lie between 20 and 40.

The mean Th/U ratio of facies 1 is 4.4 ± 0.7 within the range for marine and deltaic shales of 3.9 ± 1.1 (Myers & Wignall 1987). The mean Th/U ratio of facies 2 is 7.1 ± 2.3 with a range of 4.3–11.5.

The correlation between Th and K in facies-1 mudstones is examined in more detail in Fig. 5. The regression line of the Th–K crossplot has a gradient of 3.9 and a Th axis intercept of 4.6 ppm, i.e. at 0% there is 4.6 ppm in a non-K

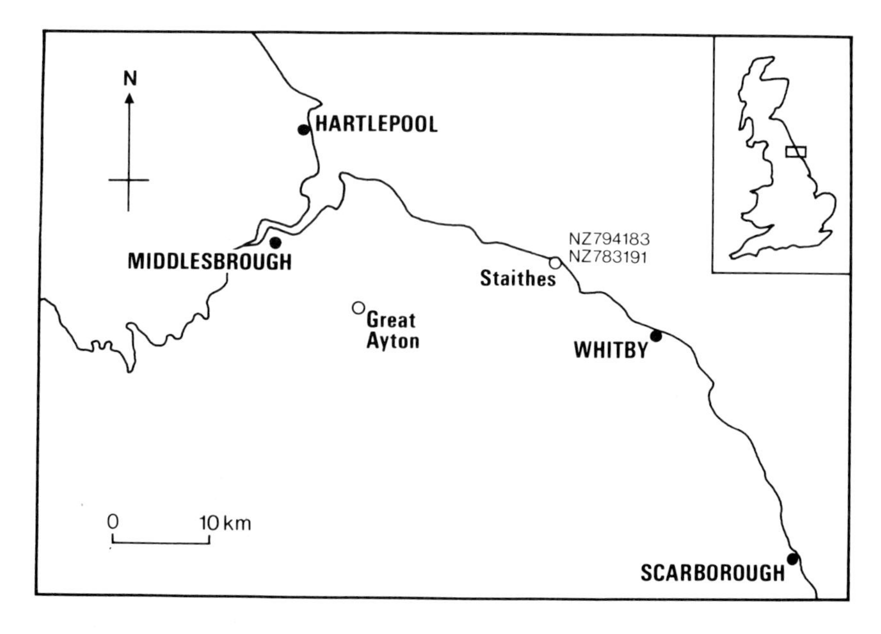


FIG. 2. Location of the Staithes section.

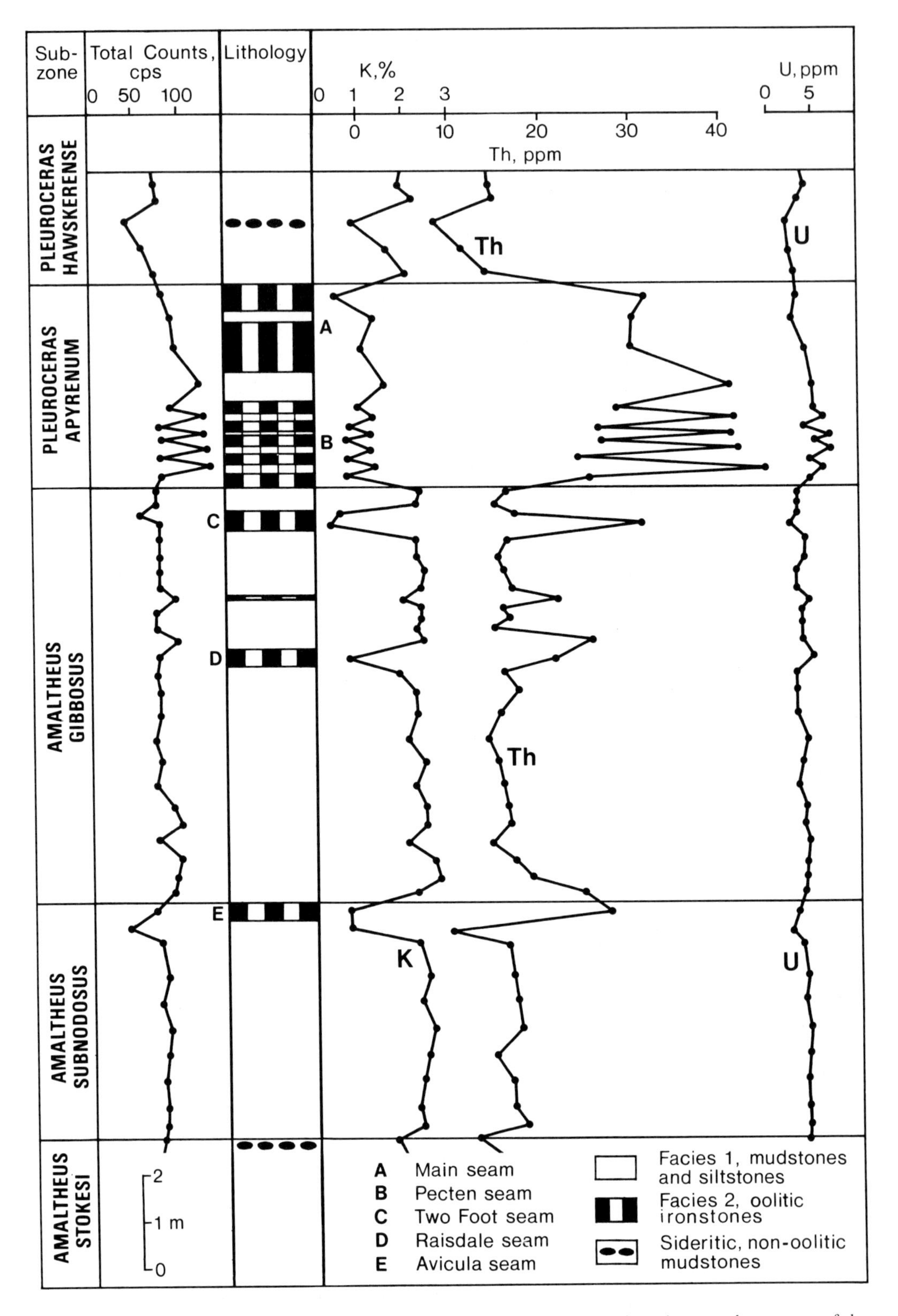


FIG. 3. Profiles of total gamma radiation, %K, ppm U and ppm Th measured on the coastal exposures of the Cleveland Ironstone Formation at Staithes, Yorkshire.

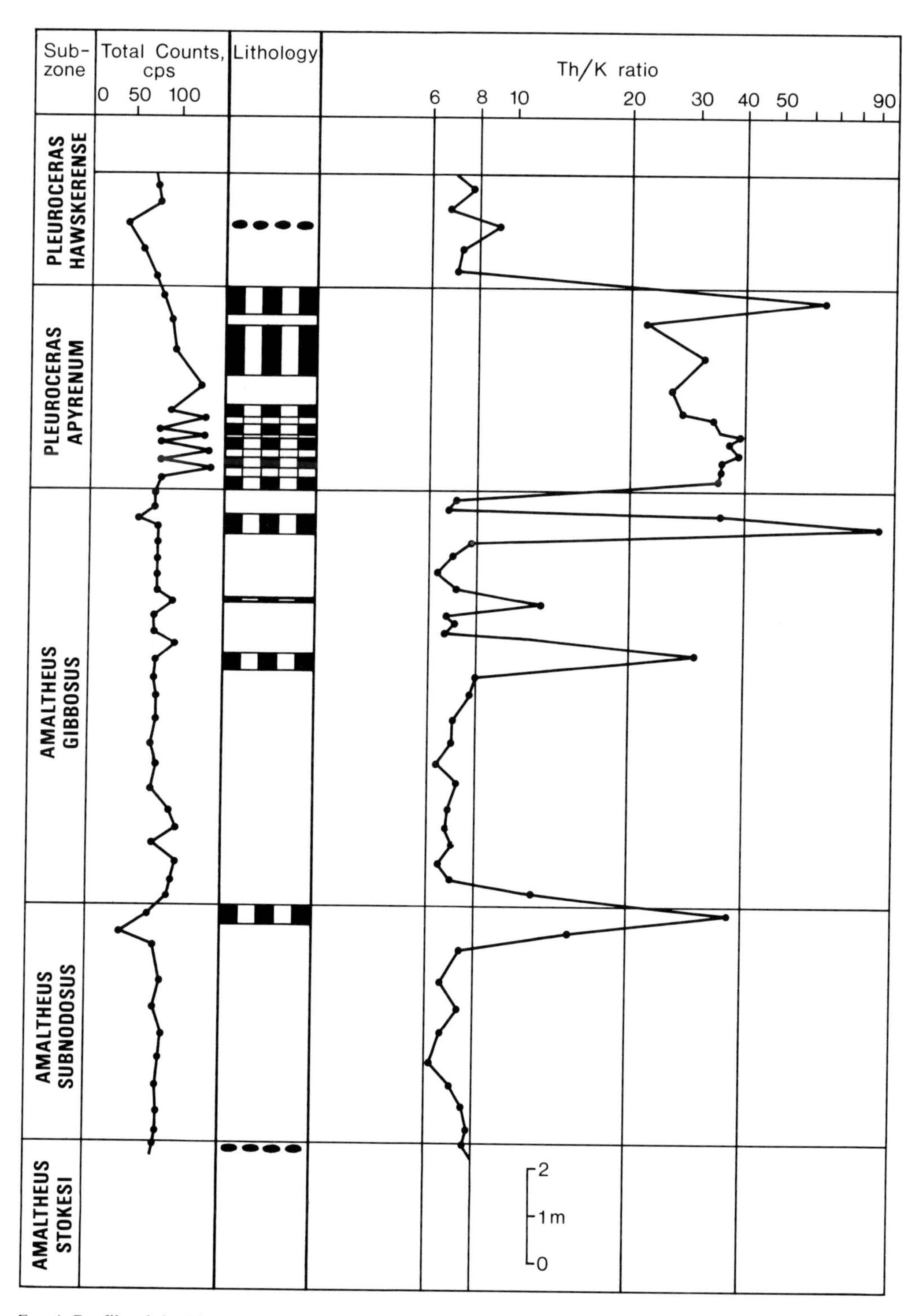


FIG. 4. Profile of the Th/K ratio through the Cleveland Ironstone Formation. Note the high Th/K ratios of the chamositic oolitic ironstones.

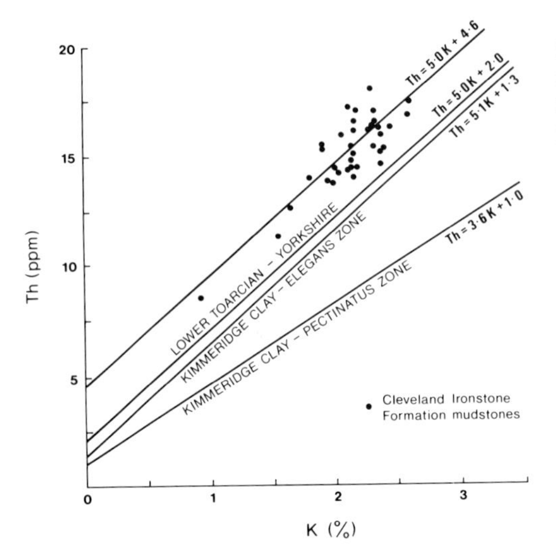


FIG. 5. Th–K cross plot of the facies-2 mudstones of the Cleveland Ironstone Formation is compared to those for other Jurassic mudstone bodies. Note the high intercept of the regression line on the Th axis. Correlation coefficients of K and Th are as follows: Cleveland Ironstone, $r = 0.84$ $n = 35$; Lower Toarcian, $r = 0.79$, $n = 62$; Kimmeridge Clay elegans zone, $r = 0.93$, $n = 31$; Kimmeridge Clay pectinatus zone, $r = 0.95$, $n = 35$.

bearing phase. This compares with a similar gradient but a much lower Th-axis intercept at 1.9 ppm in the overlying Lower Toarcian shales.

## Interpretation

The oolitic ironstones are enriched in Th and depleted in K relative to the interbedded mudstones. The Th/K ratios are similar in all the ironstones from the Avicula seam at the base, to the main seam at the top of the formation indicating that the same process of Th-enrichment occurred in all the ironstones. The fact that the Th/K ratio is unaffected by siderite cementation shows that siderite formation does not involve movement of Th and K. It also supports the view that the Th-enrichment is a primary depositional rather than diagenetic process. The Th-enrichment is unlikely to have occurred at the site of deposition because the ironstones have sharp and often erosional bases (Howard 1985) and the Th/K ratio does not increase in the sediments immediately underlying the ironstones. The ironstones, therefore, contain a transported detrital component which is enriched in Th (and depleted in K) relative to the mudstones with which they are associated.

The ironstones also have high Th/U ratios indicating that they are relatively depleted in U (or enriched in Th) compared to the interbedded mudstones. Most marine sediments have Th/U ratios in the range 3–5 (Adams & Weaver 1958, Myers 1987, Myers &Wignall 1987). Black shales are exceptions with low Th/U ratios (<2) where U has precipitated from solution under anoxic bottom water conditions (see e.g., Myers & Wignall 1987). The high Th/U ratios of the ironstones indicate highly oxidizing conditions under which U has been leached. It seems likely that the leaching of U occurred in the source-lands at the same time as the enrichment of Th and depletion of K.

## Discussion

A clue to the nature of the Th-enriched detrital component of the ironstones is present in the interbedded facies-1 mudstones. The high background level of Th associated with a non-K bearing phase shown in figure 5 can be explained by the presence of kaolinite. Catt *et al.* (1971) found that the detrital clay fraction of the middle Liassic mudstones contains up to 40% kaolinite which, with a Th content of 25–45 ppm (Herron 1986), could easily explain the excess Th. A direct correlation of Th and Al in various clay

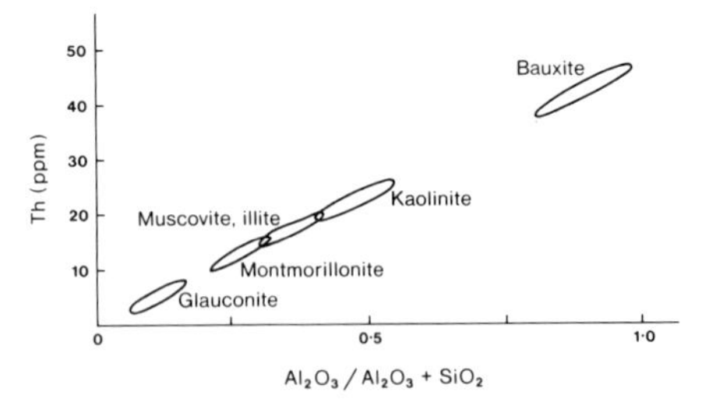


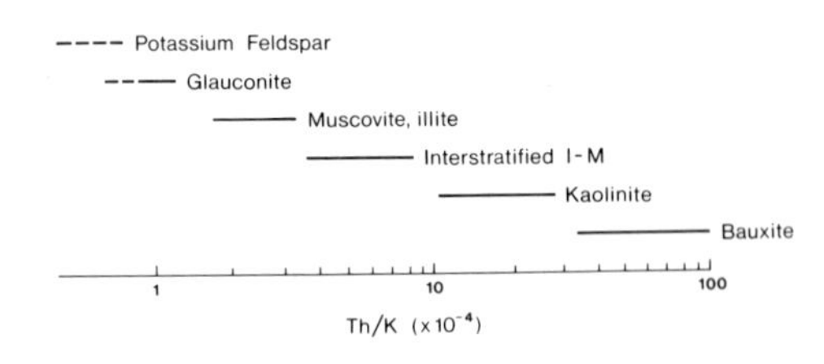


FIG. 6. Illustrating the range in Th/K ratios and the strong correlations between Th and Al in various clay minerals (after Hassan and Hossin (1975)).

minerals has been reported by Hassan & Hossin (1975) (Fig.6). This is due to the similar chemical properties of the two elements (Wedepohl 1970). The progression from illite–kaolinite–bauxite in soil days is caused by increased leaching which simultaneously enriches the clays in immobile elements such as Al and Th and depletes them in more mobile elements such as K and Si. The range of values for Th in kaolinite and bauxite/laterite (20–50ppm) is similar to those recorded in the ironstone seams. Also, $Al_2O_3/(Al_2O_3 + SiO_2)$ ratio quoted by Maynard (1986) for berthierine from the Cleveland Ironstone Formation of 0.46 is compatible with kaolinite.

The other elements which are enriched in the ironstones, namely Co, Cr, Ni, Pb, V and Zn (Catt *et al.* 1971) are known to be enriched, like Th, in residual clays (Wedepohl 1970).

In conclusion, there is strong circumstantial evidence that the ironstones originally contained concentrations of detrital kaolinitic/bauxitic material. This data agrees with models for chamositic ironstone formation which involve diagenetic alteration of detrital lateritic material.

More work is required on other ironstone sequences to confirm these trends and to study the mode of occurrence of Th in the ironstones before the full significance of this data for the origin of oolitic ironstones is understood.

# References


ADAMS, J.A.S. & RICHARDSON 1960. Thorium, uranium and zirconium concentrations in bauxite. *Economic Geology,* **55,** 1653–1675.

——& WEAVER, C.E. 1958. Thorium to uranium ratios as indicators of sedimentary process; example of the concept of geochemical facies. *Bulletin of the American Association of Petroleum Geologists,* **42,** 387–430.

BHATTACHARYYA, D.P. 1983. Origin of berthierine in ironstones. *Clays and Clay Minerals,* **31,** 173–182.

CARROLL, D. 1958. The role of clay minerals in the transportation of iron. *Geochimica et Cosmochimica Acta,* **14**

CATT, J.A, GAD, M.A., LERICHE, H.H. & LORD, A.R. 1971. Geochemistry, Micropalaeontology and origin of the Middle Lias ironstones in Northeast Yorkshire (Great Britain). *Chemical Geology,* **8,** 61–76.

CHOWNS, T.M. 1966. Depositional environment of the Cleveland Ironstone Series. *Nature,* **211,** 1286–1287.

COCHRAINE, J.K., CAREY, A.E., SCHOLKOVITZ, E.R. & SURPRENANT, L.D. 1986. The geochemistry of uranium and thorium in coastal marine sediments and pore waters. *Geochimica et Cosmochimica Acta,* **50,** 663–680.

HALLAM, A. 1966. Depositional environment of British Liassic ironstones considered in the context of their facies relationships. *Nature,* **209,** 1306–1309.

——1975. *Jurassic Environments.* Cambridge University Press, Cambridge.

HASSAN, M.N & HOSSIN, A. 1975. Contribution à l'étude des comportements du thorium et du potassium dans les roches sedimentaires. *Comptes Rendus Académie Science de Paris 28 Serie D,* 533–535.

HEIER, K. S. & BILLINGS, G. K. 1972. Potassium. *In*: WEDEPOHL, K. M. (ed) *Handbook of Geochemistry*. Vol 2. Springer Verlag, Berlin.

HERRON, M.M. 1986. Mineralogy from geochemical well logging. *Clays and Clay Minerals,* **34,** 203–213.

HOWARD, A.S. 1984. *Palaeoecology, sedimentology and depositional environments of the Middle Lias of North Yorkshire.* Unpublished PhD thesis, University of London.

——1985. Lithostratigraphy of the Staithes Sandstone and Cleveland Ironstone Formations (Lower Jurassic) of north-east Yorkshire. *Proceedings of the Yorkshire Geological Society,* **45,** 261–275.

LANGMUIR, D. & HERMAN, J.S. 1980. The mobility of thorium in natural waters at low temperatures. *Geochimica et Cosmochimica Acta* **44,** 1753–1766.

LOVBORG, L. 1971. Field determination of uranium and thorium by gamma ray spectometry, exemplified by measurements on the Ilimausaq alkaline intrusion, South Greenland. *Economic Geology,* **66,** 368–384.

MAYNARD, J.B. 1986. Geochemistry of Oolitic Iron ores an electron microprobe study. *Economic Geology,* **81,** 1473–1483.

MYERS, K.J. 1987. *Onshore-outcrop gamma ray spectrometry as a tool in sedimentological studies.* Unpublished PhD thesis, University of London.

——& WIGNALL, P.B. 1987. Understanding Jurassic organic-rich mudrocks — new concepts using gamma ray spectrometry and palaeoecology. *In:* LEGGETT, J.K. (ed) *Marine Clastic Sedimentology: New Developments and Concepts,* Graham & Trotman, London.

RAWSON, P.F., GREENSMITH, J.T. & SHALABY, S.E. 1982. Coarsening upward cycles in the uppermost Staithes and Cleveland Ironstone Formations (Lower Jurassic) of the Yorkshire Coast, England. *Proceedings of the Geological Association,* **94,** 91–93.

VAN HOUTEN, F.B. & BHATTACHARYYA, D.P. 1982. Phanerozoic oolitic ironstones — geological record and facies model. *Annual Review of Earth and Planetary Sciences,* **10,** 441–457.

——& PURUCKER, M.E. 1984. Glauconite peloids and chamositic ooids — Favourable factors, constraints and problems. *Earth Science Reviews* **20**, 211–243.

VELDE, B. 1985. Clay minerals; a physio-chemical explanation of their occurrence. *Developments in sedimentology,* **40**, Elsevier.

WEDEPOHL, K. M. (ed) 1970. *Handbook of Geochemistry*, Vol 1, Springer Verlag, Berlin.

K.J. MYERS, Department of Geology, Imperial College, London, UK.
Present address: BP Sunbury Research Centre, Sunbury-on-Thames, Middlesex, UK.

# Ironstones in the Mesozoic passive margin sequence of the Tethys Himalaya (Zanskar, Northern India): sedimentology and metamorphism

E. Garzanti, R. Haas & F. Jadoul

SUMMARY: The sedimentary succession of the Zanskar continental terrace comprises Triassic (Quartzite Series), Jurassic (Ferruginous Oolite) and Cretaceous (Giumal Sandstone) shelf siliciclastic units, all of which contain either chamositic ironstones or glauconitic greensands associated with reworked phosphorites. Ironstones are found above major unconformities at the top of shoaling arenaceous sequences, and generally mark the rapid transition from shallow-marine sands to highstand offshore pelites or pelagic foraminiferal limestones. Petrographical and sedimentological features of condensed intervals indicate deposition by transgressive fronts. During sea-level rise, high-energy waves mixed ferruginous and phosphatic grains formed in the course of earlier starved stages at low sedimentation rates, with detritus from shoreline and paralic sources reworked during stepwise coastal retreat. The timing of ironstone deposition apparently coincides with break-up stages affecting the Indian continental margin.

Ironstones are also important metamorphic markers in the Tethyan sedimentary zone, which has undergone intense fold-thrust deformation at very low to low metamorphic grade during the Tertiary Himalayan orogeny. In the Triassic iron oolites of central Zanskar, the occurrence of stilpnomelane suggests upper anchimetamorphic conditions (*c.* 300° C). In the Cretaceous greensands, however, K-rich glauconite is only peripherally replaced by incipient stilpnomelane growth, pointing to metamorphic conditions comparable to lower prehnite-pumpellyite facies (*c.* 260° C).

The sedimentary succession deposited on the northern passive continental margin of peninsular India during the Mesozoic mainly consists of carbonates, with quartzose siliciclastic intervals occurring in the Late Triassic, Middle Jurassic and mid-Cretaceous. Each of these terrigenous units contains interbedded ironstones, which are often associated with major unconformities and thus, particularly in a highly deformed region, represent invaluable stratigraphic markers and powerful keys to the understanding of sedimentary history. Because of their peculiar chemical composition, the condensed intervals are also unique indicators of metamorphic grade in passive margin carbonate-quartzarenite successions. The aim of the present paper is to describe in detail the ironstones contained in the Tethys Himalaya sequence of Zanskar (Fig. 1), and to show their potential as stratigraphic, sedimentological and metamorphic markers.

## Methods

The chemical composition of minerals was analysed with an ARL electron microprobe at the University of Innsbruch (Institut für Mineralogie und Petrographie) using standard methods. Matrix effects were corrected with the empirical Bence-Albee method and improved α-factors (after Evans, unpublished data). All Fe was analysed as $Fe^{2+}$), since electron microprobe technique does not allow routine distinction of $Fe^{2+}$ and $Fe^{3+}$. An attached energy dispersive system was used for preliminary qualitative analysis. Data are not of good quality, since measurement conditions were often difficult due to the very small size (a few microns) and the close intergrowth of the analysed minerals. Cathodoluminescence observations were carried out with a Nuclide luminoscope ELM2b, with beam voltage of 13 kV and beam current of 5 mA. The mineral content of 26 rock samples was determined by powder X-ray diffraction techniques.

## Mesozoic ironstones

### Quartzite series

The Norian-Rhaetian Quartzite Series, 200 to 250 m thick, sharply overlies an earlier Triassic pelagic to shelfal limestone succession, capped by the Zozar peritidal carbonates (Gaetani *et al.* 1985a). The sharp base of the Quartzite Series, which marks the sudden increase of terrigenous detritus, is locally characterized by graded biocalcirudites with coral-bearing lithoclasts or by calcareous sandstones with reworked phosphatic nodules. The lower part of the formation

*From* YOUNG, T. P. & TAYLOR, W. E. G. (eds), 1989, *Phanerozoic Ironstones*
Geological Society Special Publication No. 46, pp. 229-244

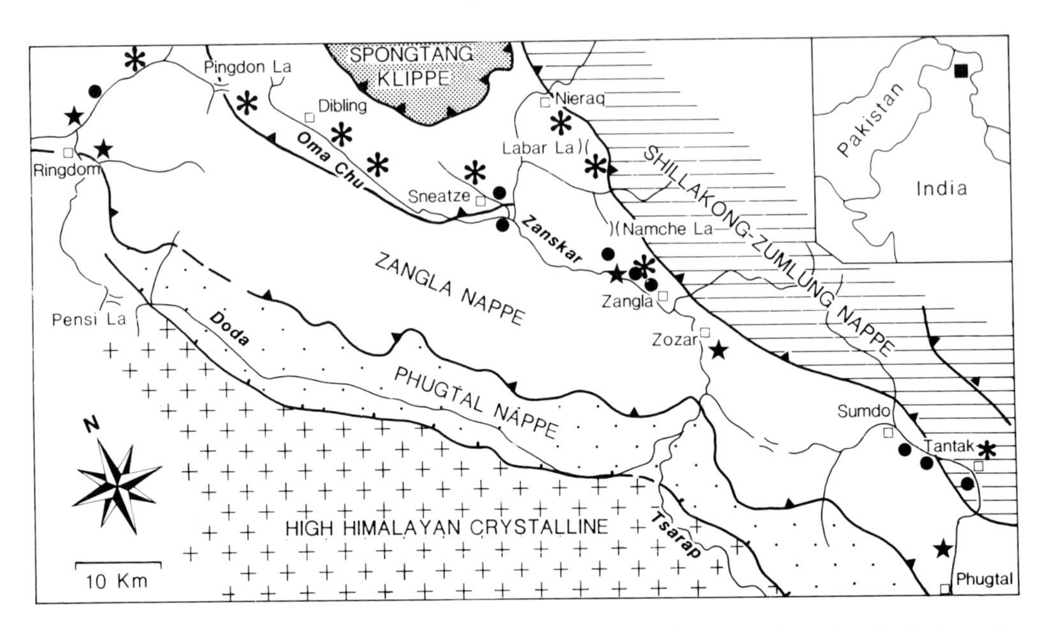


FIG. 1. Location and geologic sketch map of the studied area. Measured stratigraphic sections in the Quartzite Series (stars), Ferruginous Oolite Fm. (dots) and Giumal Sandstone (asterisks) are shown. The Zanskar synclinorium consists of several thrust sheets of Palaeozoic (Phugtal Nappe) to Mesozoic (Zangla and Shillakong-Zumlung units) Tethys Himalayan sediments, overlain by the Lingshed-Spongtang Klippe. (Gaetani *et al.* 1985b). The ironstones described in the text are virtually all from the Zangla Nappe, which contains the most proximal part of the Indian continental terrace preserved in the Zanskar Range. The Zangla Nappe is subdivided into two major thrust sheets by the Pingdon La – Oma Chu ramp-flat system (Gaetani *et al.* 1985b).

consists of lenticular biocalcarenites interbedded with very fine grained arkoses with common hummocky cross-stratification, passing upward to coarser subarkoses and locally to supermature quartzarenites with high-angle megaripple cross-bedding. Two ironstone-bearing horizons directly overlying coastal sands are found in the central part and at the top of this interval (Fig. 2). The middle part of the Quartzite Series consists of grey siltstones and burrowed biomicrites, followed by biocalcarenites interbedded with fine grained and well sorted subarkoses. At the top of the unit, thick beds of dolomitic limestones with megalodontids are associated with poorly sorted and up to medium grained quartzarenites with subrounded quartz grains and dolomite intraclasts. These layers mark the boundary with the overlying Kioto Limestone, a carbonate ramp deposit of Rhaetian to Early Dogger age.

### *Petrography of Late Triassic ironstones*

The poorly sorted bioclastic microrudites interbedded within the lower-middle Quartzite Series contain only a few isolated quartz grains (less than 10%), decreasing in maximum size from 2 mm in central Zanskar (Zangla and Zozar) to 250 μm in the more distal Phugtal section. Detrital quartz may show embayments and solution pits. Terrigenous clasts, up to some cm in size and commonly impregnated by hematite or Fe-dolomite, make up the bulk of the framework in the Zozar section. Lithoclasts consist of fine grained subarkoses with plagioclase and microcline and subordinate medium grained quartzarenites. In the Phugtal section they are less abundant, and represented by very fine grained sandstones, hybrid arenites, intra-bio-calcarenites and coarse siltstones rich in organic matter and iron oxides, which may represent either 'burrow-clasts' or pedogenic ferruginous microconcretions. Conversely, fossils become much more widespread and heterogeneous towards the east, with common echinoid plates and spines, brachiopods, pelecypods, gastropods and rarer crinoids, benthic forams (Involutinae), bryozoans or ammonites. Glaucony contained within mollusc shells or in veins and reworked phosphatic hardgrounds with borings are rare, but very well sorted iron ooids (generally 150 μm but up to 350 μm in size) may be very abundant, reaching 50% of the framework (Fig. 3a,b,c). The ooids are made of concentric laminae of chamosite (Table 1) and goethite intermixed in various proportions. Green ooids are mainly chamosite, whereas black opaque ooids consist mostly of goethite. Carbonate ooids occur only sporadically. Either detrital or biologically introduced matrix is lacking, and samples are extensively cemented by sparry calcite, often syntaxially overgrown on echinoderm fragments. Late diagenetic iron-rich calcite and siderite may also be abundant. Stilpnomelane, often in brown needles up to 200 μm long (Fig. 3d; Table 2), and migration of quartz grain boundaries is common in all stratigraphic sections.

**Ferruginous oolite formation**

The Callovian Ferruginous Oolite Fmn., which unconformably overlies the Kioto Limestone (Fig. 4), can be subdivided into four lithozones

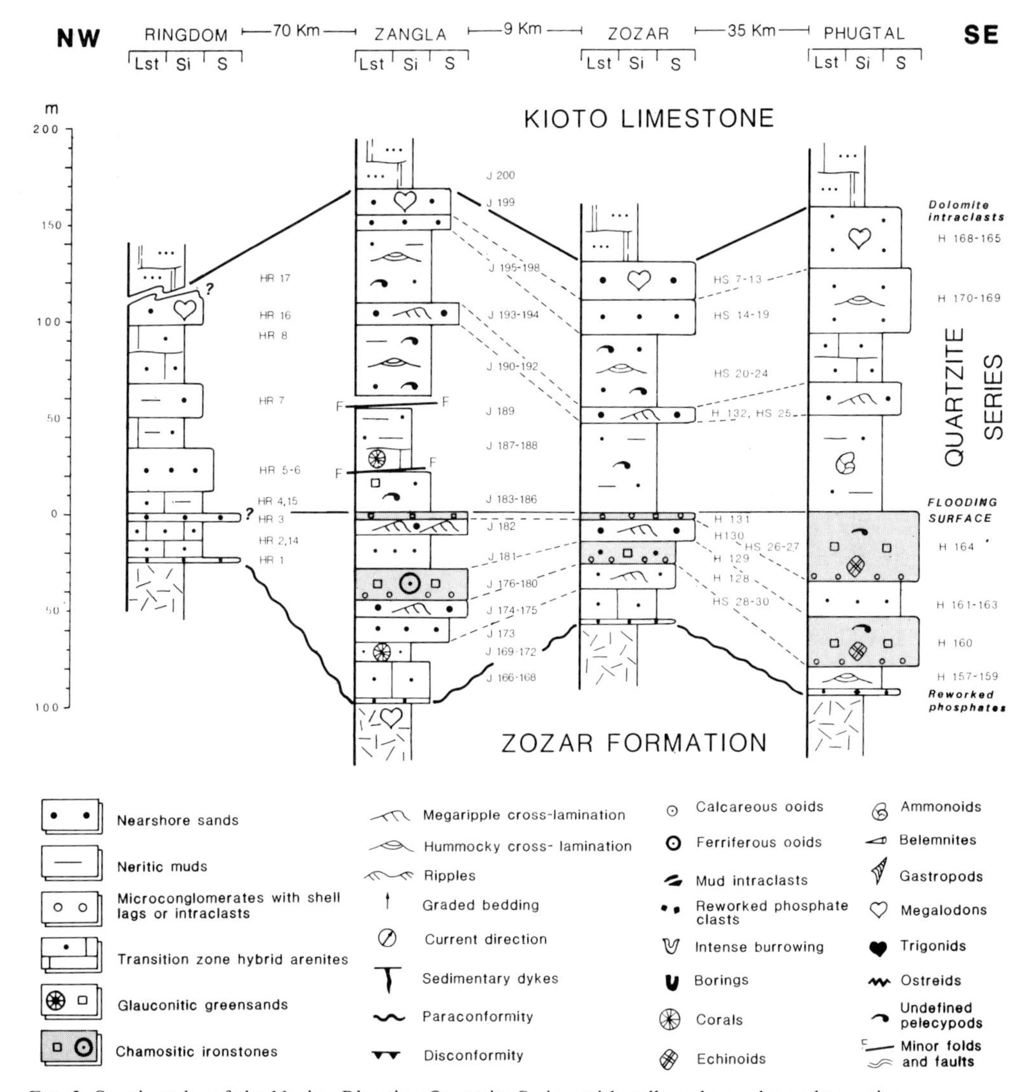


FIG. 2. Stratigraphy of the Norian-Rhaetian Quartzite Series, with collected samples and tentative lithostratigraphic correlations. Minor folding and faulting hamper precise reconstruction of the middle part of the Xangla section. The original stratigraphic thickness has been tectonically reduced in the overturned Zozar section, and particularly in the Ringdom section, where strong anchimetamorphic deformation prevents detailed sedimentological observations. Lst, limestone; Si, siltstone; S, sandstone.

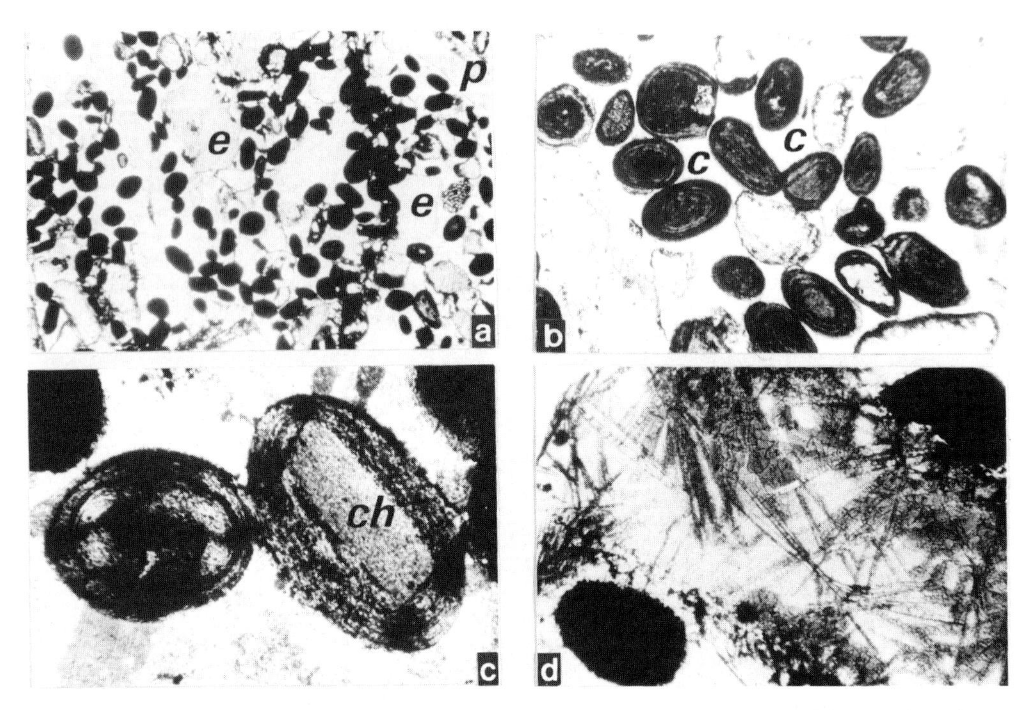


FIG. 3. Petrography of the Quartzite Series ironstones (J 177, Zangla section). (a) Bimodally sorted arenite with fine grained opaque iron ooids and coarser echinoid plates(e) and bivalves (p) (x 22, 1N). Ooids are calcite-cemented (c) and made of concentric laminae of chamosite and goethite often coating a chamositic core (ch) (b: x 68, 1N; c: x 174, 2N). (d) Stilpnomelane in well developed needles (x 174, 1N).

(Bassoullet *et al.* 1983, Jadoul *et al.* 1985). Above the basal unconformity, locally encrusted by large ostreids and locally showing borings or sedimentary dykes infilled by reddish sandy mud (Fig. 5a), lies an ironstone interval up to several metres thick (lithozone A), characterized by abundant belemnites, pelecypods or ammonites, comprising paraconglomeratic deposits with coral-bearing lithoclasts. These layers are sharply overlain by red or green pelites, passing gradually upward to fine grained and moderately sorted bioclastic quartzarenites showing graded bedding, parallel or hummocky lamination and megaripple cross-bedding at the top (lithozones B and C). The formation is capped by a widespread, 2 to 5 m thick, fossiliferous iron oolitic 'roofbed' with clear-cut lower and upper contacts and containing medium grained subarkoses with rare volcanic rock fragments (lithozone D). The Ferruginous Oolite Fmn. is followed by the Late Jurassic Spiti Shales and then by the mid-Cretaceous Giumal Sandstone.

*Petrography of Callovian chamositic ironstones.* The condensed intervals at the base and top of the Ferruginous Oolite Fmn. are characterized by the abundance of yellowish chamositic or darker chamositic/goethitic ooids locally oxidized to limonite (Fig. 5b,c). Iron ooids are spheroidal to ellipsoidal and have median diameter in the medium sand range (250 to 400 μm) and maximum diameter in the coarse sand range (600 to 1000 μm). In lithozones A and D, ooids are well sorted and cemented by calcite or ankerite (Table 3), whereas in lithozone B they are moderately sorted, finer-grained and dispersed in quartzose silt (median diameter 60 μm) or in a ferruginous groundmass (Fig. 5d). Quartz may be found at the core of, or dispersed within, the iron ooids, which may be locally replaced by carbonates or peripherally by stilpnomelane. The quartzose siliciclastic fraction is absent to common (up to 40% of the framework), and well sorted quartz grains (median diameter up to 230 μum and maximum diameter up to 700 μm) may occur together with larger Fe-ooids (type-5 textural inversion of Folk 1980; p.104). Phosphate clasts are sporadic. Large ostreid shells and brachiopods are abundant in the coarser and better washed samples of lithozones A and D. Instead, in the highly bioturbated siltstones impregnated by iron oxides or

TABLE 1. *Chemistry of chamosites from the Ferruginous Oolite Fm.*

| Chamosites | J 100 | J 104 | J 177 |
|---|---|---|---|
| $SiO_2$ | 22.6 | 21.93 | 23.42 |
| $TiO_2$ | 0.05 | 0.19 | 0.14 |
| $Al_2O_3$ | 20.37 | 19.92 | 17.97 |
| $Cr_2O_3$ | 0.11 | 0.04 | 0.10 |
| FeO[1] | 37.49 | 36.21 | 35.80 |
| MnO | 0.02 | 0.00 | 0.01 |
| MgO | 4.76 | 4.64 | 6.57 |
| CaO | 0.36 | 0.56 | 0.23 |
| $Na_2O$ | 0.01 | 0.07 | 0.04 |
| $K_2O$ | 0.00 | 0.00 | 0.20 |
| Total | 85.23 | 83.56 | 84.48 |

(J 100, lithozone D; J 104, lithozone B, Sneatze sections) and from the Quartzite Series (J 177) Zangla section. [1]Total iron as FeO. The given weight percentages of the constituent oxides result from the mean of 3 (J 104) to 8 (J 100 and J 177) selected electronmicroprobe analyses. Data are not of good quality due to the very small size of crystals; since the total is often low, stoichiometry was not calculated.

TABLE 2. *Chemistry of stilpnomelanes from the Quartzite Series ironstones.*

| Stilpnomelanes | J 177 |
|---|---|
| $SiO_2$ | 39.17 |
| $TiO_2$ | 0.01 |
| $Al_2O_3$ | 5.88 |
| $Cr_2O_3$ | 0.00 |
| FeO[1] | 31.53 |
| MnO | 0.05 |
| MgO | 3.40 |
| CaO | 0.89 |
| $Na_2O$ | 0.22 |
| $K_2O$ | 2.06 |
| Total | 83.21 |

(J 177; 3 mineral analyses averaged). [1]Total iron as FeO.

fluorapatite of lithozone B, calcareous and siliceous sponge spicules commonly occur, along with subordinate radiolarians, benthic forams and ostracods.

## Giumal sandstone

The up to 200 m thick lower part of the Giumal Sandstone consists of several coarsening-upward cycles. Cyclothems are up to a few tens of metres thick and composed of burrowed grey siltstones passing upward to fine grained subarkoses and very coarse quartzarenites. Black shale tongues with Late Aptian planktonic forams occur in the upper portion of this member (Gaetani *et al.* 1985a), which is overlain by 110–130 m of dark pelites with thin intercalations of very fine grained parallel-laminated sandstones. In the inner continental terrace (Dibling area), the 100 m thick upper Giumal starts with very coarse and cross-bedded quartzarenite bars, followed by dark pelites interbedded with immature volcanic arenites and belemnite-bearing microrudites. These layers pass upward with sharp contact to 10 m thick black glauconitic arenites. This laterally continuous ironstone interval, showing large-scale and high-angle tangential cross-lamination well displayed by aligned centimetric nodules of authigenic hematite, marks the top of the Giumal Sandstone in the outer continental

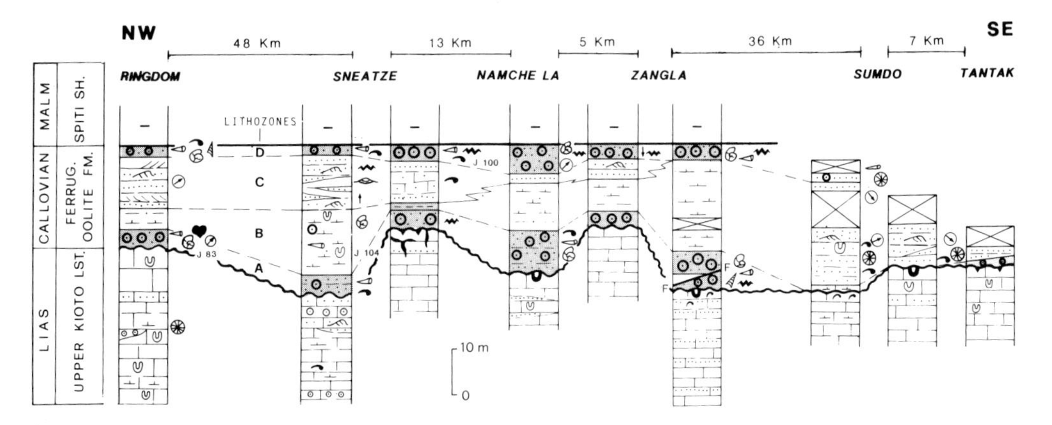


FIG. 4. Stratigraphy of the Callovian Ferruginous Oolite Formation, with tentative lateral correlation of A,B, C and D lithozones. Note that lithologies are diachronous and that the sediments overlying the unconformity become older from NW to SE. All stratigraphic sections were measured in the Zangla Nappe but for the easternmost Tantak section, which belongs to the Zumlung Nappe. See Fig. 2 for legend.

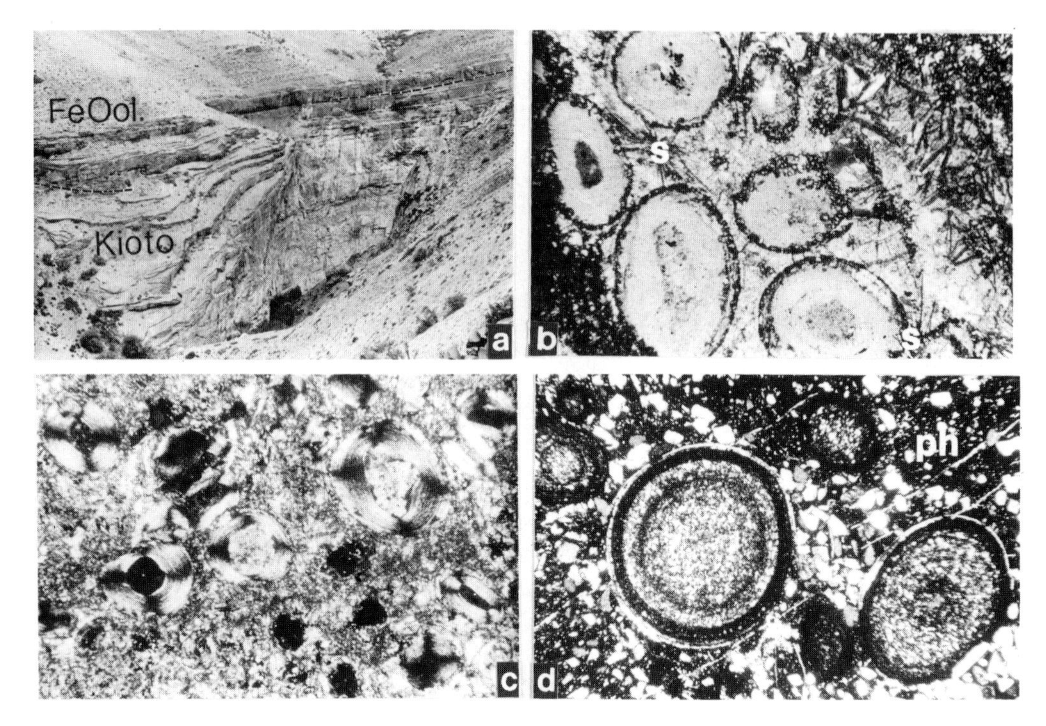


FIG. 5. Petrography of the Ferruginous Oolite Formation. (a) The Callovian F.O. Fm. onlaps onto an unconformity which corresponds to a time-gap of at least 10 Ma. (b) Chamositic ooids are calcite-cemented and peripherally corroded by stilpnomelane (s; J 83, lithozone A, Ringdom section; x 43, 1N) or scattered within a ferruginous-carbonatic groundmass (J 83; c: x 24, 2N). (d) Coarse chamositic-goethitic ooids with diffuse silica and interspersed among a quartzose silt impregnated by fluorapatite (ph; J 104, Lithozone B, Sneatze section; x 43, partly crossed Nicols).

terrace (Nieraq area; Fig. 6). The unconformable boundary with the overlying Fatu La pelagic limestones, corresponding to the transition between the Late Albian *B. breggiensis* and *R. appenninica* Biozones (M. Caron and I. Premoli Silva, personal communication), is overlain by micrites containing reworked phosphatic clasts in the Nieraq, Zumlung and Shillakong tectonic units (Baud *et al.* 1982, 1987; Bassoullet *et al.* 1983, Fig. 6). In the proximal Zanskar terrace, clastic sedimentation continues with belemnite-bearing quartzose arenites up into the Cenomanian. This interval comprises up to 2 m thick channelized rudites, with very abundant reworked phosphatic nodules yielding ammonites of Late Cenomanian age (*Protacanthoceras;* identification by A. Tintori), and marks the transition to early Turonian grey Globotruncana limestones of the Chikkim Formation (Fig. 7a).

*Petrography of Late Albian glauconitic greensands*

The upper Giumal glauconitic arenites are medium to coarse grained, well sorted and up to supermature in the Pingdon La section, coarse and poorly sorted at Sneatze (transition zone) and fine grained and moderately sorted in the Nieraq area. Detrital matrix is usually absent, but locally significant amounts were introduced by burrowing activity. The siliciclastic fraction generally consists of more than 90%, chiefly monocrystalline quartz, often highly rounded and locally displaying lobate outlines with embayments and solution pits. Feldspars are a minor component (up to 10%), often with perthitic orthoclase prevailing over microcline and plagioclase. Volcanic lithic fragments are subordinate in the Dibling area, but they become more abundant in the outer continental terrace, reaching up to 50% of terrigenous detritus at Sneatze. In the Pingdon La section, glauconitic peloids, up to 0.8 to 1 mm in size, commonly with phosphatized cores are widespread, constituting more than 50% of the framework at the base of the ironstone interval (Fig. 7b). The high K content (8-9 wt%; Table 4), the up to coarse sand size and the lobate to botryoidal shape suggest that the glaucony grains are highly evolved, and thus required

TABLE 3. *Selected electronmicroprobe analyses of carbonate cements from the Ferruginous Oolite Fmn.*

| Carbonates | H 39 | J 100 | J 104 |
|---|---|---|---|
| FeO | 1.58 | 0.53 | 0.21 |
| MnO | 0.22 | 0.03 | 0.00 |
| MgO | 0.30 | 0.68 | 0.26 |
| CaO | 54.67 | 55.23 | 55.45 |
| $CO_2$(calc.) | 44.34 | 44.42 | 43.93 |
| Total | 101.11 | 100.89 | 99.85 |
| Atoms per formula unit based on 6 oxygens, assuming stoichometric $CO_2$ | | | |
| $Fe^{2+}$ | 0.044 | 0.015 | 0.006 |
| Mn | 0.006 | 0.001 | — |
| Mg | 0.015 | 0.034 | 0.013 |
| Ca | 1.935 | 1.951 | 1.981 |
| C (calc.) | 2.000 | 2.000 | 2.000 |

(J 100, lithozone D; J 104, lithozone B, Sneatze sections) and from a topmost Giumal reworked phosphorite (H 39, Pingdon La). [1]Total iron as FeO.

considerable (up to a hundred thousand years) sedimentary breaks for their formation (Odin & Matter 1981).

Glauconitized volcanic detritus is particularly abundant in the Nieraq area. Fossils are absent, but for a few planktonic forams found in a burrow in the Nieraq section (*Ticinella* sp.; M. Caron, personal communication) and sporadic belemnites or 'fossiloid spars'. Iron ooids were never recorded, but phosphates and cherty fragments are often present in small amounts.

Syntaxial overgrowths on quartz grains occur in the coarse samples of the proximal Pingdon La section, and are welded to form patches of cement in the upper part of the ironstone, where detrital quartz also increases in abundance. Secondary hematitic nodules or pigment are very abundant in the Nieraq area, whereas pyrite is common in the Pingdon La section, where the rims of glauconitic peloids show incipient stilpnomelane growth. Green or brown stilpnomelane becomes abundant in the Nieraq area, in aggregates corroding detrital quartz or calcite. Stilpnomelane is associated with chlorite patches or abundant authigenic tourmaline in blue acicular and green prismatic crystals. Glauconite in highly cracked grains cemented by microquartz is preserved in the Zumlung Nappe at Tantak. Quartz has invariably undergone strong post-depositional deformation, with the development of undulose extinction and the incipient migration of grain boundaries where detrital grains are not separated by soft glaucony peloids.

*Petrography of Cenomanian phosphorites*

The topmost Giumal contains quartzose and bioclastic arenites, often very rich in reworked phosphatic nodules up to 5 cm in size. In these very coarse grained and commonly poorly sorted layers, washing is good and detrital matrix is absent, but intrabasinal pseudomatrix may occur due to the abundance of squashed clay intraclasts (Fig. 7c). Up to very coarse and highly rounded quartz grains form from 40% to 95% of the framework, and often show embayments or lobate outlines (Fig. 7c,d). Phosphate clasts are widespread (Fig. 7e,f), and consist of siltite/arenite fragments impregnated by a phosphatic 'matrix', phosphatized glauconite (glaucoapatite), bone material, ooids with concentric laminae, reworked early diagenetic nodules to fully developed hardgrounds with calcite-filled borings, or recycled burrow-fills ('burrow-clasts'). Belemnites are common, and ammonite phragmocones are locally found encased in phosphatic pebbles. Echinoid plates and planktonic forams become abundant at the transition with the Chikkim Limestone. Chamositic ooids are sporadic and mainly occur either within phosphatic clasts or as phosphate-cemented ooidal intraclasts. Reworked siliceous crusts, often phosphatized glauconite and glauconitized microlitic and felsitic volcanic rock fragments, are also commonly recorded.

Some layers show at least three generations of calcium carbonate cements. The early precipitation of discontinuous calcite fringes was followed by dissolution and cement collapse. The resulting, often oversized, intergranular pores were subsequently filled by two other generations of respectively bright orange-yellow and non-luminescent calcite (Fig. 8).

## Origin of ironstones and phosphorites

The Himalayan ironstones lie within siliciclastic units deposited in storm-dominated shallow-marine environments, as shown by the frequent association with hummocky cross-laminated very fine grained sands. Up to several metres thick ironstone layers, showing sharp or scoured bases and remarkable lateral continuity, are characterized by the following textural and mineralogical features:

(a) average and maximum grain size much coarser than in most adjacent terrigenous beds, commonly reaching respectively very coarse sand and pebble size;
(b) good to poor sorting; also common bimodal sorting owing to the occurrence of coarse

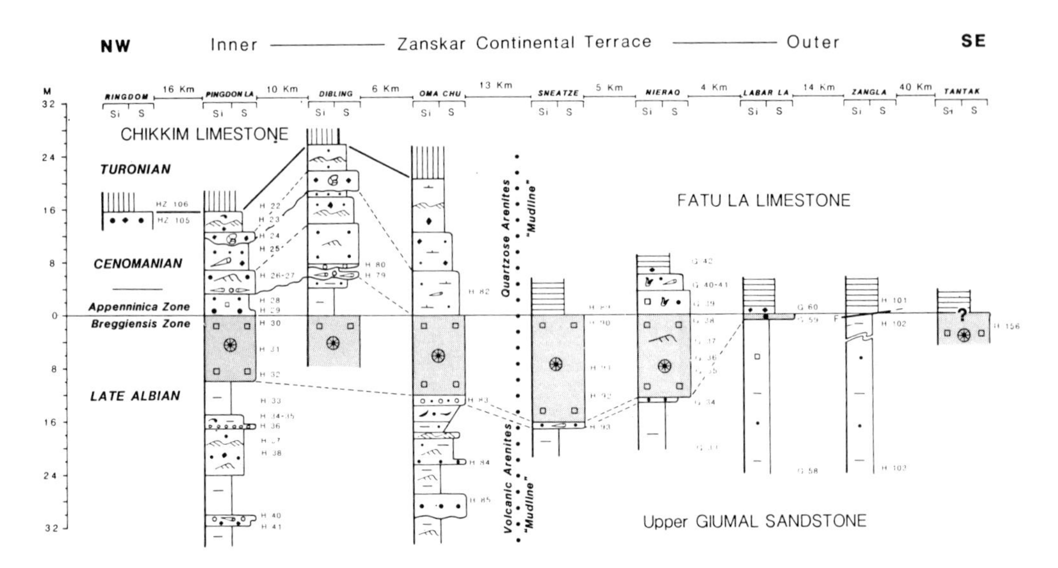


Fig. 6. Stratigraphy of the upper Giumal Sandstone. The transgression of pelagic carbonates onto the Zanskar shelf occurs through a sequence of flooding events, testified by a widespread glauconitic condensed interval in the Late Albian and by reworked phosphorites in the Cenomanian. The existence from the Albian onward of a lineament separating the continental terrace into a sandy and a muddy part (mudline) and the sharp petrographic changes from lower Giurnal subarkoses to upper Giumal volcanic arenites and to Cenomanian topmost Giumal quartzarenites point to significant tectonic control at this stage. Distance is along a projection line slightly oblique to paleo-depositional strike, which ran NNW–SSE. Datum horizon is dated micropaleontologically only in the Nieraq, Labar La and Zangla sections, whereas age control is poor in the inner continental terrace. Reconstruction of the Tantak section, which belongs to the Zumlung Nappe, is hypothetical. Si, silt; S, sand. See Fig. 2 for legend.

lithoclasts (Fig. 7e) or due to mixing of iron ooids with either a coarser-grained pebble/fossil hash (Fig. 3a) or finer-grained siliciclasts (Fig. 5d);

(c) good washing, with the absence of detrital matrix in non-burrowed samples and the abundance of primary cements in coarser arenites;

(d) occurrences of either chamosite/goethite ooids (Fig. 3b,c; 5b,c,d) or glauconitic peloids (Fig. 7b);

(f) occurrence of apatite, mainly as reworked nodules or impregnated silty to sandy clasts (Fig. 7 e,f):

(g) occurrence of iron oxides, either as pigment in arenite clasts, diagenetic hematite nodules or oxidized pyrite;

(h) occurrence of reworked terrigenous (siltite/arenite) lithoclasts;

(i) sporadic occurrence of cherty fragments and of authigenic tectosilicates within iron ooids or micritic intraclasts;

(l) occurrence of well-rounded quartz grains with corrosion pits and embayments, most likely due to solution in highly-weathered soil profiles (Fig. 7c,d; Cleary & Conolly 1972). A volcanic origin for most of the embayed quartz is unlikely, since polycrystalline embayed grains are also recorded and felsic volcanic detritus is very scanty in the whole Mesozoic succession.

These characteristics testify to:

(1) concentration of iron, silica and phosphorus on a shallow-marine continental shelf, with formation of sedimentary layers rich in phosphates and either iron ooids or glauconite during long intervals of minimal aggradation (McGhee & Bayer 1984; Gygi 1986);

(2) erosion of the shelf floor and mixing of material derived from neritic, shoreline and paralic sources, as shown by the occurrence of heterogeneous bioclasts, black pebbles and textural inversion phenomena (Goldbery 1979; Taira & Scholle 1979; Folk 1980; Flügel 1982, p.165);

(3) deposition in rapidly deepening shelfal environments, as shown by vertical transition to ammonite-bearing pelites or pelagic limestones.

The Zanskar ironstones mostly occur at the top of coarsening-upward sandy sequences and testify to relatively long periods of reduced influx of sediment during incipient transgressions (Van Houten & Purucker 1984). Shelfal environments were enriched in iron due to 'flushing' of Fe-rich paralic settings during lowstand or stillstand stages, and either glauconite pellets or chamosite ooids accumulated on starved depositional surfaces during subsequent sea-level rise (Mallinson & Otte 1986). The occurrence of soil material in many ironstone and phosphorite layers suggests that iron enrichment partly resulted from

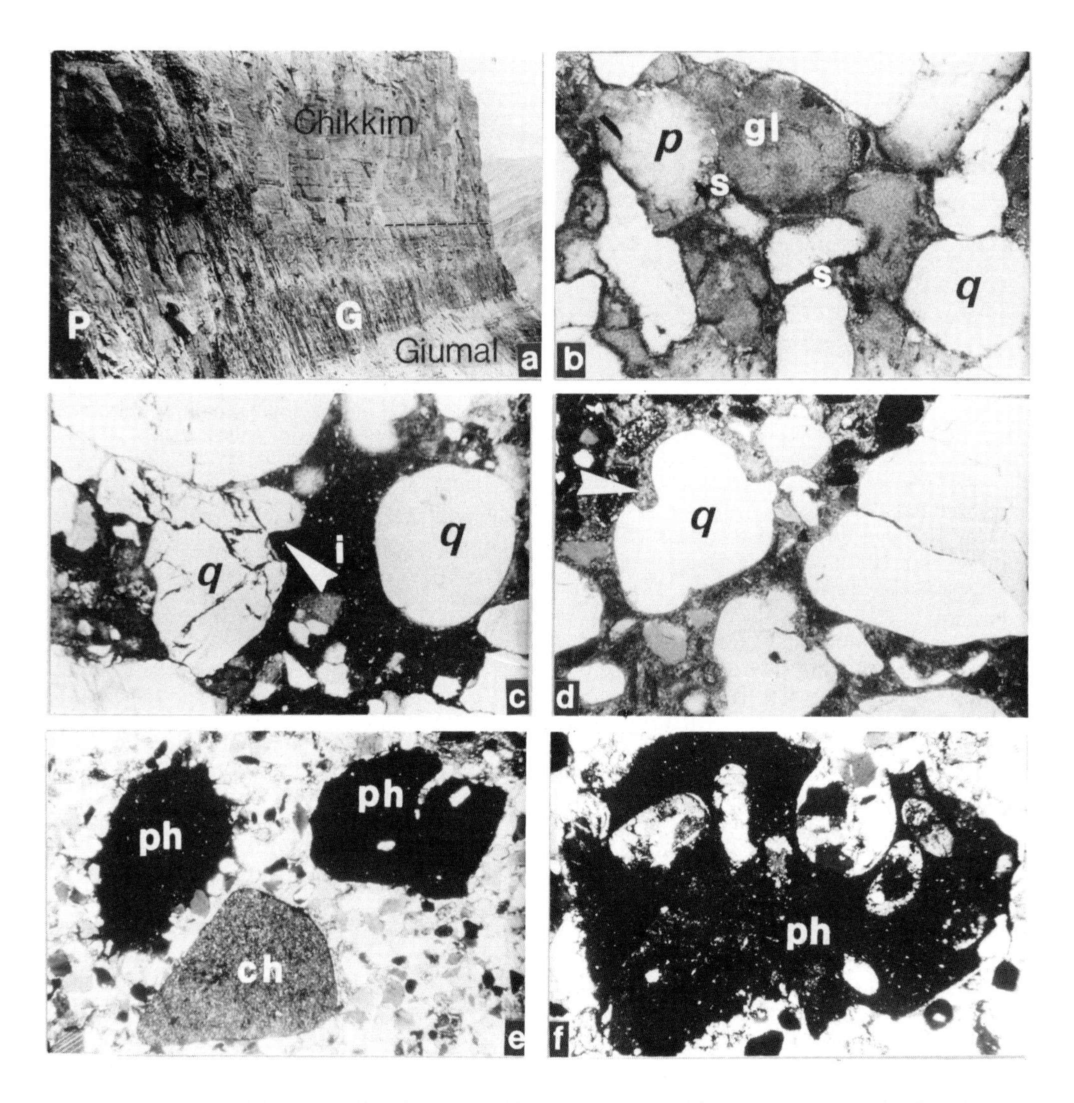


FIG. 7. Petrography of the upper Giumal condensed intervals. (a) Transition between the upper Giumal and the Chikkim Limestone in the Pingdon La section. G = 10 m thick glauconitic greensands (H 30–32); P = channelized reworked phosphorites with *Protacanthoceras* ammonites (H 24). (b) Coarse grained and well sorted glauconitic (gl) and quartzose (q) sands. Glaucony peloids show often phosphatized core (p) and are only peripherally replaced by incipient stilpnomelane (s) growth (H 31; x 57, 1N). the topmost Giumal arenites are characterized by well rounded and spherical quartz grains commonly showing solution pits, lobate outlines or embayments (arrows), and set in a muddy 'matrix' which may have been produced by squashing of mud intraclast (i) (H 79; c: x 27, 1N; d: x 22, 2N). (e) Bimodally sorted arenites with coarse cherty (ch) or phosphatic (ph) lithoclasts mixed with much finer quartzose grains (H 80; x 20, 2N). (f) Reworked phosphatic hardgrounds often show numerous borings (x 41, 2N).

TABLE 4. *Chemistry of glauconites from the uppermost Giurnal sandstone*

| Glauconites | H 39 | H 90 |
|---|---|---|
| $SiO_2$ | 46.01 | 42.35 |
| $TiO_2$ | 0.39 | 0.76 |
| $Al_2O_3$ | 13.51 | 13.53 |
| $Cr_2O_3$ | 0.10 | 0.11 |
| FeO[1] | 12.14 | 21.34 |
| MnO | 0.01 | 0.00 |
| MgO | 8.58 | 4.11 |
| CaO | 0.33 | 0.16 |
| $Na_2O$ | 0.03 | 0.06 |
| $K_2O$ | 8.97 | 8.09 |
| Total | 90.07 | 90.51 |

(H 90, Sneatze section; 3 analyses averaged) and from the Cenomanian reworked phosphorites (H 39, Pingdon La; 7 analyses averaged). [1]Total iron as FeO.

erosion of subtropical soils, with redeposition of ferruginous microconcretions and duricrust fragments in agitated marine environments. As the transgressive front advanced across the seafloor, ferruginous grains were mixed by high-energy storm waves with reworked phosphatic hardgrounds, coarse shell debris and arenite pebbles provided by stepwise shoreface retreat, and concentrated in accreting sandwaves.

The scenario envisaged for the Himalayan condensed intervals is thus one of rapid sea-level rise, when river mouths are flooded and turned into sediment-trapping estuaries, and siliciclastic detritus is mostly supplied by erosional coastal retreat and reworking of the shelf floor (Swift 1986). Ironstones were deposited at water depths of less than a few tens of metres, and generally above wave base (evidenced by well washed Triassic and Jurassic iron oolites or megaripple cross-bedded Cretaceous greensands). Iron ooids may be also found scattered in the overlying muddy sediments, which were deposited in deeper and quieter waters (spiculitic ironstones of the Ferruginous Oolite Fm., lithozone B). The abrupt and often scoured bases of the ironstones and reworked phosphorites are interpreted as ravinement surfaces, whereas the upper contacts represent flooding surfaces, marking the transition from coastal sands to deeper-water highstand muds.

## Ironstones and passive margin history

The Zanskar sedimentary succession began in the Late Permian, at the end of the rifting phase which led to the opening of Neotethys (Gaetani *et al.* 1985a; Garzanti 1986). The basal marine transgression is marked by abundant glauconite and phosphates contained in the bioclastic sublitharenites of the neritic Kuling Formation (Nicora *et al.* 1984).

### Triassic

During Scythian and Anisian times, sedimentation of the Tamba-Kurkur nodular limestones in a pelagic environment testifies to rapid thermotectonic subsidence immediately after break-up. In the Ladinian and Carnian, increasing fine terrigenous yield led to deposition of a thick shallowing-upward marly succession (Hanse Fmn.), followed by the Zozar peritidal carbonates. The first Mesozoic clastic episode recorded by the Zanskar passive margin is in the Norian, when the Quartzite Series was deposited in storm-dominated shallow-water environments. The ironstone-bearing intervals contained in the lower-middle part of the formation separate an underlying shoaling sequence, capped by cross-bedded sandstones, from deeper-water pelites (Fig. 2), and testify to reworking of older littoral deposits and to sediment starvation during rapid sea-level rise. The upper part of the Quartzite Series is a regressive megasequence, with neritic muds passing gradually upward to sublittoral arenites and megalodon limestones.

### Jurassic

Subtidal carbonate sedimentation resumed in the latest Triassic, and continued until the death of the Kioto platform in the Early Dogger (Baud *et al.* 1982; Jadoul *et al.* 1985). This event is marked by a major unconformity, pointing to generalized shelf erosion during a pre-Callovian lowstand. The occurrence of huge carbonate olistoliths of Norian to Early Dogger age in the Lamayuru continental rise succession (Bassoullet *et al.* 1981) may testify to slope instability phenomena at this stage, with gigantic megaslumps triggered by the collapse of the outer continental terrace. The disconformity is progressively onlapped by the Ferruginous Oolite Fmn. during the Callovian (Fig. 4). The formation is characterized by fossiliferous ironstone intervals at the base and top, enclosing a coarsening-upward shale-sandstone sequence deposited on a storm-controlled inner shelf. Several depositional sequences may be represented, as shown by sharp vertical sedimentologic and petrographic changes (Garzanti 1986). The chamositic ironstones are interpreted

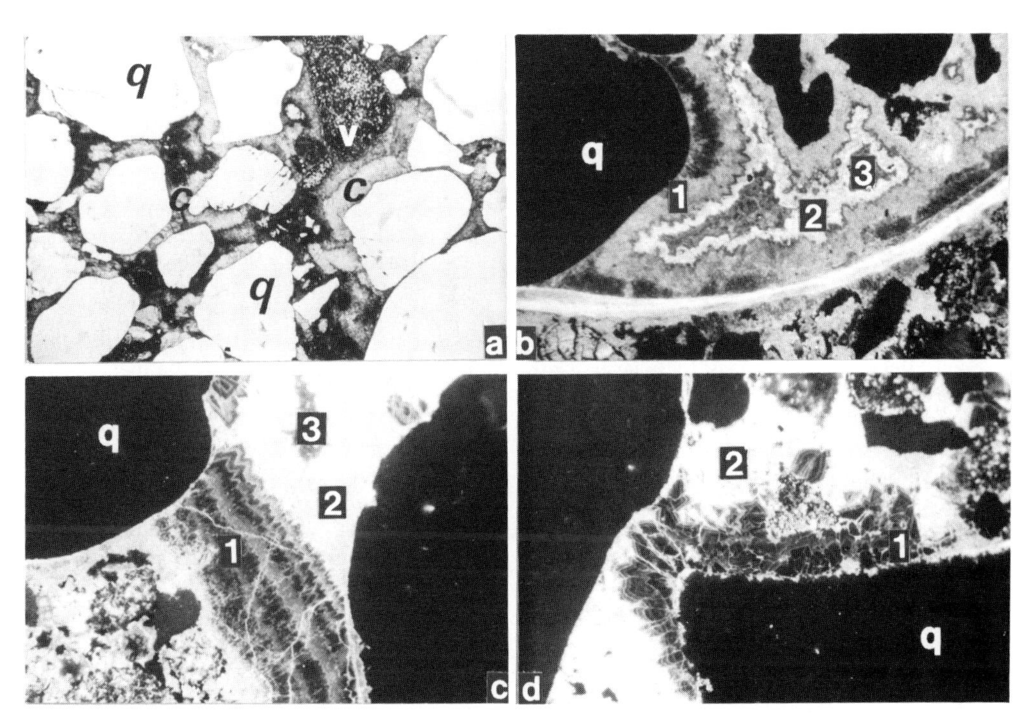


FIG. 8. Multiphase carbonate cements in the topmost Giumal Sandstone (H 79). (a) Microstalactite-like calcite cements (c) in discontinuous fringes around quartz (q) and volcanic (v) grains (x 18, 'N). Cathodoluminescence observations allow to recognize at least three generations of calcium carbonate cements (all CL photos x 48). (b) A first generation of dull bladed calcite (1) is followed by bright luminescent calcite (2) and then by non-luminescent pore-filling spar (3). (c) Alternating non-luminescent and dull isoriented fringes of phase (1) tend to cement detrital quartz grains (q) and show microfractures filled by bright calcite of phase (2). (d) Detail of early cements (1), which are fractured and corroded by later bright calcite probably formed in more reducing phreatic marine environments (2).

as transgressive deposits associated with unconformities and characterized by low sedimentation rates, as testified by the condensed ammonite fauna (Emery 1968; Sellwood 1978). Transgressive conditions are consistent with the relatively coarse grain size and poor sorting of these sands, deposited in high-energy open shelf environments (Abbott 1985). The deposition of the Ferruginous Oolite Formation coincides with the global Callovian sea-level rise (Hallam 1981; Vail & Todd 1981), linked to a major kinematic reorganization in the Tethyan domain (Dercourt *et al.* 1986; Savostin *et al.* 1986). Next, in the Late Jurassic, the slowly subsiding Zanskar margin was characterized by widespread pelitic sedimentation.

## Cretaceous

Lowermost Cretaceous glauconitic ironstones are commercially exploited in the Salt Range equivalent of the Upper Spiti Shale (Hallam & Maynard 1987). In the sedimentary succession of the Zangla Nappe, however, only the Jurassic lower part of the Spiti Shale is documented, and the Neocomian is possibly missing (Gaetani *et al.*, 1985a). Probably only in the Aptian does the Giumal Sandstone record the multiphase progradation of clastic detritus into neritic depositional environments.

In the Late Albian, ironstones rich in glauconite and hematite mark the sharp transition to the pelagic Fatu La Limestone in the outer continental terrace. If the abundance of iron oxides at Nieraq is not a late weathering feature, it might suggest that glauconitic greensands have been exposed and oxidized shortly after deposition, during development of a type-1 unconformity (975 Ma, Haq *et al.* 1987). Warm and humid climates with mature soils formed on the Indian coastal plains and the probably still active basaltic volcanism (Gaetani *et al.* 1985a) favoured the concentration of iron minerals on the stranded

Zanskar shelf, during one of the most favourable periods in earth history for glauconitic sedimentation (Odin & Matter 1981; Van Houten & Purucker 1984).

In the Cenomanian, clastic sedimentation becomes restricted to more proximal areas because of progressive flooding of the shelf. During the stepwise advance of the transgressive front, repeated interruptions permitted increased burrowing and development of hematitized or phosphatized crusts, which were subsequently eroded and deposited as reworked phosphorites. The complex polyphase early cementation shown by some of these layers, with an alternation of phreatic marine to possibly vadose precipitation and dissolution-collapse episodes, is also ascribed to sea-level fluctuations and the development of unconformities (Harris *et al.* 1986). Eventually, siliciclastic detritus disappeared and pelagic oozes encroached onto the whole Zanskar continental terrace in the Turonian, during perhaps the most pronounced sea-level highstand of the Phanerozoic (Vail *et al.* 1977; Haq *et al.* 1987).

The lateral facies change between the grey Chikkim and the multicoloured Fatu La Globotruncana limestones occur in the lower Oma Chu drainage area (Kelemen & Sonnenfeld 1983). This lateral transition zone corresponds to a sharp increase of iron oxidation both in the Late Albian Giumal ironstone and in the overlying foraminiferal oozes, testifying to more restricted circulation in the inner continental terrace, generally characterized by anoxic environments (Berner 1981). At earlier times, during deposition of the upper Giumal Sandstone, estuarine coarse quartzarenite bars passed laterally to neritic fine grained volcanic arenites and pelites across the same narrow zone (Fig. 6; 'mudline' of Stanley *et al.* 1983), which may be thus interpreted as a structurally controlled lineament.

In the Campanian, pelagic carbonate sedimentation was replaced by the Kangi La Marls, overlain by the Marpo carbonate ramp in the Late Maastrichtian (Nicora *et al.* 1986). After a regressive terrigenous episode close to the Cretaceous-Tertiary boundary, another ironstone with glaucony and phosphates is found in the mid-Paleocene (see Nicora *et al.* (1986) for detailed descriptions). This condensed interval marks the rapid transition from the coastal Stumpata Quartzarenite to open shelf and deeper-water limestones with planktonic forams, and thus testifies to a major flooding event. Finally, during the Eocene, the Zanskar continental terrace was involved in the India–Eurasia collision, and underwent intense fold-thrust deformation during the Himalayan orogeny.

## Metamorphism of the Zanskar sediments

The Tethys Himalayan Zone is bounded to the south by the mesograde metamorphic nappes of the High Himalayan Crystalline zone (Fig. 1), where all Barrovian zones from biotite to sillimanite are represented in the Tertiary regional metamorphism (Honegger *et al.* 1982). The tectonic contact between the metamorphic and the sedimentary rocks corresponds, according to Herren (1987), to a late shear zone with a normal sense of displacement, where the transition from upper amphibolite to lower greenschist facies occurs within 200 m. In the Tethys Himalayan sediments, metamorphic grade decreases from lower greenschist facies in western Zanskar (Ringdom area), to anchimetamorphic conditions in central Zanskar (Zangla-Zozar area), as shown by quartz recrystallization features and the colour alteration index of Triassic conodonts (Baud *et al.* 1984).

### Metamorphism of Mesozoic ironstones

Glauconitic greensands are particularly useful metamorphic indicators at very low grade. Glauconite is stable up to lower anchizonal conditions (Zone I of Frey *et al.* 1973), whereas in the middle anchizone, approximately corresponding to the transition from zeolite to prehnite-pumpellyite facies, it reacts with quartz to form stilpnomelane (Zone II). Next, at thermal conditions approaching low metamorphic grade, biotite appears (Winkler 1976, pp. 206–210; Zone III).

#### *Quartzite series*

The occurrence of stilpnomelane in the ironstones interbedded within the Quartzite Series (Table 2) points to conditions comparable to upper Zone II in the Zangla area (Frey *et al.* 1973). Quartz and phyllosilicate recrystallization features suggest correlation with the 'zone of quartzite-like structures and hydromica-chloritic cements' (Kossovskaya & Shutov 1958; Frey 1970). Anchimetamorphic temeratures in central Zanskar were thus around 300°C, which are consistent with the dark grey colour of conodonts in the underlying Triassic units (Baud *et al.* 1984). In the western Zanskar Ringdom section, incipient growth of biotite, as well as the abundance of authigenic muscovite in lamellae up to 20 μm in width and extensive quartz recrystallization with newly-grown crystals exceeding 50 μm in size, testify to epimetamorphic or lower greenschist conditions (Frey 1970; Young 1976), at temperatures above 350°C ('late metagenesis' of Kossovskaya & Shutov 1970).

*Ferruginous Oolite Formation*

In the Ringdom section, extensive migration of quartz grain boundaries with newly-grown crystals 20 to 30 μm in size and formation of abundant stilpnomelane testify to conditions approaching lower greenschist facies. No metamorphic indicators were found in the other stratigraphic sections.

*Giumal Sandstone*

In the condensed intervals found at the top of the Giumal Sandstone in the Dibling area, K-rich (8–9 wt.%; Table 4) glauconitic mica is preserved, and stilpnomelane growth is confined to the rims of glaucony grains. Metamorphic conditions were thus comparable to the beginning of zone II (Frey *et al.* 1973), at temperatures around 260° C, as also shown by the common occurrence of deformation lamellae in quartz grains and by the incipient migration of crystal boundaries (Völl 1982). At Nieraq, dissolution cleavage, incipient breakdown of detrital biotite, extensive stilpnomelane growth and newly-formed crystals reaching 25 μm in size in quartz-filled fractures are observed, suggesting temperatures closer to 300°C. In the Ringdom section, newly-grown quartz crystals become widespread and reach 40 μm in size, testifying to conditions approaching lower greenschist facies (temperatures close to 350°C).

## Conclusions

In the passive margin succession of the Zanskar Tethys Himalaya, chamositic ironstones or glauconitic greensands associated with reworked phosphorites are recorded in the Late Permian, Late Triassic, Middle Jurassic, mid-Cretaceous and mid-Paleocene. These condensed intervals are of key importance for stratigraphic correlation and facies analysis, since they were deposited in the process of shelf formation during major widespread transgressions ('transgressive system tracts' of Haq *et al.* 1987) as suggested by petrographic and sedimentological features. When sea level was rising at an increasing rate, high-energy waves eroded the shelf floor and mixed iron ooids, glauconite and reworked phosphatic hardgrounds, formed during earlier starved stages at low sedimentation rates, with coarse shell debris, arenite pebbles and soil material provided by ravine-forming processes during stepwise shoreline retreat.

Laterally continuous ironstones mark the rapid transition from coastal clastics to offshore deposits. Even minor condensed intervals, found at top of shoaling arenaceous sequences as for the lower-middle Quartzite Series, are overlain by finer-grained and deeper-water sediments, and thus testify to significant flooding events. Ferruginous oolites are most abundant in the Callovian, a period of generalized transgression all along the Himalayas, when they onlap onto a disconformity corresponding to a time-gap of at least 10 Ma (Jadoul *et al.* 1985). Glauconitic arenites and reworked phosphorites are instead best developed at the top of the Giumal Sandstone, and strongly recall the coeval reworked hardgrounds found in the passive margin succession of the Helvetic Alps, the formation of which involved complex multiphase early diagenesis in periods of starved sediment supply and rapid eustatic changes (Ouwehand 1986; Föllmi & Ouwehand 1987).

The Zanskar ironstones closely compare with the transgressive arenites found at the base of depositional megasequences in other rifted-margins (Abbott 1985; Kirk 1985), and show a surprising correspondence with the major break-up episodes which characterize the complex spreading history of the Gondwana fragments (Audley-Charles 1984; Rowley *et al.* 1986). It may be noted the distribution of ironstones, which often contain significant volcanic detritus, coincides with major basaltic volcanic episodes affecting the Indian margin in the Late Permian (Honegger *et al.* 1982), in the Late Triassic (Reuber *et al.* 1987), in the mid-Late Jurassic (Kanwar & Bhandari 1979; Reuber *et al.* 1988), in the mid-Cretaceous (Gaetani *et al.* 1985a) and in the Paleocene (Courtillot *et al.* 1986).

During the Tertiary Himalayan orogeny, the Zanskar continental terrace was dismembered into several thrust sheets which were stacked and deformed at very low to low metamorphic grade. In the chamositic ironstones interbedded within the Triassic Quartzite Series, the occurrence of stilpnomelane and quartz recrystallization features suggest temperatures increasing from about 300°C in central Zanskar to more than 350° C in the western Ringdom area. In the Cretaceous Giumal greensands, however, at a higher structural and stratigraphic level, K-rich glauconite is preserved and stilpnomelane growth is confined to the rims of glaucony grains, pointing to metamorphic conditions comparable to lower prehnite-pumpellyite facies, at temperatures between 270°(Dibling area) and 300°C (Nieraq area).

ACKNOWLEDGEMENTS: We are grateful to P. Ouwehand for very careful critical comments and cathodoluminescence observations; to D. Fontana for assistance with SEM and to E. Mutti, F. Fonnesu, I. Premoli, A. Greco, E. Herren, K. Honegger, J.A. Dockal, W.B. Harris and P. R. Vail for elucidations and helpful discussions. A. Baud, G. Mascle, R. Casnedi, E. Fois, M. Gaetani, A. Nicora and A. Tintori were with us in the field; M. Carmine, G. Arosio and F. Cerizzi helped in various ways. S. Antico made the drawings and C. Malinverno excellent thin sections.

R. Haas performed the X-ray and microprobe analyses; F. Jadoul made cathodoluminescence observations and studied the Ferruginous Oolite Fm. in the field; E. Garzanti is responsible for the field and petrographic study of the sandstone units and for the sedimentological and metamorphic interpretation.

# References

ABBOTT, W.O. 1985. The recognition and mapping of a basal transgressive sand from outcrop, subsurface and seismic data. *In:* BERG, O.R. & WOOLVERTON, D.G. (eds) *Seismic stratigraphy* II. Memoir of American Association of Petroleum Geologists, **39**, 157–167.

AUDLEY-CHARLES, M.G. 1984. Cold Gondwana, warm Tethys and the Tibetan Lhasa block. *Nature,* **310**, 165.

BASSOULLET, J.P. COLCHEN, M., JUTEAU, Th., MARCOUX, J., MASCLE, G. & REIBEL, G. 1983. Geological studies in the Indus suture zone of Ladakh (Himalayas). *In:* GUPTA, V.J. (ed.) *Stratigraphy and structure of Kashmir and Ladakh Himalaya. Contributions to Himalayan geology,* **2**, Hindustani, Delhi, 96–124.

——,——, MARCOUX, J. & MASCLE, G. 1981. Les masses calcaires du flysch triasico-jurassique de Lamayuru (zone de la suture de l'Indus, Himalaya de Ladakh): klippes sédimentaires et éléments de plate-forme remaniés. *Rivista Italiana di Palaeontologia e Stratigrafia, 1980,* 4, 825–844.

BAUD, A. *et al.* 1982. Le contact Gondwana-peri-Gondwana dans le Zanskar oriental (Ladakh, Himalaya). *Bulletin Société Géologique de France,* **24**, 241–361.

——, GAETANI, M., GARZANTI, E., FOIS, E., NICORA, A. & TINTORI, A. 1984. Geological observations in southeastern Zanskar and adjacent Lahul area (northwestern Himalaya). *Eclogae Geologicae Helvetiae,* **77**, 171–197.

——, GARZANTI, E. & MASCLE, G. 1987. Evolution des facies durant le Crétacé superieur au Zanskar (NW Himalaya). *2ème Workshop in Himalayan Geology,* Nancy, p.2.

BERNER, R.A. 1981. New geochemical classification of sedimentary environments. *Journal of Sedimentary Petrology,* **51**, 359–365.

CLEARY, W.J. & CONOLLY, J.R. 1972. Embayed quartz grains in soils and their significance. *Journal of Sedimentary Petrology,* **42**, 899–904.

COURTILLOT, V., HESSE, J., VANDAMME, D., MONTIGNY, R., JAEGER, J.J. & CAPPETTA, H. 1986. Deccan flood basalts at the Cretaceous/Tertiary boundary? *Earth and Planetary Science Letters,* **80**, 361–374.

DERCOURT, J., *et al.* 1986. Geological evolution of the Tethys belt from the Atlantic to the Pamirs since the Lias. *Tectonophysics,* **123**, 241–315.

EMERY, K.O. 1968. Relict sediments on continental shelves of the world. *Bulletin of the American Association of Petroleum Geologists,* **52**, 445–464.

FLÜGEL, E. 1982. *Microfacies analysis of limestones.* Springer Verlag, Heidelberg.

FOLK, R.L. 1980. *Petrology of sedimentary rocks.* Hemphill's Austin.

FÖLLMI, K. & OUWEHAND, P.J. 1987. Die Garschella Formation: Neue stratigraphische Daten aus der 'mittleren' Kreide des ostschweizerischen und vorarlberger Helvetikums. *Eclogae Geologicae Helvetiae,* **80**, 141–191.

FREY, M. 1970. The step from diagenesis to metamorphism in pelitic rocks during alpine orogenesis. *Sedimentology,* **15**, 261–279.

——,HUNZIKER, J.C., ROGGWILLER, P. & SCHINDLER, C. 1973. Progressive neidriggradige Metamorphose glaukonitführender Horizonte in den helvetischen Alpen der Ostschweiz. *Contributions to Mineralogy and Petrology,* **39**, 185–218.

GAETANI, M., CASNEDI, R., FOIS, E., GARZANTI, E., JADOUL, F., NICORA, A. & TINTORI, A. 1985a. Stratigraphy of the Tethys Himalaya in Zanskar, Ladakh — Initial report. *Rivista Italiana di Paleontologia e Stratigrafia,* **91**, 443–478.

——,GARZANTI, E. & JADOUL, F. 1985b. Main structural elements of Zanskar, NW Himalaya. *Rediconti della Societa Geologica Italiana,* **8**, 3–8.

GARZANTI, E. 1986. *Storia sedimentaria del margine continentale settentrionale della placca indiana (Ladakh, India).* Ph.D. Thesis, Universita di Milano.

GOLDBERY, R. 1979. A textural inversion phenomenon within Lower Jurassic red beds of the Ardon Formation, Makhtesh Ramon, (Israel). *Journal of Sedimentary Petrology,* **49**, 891–900.

GYGI, R.A. 1986. Eustatic sea level changes of the Oxfordian (Late Jurassic) and their effect documented in sediments and fossil assemblages of an epicontinental sea. *Eclogae Geologicae Helvetiae,* **79**, 455–491.

HALLAM, A. 1981. A revised sea-level curve for the early Jurassic. *Journal of the Geological Society, London,* **138,** 735–743.
——& MAYNARD, J.B. 1987. The iron ores and associated sediments of the Chichali formation (Oxfordian to Valanginian) of the Trans-Indus Salt Range, Pakistan. *Journal of the Geological Society, London,* **144,** 107–114.
HAQ, B., HARDENBOL, J. & VAIL, P.R. 1987. Chronology of fluctuating sea levels since the Triassic (250 My ago to the present) *Science,* **235,** 1156–1167.
HARRIS, W.B., ZULLO, V.A. & OTTE, L.J. 1986. Eocene carbonate facies of the North Carolina coastal plain. In: TEXTORIS, D. (ed.). *SEPM Midyear Meeting, Raleigh, Field guidebooks,* 253–332.
HERREN, E. 1987. The Zanskar Shear Zone: northeast-southwest extension within the Higher Himalayas (Ladakh, NW Himalaya, India). *Geology,* **15,** 409–413.
HONEGGER, K., DIETRICH, V., FRANK, W., GANSSER, A., THONI, M. & TROMMSDORFF, V. 1982. Magmatism and metamorphism in the Ladakh Himalayas (the Indus-Tsangpo suture zone): *Earth and Planetary Science Letters,* **60,** 253–292.
JADOUL, F., FOIS, E., TINTORI, A. & GARZANTI, E. 1985. Preliminary results on Jurassic stratigraphy in Zanskar (NW Himalaya). *Rendiconti della Societa Geologica Italiana,* **8,** 9–13.
KANWAR, S.S. & BHANDARI, A.K. 1979. Stratigraphy, structure and sedimentation of part of Lahaul and Spiti District, Himachal Pradesh. *Miscellaneous Publications of the Geological Survey of India, (1976),* **41,** 169–178.
KELEMEN, P.B. & SONNENFELD, M.D. 1983. Stratigraphy, structure, petrology and local tectonics, Central Ladakh, NW Himalaya. *Bollettino Svizzero di Mineralogia e Petrografia,* **63,** 267–287.
KIRK, R.B. 1985. A seismic stratigraphic case history in the eastern Barrow Subbasin, North west shelf, Australia. In: BERG, O.R. & WOOLVERTON, D.G. (eds.) *Seismic stratigraphy II. Memoir of the American Association of Petroleum Geologists,* **39,** 183–207.
KOSSOVSKAYA, A.G. & SHUTOV, V.D. 1958. Zonality in the structure of terrigenous deposits in platform and geosynclinal regions. *Eclogae Geologicae Helvetiae,* **51,** *656–666.*
——*1970. Main aspects of the epigenesis problem. Sedimentology,* **15,** 11–40.
MALLINSON, D.J. & OTTE, L.J. 1986. A model for the occurrence of glauconia on the modern and Miocene North Carolina continental shelf/slope (Abstract) *SEPM Midyear Meeting, Raleigh,* **70.**
McGHEE, G.R. & BAYER, U. 1984. The local signature of sea-level changes. *In*: BAYER, U. & SEILACHER, A. (eds) *Sedimentary and evolutionary cycles.* Lecture Notes in Earth Sciences, Springer Verlag, Heidelberg. 98–112.
NICORA, A., GAETANI, M. & GARZANTI, E. 1984. Late Permian to Anisian in Zanskar (Ladakh, Himalaya). *Rendiconti della Societa Geologica Italiana,* **7,** 27–30.
——,GARZANTI, E. & FOIS, E. 1986. Maastrichtian to Paleocene evolution of the Zanskar continental shelf (NW Himalaya, India). *Rivista Italiana di Paleontologia e Stratigrafia,* 439–495.
ODIN, G.S. & MATTER, A. 1981. De glauconiarum origine. *Sedimentology,* **28,** 611–641.
OUWEHAND, P.J. 1986. Werden Phosphorite wirklich frühdiagenetisch gebildet? Beispiele aus der helvetischen 'Mittelkreide' der Ostschweiz (Abstract) *Deutschsprachiges Sedimentologen-Treffen, Freiburg BRD,* **4.**
REUBER, I., BOILLET, G., COLCHEN, M. & MASCLE, G. 1988. Modèle d'évolution de la marge indienne de la Tèthys en Himalaya du Ladakh du Permien à l'Aptian-Albien. *Abstract Himalayan Workshop, Lansanne,* 30.
——, COLCHEN, M. & MEVEL, C. 1987. The geodynamic evolution of the South-Tethyan margin in Zanskar, NW Himalaya, as revealed by the Spongtang ophiolitic mélanges. *Geodinamica Acta,* **1,** 283–296.
ROWLEY, D.B., LOTTES, A.L., NIE, S.Y. & ZIEGLER, A.M. 1986. Tectonic evolution of the Himalayas within the context of Gondwanan-Eurasian relative motions. *Bulletin of the Geological Society of America, Abstracts with programs,* 735.
SAVOSTIN, L.A., SIBUET, J.C. ZONENSHAIN, L.P., LE PICHON, X. & ROULET, M.J. 1986. Kinematic evolution of the Tethys belt from the Atlantic Ocean to the Pamirs since the Triassic. *Tectonophysics,* **123,** 1–35.
SELLWOOD, B.W. 1978. Jurassic. *In*: MCKERROW, W.S. (ed.). *Ecology of fossils.* M.I.T. Press, Cambridge, 204–279.
STANLEY, D.J., ADDY, S.K. & BEHRENS, E.W. 1983. The mudline: variability of its position relative to shelfbreak. *SEPM Special Publication* **33,** 279–298.
SWIFT, D.J.P. 1986. Shelf sands and sandstone types. (Abstract) *SEPM Mid-year Meeting, Raleigh,* 108.
TAIRA, A. & SCHOLLE, P.A. 1979. Origin of bimodal sands in some modern environments. *Journal of Sedimentary Petrology,* **49,** 777–786.
VAIL, P.R., HARDENBOL, J. & TODD, R.G. 1984. Jurassic unconformities, chronostratigraphy and sea-level changes from seismic stratigraphy and biostratigraphy. *In*: SCHLEE, J. (ed.) *Interregional unconformities and hydrocarbon accumulation. Memoir of the American Association of Petroleum Geologists,* **36,** 129–144.
——,MITCHUM, R.M. & THOMPSON, S. 1977. Global cycles of relative changes in sea level. *In*: PAYTON CE (ed.) *Seismic stratigraphy — applications to hydrocarbon exploration. Memoir of the American Association of Petroleum Geologists,* **26,** 83–97.
——& TODD, R.G. 1981. Northern North Sea Jurassic unconformities, chronostratigraphy and sea-level changes from seismic stratigraphy. *In*: ILLING, L. & HOBSON, G. (eds.) *Petroleum geology of the*

*continental shelf of the northwest Europe.* Heyden, Institute of Petroleum, London, 216–235.

VAN HOUTEN, F.B. & PURUCKER, M.E. 1984. Glauconitic peloids and chamositic ooids — Favorable factors, constraints and problems. *Earth Science Reviews,* **20**, 211–243.

VÖLL, G. 1982. Bewegung von Korngrenzen in Gesteinen. Vortrag, DFG 'Gesteinskinetic-Sitzung', 24.4.82, Tubingen.

WINKLER, H.G.F. 1976. *Petrogenesis of metamorphic rocks* (4th Edition) Springer Verlag, Heidelberg.

YOUNG, S.W. 1976. Petrographic textures of detrital polycrystalline quartz as an aid to interpreting crystalline source rocks. *Journal of Sedimentary Petrology,* **46**, 595–603.

E. GARZANTI, Dip. Scienze della Terra, Via Mangiagalli 34, Milano, Italy.
R. HAAS, Geologisches Institut, ETH Zentrum, Zurich, Switzerland.
F. JADOUL, Dip. Scienze della Terra, Via Mangiagalli 34, Milano, Italy.

# Geographical Index

# Index

## Journal of the Geological Society

Chief Editor: M.J. Le Bas

Published bi-monthly, the Journal of the Geological Society enjoys a very wide international circulation and has been published continuously since 1845. It is a leading international organ for significant research in all branches of the geological sciences. It is the medium of choice for publication of work by internationally recognized authorities.

Papers are accepted from both Fellows and non-Fellows. A Short Papers section has been introduced for rapid publication of topical information.

## Quarterly Journal of Engineering Geology

Scientific Editor: A.B. Hawkins

This is a journal of the Geological Society with a wide international circulation. Original papers are accepted from Fellows and non-Fellows working in the UK or overseas. The journal is designed for papers which deal with any of the subjects within the field of geology as applied to civil engineering, mining practices and water resources, including rock mechanics and geotechnics. Its wide scope also includes applied sedimentology, pedology, geohydrology and the engineering application of geophysics. Case histories and review articles are also published.

For both the Journal of the Geological Society and the Quarterly Journal of Engineering Geology, typescripts for publication, editorial enquiries and correspondence should be sent to: The Editorial Department, Geological Society Publishing House, Unit 7, Brassmill Enterprise Centre, Brassmill Lane, Bath BA1 3JN. Tel 0225-445046.

Enquiries and business correspondence relating to sales and subscription matters should be sent to: Journals Subscription Department, Geological Society, Burlington House, Piccadilly, London W1V 0JU. Tel: 01-434-9944.

Fellows and other members of the Geological Society enjoy substantial reduction on the price of books published by the Society.

Fellowship dues include a subscription to one of the above mentioned journals or the Journal of Marine and Petroleum Geology (published jointly with Butterworths). Several other journals can be subscribed to by Fellows at substantially reduced cost where the Society has an agreement. Fellowship is open to geologists worldwide; further information is available from the Executive Secretary, Geological Society, Burlington House, Piccadilly, London W1V 0JU.

## Geology Today

Edited by J.H. McD. Whitaker and P.J. Smith and published bi-monthly by Blackwell Scientific Publications Ltd., under the co-sponsorship of The Geological Society and The Geologists' Association.

This journal serves professional geologists wishing to keep abreast of developments in areas outside their own field, as well as amateur geologists, students and their instructors. The Editorial Board selects several articles for each issue on various topics of current interest in both pure and applied geology, to provide both specialists and non-specialists with clear and readable information covering a wide scope. The journal also features shorter notes, news of meetings, exhibitions and field excursions, contributions from foreign correspondents and book reviews.